Human Cell Culture Protocols

METHODS IN MOLECULAR MEDICINE™

John M. Walker, SERIES EDITOR

Human Cell Culture Protocols, edited by *Gareth E. Jones,* 1996
Antisense Therapeutics, edited by *Sudhir Agrawal,* 1996
Vaccine Protocols, edited by *Andrew Robinson, Graham H. Farrar, and Christopher N. Wiblin,* 1996
Prion Diseases, edited by *Harry F. Baker and Rosalind M. Ridley,* 1996
Molecular Diagnosis of Cancer, edited by *Finbarr Cotter,* 1996
Molecular Diagnosis of Genetic Diseases, edited by *Rob Elles,* 1996
Herpes Simplex Virus Protocols, edited by *Moira S. Brown and Alasdair MacLean,* 1996
***Helicobacter pylori* Protocols,** edited by *Christopher L. Clayton and Harry T. Mobley,* 1996
Lectins in Medical Research, edited by *Jonathan M. Rhodes and Jeremy D. Milton,* 1996
Gene Therapy Protocols, edited by *Paul Robbins,* 1996

METHODS IN MOLECULAR MEDICINE™

Human Cell Culture Protocols

Edited by
Gareth E. Jones
King's College, University of London, UK

Humana Press ✱ Totowa, New Jersey

© 1996 Humana Press Inc.
999 Riverview Drive, Suite 208
Totowa, New Jersey 07512

All rights reserved. No part of this book may be reproduced, stored in a retrieval system, or transmitted in any form or by any means, electronic, mechanical, photocopying, microfilming, recording, or otherwise without written permission from the Publisher. Methods in Molecular Medicine™ is a trademark of The Humana Press Inc.

All authored papers, comments, opinions, conclusions, or recommendations are those of the author(s), and do not necessarily reflect the views of the publisher.

This publication is printed on acid-free paper. ∞
ANSI Z39.48-1984 (American Standards Institute) Permanence of Paper for Printed Library Materials.

Cover illustration: Fig. 2 from Chapter 7, "Culture of Human Normal Brain and Malignant Brain Tumors for Cellular, Molecular, and Pharmacological Studies," by Francis Ali-Osman.

Photocopy Authorization Policy:
Authorization to photocopy items for internal or personal use, or the internal or personal use of specific clients, is granted by Humana Press Inc., provided that the base fee of US $5.00 per copy, plus US $00.25 per page, is paid directly to the Copyright Clearance Center at 222 Rosewood Drive, Danvers, MA 01923. For those organizations that have been granted a photocopy license from the CCC, a separate system of payment has been arranged and is acceptable to Humana Press Inc. The fee code for users of the Transactional Reporting Service is: [0-89603-335-X/96 $5.00 + $00.25].

Printed in the United States of America. 10 9 8 7 6 5 4 3 2 1

Library of Congress Cataloging in Publication Data

Main entry under title:

Methods in molecular medicine™.

Human cell culture protocols / edited by Gareth E. Jones.
    p. cm. — (Methods in molecular medicine™)
    Includes index.
    ISBN 0-89603-335-X (alk. paper)
    1. Human cell culture—Laboratory manuals. I. Jones, Gareth E., 1947– . II. Series.
    [DNLM: 1. Cells, Cultured. 2. Cytological Techniques. QS 525 H9175 1996]
    QH585.2.H85 1996
    DNLM/DLC   96-4625
    for Library of Congress   CIP

Preface

Cell culture is now a routine approach for the preparation of cells from a variety of sources as an adjunct or replacement for animal work. In an earlier era, the methodology seemed shrouded in mystery, a useful device for sustaining exclusivity by some of the pioneers, notably Alexis Carrel. Although he was undoubtedly responsible for many early developments, his claims for the extreme difficulty of the tissue culture techniques put off a lot of his contemporaries and interest in tissue culture declined markedly for many years even after he gave up his own studies. Nevertheless, not everyone succumbed to the inertia engendered by Carrel's domination, and the work of others such as Harrison and Rous was sufficiently innovative to attract a small stream of followers. By the outbreak of World War II, increased knowledge of cell metabolism had brought about significant improvements in the formulation of culture media and fully defined media were being produced by the 1960s.

Despite technical advances, such as using enzymes or defined or near-defined media, cell or tissue culture was still a time-consuming technique at this time, largely because of the huge chore of weekly media preparation. Fortunately the growing importance of cell culture drove the development of a whole industry devoted to supplying materials for tissue culture, so that by today there are companies providing 80 varieties of media, along with culture flasks, pipets, enzyme solutions, and even cultures of cells. Much of the drudge facing the worker in cell culture has now disappeared, and there are now many books published devoted entirely to the detailed description of specialist culture techniques. So why produce this volume?

The origins of *Human Cell Culture Protocols* lie in the repeated requests of clinical scientists and their allies for advice on cell culture. Despite the great advances described above, it is fair to say that the bulk of the available help is directed to those working with nonhuman material. Even where human cells are described, there has been a tendency to concentrate on a few well-established cell lines, such as HeLa or A431 cells. Though such sources can be very useful for certain work, no one would pretend that such lines are representative of normal human tissue. Scientists closely involved with clinical research have largely been obliged to adapt many of the techniques developed for the culture of such cell lines, introducing many innovative approaches to deal with the particular problems of establishing cultures from biopsied human samples. Very great progress has been made over the last decade, driven by the demands of pharmaceutical companies as much as the academic interest in the cellular basis of human development and disease. These techniques are published, but not in a convenient form for rapid retrieval, since they are usually embedded in the methods sections of research papers. *Human Cell Culture Protocols* is the first attempt to provide a catalog of protocols devoted entirely to human cell culture.

It has been estimated that there are some 200 distinct cell types in the human body, though such a classification is somewhat arbitrary. Arbitrary or not, the present book could not possibly cover this range so I have selected contributors who were able to provide detailed protocols for a selection of the major tissue groupings. Connective tissue cells are well catered for, from fibroblasts through chondrocytes to osteocytes. Blood cells are similarly explored in detail with chapters covering NK-cells, monocytes, B-lymphocytes, and T-lymphocytes. Several examples of epithelial types are described; keratinocytes and airway cells being standard tissues used by many clinical investigators, but I also include less well-known material, such as thymic epithelial cells, which have unique functions in relation to normal organ function. Muscle is another major area of clinical research, and there are chapters detailing the culture of two very different smooth muscle sources, namely uterine and aortic smooth muscle. Skeletal muscle culture is also described, but I have not included a chapter on cardiac myocytes, since no satisfactory method yet exists for this material. Other major organs that had to be included were liver and kidney, as well as endothelia and some of the more esoteric tissues that nevertheless have considerable clinical significance. Among the selected range included in this volume you will find melanocytes, Langerhans' cells, trophoblasts, and cells of the conjunctiva.

The choice of chapters owes much to the advice of many clinical colleagues and friends, but I feel that coverage of neuronal tissue (including their support cells) must await a future edition. To provide some compensation, I have provided, in chapters prepared by Manuel Vega, protocols that demonstrate the power of modern molecular techniques applied to genetic disorders, in this case the disorder being cystic fibrosis. Finally I thought it wise to include a chapter on the detection of mycoplasma in cell cultures. These agents are pervasive in human material and, unlike most bacteria and common fungal contamination, not easy to eliminate once they spread throughout your cultures.

I would like to conclude by thanking Clare Wise for her ready assistance with proofreading and Professor John Walker for his steady encouragement throughout my labors. Thanks also to the staff of Humana Press for their tolerance of my sometimes tardy responses to their requests for chapter revisions, my only excuse is the constant wail of all academics—too busy!

Gareth E. Jones

Contents

Preface v
Contributors xi

1 Establishment and Maintenance of Normal Human Keratinocyte Cultures,
  Claire Linge 1
2 Cultivation of Normal Human Epidermal Melanocytes,
  Mei-Yu Hsu and Meenhard Herlyn 9
3 Cultivation of Keratinocytes from the Outer Root Sheath of Human Hair Follicles,
  Alain Limat and Thomas Hunziker 21
4 Isolation and Short-Term Culture of Human Epidermal Langerhans' Cells,
  Corinne Moulon 33
5 Growth and Differentiation of Human Adipose Stromal Cells in Culture,
  Brenda Strutt, Wahid Khalil, and Donald Killinger 41
6 Human Fetal Brain Cell Culture,
  Mark P. Mattson 53
7 Culture of Human Normal Brain and Malignant Brain Tumors for Cellular, Molecular, and Pharmacological Studies,
  Francis Ali-Osman 63
8 Cell Culture of Human Brain Tumors on Extracellular Matrices: *Methodology and Biological Applications,*
  Manfred Westphal, Hildegard Nausch, and Dorothea Zirkel 81
9 Isolation and Culture of Human Umbilical Vein Endothelial Cells,
  David M. L. Morgan 101
10 Human Thymic Epithelial Cell Cultures,
  Anne H. M. Galy 111
11 Generation and Cloning of Antigen-Specific Human T-Cells,
  Hans Yssel 121

12 Protection of Purified Human Hematopoietic Progenitor Cells by Interleukin-1β or Tumor Necrosis Factor-α,
Robert J. Colinas *137*

13 Human Mononuclear Phagocytes in Tissue Culture,
Yona Keisari *153*

14 Purification of Peripheral Blood Natural Killer Cells,
Ian M. Bennett and Bice Perussia *161*

15 Cystic Fibrosis Airway Epithelial Cell Culture,
Manuel A. Vega *173*

16 Gene Transfer into Cystic Fibrosis Airway Epithelial Cells,
Manuel A. Vega *185*

17 Genetic Analysis of Cystic Fibrosis Airway Epithelial Cells,
Manuel A. Vega *193*

18 Human Tracheal Gland Cells in Primary Culture,
Marc D. Merten *201*

19 Human Chondrocyte Cultures as Models of Cartilage-Specific Gene Regulation,
Mary B. Goldring *217*

20 Isolation and Culture of Bone-Forming Cells (Osteoblasts) from Human Bone,
James A. Gallagher, Roger Gundle, and Jon N. Beresford *233*

21 The Isolation of Osteoclasts from Human Giant Cell Tumors and Long-Term Marrow Cultures,
Catherine A. Walsh, John A. Carron, and James A. Gallagher *263*

22 In Vitro Cellular Systems for Studying OC Function and Differentiation: *Primary OC Cultures and the FLG 29.1 Model,*
Donatella Aldinucci, Julian M. W. Quinn, Massimo Degan, Senka Juzbasic, Angela De Iuliis, Salvatore Improta, Antonio Pinto, and Valter Gattei *277*

23 Isolation, Purification, and Growth of Human Skeletal Muscle Cells,
Grace K. Pavlath *307*

24 Cultures of Proliferating Vascular Smooth Muscle Cells from Adult Human Aorta,
Heide L. Kirschenlohr, James C. Metcalfe, and David J. Grainger *319*

Contents

25 Human Myometrial Smooth Muscle Cells and Cervical Fibroblasts in Culture: *A Comparative Study,*
    Françoise Cavaillé, Dominique Cabrol, and Françoise Ferré 335

26 Primary Culture of Human Antral Endocrine Cells,
    Alison M. J. Buchan 345

27 Primary Cell Culture of Human Enteric Neurons: *Submucosal Plexus,*
    Eric A. Accili and Alison M. J. Buchan 357

28 Isolation and Culture of Human Hepatocytes,
    Martin K. Bayliss and Paul Skett 369

29 Culture of Human Pancreatic Islet Cells,
    Stellan Sandler and Décio L. Eizirik 391

30 Isolation and Culture of Human Renal Cortical Cells with Characteristics of Proximal Tubules,
    Vicente Rodilla and Gabrielle M. Hawksworth 409

31 Glomerular Epithelial and Mesangial Cell Culture and Characterization,
    Heather M. Wilson, Keith N. Stewart, and Alison M. MacLeod ... 419

32 Enzymatic Isolation and Serum-Free Culture of Human Renal Cells: *Retaining Properties of Proximal Tubule Cells,*
    John H. Todd, Kenneth E. McMartin, and Donald A. Sens 431

33 Culture of Human Renal Medullary Interstitial Cells,
    Klara Tisocki and Gabrielle M. Hawksworth 437

34 Culture Technique of Human Pituitary Adenoma Cells,
    Masanori Kabuto 447

35 Preparation of Human Trophoblast Cells for Culture,
    Suren R. Sooranna 457

36 Superfused Microcarrier Cultures of BeWo Choriocarcinoma Cells: *A Dynamic Model for Studies of Human Trophoblast Function,*
    Suren R. Sooranna and Bryan M. Eaton 465

37 Corneal Organ Culture,
    Robert W. Lambert, Normand R. Richard, and Janet A. Anderson 477

38 Keratocyte Cell Culture,
    Normand R. Richard, Robert W. Lambert, and Janet A. Anderson 489

39 Conjunctiva: *Organ and Cell Culture,*
    Monica Berry, Roger B. Ellingham, and Anthony P. Corfield 503

40 Establishment and Maintenance of In Vitro Cultures of Human Retinal Pigment Epithelium,
Eva L. Feldman, Monte A. Del Monte, Martin J. Stevens, and Douglas A. Greene 517

41 Demonstration of Mycoplasma Contamination in Cell Cultures by a Mycoplasma Group-Specific Polymerase Chain Reaction,
Frank van Kuppeveld, Jos van der Logt, Karl-Erik Johansson, and Willem Melchers 525

Index 539

Contributors

ERIC A. ACCILI • *Department of Physiology and General Biochemistry, University of Milan, Italy*
DONATELLA ALDINUCCI • *The Leukemia Unit, Division of Medical Oncology, Avino, Italy*
FRANCIS ALI-OSMAN • *Department of Experimental Pediatrics, Section of Molecular Therapeutics, University of Texas M. D. Anderson Cancer Center, Houston, TX*
JANET A. ANDERSON • *National Vision Research Institute, San Diego, CA*
MARTIN K. BAYLISS • *Department of Bioanalysis and Drug Metabolism, Glaxo Wellcome Research and Development, Hertfordshire, UK*
IAN M. BENNETT • *Thomas Jefferson University, Philadelphia, PA*
JON N. BERESFORD • *School of Pharmacy and Pharmacology, University of Bath, Avon, UK*
MONICA BERRY • *Department of Ophthalmology, Bristol Eye Hospital, University of Bristol, UK*
ALISON M. J. BUCHAN • *Department of Physiology, University of British Columbia, Vancouver, Canada*
DOMINIQUE CABROL • *INSERM U., Paris, France*
JOHN A. CARRON • *Department of Human Anatomy and Cell Biology, University of Liverpool, UK*
FRANÇOISE CAVAILLÉ • *INSERM U., Paris, France*
ROBERT J. COLINAS • *Wadsworth Center, New York State Department of Health, Albany, NY*
ANTHONY P. CORFIELD • *Department of Medicine, Bristol Royal Infirmary, Bristol, UK*
MASSIMO DEGAN • *The Leukemia Unit, Division of Medical Oncology, Avino, Italy*
ANGELA DE IULIIS • *The Leukemia Unit, Division of Medical Oncology, Avino, Italy*
MONTE DEL MONTE • *Departments of Ophthalmology and Pediatrics, University of Michigan Medical School, Ann Arbor, MI*

BRYAN M. EATON • *Department of Obstetrics and Gynecology, Chelsea and Westminister Hospital, London, UK*

DÉCIO EIZIRIK • *Department of Medical Cell Biology, Uppsala University, Uppsala, Sweden*

ROGER B. ELLINGHAM • *Department of Ophthalmology, Bristol Eye Hospital, University of Bristol, UK*

EVA FELDMAN • *Department of Neurology, University of Michigan, Ann Arbor, MI*

FRANÇOISE FERRÉ • *INSERM U., Paris, France*

JAMES A. GALLAGHER • *Department of Human Anatomy and Cell Biology, Bone Cell Research Group, University of Liverpool, UK*

ANNE GALY • *Karmanos Cancer Center, Detroit, MI*

VALTER GATTEI • *The Leukemia Unit, Division of Medical Oncology, Avino, Italy*

MARY B. GOLDRING • *Arthritis Research Laboratory, Massachusetts General Hospital and Division of Medical Sciences, Harvard Medical School, Boston, MA*

DAVID J. GRAINGER • *Department of Biochemistry, University of Cambridge, UK*

DOUGLAS A. GREENE • *Department of Internal Medicine, Michigan Diabetes Research and Training Center, University of Michigan Medical School, Ann Arbor, MI*

ROGER G. GUNDLE • *Department of Orthopaedic Surgery, The Nuffield Orthopaedic Centre, Headington, Oxford, UK*

GABRIELLE M. HAWKSWORTH • *Department of Medicine and Therapeutics, University of Aberdeen, Scotland*

MEENHARD HERLYN • *The Wistar Institute, Philadelphia, PA*

MEI-YU HSU • *The Wistar Institute, Philadelphia, PA*

THOMAS HUNZIKER • *Dermatological Clinic, University of Bern, Switzerland*

SALVATORE IMPROTA • *The Leukemia Unit, Division of Medical Oncology, Avino, Italy*

KARL-ERIK JOHANSSON • *The National Veterinary Institute, Uppsala, Sweden*

SENKA JUZBASIC • *The Leukemia Unit, Division of Medical Oncology, Avino, Italy*

MASANORI KABUTO • *Department of Neurosurgery, The Fukui Medical School, Fukui, Japan*

YONA KEISARI • *Department of Human Microbiology, Sackler Faculty of Medicine, Tel-Aviv University, Tel-Aviv, Israel*

Contributors

WAHID KHALIL • *Lawson Research Institute, St. Joseph's Health Centre, London, Ontario, Canada*

DONALD KILLINGER • *Lawson Research Institute, St. Joseph's Health Centre, London, Ontario, Canada*

HEIDE L. KIRSCHENLOHR • *Department of Biochemistry, University of Cambridge, UK*

FRANK VAN KUPPEVELD • *Department of Medical Microbiology, University of Nijmegen, The Netherlands*

ROBERT W. LAMBERT • *National Vision Research Institute, San Diego, CA*

ALAIN LIMAT • *Dermatological Clinic, University of Bern, Switzerland*

CLAIRE LINGE • *Mount Vernon Hospital, Northwood, Middlesex, UK*

JOS VAN DER LOGT • *Department of Medical Microbiology, University of Nijmegen, The Netherlands*

ALISON M. MACLEOD • *Department of Medicine and Therapeutics, University of Aberdeen, Scotland*

MARK P. MATTSON • *Department of Anatomy and Neurobiology, Sanders-Brown Research Center on Aging, University of Kentucky, Lexington, KY*

KENNETH E. MCMARTIN • *Department of Pharmacology, Louisiana State University Medical Center, Shreveport, LA*

WILLEM MELCHERS • *The National Veterinary Institute, Uppsala, Sweden*

MARC D. MERTEN • *Groupe de Recherche sur les Glandes Exocrines, Faculté de Médicine, Marseille, France*

JAMES C. METCALFE • *Department of Biochemistry, University of Cambridge, UK*

DAVID M. L. MORGAN • *Vascular Biology Research Centre, Biomedical Sciences Division, King's College, London, UK*

CORINNE MOULON • *Max Planck Institute of Immunobiology, Freiburg, Germany*

HILDEGARD NAUSCH • *Department of Neurosurgery, Laboratory for Brain Tumor Biology, Eppendorf University Hospital, Hamburg, Germany*

GRACE K. PAVLATH • *Department of Pharmacology, Emory University, Atlanta, GA*

BICE PERUSSIA • *Thomas Jefferson University, Philadelphia, PA*

ANTONIO PINTO • *The Leukemia Unit, Division of Medical Oncology, Avino, Italy*

JULIAN M. W. QUINN • *Nuffield Department of Pathology and Bacteriology, John Radcliffe Hospital, Oxford, UK*

NORMAND R. RICHARD • *National Vision Research Institute, San Diego, CA*
VICENTE RODILLA • *Department of Medicine and Therapeutics, University of Aberdeen, Scotland*
STELLAN SANDLER • *Department of Medical Cell Biology, Uppsala University, Uppsala, Sweden*
DONALD A. SENS • *Department of Pathology, Robert C. Byrd Health Sciences Center, West Virginia University, Morgantown, WV*
PAUL SKETT • *Department of Biomedical and Life Sciences, University of Glasgow, Scotland*
SUREN R. SOORANNA • *Department of Obstetrics and Gynecology, Chelsea and Westminister Hospital, London, UK*
MARTIN J. STEVENS • *Department of Internal Medicine, Michigan Diabetes Research and Training Center, University of Michigan Medical School, Ann Arbor, MI*
KEITH N. STEWART • *Department of Medicine and Therapeutics, University of Aberdeen, Scotland*
BRENDA STRUTT • *Lawson Research Institute, St. Joseph's Health Centre, London, Ontario, Canada*
KLARA TISOCKI • *Department of Medicine and Therapeutics, University of Aberdeen, Scotland*
JOHN H. TODD • *Department of Pathology, Robert C. Byrd Health Sciences Center, West Virginia University, Morgantown, WV*
MANUEL A. VEGA • *Department of Biology, Human Gene Therapy Working Group, Universidad Nacional del Sur, Bahia Blanca, Argentina*
CATHERINE A. WALSH • *Department of Human Anatomy and Cell Biology, University of Liverpool, UK*
MANFRED WESTPHAL • *Department of Neurosurgery, Laboratory for Brain Tumor Biology, Eppendorf University Hospital, Hamburg, Germany*
HEATHER M. WILSON • *Department of Medicine and Therapeutics, University of Aberdeen, Scotland*
HANS YSSEL • *Department of Human Immunology, DNAX Research Institute, Palo Alto, CA*
DOROTHEA ZIRKEL • *Department of Neurosurgery, Laboratory for Brain Tumor Biology, Eppendorf University Hospital, Hamburg, Germany*

1

Establishment and Maintenance of Normal Human Keratinocyte Cultures

Claire Linge

1. Introduction

Keratinocytes are the major cellular component of the epidermis, which is the stratified squamous epithelia forming the outer-most layer of skin. The keratinocytes lie on a basement membrane and are organized into distinct cell layers which differ morphologically and biochemically. These regions from the basement membrane outward are the basal, spinous, granular, and cornified layers. Cellular proliferation takes place mainly in the basal layer. On division, keratinocytes give rise to either replacement progenitor cells and/or cells that are committed to undergo the process of terminal differentiation. These latter cells leave the basal layer and gradually migrate upward, simultaneously progressing along the differentiation pathway as they go. Finally they reach the outer surface of the epidermis in the form of fully mature functional cells, the corneocytes. The function of these mature cells is the protection of the underlying viable tissues from the external milieu.

The reasons for studying keratinocytes are many-fold, and include investigation of the pathogenesis of keratinocyte-related diseases and also examination of the control mechanisms of proliferation and differentiation. The development of a long-term in vitro keratinocyte system, allowing precise experimental manipulation of these cells, has been instrumental in the rapid advances made in these fields over the last two decades.

Initial attempts to grow keratinocytes were limited to the use of organ and explant cultures *(1)*. Using these techniques, whole pieces of skin can be kept alive in the short term, and growth is confined to the tissue fragment or onto the plastic surrounding the explant. However, these cultures have an extremely short life-span and also limited application, since mixed cultures of keratinocytes and fibroblasts are obtained. Fibroblasts present a major problem where keratinocyte culture is concerned. Even small amounts of fibroblast contamination can lead to their overgrowth of keratinocyte cultures. This is because of the high proliferation rate of fibroblasts compared with that of keratinocytes even under optimum conditions for keratinocyte growth.

The greatest advance in the development of a long-term keratinocyte culture method came in 1975, when Rheinwald and Green reported the serial cultivation of pure cultures of keratinocytes from a single-cell suspension of epidermal cells *(2)*. This was achieved by growing the cells in serum-containing medium on a mesenchymal feeder cell layer (irradiated mouse 3T3 cells). Using this feeder layer of viable, yet nonproliferating, mesenchymal cells, reduced fibroblast contamination and growth vastly, if not completely, but enhanced the proliferation of keratinocytes. The longevity of these keratinocyte cultures was further improved on the addition of a variety of mitogens discovered to be important for the health and growth of keratinocytes. A list of these cytokines and the relevant references are given in Section 2. The most vital of these mitogens is epidermal growth factor (EGF) *(3)*.

Since the introduction by Rheinwald and Green of a method of long-term culture of keratinocytes, alternative culture methods have been developed, each being designed for specific experimental requirements. The degree to which the pattern of keratinocyte differentiation in vivo is reproduced in vitro depends on the conditions and methods of culture. Keratinocyte cultures can vary from undifferentiated monolayers *(4)* under low calcium conditions (<0.06 mM) to fully differentiated stratified multilayers achieved when grown in skin-equivalent cultures *(5–8)*. Skin equivalents have been fashioned by growing keratinocytes on the following:

1. Collagen disks: Collagen-based, thin permeable membranes produced by ICN Flow.
2. ECM gels: Usually collagen-based, but can contain other extracellular matrix (ECM) constituents, such as laminin or fibronectin.
3. Dermal equivalent: ECM gels that contain viable mesenchymal cells, such as dermal fibroblasts or fibroblastic cell lines.
4. DED: De-epidermized dermis. Produced by multiple freeze–thawing of a piece of skin, after which the dead epidermis can simply be peeled off. This treatment kills off all endogenous cells, leaving an uninhabited connective tissue skeleton, which can then be reseeded with cells of the experimenters choice.

The more closely the culture conditions duplicate the tissue environment (i.e., acidic pH, collagenous substrate, presence of mesenchymal cells, air interface, etc.), the more complete the expression of epidermal differentiation characteristics.

The method detailed in this chapter is adapted from that of Rheinwald and Green, and allows the long-term maintenance of keratinocytes in culture supplying a stock of healthy cells, which can be used either as they are or in any of the alternative methods mentioned for experimentation.

2. Materials

1. 3T3 cells: available from the European Collection of Animal Cell Cultures (ECACC #88031146).
2. 3T3 medium: Dulbecco's modified Eagle's medium (DMEM) supplemented with: 10% fetal calf serum (FCS), 4 mM L-glutamine, 100 U/mL penicillin, 100 µg/mL streptomycin. The shelf life of this medium is approx 4–6 wk. Owing to the instability of L-glutamine at 4°C, however, fresh L-glutamine can be added. Media supplements are stored as concentrated stocks at –20°C, and media (both basic and supplemented) are stored at 4°C. Note: All reagents are available from tissue-culture retailers. FCS should be batch tested to obtain optimum serum for cell growth.
3. Freshly isolated normal human skin in the form of abdominal or breast reductions or circumcisions.
4. Skin transport medium: DMEM supplemented with: 10% FCS, 100 U/mL penicillin, 100 µg/mL streptomycin, 2.5 µg/mL fungizone, 50 µg/mL gentamicin. Media supplements are stored as concentrated stocks at –20°C, and media (both basic and supplemented) are stored at 4°C.
5. Keratinocyte growth medium (KGM): made up of a 3:1 (v/v) mixture of Ham's F12 and DMEM media, supplemented with: 10% FCS, 4 mM L-glutamine, 100 U/mL penicillin, 100 µg/mL streptomycin, 0.4 µg/mL hydrocortisone *(9)*, $10^{-10}M$ cholera enterotoxin *(10)*, 5 µg/mL transferrin *(11)*, $2 \times 10^{-11}M$ liothyronine *(11)*, $1.8 \times 10^{-4}M$ adenine *(12)*, 5 µg/mL insulin *(11)*, and 10 ng/mL EGF *(9)*. This media should be used fresh if possible, but has a shelf life of approx 1 wk. Media supplements are stored as concentrated stocks at –20°C, and media (both basic and supplemented) are stored at 4°C. Stock supplements are usually made up in phosphate-buffered saline (PBS) containing a carrier protein, such as 0.1% bovine serum albumin. Certain media supplements require dissolving as follows: Na tri-iodothyronine dissolves initially in 1 part of HCl and 2 parts ethanol, adenine dissolves in NaOH, pH 9.0, insulin dissolves in 0.05M HCl, and hydrocortisone dissolves in EtOH.
6. PBS: All PBS referred to in this text lacks calcium and magnesium ions and is made up of the following: 1% (w/v) NaCl, 0.025% KCl, 0.144% Na_2HPO_4, and 0.025% KH_2PO_4. This solution is pH-adjusted to 7.2, autoclaved at 121°C (15 psi) for 15 min, and stored at room temperature.
7. Trypsinization solution: 1 vol of trypsin stock is added to 4 vol of EDTA stock and used immediately. Trypsin stock: trypsin (Difco, Detroit, MI 1:250) is made up of 0.25% (w/v) Tris-saline, pH 7.7 (0.8% NaCl, 0.0038% KCl, 0.01% Na_2HPO_4, 0.1% dextrose, 0.3% trizma base). Stocks are filter sterilized and stored aliquoted at –20°C. EDTA stock: EDTA is made up of 0.02% (w/v) EDTA in Ca- and Mg-free PBS, autoclaved at 121°C (15 psi) for 15 min, and stored at room temperature.

8. The following sterile equipment is required: forceps, scalpel, iris scissors, hypodermic needles, medical gauze, tissue-culture flasks, and Petri dishes.
9. Mitomycin-C stock solution: dissolve in sterile H_2O to a concentration of 400 µg/mL. Store at 4°C in the absence of light. Solution is stable for 3–4 mo.
10. Dispase medium: 3T3 medium containing 2 mg/mL Dispase (Boehringer Mannheim, Mannheim, Germany) and filter sterilized. Use immediately.

3. Methods
3.1. Routine Maintenance of 3T3 Cell Line

These adherent cells are grown in 3T3 media at 37°C to near confluence (*see* Note 1) and passaged as follows:

1. Remove media from flask, and wash cells with an equivalent volume of PBS.
2. Add the trypsinization mixture to the flask at approx 1.5 mL/25 cm^2 surface area, and incubate at 37°C for approx 5 min, or until all cells have rounded up.
3. Add 4 vol of medium to deactivate the trypsin and EDTA, and disperse the cells with repeated pipeting.
4. Estimate the cell number using a hemocytometer, and pellet the cells at approx 300*g* for 5 min.
5. Resuspend the cells in fresh media, and seed into flasks or Petri dishes at approx 3×10^3 cells/cm^2 of surface area. Note: Density of cells at seeding can be varied depending on when confluence is required.
6. Cells should reach confluence in approx 3–5 d.

3.2. Production of Feeder Layers

1. Select flasks of exponentially growing 3T3, which have no more than 50% of the flask's surface area covered by cells, replace the media, and incubate for a further 24 h.
2. Add approx 1–10 µg of mitomycin-C/mL of medium (*see* Note 2), and incubate for a further 12 h.
3. Wash the flask three times with fresh medium. Incubate the cells with the final wash for approx 10–20 min at 37°C.
4. Harvest the cells immediately by trypsinization in the usual manner (detailed in Section 3.1.) and seed in fresh flasks at approx 2.5×10^4 cells/cm^2 in keratinocyte media (1 mL media/5 cm^2 of plastic surface area).
5. Incubate at 37°C for approx 12 h to allow the cells to adhere and spread before seeding with keratinocytes.

3.3. Initiation of Keratinocyte Cultures

1. Place skin sample directly from patient (*see* Note 3) as sterilely as possible into a Universal containing a covering volume of skin transport media at 4°C (*see* Note 4).
2. Before processing, remove skin from the transport media, submerge briefly in alcohol three times, and shake dry in the tissue-culture hood.

3. Place skin into a shallow sterile container (a 10-cm Petri dish is perfect for small skin samples), and using fine forceps and iris scissors, trim away the hypodermis, i.e., the adipose and loose connective tissue, until only the epidermis and the relatively dense dermis remain (*see* Notes 5 and 6).
4. Flatten the skin, epidermis down, onto the surface of the Petri dish and using a sterile scalpel, cut the skin into long 2–3-mm thin strips.
5. Place the strips into a Universal containing at least a covering amount of dispase medium, and incubate either overnight at 4°C or for 2–4 h at 37°C.
6. After the incubation, remove the strips of skin from the dispase media, dab excess media off on the inside of the lid of a 10-cm Petri dish, and place the relatively media-free strips into the Petri dish. Peel the epidermis away from the dermis with two sterile hypodermic needles. The epidermis is a semiopaque thin layer, whereas the connective tissue of the dermis will have absorbed fluid and will appear as a thick swollen slightly gelatinous layer. This should come away easily. If sections remain attached, then either the strips were too thick or further incubation in dispase is required. (Note: This should not be a problem after incubation overnight at 4°C.)
7. Place the epidermis strips only into 5 mL of trypsin stock solution, and shake rapidly for 1 min. Add 15 mL of DMEM/10% FCS to inactivate the trypsin. Remove upper epidermal layer pieces by pipeting through sterile gauze into a sterile universal.
8. Pellet the single-cell suspension by centrifugation at approx $300g$ for 5 min, resuspend in keratinocyte growth media, and count. Seed at approx $2–5 \times 10^4$ viable cells/cm^2 onto the preplated feeder layers.

3.4. Routine Culture of Keratinocyte Strains

1. Regularly change the medium twice per week.
2. With time, the 3T3 feeder cells will begin to die and detach from the flask. Replace these with fresh feeder cells as necessary (*see* Note 7).
3. The cultures should reach confluence within 10–14 d. It is important to passage the keratinocytes before they reach confluence, i.e., when they cover approx 70–80% of the surface area of the flask or Petri dish (*see* Note 8).
4. To passage the keratinocytes, proceed as described for passaging of 3T3 cells in Section 3.1., steps 1–4, with the exception that keratinocytes will take longer to trypsinize (10–15 min) and will require vigorous agitation of the flask to detach the rounded up cells from the surface.
5. Once counted, seed the keratinocytes onto fresh feeder layers at a density of approx $5–50 \times 10^3$ viable cells/cm^2. The density seeded depends on when confluence is required. Healthy secondary keratinocyte cultures should reach confluence within 7–10 d (*see* Note 9).

4. Notes

1. The 3T3 cell line is an undemanding cell line to maintain in culture, and few problems should be encountered by a competent tissue culturist. The only thing

to note is that cells that have become overconfluent and begun to pile up (i.e., form foci) should not be used either for the continuation of stocks or for the production of feeder cells, since the cells appear to transform further and can become resistant to irradiation or mitomycin-C treatment, maintaining their proliferative ability and thereby overrunning keratinocyte cultures.
2. The exact concentration of mitomycin-C required to produce viable, yet nonproliferating 3T3 cells should be titrated, since it varies with batch. Alternatively, feeder cells can be produced by irradiation with approx 6000 rads using a γ-irradiator (cobalt^{60}). This can be performed on 3T3 cells that are either attached to the flask surface or in suspension (depending on the size of your irradiator). The exact dose of radiation required to produce viable, yet nonproliferating cells must be titrated.
3. Generally, keratinocyte cultures from younger patients (<16 yr old) have a greater growth potential. It is advised that cultures should not be initiated from patients older than 60 yr.
4. Skin samples remain viable for up to 20 h when stored in skin transport media at 4°C.
5. The density of the hypodermis varies with the biopsy site. For foreskins, the hypodermis is particularly loose and therefore easily dissectible, whereas skin taken from the back has an extremely dense hypodermis, which proves difficult to remove. In the latter case, just remove as much extraneous connective tissue as possible.
6. Skin is often contaminated with bacteria or yeast. Submerging the skin sample in alcohol before processing should kill most forms of contamination. However, pockets of bacteria which have become trapped in sweat or sebaceous pores may be present. Foreskins are particularly prone to blocked pores. Fortunately, once the skin is stretched upside down across the Petri dish, the presence of blocked pores is usually obvious. The affected areas of skin should be carefully dissected out and discarded, taking particular care not to cut into the blocked pore.
7. An adequate feeder layer density is extremely important for the continued growth of keratinocytes and the reduction of fibroblast growth. A good feeder layer should cover approx 70% of the surface area.
8. In order to maintain healthy cultures of rapidly growing keratinocytes (*see* Fig. 1), it is imperative that keratinocyte cultures are passaged well before full confluence is reached, when approx 70% of the flask's surface area is covered with keratinocytes. If this is not done, then the underlying proliferative keratinocytes will begin to die off or differentiate. This is presumably owing to the relatively impermeable multiple layers of differentiating keratinocytes reducing the nutrients available to the basal layer.
9. Keratinocyte stocks can be successfully stored in liquid nitrogen. Only cultures of rapidly growing keratinocytes should be chosen for freezing, i.e., select flasks where only 50% of the surface area is covered by keratinocyte colonies. Trypsinize, count, and pellet the cells as usual. Resuspend the cells at approx $1–5 \times 10^6$ cells/mL in 90% FCS and 10% dimethyl sulfoxide, place into cryotubes

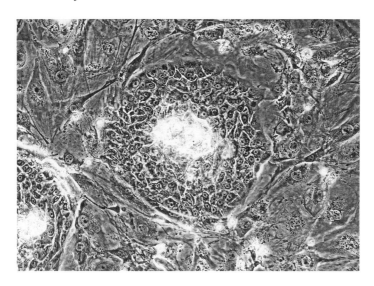

Fig. 1. A 7-d-old primary culture showing a healthy keratinocyte colony (center) surrounded by dying feeder cells. Note the symmetrical appearance of the colony, the smooth rounded edges of which are typical of rapidly growing keratinocyte colonies. The phase-bright debris located at the center of the colony is commonly seen, particularly in primary cultures, and is thought to be owing to cellular aggregation of terminally differentiating cells to the proliferating cells before the latter adhere to the plastic and begin to grow. Magnification 200×.

immediately, insulate tubes (wrap in multiple tissue layers or place within polystyrene container), and freeze overnight at –80°C, before placing into liquid nitrogen.

References

1. Cruickshank, C. N., Cooper, J. R., and Hooper, C. (1960) The cultivation of cells from adult epidermis. *J. Invest. Dermatol.* **34,** 339–342.
2. Rheinwald, J. G. and Green, H. (1975) Serial cultivation of strains of human epidermal keratinocytes: the formation of keratinizing colonies from single cells. *Cell* **6,** 331–344.
3. Rheinwald, J. G. and Green, H. (1977) Epidermal growth factor and the multiplication of cultured human epidermal keratinocytes. *Nature* **265,** 421–424.
4. Boyce, S. T. and Ham, R. G. (1983) Calcium regulated differentiation of normal epidermal keratinocytes in chemically defined clonal culture and serum-free serial culture. *J. Invest. Dermatol.* **81,** 33s–40s.
5. Bell, E., Sher, S., Hull, B., Merril, C., Rosen, S., Chamson, A., Asselineau, D., Dubertret, L., Coulomb, B., Lapiere, C., Nusgens, B., and Neveux, Y. (1983) The reconstitution of living skin. *J. Invest. Derm.* **81,** 2s–10s.

6. Pruneiras, M., Regnier, M., and Woodley, D. (1983) Methods of cultivation of keratinocytes at an air liquid interface. *J. Invest. Dermatol.* **81,** 28s–33s.
7. Boyce, S. T., Christianson, D. J., and Hansborough, J. F. (1988) Structure of a collagen-glycosaminoglycan dermal skin substitute optimised for cultured human epidermal keratinocytes. *J. Biomed. Mater. Res.* **22,** 939–957.
8. Yannas, I. V., Lee, E., Orgill, D. P., Skrabut, E. M., and Murphy, G. F. (1989) Synthesis and characterisation of a model extracellular matrix that induces partial regeneration of adult mammalian skin. *Proc. Natl. Acad. Sci. USA* **86,** 933–937.
9. Rheinwald, J. G. (1980) Serial cultivation of normal human epidermal keratinocytes. *Methods. Cell Biol.* **21,** 229–254.
10. Green, H. (1978) Cyclic AMP in relation to proliferation of the epidermal cell: a new view. *Cell* **15,** 801–811.
11. Watt, F. M. and Green, H. (1981) Involucrin synthesis is correlated with cell size in human epidermal cultures. *J. Cell Biol.* **90,** 738–742.
12. Wu, Y. J., Parker, L. M., Binder, N. E., Beckett, M. A., Sinard, J. H., Griffiths, C. T., and Rheinwald, J. G. (1982) The mesothelial keratins: a new family of cytoskeletal proteins identified in cultured mesothelial cells and non-keratinising epithelia. *Cell* **31,** 693–703.

2

Cultivation of Normal Human Epidermal Melanocytes

Mei-Yu Hsu and Meenhard Herlyn

1. Introduction

An important approach in studies of normal, diseased, and malignant cells is their growth in culture. The isolation and subsequent culture of human epidermal melanocytes has been attempted since 1957 *(1–5)*, but only since 1982 have pure normal human melanocyte cultures been reproducibly established to yield cells in sufficient quantity for biological, biochemical, and molecular analyses *(6)*. Selective growth of melanocytes, which comprise only 3–7% of epidermal cells in normal human skin, was achieved by suppressing the growth of keratinocytes and fibroblasts in epidermal cell suspensions with the tumor promoter 12-*O*-tetradecanoyl phorbol-13-acetate (TPA) and the intracellular cyclic adenosine 3', 5' monophosphate (cAMP) enhancer cholera toxin, respectively, which both also act as melanocyte growth promoters. Recent progress in basic cell-culture technology, along with an improved understanding of culture requirements, has led to an effective and standardized isolation method, and special culture media for selective growth and long-term maintenance of human melanocytes. The detailed description of this method is aimed at encouraging its use in basic and applied biological research.

2. Materials

1. Normal skin-transporting medium: The medium for collecting normal skin is composed of Hank's balanced salt solution (HBSS without Ca^{2+} and Mg^{2+}; Gibco-BRL [Grand Island, NY], #21250-089) supplemented with penicillin (100 U/mL; USB [Cleveland, OH], #199B5), streptomycin (100 µg/mL; USB, #21B65), gentamicin (100 µg/mL; BioWittaker [Walkersville, MD], #17-518Z), and fungizone (0.25 µg/mL; JRH Biosciences [Lenexa, KS], #59-604-076). After sterilization through a 0.2-µm filter, the skin-transporting medium is transferred into sterile containers in 20-mL aliquots and stored at 4°C for up to 1 mo.

From: *Methods in Molecular Medicine: Human Cell Culture Protocols*
Edited by: G. E. Jones Humana Press Inc., Totowa, NJ

2. Epidermal isolation solution: Dilute 0.5 mL of 2.5% trypsin solution (Bio-Wittaker, #17-160E) with 4.5 mL of HBSS without Ca^{2+} and Mg^{2+} at pH 7.4 to yield a final trypsin concentration of 0.25%.
3. Cell-dispersal solution: 1.25 U/mL dispase (neutral protease, grade II; Boehringer Mannheim [Indianapolis, IN], #295-825), 0.1% (w/v) hyaluronidase (type 1S from bovine testis; Sigma [St. Louis, MO], #H3506), and 10% heat-inactivated fetal calf serum (FCS; Sigma, #F2442) in MCDB 153 medium (Sigma, #M7403) supplemented with 2 mM $CaCl_2$ and mixed with Leibovitz's L-15 at a 4:1 (v/v) ratio.
4. Bovine pituitary extract *(7,8)*: The following should be prepared before extraction:
    a. Weigh out bovine pituitary glands (Pel-Freeze Biologicals [Rogers, AR], #57133-2) into 25–30-g batches. Place in Ziploc® bags, and store at –70°C.
    b. Thirty liters of cold (4°C) 1X phosphate-buffered saline (PBS) without Ca^{2+} and Mg^{2+}.
    c. One liter of cold 0.15M NaCl saline solution.
    d. One liter of 0.2 mM EDTA solution.
    e. Prechill high-speed centrifuge rotor at 4°C.
    f. Boil 6000–8000 Dalton dialysis tubing (10 strips about 2 ft in length; Spectrum Medical Industries [Los Angeles, CA], #132655) twice in ddH_2O and once in EDTA solution (prepared in Section 2., step 4d). Boil for 20 min each time. Leave tubing in beaker filled with 1X PBS, and store at 4°C for up to 3 d.

    The bovine pituitary extract is prepared as follows (steps g–p are performed in a cold room):
    g. Thaw and rinse pituitary glands in ddH_2O, handling each batch of preweighed pituitary glands separately.
    h. Pulse-blend thawed pituitary glands in a blender containing 2.38 mL cold saline solution/g of pituitary gland to break up large pieces. Pulses should not exceed 30 s because the temperature must remain low. Transfer the resulting mixture into a 2-L flask.
    i. Repeat steps g and h for all batches of pituitary glands.
    j. Stir the pooled mixture in the 2-L flask for 90 min.
    k. Pour the mixture into plastic centrifuge bottles.
    l. Spin at 12,000g for 45 min at 4°C in a prechilled rotor to remove debris.
    m. Pipet supernatant from bottles into dialysis tubing, and begin dialysis against cold 1X PBS in a cold room.
    n. Change the buffer three times in 3 d.
    o. Filter sequentially through low protein-binding filters of 0.45 µm (Millipore [Marlborough, MA], SLHV025 LS) to eliminate any fragments and 0.2 µm (Millipore, SLGV025 LS) to sterilize the filtrate. Prepare 5-mL aliquots, and store at –70°C for up to 6 mo. Once thawed, pituitary extract should be diluted in medium immediately.
    p. Determine the protein concentration of the extract using a protein assay kit (Pierce [Rockford, IL], BCA Protein Assay Reagent, #23225H), and titrate the optimal concentration in medium (approx 40 µg/mL).

5. Melanocyte growth medium (MGM): The following stock solutions are required:
   a. Insulin (Sigma, #I5500), 5 mg/mL stock. Dissolve 0.5 g of crystalline insulin in 1 mL of $0.01M$ HCl solution, and bring the volume up to 100 mL with ddH_2O. Filter-sterilize, prepare single-use aliquots in sterile vials, and store at $-70°C$ for up to 6 mo. Do not thaw and refreeze. Use 0.5 mL/500 mL of medium to give a final insulin concentration of 5 µg/mL.
   b. Epidermal growth factor (EGF; Sigma, #E4127), 5 µg/mL stock. Suspend 0.1 mg of EGF powder in 20 mL of HBSS without Ca^{2+} and Mg^{2+}. Sterilize through a 0.2-µm filter, prepare 1-mL aliquots, and store at $-70°C$. Use 0.5 mL/500 mL of medium to give a final EGF concentration of 5 ng/mL. Avoid repeated freezing and thawing.
   c. TPA (Chemicals for Cancer Research [Chanhassen, MN], #8005), 0.25 mg/mL stock. Dissolve 10 mg of TPA in 40 mL of 100% ethanol, aliquot, seal with Parafilm®, and store at $-20°C$. Use 20 µL/500 mL of medium to give a final TPA concentration of 10 ng/mL.
   MGM is prepared as follows: Mix MCDB 153 (Sigma, #M7403) supplemented with 2 mM of $CaCl_2$, with Leibovitz's L-15 (Gibco-BRL, #41300-070) at a 4:1 ratio (v/v), and add 2% heat-inactivated FCS (Sigma, #F2442), 5 µg/mL of insulin, 5 ng/mL of EGF, 10 ng/mL of TPA, and bovine pituitary extract to yield 40 µg/mL of pituitary protein in the medium. Store the MGM at 4°C for up to 8 d.
6. Trypsin-versene solution: Make a 5X stock by mixing 0.5 mL of trypsin solution (2.5%; BioWittaker, #17-160E) with 100 mL of versene solution composed of 0.1% EDTA (Fisher [Pittsburgh, PA], #02793-500) in Ca^{2+}- and Mg^{2+}-free PBS (pH 7.4). To prepare trypsin-versene solution, dilute 5X stock with HBSS (Ca^{2+}-, Mg^{2+}-free) to give a final concentration of 0.0025% trypsin and 0.02% versene.
7. Cell-preservative medium: Prepare 5% (v/v) dimethyl sulfoxide (DMSO; Sigma, #D2650) in 95% heat-inactivated FCS as needed.

3. Methods
3.1. Day 1

1. Prepare the following in a laminar flow hood: one pair of sterile forceps, curved scissors, and surgical scalpel blade; 5 mL of epidermal isolation solution (*see* Section 2., step 2) in a sterile centrifuge tube; 10 mL of Ca^{2+}- and Mg^{2+}-free HBSS in a sterile nontissue-culture Petri dish; and 10 mL of 70% ethanol in a separate sterile Petri dish.
2. Soak the skin specimens in 70% ethanol for 30 s. Transfer skin to another Petri dish containing HBSS to rinse off ethanol (*see* Notes 1 and 2).
3. Cut skin-ring open, and trim off fat and subcutaneous tissue with scissors (*see* Note 3).
4. Cut skin into pieces (approx 2 × 3 mm²) using the surgical scalpel blade with one-motion cuts (*see* Note 4).
5. Transfer the pieces into the tube containing epidermal isolation solution. Cap, invert, and incubate the tube in the refrigerator at 4°C for 18–24 h (*see* Note 5).

3.2. Day 2

1. Prepare the following in a laminar flow hood: one pair of sterile forceps and a surgical scalpel blade; 10 mL of Ca^{2+}- and Mg^{2+}-free HBSS in a sterile, nontissue-culture Petri dish; two empty sterile Petri dishes; and 5 mL of cell-dispersal solution in a 15-mL centrifuge tube.
2. Pour tissue in epidermal isolation solution into one of the empty Petri dishes. Transfer tissue pieces to the Petri dish containing HBSS. Separate epidermis (thin translucent layer) from dermis (thick opaque layer) using the forceps. Hold the dermal part of the skin piece with the forceps, and gently slide the epidermal side on the dry surface of a nontissue-culture Petri dish. The epidermis should stick to the Petri dish. Discard the dermis immediately (*see* Note 6).
3. Add a drop of cell-dispersal solution on the resulting epidermis to avoid drying and to neutralize trypsin while isolating the epidermis from the remaining skin pieces. Repeat procedure in steps 2 and 3 for each piece of tissue (*see* Note 6).
4. Transfer the collected epidermal sheets with cell-dispersal solution from the Petri dish to the centrifuge tube. Incubate the tube at 37°C for 2–4 h depending on cell disaggregation. Vortex the tube vigorously to release single cells from epidermal sheets. Wash the resulting single-cell suspension three times with Ca^{2+}- and Mg^{2+}-free HBSS. Centrifuge for 5 min at 2000 rpm at room temperature. Aspirate the supernatant, which may contain remaining stratum corneum. Resuspend the pellet with MGM (*see* Note 7).
5. Plate the resulting epidermal cell suspension at approx 2×10^5 cells/cm^2 in the tissue-culture vessel. Incubate at 37°C in 5% CO_2/95% air for 48–72 h.

3.3. After 2 Days

1. Wash culture with MGM to remove nonadherent cells. Medium change should be performed twice each week. If the culture is contaminated with fibroblasts, start treatment with MGM containing 200 µg/mL of geneticin (G418; Gibco-BRL, #11811) for 2–3 d. Seventy percent confluent primary cultures can be obtained in 2 wk (*see* Note 8).
2. Subcultivation: Primary cultures established from foreskins usually reach 70% confluence within 12 d. Cultures are treated with trypsin-versene solution (*see* Section 2., step 6) at room temperature for 1 min, harvested with Leibovitz's L-15 containing 10% heat-inactivated FCS, centrifuged at 2000 rpm for 3 min, resuspended in MGM, reinoculated at 10^4 cells/cm^2, and serially passaged. Medium is changed twice each week.
3. Cryopreservation: Melanocyte suspensions harvested by trypsin-versene and Leibovitz's L-15 containing 10% FCS are centrifuged at 2000 rpm for 5 min and resuspended in cell-preservative medium (*see* Section 2., step 7) containing 5% DMSO as a cryopreservative. Cells are normally suspended at a density of 10^6/mL and transferred to cryotubes. The tubes are then placed in a plastic sandwich box (Nalgene™ Cryo 1°C Freezing Container; Nalge [Rochester, NY], #5100-0001), which is immediately transferred to a –70°C freezer. The insulation of the box ensures gradual cooling of the cryotubes and results in over 80%

viability of the cells on thawing. After overnight storage in the freezer, the cryotubes are placed in permanent storage in liquid nitrogen.
4. Thawing: The melanocyte suspension is thawed by placing the cryotube in a water bath at 37°C. When the cell-preservative medium is almost, but not totally defrosted, the outside of the tube is wiped with 70% ethanol. The cell suspension is then withdrawn, quickly diluted in MGM at room temperature, centrifuged, and resuspended in fresh MGM. Cell viability is determined by Trypan Blue exclusion. The resulting melanocytes are then seeded at a density of $10^4/cm^2$.

3.4. Results

3.4.1. Minimal Growth Requirements

Earlier studies of normal melanocytes *(6,9,10)* were done using media containing 5–15% FCS, which provides a host of poorly characterized growth-promoting activities. Deprivation of serum and brain tissue extracts from media has led to the delineation of four groups of chemically defined melanocyte mitogens.

1. Peptide growth factors, including basic fibroblast growth factor (bFGF; *11,12*), which is the main growth-promoting polypeptide in bovine hypothalamus and pituitary extracts; insulin/insulin-like growth factor-1 (IGF-1; *13*); EGF *(14,15)*; transforming growth factor-α (TGF-α; *16*); endothelins (ET; *17*); and hepatocyte growth factor/scatter factor (HGF/SF; *18,19*).
2. Calcium, since reduction of Ca^{2+} concentrations in MGM from an optimal 2.0 to 0.03 mM reduces cell growth by approx 50% *(20)*, and cation-binding proteins, such as tyrosinase at $10^{-11}M$, and ceruloplasmin at 0.6 U/mL *(20)*.
3. Enhancers of intracellular levels of cAMP, including α-melanocyte-stimulating hormone (α-MSH) at 10 ng/mL *(21)*; forskolin at $10^{-9}M$ *(20)*, follicle-stimulating hormone (FSH) at $10^{-7}M$ *(22)*; and cholera toxin at $10^{-12}M$ *(6,20,23,24)*.
4. Activators of protein kinase C (PKC), such as TPA *(25)*, which is lipophilic and cannot be removed by simple washing, and 20-oxo-phorbol-12,13-dibutyrate (PDBu) at $10^{-6}M$ *(25)*, which is a similar derivative, but more hydrophilic. Recent data suggest that the tigliane class phorbol compounds, such as 12 deoxyphorbol,13 isobutyrate (DPIB) and 12 deoxyphorbol,13 phenylacetate (DPPA), which possess diminished tumor-promoting activity, are able to activate PKC as well as stimulate melanocyte proliferation *(26)*.

3.4.2. Morphology

Human epidermal melanocytes grown in MGM normally exhibit a bi- or tripolar morphology with varying degrees of pigmentation (Fig. 1). Dendricity may increase at higher passage levels.

3.4.3. Expression of Antigens

Extensive studies have been done to characterize the antigenic phenotype of malignant melanoma cells *(27)*. On the other hand, very few attempts have

Fig. 1. Morphology of normal human epidermal melanocytes grown in medium supplemented with bFGF (pituitary extract), serum, and phorbol ester.

been made to produce monoclonal antibodies (MAbs) to normal melanocytes *(15,28)*. Cultured melanocytes share with melanoma cells the expression of a variety of cell-surface antigens (melanoma-associated antigens), including p97 melanotransferrin, integrin β_3 subunit of the vitronectin receptor, gangliosides GD_3 and 9-O-acetyl GD_3, chondroitin sulfate proteoglycan *(15)*, and Mel-CAM/MUC18 *(29)*. However, these antigens are not expressed by normal melanocytes in situ *(30)*. Table 1 summarizes the expression of antigens on melanocytes *in situ* and in culture. The observed divergent antigenic phenotype in culture and *in situ* suggests a role for epidermal microenvironmental signals in controlling the melanocytic phenotype. Indeed, accumulating evidence indicates that undifferentiated keratinocytes can control proliferation, morphology, pigmentation, and antigen expression of melanocytes in coculture *(28,32–35)*. However, the underlying mechanisms responsible for keratinocyte–melanocyte interactions remain unclear.

3.4.4. Growth Characteristics

Melanocytes from neonatal foreskin can be established with a success rate of 80% and have a maximum life-span of 60 doublings, with a doubling time of 2–6 d. Heavily pigmented cells isolated from black individuals have a shorter doubling time and tend to senesce after 20–30 doublings. By contrast, epider-

Table 1
Expression of Antigens on Melanocytes *In Situ* and in Culture[a]

Antigens	*In situ*	In culture[b]
CD26	++++	++++
gp145	++	++
TRP-1	++++	++++
E-cadherin	++++	++++
α-Catenin	++++	++++
β-Catenin	++++	++++
Integrin $β_3$ subunit	–	++++
Tenascin	–	+
Fibronectin	–	++++
Chondroitin sulfate proteogylcan	±	++++
p97 Melanotransferrin	–	++++
NGF-receptor (p75)	±	++++
9-*O*-acetyl GD_3	±	+++
GD_3	±	++++
HLA-DR	–	–
Mel-CAM (MUC18)	–	++++

[a]–, Lack of expression; ±, 0–20%; +, 20–40%; ++, 40–60%; +++, 60–80%; ++++, 80–100%.
[b]Results were obtained with melanocytes grown in MGM containing TPA at passages 2–20, with the exception of gp145, which is expressed more strongly on melanocytes cultured in the absence of TPA (*31* and unpublished data).

mal melanocytes from adult skin only grow in about 10% of cases and for no more than 10 doublings with a doubling time of 7–14 d. The cells do not grow beyond 70% confluency and exhibit signs of growth arrest by contact inhibition. Normal melanocytes do not proliferate anchorage independently in soft agar and are nontumorigenic in athymic nude mice *(14)*.

4. Notes

1. Tissue source and collection: The sources of tissue for melanocyte cultures are human neonatal foreskins obtained from routine circumcision and normal adult skin acquired from reduction mammoplasty. At the time of excision, the skin is placed into a sterile container with 20 mL of normal skin-transporting medium (*see* Section 2., step 1) supplied in advance and kept near the surgical area at 4°C. Specimens are delivered immediately to the tissue-culture laboratory or stored at 4°C. Neonatal foreskins can be kept for up to 48 h, and normal adult skin, for up to 24 h. However, the fresher the specimens, the higher the yield of live cells on isolation.
2. Sterilization of skin specimens: Reduce contamination by a short treatment (30 s) of intact skin with 70% ethanol in a laminar flow hood. After sterilization, rinse samples with HBSS.

3. Preliminary tissue preparation: Place tissues in a 100-mm nontissue-culture Petri dish, and remove most of the subcutaneous fat and membranous material with curved scissors.
4. Adjustment of tissue size for trypsinization: To improve reagent penetration, cut the skin samples into small pieces (approx 2×3 mm^2) rinsed in HBSS.
5. Trypsinization: Since the first report in 1941 *(24)*, epidermal cell suspensions have usually been prepared using enzymes (most commonly, trypsin). Pieces of skin are incubated in epidermal isolation solution for up to 24 h at 4°C.
6. Separation of epidermis from dermis: After incubation with trypsin, the epidermal isolation solution is replaced by HBSS. As originally described *(24)*, crude trypsin splits epidermis from the dermis along the basement membrane. Since melanocytes are located just above the basement membrane, removal of this lowest layer of epidermal cells requires some effort. Each piece of skin is held with forceps with dermal side up. The epidermal sheet is detached by sliding the specimen onto the dry surface of a nontissue-culture Petri dish. To prevent the epidermis from drying and to stop trypsinization, a drop of cell-dispersal solution containing 10% FCS can be added to the resulting epidermal sheet. To avoid potential sources of fibroblast contamination, dermal pieces should be discarded immediately once they are separated from the epidermis, and the forceps used to hold the dermis should never touch the epidermal sheets. Contaminated dermis is easily recognized by its white color in contrast to the yellowish-brown color of the epidermis. Isolated epidermal sheets in cell-dispersal solution are then transferred to a centrifuge tube.
7. Cell dispersal techniques: A single-cell suspension is prepared by enzymatic treatment. In the centrifuge tube, clumps of epidermal tissue are dissociated by cell-dispersal solution at 37°C for 2–4 h. The resulting single-cell suspension is washed three times with Ca^{2+}- and Mg^{2+}-free HBSS to remove enzymes. Cells are then pelleted by centrifugation at 2000 rpm for 5 min and resuspended in MGM. The epidermal cell suspension is then seeded in single wells of 24-well plastic tissue-culture plates and incubated at 37°C in an atmosphere of 5% CO_2 and 95% air.
8. Selective growth: Most methods for growing pure cultures of melanocytes from epidermal cell suspensions depend on optimal conditions that enable melanocytes, but not keratinocytes, to attach to a substrate and proliferate. These conditions include high oxygen tension *(36)*, high seeding density *(37)*, and the presence of sodium citrate *(38)*, phorbol esters *(6)*, and 5-fluorouracil *(39)*.

   The presence of phorbol esters (TPA) not only suppresses the growth of keratinocytes, but also promotes melanocyte growth. However, despite the potency of TPA in stimulating melanocytes, even minimal fibroblast contamination will eventually result in overgrowth by these more rapidly dividing cells. Fibroblast contamination can be eliminated by treatment of cultures with MGM containing 200 mg/mL of geneticin (G418; Gibco-BRL, #11811) for 2–3 d.
9. Pitfalls and alternatives (*see* Table 2): The presence of TPA in the medium has been shown to reduce the numbers of melanosomes in human melanocytes in culture and to delay the onset of melanization *(6)*. Thus, although this reagent

Table 2
Phenotype of Neonatal Foreskin Melanocytes in Culture[a,b] (31)

Culture conditions	Growth			Morphology[c]			Pigmentation			Antigen expression					
										NGFR			gp145		
	Passage			Passage			Passage			Passage			Passage		
	1	5	8	1	5	8	1	5	8	1	5	8	1	5	8
TPA	+++	+++	+++	D	D	D	+++	+++	+++	++	+++	+++	0	0	+
No TPA	+++	+	0	S	F	F	+	0	0	++++	0				+++

[a]All cultures were maintained in the same base medium: 4 parts MCDB 153 with 1 part L-15, supplemented with insulin, EGF, bovine pituitary extract (BPE), and 2% FCS.
[b]+ to ++++, Degree of growth, pigmentation, or antigen expression; 0, no growth or >14 d doubling, no pigmentation, and no expression of antigen.
[c]S, spindle; F, flat, polygonal; D, dendritic.

supports long-term culture of human melanocytes, it may have limited use in studies of melanocyte differentiation. When melanocytes are established in medium without TPA, they grow at doubling times of 4–7 d for the first 2–3 passages and senesce by passage 5. Initially, they assume a spindle morphology, which changes by passages 3–5 to a flat, polygonal morphology *(40)*. The flat polygonal cells are unpigmented and proliferate slowly. Concomitant with the morphological and proliferative changes, there is a decrease in expression of the nerve growth factor (NGF) receptor and an increase in expression of gp145 (Table 2).

There are other alternative media for melanocyte culture. TIP medium, a TPA-containing medium, consists of 85 nM TPA, 0.1 mM isobutylmethyl xanthine (IBMX), and 10-20 µg protein/mL placental extract in Ham's F-10 medium supplemented with 10% newborn calf serum *(41)*. TPA-free medium *(42)*, composed of Ca^{2+}-free M199 medium supplemented with 5–10% chelated FCS, 10 µg/mL insulin, 10 ng/mL EGF, $10^{-9}M$ triiodothyronine, 10 µg/mL transferrin, $1.4 \times 10^{-6}M$ hydrocortisone, $10^{-9}M$ cholera toxin, and 10 ng/mL bFGF *(42)*, can also support short-term culture of melanocytes.

References

1. Hu, F., Staricco, R. J., Pinkus, H., and Fosnaugh, R. (1957) Human melanocytes in tissue culture. *J. Invest. Dermatol.* **28**, 15–32.
2. Karasek, M. and Charlton, M. E. (1980) Isolation and growth of normal human skin melanocytes. *Clin. Res.* **28**, 570A.
3. Kitano, Y. (1976) Stimulation by melanocyte stimulating hormone and dibutyryl adenosine 3', 5'-cyclic monophosphate of DNA synthesis in human melanocytes in vitro. *Arch. Derm. Res.* **257**, 47–52.
4. Mayer, T. C. (1982) The control of embryonic pigment cell proliferation in culture by cyclic AMP. *Dev. Biol.* **94**, 509–614.
5. Wilkins, L. M. and Szabo, G. C. (1981) Use of mycostatin-supplemented media to establish pure epidermal melanocyte culture (abstract). *J. Invest. Dermatol.* **76**, 332.
6. Eisinger, M. and Marko, O. (1982) Selective proliferation of normal human melanocytes in vitro in the presence of phorbol ester and cholera toxin. *Proc. Natl. Acad. Sci. USA* **79**, 2018–2022.
7. Kano-Sueoka, T., Campbell, G. R., and Gerber, M. (1977) Growth stimulating activity in bovine pituitary extract specific for a rat mammary carcinoma cell line. *J. Cell Physiol.* **93**, 417–424.
8. Tsao, M. C., Walthall, B. J., and Ham, R. G. (1982) Clonal growth of normal human epidermal keratinocytes in a defined medium. *J. Cell Physiol.* **110**, 219–229.
9. Herlyn, M., Herlyn, D., Elder, D. E., Bondi, E., LaRossa, D., Hamilton, R., Sears, H., Balaban, G., Guerry, D., Clark, W. H., and Koprowski, H. (1983) Phenotypic characteristics of cells derived from precursors of human melanoma. *Cancer Res.* **43**, 5502–5508.
10. Herlyn, M., Thurin, J., Balaban, G., Bennicelli, J., Herlyn, D., Elder, D. E., Bondi, E., Guerry, D., Nowell, P., Clark, W. H., and Koprowski, H. (1985) Characteristics of cultured human melanocytes isolated from different stages of tumor progression. *Cancer Res.* **45**, 5670–5676.

11. Halaban, R., Ghosh, S., and Baird, A. (1987) bFGF is the putative natural growth factor for human melanocytes. *In Vitro* **23**, 47–52.
12. Halaban, R., Kwon, B. S., Ghosh, S., Delli Bovi, P., and Baird, A. (1988) bFGF as an autocrine growth factor for human melanomas. *Oncogene Res.* **3**, 177–186.
13. Rodeck, U., Herlyn, M., Menssen, H. D., Furlanetto, R. W., and Koprowski, H. (1987) Metastatic but not primary melanoma cell lines grow in vitro independently of exogenous growth factors. *Int. J. Cancer* **40**, 687–690.
14. Herlyn, M., Rodeck, U., Mancianti, M. L., Cardillo, F. M., Lang, A., Zross, A. H., Jambrosic, J., and Koprowski, H. (1987) Expression of melanoma-associated antigens in rapidly dividing human melanocytes in culture. *Cancer Res.* **47**, 3057–3061.
15. Herlyn, M., Clark, W. H., Rodeck, U., Mancianti, M. L., Jambrosic, J., and Koprowski, H. (1987) Biology of tumor progression in human melanocytes. *Lab. Invest.* **56**, 461–474.
16. Pittelkow, M. R. and Shipley, G. D. (1989) Serum-free culture of normal human melanocytes: growth kinetics and growth factor requirements. *J. Cell Physiol.* **140**, 565–576.
17. Imokawa, G., Yada, Y., and Miyagishi, M. (1992) Endothelins secreted from human keratinocytes are intrinsic mitogens for human melanocytes. *J. Biol. Chem.* **267**, 24,675–24,680.
18. Halaban, R., Rubin, J. S., Funasaka, Y., Cobb, M., Boulton, T., Faletto, D., Rosen, E., Chan, A., Yoko, K., and White, W. (1992) Met and hepatocyte growth factor/scatter factor signal transduction in normal melanocytes and melanoma cells. *Oncogene* **7**, 2195–2206.
19. Matsumoto, K., Tajima, H., and Nakamura, T. (1991) Hepatocyte growth factor is a potent stimulator of human melanocyte DNA synthesis and growth. *Biochem. Biophys. Res. Commun.* **176**, 45–51.
20. Herlyn, M., Mancianti, M. L., Jambrosic, J., Bolen, J. B., and Koprowski, H. (1988) Regulatory factors that determine growth and phenotype of normal human melanocytes. *Exp. Cell Res.* **179**, 322–331.
21. Abdel-Malek, Z. A. (1988) Endocrine factors as effectors of integumental pigmentation. *Dermatol. Clin.* **6**, 175–184.
22. Adashi, E. Y., Resnick, C. E., Svoboda, M. E., and Van Wyk, J. J. (1986) Follicle-stimulating hormone enhances somatomedin C binding to cultured rat granulosa cells. *J. Biol. Chem.* **261**, 3923–3926.
23. Gilchrest, B. A., Vrabel, M. A., Flynn, E., and Szabo, G. (1984) Selective cultivation of human melanocytes from newborn and adult epidermis. *J. Invest. Dermatol.* **83**, 370–376.
24. Medawar, P. B. (1941) Sheets of pure epidermal epithelium from human skin. *Nature* **148**, 783.
25. Niedel, J. E. and Blackshear, P. J. (1986) Protein kinase C, in *Phosphoinositides and Receptor Mechanisms* (Putney, J. W., Jr., ed.), Liss, New York, pp. 47–88.
26. Cela, A., Leong, I., and Krueger, J. (1991) Tigliane-type phorbols stimulate human melanocyte proliferation: potentially safer agents for melanocyte culture. *J. Invest. Dermatol.* **96**, 987–990.

27. Herlyn, M. and Koprowski, H. (1988) Melanoma antigens: immunological and biological characterization and clinical significance. *Ann. Rev. Immunol.* **6**, 283–308.
28. Houghton, A. N., Eisinger, M., Albino, A. P., Cairncross, J. G., and Old, L. J. (1982) Surface antigens of melanocytes and melanomas: markers of melanocyte differentiation and melanoma subsets. *J. Exp. Med.* **156**, 1755–1766.
29. Shih, I.-M., Elder, D. E., Hsu, M.-Y., and Herlyn, M. (1994) Regulation of Mel-CAM/MUC18 expression on melanocytes of different stages of tumor progression by normal keratinocytes. *J. Am. Pathol.* **145**, 837–845.
30. Elder, D. E., Rodeck, U., Thurin, J., Cardillo, F., Clark, W. H., Stewart, R., and Herlyn, M. (1989) Antigenic profile of tumor progression stages in human melanocytes, nevi, and melanomas. *Cancer Res.* **49**, 5091–5096.
31. Valyi-Nagy, I. and Herlyn, M. (1991) Regulation of growth and phenotype of normal human melanocytes in culture, in *Melanoma 5, Series on Cancer Treatment and Research* (Nathanson, L., ed.), Kluwer Academic, Boston, MA, pp. 85–101.
32. Scott G. A. and Haake, A. R. (1991) Keratinocytes regulate melanocyte number in human fetal and neonatal skin equivalents. *J. Invest. Dermatol.* **97**, 776–781.
33. DeLuca, M., D'Anna, F., Bondanza, S., Franzi, A. T., and Cancedda, R. (1988) Human epithelial cells induce human melanocyte growth in vitro but only skin keratinocytes regulate its proper differentiation in the absence of dermis. *J. Cell Biol.* **107**, 1919–1926.
34. Valyi-Nagy, I., Hirka, G., Jensen, P. J., Shih, I.-M., Juhasz, I., and Herlyn, M. (1993) Undifferentiated keratinocytes control growth, morphology, and antigen expression of normal melanocytes through cell-cell contact. *Lab. Invest.* **69**, 152–159.
35. Herlyn, M. and Shih, I.-M. (1994) Interactions of melanocytes and melanoma cells with the microenvironment. *Pigment Cell Res.* **7**, 81–88.
36. Riley, P. A. (1975) Growth inhibition in normal mammalian melanocytes in vitro. *Br. J. Dermatol.* **92**, 291–304.
37. Mansur, J. D., Fukuyama, K., Gellin, G. A., and Epstein, W. L. (1978) Effects of 4-tertiary butyl catechol on tissue cultured melanocytes. *J. Invest. Dermatol.* **70**, 275–279.
38. Prunieras, M., Moreno, G., Dosso, Y., and Vinzens, G. (1976) Studies on guinea pig skin cell cultures: V. Co-cultures of pigmented melanocytes and albino keratinocytes, a model for the study of pigment transfer. *Acta Dermatovenereol.* **56**, 1–9.
39. Tsuji, T. and Karasek, M. (1983) A procedure for the isolation of primary cultures of melanocytes from newborn and adult human skin. *J. Invest. Dermatol.* **81**, 179,180.
40. Herlyn, M., Clark, W. H., Rodeck, U., Mancianti, M. L., Jambrosic, J., and Koprowski, H. (1987) Biology of tumor progression in human melanocytes. *Lab. Invest.* **56**, 461–474.
41. Halaban, R., Langdon, R., Birchall, N., Cuono, C., Baird, A., Scott, G., Moellmann, G., and McGuire, J. (1988) Basic fibroblast growth factor from keratinocytes is a natural mitogen for melanocytes. *J. Cell Biol.* **107**, 1611–1619.
42. Tang, A., Eller, M. S., Hara, M., Yaar, M., Hirohashi, S., and Gilchrest, B. A. (1994) E-cadherin is the major mediator of human melanocyte adhesion to keratinocytes in vitro. *J. Cell Sci.* **107**, 983–992.

3

Cultivation of Keratinocytes from the Outer Root Sheath of Human Hair Follicles

Alain Limat and Thomas Hunziker

1. Introduction

The outer root sheath (ORS) of hair follicles is a multilayered tissue made up predominantly by undifferentiated keratinocytes *(1,2)*. Although the functions of the ORS cells for hair growth are not established, it is known that the ORS cells can contribute to the regeneration of the epidermis, as during healing of superficial wounds where the ORS cells migrate out of the follicle to repopulate the denuded area *(3,4)*. Recent studies also suggest that stem cells for various epithelial cell populations of the skin are located in the ORS tissue *(5,6)*.

Because ORS cells can be regarded as undifferentiated epidermal keratinocytes *(1,2,7,8)*, they represent a source of easily and repeatedly available keratinocytes, avoiding the dependency on surgery or suction blister material. Moreover, the use of ORS cells is especially suited if cocultures with autologous cells (e.g., peripheral blood mononuclear cells) are performed *(9)*.

In the past, several methods for the cultivation of human ORS cells have been described, most of which were based on explanting plucked anagen hair follicles on different growth substrata, such as collagen *(10)*, bovine eye lens capsules *(11)*, or collagen gels populated with fibroblasts *(12)*.

We have developed a simple technique for the cultivation of ORS cells that yields substantially higher cell numbers and in a shorter time as compared to the explant techniques *(13,14)*. Our laboratory routine enables the initiation of primary cultures starting with ORS cells released from two hair follicles per culture dish. The main steps of this technique comprise the plucking of scalp hair follicles, the dissociation of the ORS cells from the follicle, and the plating of the dissociated ORS cells in a growth-supportive medium on a preformed feeder layer *(14,15)*. Because the culture is started with low cell numbers, a crucial point of the protocol for the primary cultivation is the use of feeder layers made of postmitotic human dermal fibroblasts *(14)*. We have found that the use of postmitotic human dermal fibroblasts instead of the conventional 3T3-feeder system *(16)* has a number of advantages, for instance, a higher reproducibility in the preparation of the feeder layers, which can be stored for several weeks in the CO_2 incubator or cryopreserved in liquid nitrogen without loss of the growth-promoting properties for epithelial cells *(14,15)*.

2. Materials
2.1. Tissue-Culture Facility

Most items needed to isolate and cultivate ORS cells from plucked hair follicles belong to the standard equipment of a cell-culture laboratory:

1. A tissue-culture cabinet (preferentially with vertical air flow).
2. A stereomicroscope.
3. A humidified, carbon dioxide (5% in air) incubator set at 37°C.
4. An inverted, phase-contrast microscope (magnification: 100 and 200×).
5. A bench-top centrifuge (e.g., Heraeus Sepatech, Osterode, Germany).
6. Fine tweezers (curved and rectilinear ones).
7. Gross forceps.
8. Fine scissors (curved and rectilinear ones).
9. Miniscalpels (Opthalmic Knife 45°, Alcon Surgical, Fort Worth, TX).
10. 35-, 60-, and 100-mm tissue-culture and bacteriological dishes (e.g., Falcon; Becton Dickinson, Bedford, MA).
11. Pasteur pipets.

2.2. Tissue-Culture Reagents and Solutions

1. Phosphate-buffered saline (PBS) (Dulbecco) with Ca^{2+} and Mg^{2+} (e.g., Seromed L 1813; Biochrom Berlin, Germany).
2. PBS (Dulbecco) without Ca^{2+} and Mg^{2+} (e.g., Seromed L 182-01).
3. Rinsing medium: DMEM (e.g., Seromed F 0435) buffered with 0.25 mM HEPES, pH 7.2 (e.g., Seromed L 1613) and containing 10% fetal calf serum (FCS) (e.g., Seromed or Gibco, Life Technologies, Gaithersburg, MD) and 40 U/mL penicillin and 40 μg/mL streptomycin (e.g., both from Seromed A 2213).
4. Trypsin 0.1% (w/v) and EDTA 0.02% (w/v) in PBS (Dulbecco) without Ca^{2+} and Mg^{2+}. Trypsin and EDTA purchased, for example, from Seromed (L 2133 and L 2113, respectively).
5. 0.02% (w/v) EDTA in PBS (Dulbecco) without Ca^{2+} and Mg^{2+}.
6. 0.05% (w/v) Trypsin, 0.02% (w/v) EDTA in PBS (Dulbecco) without Ca^{2+} and Mg^{2+}.

Cultivation of Keratinocytes

7. Medium for primary cultures of ORS cells *(13,17)*: First, 1000 mL DMEM (e.g., Seromed F 0435) and 1000 mL Ham's F12 (e.g., Seromed F 0813) are prepared, the pH adjusted to 7.0–7.3 if necessary, and sterilized through a 0.2-μm filter.
8. For the preparation of 100 mL of culture medium, 75 mL DMEM and 25 mL Ham's F12 are mixed, and the following supplements (for preparation, *see* steps 10–13) added:
    a. 1 mL adenine solution (Boehringer Mannheim; [Mannheim, Germany] 102 067).
    b. 0.1 mL insulin (Sigma [St. Louis, MO] I 5500).
    c. 0.1 mL triiodothyronine (Sigma T 2877).
    d. 0.2 mL hydrocortisone (Sigma H 4001).
    This formulation can be stored at 4°C for up to 4 wk.
9. Portions of 100 mL of final culture medium are prepared by adding the following supplements:
    a. 1 mL glutamine (Seromed K 0280).
    b. 10 μL EGF (Sigma E 4127).
    c. 0.1 mL choleratoxin (Sigma C 3012).
    d. 1 mL penicillin/streptomycin (Seromed A 2213).
    e. 1 mL fungizone (Gibco 15290-026).
    f. 10 mL FCS (e.g., Seromed or Gibco; *see* Note 4).
    This final medium has to be used within 10 d.
10. Solution of adenine: Dissolve 182 mg adenine (Boehringer Mannheim 102 067) in 100 mL bidistilled water and 0.7 mL HCl 1N, stirred until complete dissolution. Filter sterilize and make aliquots of 1.2 mL, which are to be stored at $-20°C$.
11. Solution of insulin: Dissolve 50 mg insulin (Sigma I 5500) in 10 mL 0.005N HCl. Filter sterilize and make aliquots of 0.2 mL, which are to be stored at $-20°C$.
12. Solution of hydrocortisone: Dissolve 25 mg hydrocortisone (Sigma H 4001) in 5 mL ethanol, and make aliquots of 0.5 mL, which are to be stored at $-20°C$ (stock solution). Dilute 0.4 mL of the stock solution to 10 mL with DMEM (e.g., Seromed F 0435), filter sterilize this diluted solution, and make aliquots of 0.3 mL, which are to be stored at $-20°C$.
13. Solution of triiodothyronine: Dissolve 13.6 mg triiodothyronine (Sigma T 2877) in a minimal volume of 0.02N sodium hydroxide. Bring the volume to 100 mL with H_2O, filter sterilize, and store at $-20°C$. To prepare the final solution, add 40 μL of the concentrated solution to 1960 μL DMEM, make aliquots of 0.15 mL, and store them at $-20°C$.
14. Solution of choleratoxin: Dissolve 1 mg choleratoxin (Sigma C 3012) in 1.18 mL of bidistilled water and filter sterilize (stock solution). Dilute 0.1 mL of the stock solution to 10 mL with DMEM (e.g., Seromed F 0435) containing 10% FCS, and store at 4°C.
15. Solution of EGF: Dissolve 100 μg EGF (Sigma E 4127) in 1 mL DMEM (e.g., Seromed F 0435) containing 10% FCS, filter sterilize, and make aliquots of 15 μL which are stored at $-20°C$.
16. Media for the subcultivation of ORS cells: Keratinocyte growth medium (KGM) used for the subcultivation of ORS cells on tissue-culture plastic is based on the

formulation of MCDB 153 *(18)* and can be purchased from Clonetics Corporation (San Diego, CA) or from Promocell (Heidelberg, Germany).
17. Medium for the cultivation of the fibroblasts: DMEM (e.g., Seromed F 0435) buffered with 3.7 mg/mL $NaHCO_3$ (e.g., Seromed L 1703) and containing 10% FCS (e.g., Seromed or Gibco), 10 U/mL penicillin, and 10 µg/mL streptomycin.
18. Mitomycin C (Sigma M 0503): A stock solution is prepared by dissolving mitomycin C in PBS (Dulbecco) with Ca^{2+} and Mg^{2+} at a concentration of 100 µg/mL, which is sterilized by filtration through a 0.2-µm filter. Aliquots of 0.9 mL of stock solution are stored at –20°C. For treatment of the fibroblasts, mitomycin C is used at a final concentration of 8 µg/mL (e.g., 0.8 mL stock solution/10 mL DMEM containing 10% FCS).

3. Methods

3.1. Plucking of Hair Follicles

Usually, we isolate scalp hair follicles from the occipital region. Using the same protocol, follicles from other anatomical sites, such as beard, leg, and genital region, can be isolated. Optimal recovery of the ORS tissue during the plucking procedure is achieved by observing carefully the following protocol:

1. The hairs to be plucked are exposed by pulling up the adjacent hair. A few number of hairs (maximally 3–4) are gripped with gross sterile forceps as close as possible to the skin surface. The hairs are pulled out by a jerky movement made perpendicular to the skin surface (*see* Note 1).
2. The follicular material is then directly collected into a 60-mm bacteriological dish containing 5 mL rinsing medium, by cutting with fine sterile scissors. The remaining distal keratinized hair shaft is discarded. At least one follicle has to be prepared per final milliliter of culture medium.
3. The follicles in the anagen phase (i.e., growing phase of the hair cycle *(19)*, indicated by the visible ORS tissue; *see* Fig. 1A [p. 26] and Note 2) are selected under a dissecting microscope (stereomicroscope) and transferred into a new 60-mm bacteriological dish containing 5 mL of rinsing medium.
4. We usually remove the bulbar part as well as the distal fifth of the follicular length (corresponding to the infundibular part) using miniscalpels, which ensures that the only living cell population in the remaining follicle is constituted by ORS cells (*see* Note 3). In Fig. 1A, the sites where the cuts are applied are marked by arrowheads. The micropreparative manipulation under the stereomicroscope does not necessarily need to be performed in the sterile cabinet, but can be done in a clean place.
5. The prepared follicles are rinsed four times by consecutive transfers in 60-mm bacteriological dishes containing 5 mL of rinsing medium.

3.2. Isolation of the ORS Cells from the Follicles

1. The follicles are deposited into an empty 35-mm bacteriological dish in such a way that they are in close vicinity, though separated from each other. This guar-

antees free access of the trypsin during the subsequent disaggregation step. Some residual medium is aspirated with a Pasteur pipet.
2. The follicles are covered by a minimal volume of trypsin (0.1%)/EDTA (0.02%) solution (a droplet if the number of follicles is <5 or 1 mL for 50 or more follicles).
3. The follicles are then incubated at 37°C until detachment of the outermost ORS cells becomes visible (Fig. 1B). This detachment procedure usually takes approx 15–20 min, but its completeness has to be checked under the inverted microscope. Effective trypsinization is recognized by the fact that the ORS tissue becomes loosened and single ORS cells are visible around the follicle (Fig. 1B).
4. The trypsin is inactivated by the addition of 1 mL of culture medium (5 mL if 1 mL trypsin was used). The follicles are pipeted up and down through a Pasteur pipet several times, taking care to avoid the formation of foam. The cell suspension still containing the follicles is now transferred into a 50-mL tube. The 35-mm dish is rinsed twice with 1 mL of culture medium, which is added to the 50-mL tube.
5. The medium in the 50-mL tube is made up to 7 mL (*see* Note 4). Further release of ORS cells still adhering to the follicles is achieved by vigorous pipeting of the suspension through a 5-mL pipet at least 50 times. The suspension is then diluted with culture medium in such a way that the final volume in milliliters corresponds to the number of follicles prepared. If only few follicles are prepared, it is better to reduce the volumes during the isolation procedure in order to avoid a centrifugation step.

3.3. Primary Cultivation of ORS Cells

1. The ORS cell suspension is distributed in culture dishes containing a preformed feeder layer (*see* Section 3.5.) and the hair follicles mostly denuded from the ORS tissue are removed with fine tweezers.
2. The cultures are incubated at 37°C in air with 5% CO_2. If starting with ORS cells from 1 follicle/mL culture medium, a seeding density of about 1×10^3 cells/cm^2 is achieved, so that only few round cells are visible over the feeder layer. Spreading of ORS cells occurs only after 2–4 d, with the ORS cells being located predominantly between feeder cells (Fig. 1C). At this time, the ORS cells display the typical epitheloid morphology, with a well-discernible nucleus and a large cytoplasm. With time, colonies of ORS cells develop, which push aside the feeder cells (Fig. 1D; *see* Note 5).
3. The first medium change is done not before culture d 7.
4. Thereafter, medium is changed three times/week. The medium changes remove the detached feeder cells (*see* Note 5). As the size of the colonies increases, the ORS cells become more compactly arranged, while their apparent size decreases and the cytoplasm becomes less striking. During the first culture days, the proliferation seems rather slow, but the cell number increases rapidly as soon as the culture is in the logarithmic growth phase (from d 6–7 on). Around d 12–14, the culture is 80–100% confluent (Fig. 1E; *see* Note 6).

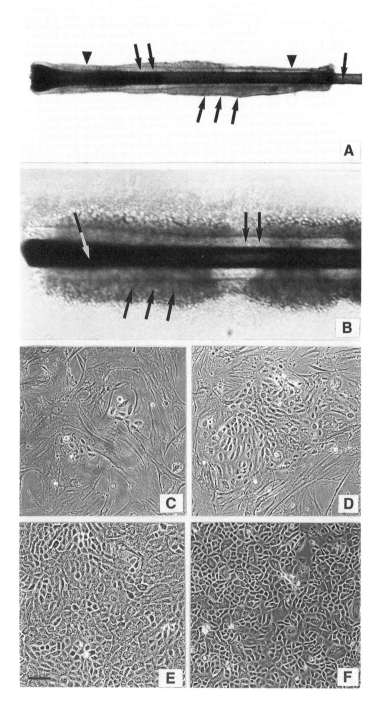

Fig. 1.

3.4. Subcultivation of ORS Cells

1. For subcultivation of the ORS cells (*see* Note 6), residual feeder cells are first selectively removed by incubation at 37°C for 2–3 min with EDTA (0.02%) in PBS without Ca^{2+} and Mg^{2+}. Effective removal of the feeder cells is obtained by vigorously pipeting the EDTA solution several times against the feeder cells.
2. The EDTA solution is then aspirated, and the ORS cells rinsed three times with PBS without Ca^{2+} and Mg^{2+} to remove all the feeder cells.
3. The ORS cells are incubated at 37°C in 0.1% trypsin/0.02% EDTA in PBS without Ca^{2+} and Mg^{2+} (for example, 0.5 mL/35-mm dish). Cell disaggregation is usually completed within 8–10 min, but has to be checked under the inverted microscope.
4. Trypsin is blocked by adding 1.5 mL of culture medium/35-mm culture dish. A single-cell suspension is obtained by vigorous pipeting through a 5-mL pipet.
5. A cell count is performed using, for example, a hematocytometer chamber (*see* Notes 7 and 8).
6. The cells are centrifuged at 250g (e.g., at 1000 rpm in a Heraeus Megafuge) for 8 min at room temperature.
7. The supernatant is aspirated and the cells resuspended in the selected media.
8. Secondary cultures of ORS cells are best performed in low (i.e., 0.15 m*M*) Ca^{2+} media (KGM) on tissue-culture plastic in the absence of feeder cells. In this case, plating densities as low as 1000 cells/cm² are easily achievable, with a maximal number of subcultures around 3–5.

Fig. 1. *(previous page)* **(A)** Phase-contrast photograph of a plucked human hair follicle that is in the growth phase (anagen) of the hair cycle. The different layers of the follicles are marked by arrows: hair shaft (simple arrow), inner root sheath (double arrow), outer root sheath (triple arrow). The arrowheads indicate where the cuts are made to remove the bulbar part (on the left side) and the infundibular portion situated *in situ* above the entry of the sebaceous gland (on the right side). **(B)** Phase-contrast photograph of a plucked human hair follicle after incubation in trypsin (0.1%)/EDTA (0.02%) for 20 min. As soon as the ORS cells become clearly detached from the hair shaft as seen here, the trypsinization is completed and can be stopped by the addition of culture medium containing FCS. Arrows mark the different layers of the follicle as in (A). **(C–E)** Time course of a primary culture of ORS cells. At d 4 (C), small islands of compactly arranged ORS cells recognizable by their epithelial morphology become visible. The ORS cells are surrounded by the fibroblast feeder cells, which are easily recognizable by their much larger size and pleiomorphic shape. At d 8 (D), the islands of ORS cells have further expanded and thereby detached the feeder cells. (E) The ORS cells have reached 80–100% confluence at d 12. At this time, the ORS cells have to be subcultured in order to avoid becoming postmitotic and to avoid entering the route of terminal differentiation. **(F)** Phase-contrast photograph of a confluent secondary culture of ORS cells in KGM (low Ca^{2+} medium). Bar: 50 µm.

3.5. Preparation of Feeder Layers Made of Postmitotic Dermal Fibroblasts (14,15)

1. Cultures of human dermal fibroblasts, either purchased from cell repositories (e.g., ATCC, Rockville, MD) or established from skin explants *(20)*, are propagated in DMEM supplemented with 10% FCS with a weekly split ratio of 1:2, best in 100-mm culture dishes (*see* Note 9). Subcultivation is done with 0.05% trypsin/0.02% EDTA in PBS without Ca^{2+} and Mg^{2+} at 37°C for approx 5 min (e.g., 1.5 mL trypsin/EDTA solution/100-mm dish). Trypsin is inactivated by the addition of 3.5 mL DMEM supplemented with 10% FCS (*see* Note 10).
2. To prepare feeder layers, confluent cultures are subcultured in a 1:4 ratio instead of the usual 1:2 ratio, and incubated overnight at 37°C in the CO_2 incubator (*see* Note 11).
3. The next day, these fibroblasts are rendered postmitotic by treatment for 5 h at 37°C in the CO_2 incubator with 8 µg/mL mitomycin C in DMEM supplemented with 10% FCS (10 mL DMEM containing 10% FCS and 8 µg/mL mitomycin C/100-mm culture). *See* Note 12.
4. After the 5 h of incubation, the cells are rinsed at least four times with PBS containing Ca^{2+} and Mg^{2+}, followed by one rinse with PBS without Ca^{2+} and Mg^{2+}.
5. The fibroblasts are detached by treatment with 0.05% trypsin/0.02% EDTA in PBS without Ca^{2+} and Mg^{2+} at 37°C for approx 3 min (e.g., 1.5 mL trypsin/EDTA solution/100-mm dish). Detachment of the cells can be speeded up by carefully agitating the culture dish.
6. The trypsin is blocked by addition of 4.5 mL of DMEM containing 10% FCS/100-mm dish.
7. The cells are thoroughly suspended and transferred to a 50-mL tube.
8. The culture dish is rinsed once more with 5 mL of medium, which is added to the 50-mL tube.
9. A cell count of the suspension is performed.
10. The suspension is diluted in DMEM containing 10% FCS in such a way as to obtain a suspension of 2×10^4 cells/mL (corresponding to $4–5 \times 10^3$ cells/cm^2 culture surface).
11. The suspension is now distributed in the selected culture dishes, usually in 35-mm culture dishes.
12. The now ready-to-use feeder layers can be stored at 37°C in the CO_2 incubator for at least 20–30 d with a weekly medium change (DMEM containing 10% FCS) until use.
13. For long-term storage, feeder cells can be kept frozen in liquid nitrogen *(15)*. For this purpose, 1×10^6 mitomycin C-treated fibroblasts are suspended in 1 mL of DMEM containing 10% FCS and 10% dimethyl sulfoxide (Serva, Heidelberg, Germany), transferred to a 1.8-mL cryotube, left inside a styropor box at –80°C for 24 h, and finally transferred into the liquid nitrogen tank.
14. A reproducible plating efficiency of 50% is obtained when recovering the frozen cells from the cryogenic storage. For preparation of feeder layers, the fibroblasts have thus to be suspended at a density of 4×10^4 cells/mL medium.

4. Notes

1. If the movement applied to pluck the follicles is not jerky enough, the ORS tissue around the hair shaft will not be completely harvested. In some donors, however, plucking of follicles yields only small amounts of ORS tissue, mainly because the skin is too tough.
2. In the human scalp, around 80% of the hairs are in the growth (anagen) phase and thus bear ORS cells (*see* Fig. 1A).
3. If ORS cells are cultivated as substitutes for epidermal keratinocytes, it is not necessary to remove the bulbar and the most distal part of the follicle. In this case, however, the presence of epithelial cells from the hair matrix or from the infundibular portion cannot be ruled out *(21)*.
4. Because of the presence of feeder cells in the primary culture, most batches of FCS will work, and therefore, do not necessarily have to be tested.
5. Selective removal of the feeder cells by the EDTA treatment is in principle possible as soon as the colonies of ORS cells have reached a critical size (more than 100 cells). The ORS cells can then be used for the selected purpose.
6. Because differentiation of ORS cells can start in the center of the colonies before the culture has reached confluence, the state of the cells has to be monitored under the inverted microscope. Thus, it is advisable to subculture the cells as long as they are in the proliferative state (morphologically recognizable by the easily recognizable nucleus surrounded by a bright cytoplasm) rather than to try to increase the cell number by waiting until the culture is confluent, but with the risk of the presence of a high percentage of postmitotic keratinocytes.
7. By initiating a primary culture with cells from two plucked hair follicles/35-mm culture dish, a yield of around 1×10^6 ORS cells is obtained within 2 wk.
8. We have found no donor age-specific differences in the growth behavior of ORS cells.
9. Thus far, we have found no marked variations in the feeder capacity between different strains of fibroblasts *(14)*.
10. The fibroblast strain selected for use as feeder cells should be first extensively propagated and then kept in aliquots in the frozen state. When used for the preparation of feeder layers, the fibroblasts from a cryotube are then used maximally for 6 wk with a split ratio of 1:2/week (*see* Section 3.5.) and treatment with mitomycin C when needed.
11. In the preparation of the feeder layers, a crucial point is the density of the fibroblasts. If this density is too high, the ORS cell colonies are hampered in their lateral expansion by the opposite tendency of the fibroblasts, which tremendously increase in cell size in the postmitotic state *(14,22)*. Conversely, the growth-stimulatory capacity of the feeder layer is reduced if the fibroblasts are present at too low a density.
12. Alternatively, the postmitotic state can be induced by irradiation of fibroblasts with X-rays *(14,15)*: Fibroblast from a confluent culture are detached by trypsinization (*see* Section 3.5.), and the cells counted and resuspended at a density of $0.5–1 \times 10^6$ cells/mL in cooled DMEM containing 10% FCS and penicil-

lin/streptomycin. Aliquots of 0.9 mL of this suspension are filled in cryotubes (e.g., Nunc, Life Technologies, Gaithersburg, MD, or Wheaton, Millville, NJ). The cryotubes are irradiated with a single dose of 7000 cGy (50 kV, 1.0-mm aluminum filter) using, for example, a Dermopan 2 (Siemens, Erlangen, Germany). Immediately after irradiation, 0.1 mL dimethyl sulfoxide (Serva) is added to the cryotubes, and the cells then frozen for the first 24 h in a styropor box at –80°C, and afterward in liquid nitrogen for prolonged storage.

13. This protocol using the aforementioned culture medium and the feeder layers made of postmitotic fibroblasts is equally suitable for primary cultures of epidermal (interfollicular) keratinocytes *(14)*.

References

1. Coulombe, P. A., Kopan, R., and Fuchs, E. (1989) Expression of keratin K14 in the epidermis and hair follicle: insights into complex programs of differentiation. *J. Cell. Biol.* **109,** 2295–2312.
2. Stark, H. J., Breitkreutz, D., Limat, A., Bowden, P., and Fusenig, N. E. (1987) Keratins of the human hair follicle: "hyperproliferative" keratins consistently expressed in outer root sheath cells in vivo and in vitro. *Differentiation* **35,** 236–248.
3. Eisen, A. Z., Holyoke, J. B., and Lobitz, W. C. (1955) Responses of the superficial portion of the human pilosebaceous apparatus to controlled injury. *J. Invest. Dermatol.* **25,** 145–156.
4. Bishop, G. H. (1946) Regeneration after experimental removal of skin in man. *Am. J. Anat.* **76,** 253–281.
5. Cotsarelis, G., Sun, T. T., and Lavker, R. M. (1990) Label-retaining cells reside in the bulge area of pilosebaceous unit: implications for follicular stem cells, hair cycle, and skin carcinogenesis. *Cell* **61,** 1329–1337.
6. Yang, J. S., Lavker R. M., and Sun, T. T. (1993) Upper human hair follicle contains a subpopulation of keratinocytes with superior in vitro proliferative potential. *J. Invest. Dermatol.* **101,** 652–659.
7. Limat, A., Breitkreutz, D., Hunziker, T., Boillat, C., Wiesmann, U., Klein, E., Noser, F., and Fusenig, N. E. (1991) Restoration of the epidermal phenotype by follicular outer root sheath cells in recombinant culture with dermal fibroblasts. *Exp. Cell. Res.* **194,** 218–227.
8. Limat, A., Breitkreutz, D., Hunziker, T., Klein, E., Noser, F., Fusenig, N. E., and Braathen, L. R. (1993) Outer root sheath (ORS) cells organize into epidermoid cyst-like spheroids when cultured inside matrigel: a light microscopic and immunohistological comparison between human ORS cells and interfollicular keratinocytes. *Cell Tissue Res.* **275,** 169–176.
9. Wyss-Coray, T., Gallati, H., Pracht, I., Limat, A., Mauri, D., Frutig, K., and Pichler, W. J. (1993) Antigen-presenting human T cells and antigen-presenting B cells induce a similar cytokine profile in specific T cell clones. *Eur. J. Immunol.* **23,** 3350–3357.
10. Imcke, E., Mayer-da-Silva, A., Detmar, M., Tiel, H., Stadler, R., and Orfanos, C. E. (1987) Growth of human hair follicle keratinocytes in vitro: ultrastructural features of a new model. *J. Am. Acad. Dermatol.* **17,** 779–786.

11. Weterings, P. J. J. M., Vermorken, A. J. M., and Bloemendal, H. (1991) A method for culturing human hair follicle cells. *Br. J. Dermatol.* **104**, 1.
12. Lenoir, M. C., Bernard, B. A., Pautrat, G., Darmon, M., and Shroot B. (1988) Outer root sheath cells of human hair follicle are able to regenerate a fully differentiated epidermis in vitro. *Dev. Biol.* **130**, 610–620.
13. Limat, A. and Noser, F. K. (1986) Serial cultivation of single keratinocytes from the outer root sheath of human scalp hair follicles. *J. Invest. Dermatol.* **87**, 485–488.
14. Limat, A., Hunziker, T., Boillat, C., Bayreuther, K., and Noser, F. (1989) Postmitotic human dermal fibroblasts efficiently support the growth of human follicular keratinocytes. *J. Invest. Dermatol.* **92**, 758–762.
15. Limat, A., Hunziker, T., Boillat, C., Noser, F., and Wiesmann, U. (1990) Postmitotic human dermal fibroblasts preserve intact feeder properties for epithelial cell growth after long-term cryopreservation. *In Vitro Cell. Dev. Biol.* **26**, 709–712.
16. Rheinwald, J. G. and Green, H. (1975) Serial cultivation of strains of human epidermal keratinocytes: the formation of keratinizing colonies from single cells. *Cell* **6**, 331–343.
17. Wu, Y. J., Parker, L. M., Binder, N. E., Beckett, M. A., Sinard, J. H., Griffiths, C. T., and Rheinwald, J. G. (1982) The mesothelial keratins: a new family of cytoskeleton proteins identified in cultured mesothelial cells and nonkeratinizing epithelia. *Cell* **31**, 693–703.
18. Boyce, S. T. and Ham, R. G. (1983) Calcium-regulated differentiation of normal human epidermal keratinocytes in chemically defined clonal culture and serum-free serial culture. *J. Invest. Dermatol.* **81**, 33s–40s.
19. Parakkal, P. F. (1990) Catagen and telogen phases of the growth cycle, in *Hair and Hair Diseases* (Orfanos, C. E. and Happle, R., eds.), Springer Verlag, Berlin, pp. 99–116.
20. Sly, W. S. and Grubb, J. (1979) Isolation of fibroblasts from patients, in *Methods in Enzymology*, vol. 58 (Jakoby, W. B. and Pastan, I. H., eds.), Academic, London, pp. 440–450.
21. Limat, A., Breitkreutz, D., Thiekoetter, G., Noser, F., Hunziker, T., Braathen, L. R., and Fusenig, N. E. (1993) Phenotypic modulation of human hair matrix cells (trichocytes) by environmental influence in vitro and in vivo. *Epithelial Cell Biol.* **2**, 55–65.
22. Bayreuther, K., Rodemann, H. P., Hommel, R., Dittmann, K., Albiez, M., and Francz, P. (1988) Human skin fibroblasts in vitro differentiate along a terminal cell lineage. *Proc. Natl. Acad. Sci. USA* **85**, 5112–5116.

4

Isolation and Short-Term Culture of Human Epidermal Langerhans' Cells

Corinne Moulon

1. Introduction

The human epidermis consists of a heterogeneous population of cells including keratinocytes in various stages of differentiation, Merkel cells, melanocytes, and Langerhans' cells (LC). The latter account for only around 1–3% of the epidermal population, but they represent the major antigen-presenting cells of the epidermis *(1)*.

Most of the studies concerning LC have been performed in the mouse model, and murine epidermal LC represent the best characterized nonlymphoid dendritic cells. It has thus been shown that during short-term in vitro incubation (2–3 d), these cells undergo modifications of both their phenotypical characteristics, as well as their functional capacities. During culture, LC enlarge, develop numerous cytoplasmic sheets, express more class I and II MHC antigens and adhesion molecules, and lose Fc receptors and Birbeck granules *(2)*. At the same time, their T-cell stimulatory capacities develop, but their processing abilities for native protein are greatly reduced *(3–6)*. All these modifications, which are mediated by the cytokine granulocyte macrophage-colony stimulating factor (GM-CSF) *(7,8)*, make them fully resemble blood and lymphoid dendritic cells *(9)*. As a result of these differences between fresh and cultured LC in vitro, and the in vivo experiments *(10)*, it has been proposed that this short-term in vitro incubation mimics what occurs in vivo during the migration of epidermal LC to draining lymph nodes *(11)*. Epidermal LC have thus been cultured in vitro as a model for studying the relationship between LC within the epidermis and within draining lymph nodes.

The data available concerning human LC are much more limited, and this is the result of several points: First, only small numbers of LC are available from skin (1–3%), so that large amounts of material are required to generate sufficient cells to be able to use them in functional assays. Another limitation is that, like their murine counterparts *(3)*, these cells do not divide in vitro and no human LC lines exist yet. The only "culture" that is possible with these cells actually consists of a short-term incubation of 2–3 d as reported for murine LC. Despite these problems, it has been shown that on in vitro incubation, human LC exhibit analogous phenotypic changes to those described in the murine system *(12,13)*. In addition, downregulation of CD1a molecule, an antigen that is not expressed on murine LC, is observed. Concerning the functional data, there seem to be some discrepancies between murine and human systems. Although the functional immunostimulatory capacities of LC are considerably enhanced on culture *(14–17)*, fresh human LC are also able to stimulate a primary response *(14,16,18)*. Moreover, GM-CSF greatly enhances the viability of LC during incubation, but it does not functionally have as potent an effect on human as on murine LC *(19)*.

In this chapter, the method currently used in my laboratory for preparation of epidermal cell (EC) suspensions from normal human skin is described. These bulk EC suspensions can be used as a source of antigen-presenting cells to activate T- and B-cells. However, to perform detailed functional studies, partially enriched or purified LC preparations are needed.

Different techniques, varying in time and complexity, have been utilized to achieve purification of human LC *(18,20–24)*. One of these techniques, based on differences in buoyant density between LC and other EC, is presented here (Note 1). In contrast to other methods (Note 2), the LC population obtained following this procedure is unlabeled and thus suitable for functional and phenotypical assays. Finally, conditions used for the short-term culture of these LC suspensions are described.

2. Materials

2.1. Skin Samples

Normal human skin is obtained from patients undergoing reconstructive plastic surgery of the breast or abdomen. It can also be obtained from human cadavers if taken <24 h from the time of death.

2.2. Instruments

1. Metallic instruments: Dissection box containing a scalpel, a big pair of scissors, a fine curved pair of scissors, and pairs of curved and straight forceps. All of them have to be sterile.
2. Preparation of dermo–epidermal slices: A cork dissection board, sterile gauze, dissection pins, and 10-cm Petri dishes are required. A skin-parer or corn-cutter (e.g., Credo, Solingen, Germany) and blades are used for producing the cutaneous slices. They are autoclaved before use.

2.3. Solutions and Reagents

1. Hank's balanced salt solution (HBSS; Gibco, Grand Island, NY): without calcium and magnesium supplemented with 50 µg/mL gentamicin.
2. Trypsin: Prepare a stock solution of 0.25% trypsin (Powder trypsin from Difco Laboratories, Detroit, MI) in HBSS. Filter and store aliquots at –20°C. Repeated cycles of freezing and thawing will reduce the potency of the solution (Note 3).
3. Fetal calf serum (FCS): Heat inactivated (56°C for 30 min).
4. Trypan Blue: 0.16% solution in PBS. Filter and store the solution at –20°C.
5. Culture medium (called complete RPMI medium): RPMI-1640 medium (Gibco) supplemented with 10% heat-inactivated human AB serum, 2 mM L-glutamine, 50 µg/mL gentamicin, and 1 µg/mL indomethacin (Sigma, St. Louis, MO).
6. Lymphoprep (Nycomed Pharma SA, density = 1.077 g/cm^3): Store at –4°C. For the third step, prepare Lymphoprep diluted with distilled water: 1.6 mL for 3.4 ml Lymphoprep. Make the required volume of solution just before use.
7. Human recombinant GM-CSF (Genzyme, Boston, MA): Store aliquots at –80°C. Use at a concentration of 200 U/mL.

3. Methods
3.1. Preparation of Epidermal Cell Suspensions

Epidermal cell suspensions are obtained by a standard trypsinization technique, which allows the separation of dermis from epidermis and disrupts these epidermal sheets into single cells (Note 4).

1. On receiving the skin sample, carefully remove the underlying subcutaneous fat with a scalpel. In the case of a large sample, it is better to cut it into smaller pieces (of 100 cm^2 maximum) since this improves the EC yield (Note 5).
2. Soak the cutaneous samples in gentamicin-supplemented HBSS for 30 min before processing.
3. Prepare Petri dishes containing 20 mL of HBSS with 0.05% trypsin and gentamicin, and store them at 4°C until use.
4. Cover the dissection board with sterile gauze and pin out the skin. It is important to ensure that the sample is kept very taut.
5. Wipe the skin, and use a corn-cutter to slice off thin sheets (around 0.15 mm) consisting of the epidermis and the papillary dermis, changing the blade as required. Spread out the resulting cutaneous slices dermal side down on the previously prepared Petri dishes containing 0.05% trypsin. This is the most crucial step of the procedure, since the cutaneous slices should not be too thick to prevent the trypsin from cleaving the dermis from the epidermis. However, slices that are too thin will result in excessive trypsin digestion of the epidermis and a loss of epidermal cells. The ideal thickness of the slices is one where the dermo–epidermal slices stretch out by themselves over the trypsin.
6. Incubate the Petri dishes either for 1 h in 37°C incubator or overnight at 4°C.
7. Prepare Petri dishes containing 10 mL of HBSS supplemented with 10% FCS to stop trypsin action.

8. Taking one cutaneous slice at a time, separate the epidermis from the underlying dermis using extrafine forceps (pairs of curved and straight forceps are best). Trypsinization is optimal when the epidermis separates easily from the dermis in one piece.
9. Pool the epidermal sheets into the Petri dishes containing FCS, and mince them up with a fine curved pair of scissors. A single cell suspension is obtained by vigorously pipeting the epidermal sheets up and down several times using a 5- or 10-mL pipet.
10. Sieve the suspension through a sterile gauze to remove residual clumps, debris, and stratum corneum.
11. Wash the suspension three times with HBSS supplemented with 10% FCS by centrifugation at 300g for 10 min, and resuspend the EC in HBSS supplemented with FCS.
12. Assess the viability, which should be superior to 80%, by the trypan blue exclusion test (Note 6).

3.2. Enrichment of LC

Enrichment of up to 80% LC is achieved by three successive density gradient centrifugation steps (Note 7).

1. Adjust the concentration of the EC suspension to 5×10^6 cells/mL.
2. Apply the suspension onto Lymphoprep in 15-mL tubes with a 2:1 ratio of EC suspension:Lymphoprep.
3. Centrifuge for 20 min at 400g.
4. Carefully harvest the low-density fraction at the interface, collect in a 15-mL tube, and wash twice with FCS-supplemented HBSS.
5. Enumerate the cells, estimate the percentage of LC using phase-contrast microscopy (Note 8), and proceed to the second step following the same procedure (steps 1–5).
6. The last centrifugation step is the same (steps 1–5), except that the Lymphoprep used is diluted with distilled water. Depending on the number of EC, this step can be done using 5-mL instead of 15-mL tubes.

3.3. Short-Term Culture of LC

This incubation can also be performed using bulk or partially enriched EC suspensions.

1. EC suspensions are plated in complete RPMI medium containing 200 U/mL of GM-CSF at a concentration of 5×10^5 cells/mL for highly enriched LC suspensions, or at 10^6 cells/mL for partially enriched EC suspensions.
2. Incubate the plates for 48–72 h.
3. Recover the nonadherent fractions, which contain all LC.
4. Proceed to a Lymphoprep gradient centrifugation, as explained before, to remove the dead cells and the remaining keratinocytes.

4. Notes

1. It should be noted that this technique has some drawbacks. Indeed, according to the cutaneous sample, LC recovery varies considerably. Furthermore, gradient centrifugation of fresh EC suspensions selects for the subset of LC that are low density. In some experiments, this might lead to an important loss of material.
2. Positive selection procedures may be used for the purification of LC. However, these techniques will involve binding a ligand (e.g., monoclonal antibodies [MAbs]) to cell-surface molecules (usually HLA-DR or CD1a molecules in the case of LC), which may interfere with functional experiments. FACS sorting, panning, and MACS can be recommended depending on the kind of experiments you want to perform.
3. A critical factor in the preparation of the EC suspensions is the trypsin used. The trypsinization technique indicated in this method works well (usually better at 1 h at 37°C than overnight at 4°C). Since I have noted considerable lot-to-lot variation in trypsin efficiency, it is, however, recommended to establish the optimal conditions for trypsinization with every new batch of this reagent. Furthermore, studies that investigate trypsin-sensitive molecules on LC should ideally use as low concentrations of this enzyme as possible. It has been found that a concentration of 0.05%, five times lower than usually described, provides the best balance between poor tissue digestion and excessive molecular damage.
4. The method described here for preparation of the dermo–epidermal slices is cheap, but rather time consuming, and requires some practice to find the right thickness for the cutaneous slices. Keratome sets (which allow the production of regular slices of 0.1 mm) are commercially available, but at considerable expense.
5. Depending on when the skin samples are received, it is not always possible to proceed immediately to the preparation of EC suspensions. Skin samples may be stored in HBSS with gentamicin for 24 h at 4°C before processing without affecting the LC recovery.
6. Epidermal cell suspensions can be stored, if necessary, in liquid nitrogen (using dimethyl sulfoxide at a final concentration of 10% in RPMI-1640 medium supplemented with 20% FCS). Once thawed, the epidermal suspensions retain their original capacities and can be used for further LC enrichment. However, some cell loss will occur.
7. Routinely, the enrichment procedure leads to a suspension containing about 8–15% LC after the first centrifugation step and 30–50% after the second step. The last round of centrifugation usually provides an EC population containing 80–95% LC.
8. Initially, enrichment is best assessed by antibody staining of LC (using anti-HLA-DR or anti-CD1a MAbs). However, with experience, intermediate preparations (after the first and second steps) may be examined by phase-contrast microscopy to estimate the LC proportions (keratinocytes appear rounded, whereas in contrast, LC exhibit short cell processes, which make the cell recognizable).

References

1. Teunissen, M. B. M. (1992) Dynamic nature and function of epidermal Langerhans cells in vivo and in vitro: a review, with emphasis on human Langerhans cells. *Histochemistry J.* **24**, 697–716.
2. Romani, N., Schuler, G., and Fritsch, P. (1991) Identification and phenotype of epidermal Langerhans cells, in *Epidermal Langerhans Cells* (Schuler, G., ed.), CRC, Boca Raton, FL, pp. 49–86.
3. Inaba, K., Schuler, G., Witmer, M. D., Valinsky, M. D., Atassi, B., and Steinman, R. M. (1986) The immunologic properties of purified Langerhans cells: distinct requirements for stimulation of unprimed and sensitized T lymphocytes. *J. Exp. Med.* **164**, 605–613.
4. Schuler, G. and Steinman, R. M. (1985) Murine epidermal Langerhans cells mature into potent immunostimulatory dendritic cells in vitro. *J. Exp. Med.* **161**, 526–546.
5. Romani, N., Koide S., Crowley, M., Witmer-Pack, M., Livingstone, A. M, Fathman, C. G., Inaba K., and Steinman R. M. (1989) Presentation of exogenous protein antigens by dendritic cells to T cell clones. Intact protein is presented best by immature, epidermal Langerhans cells. *J. Exp. Med.* **169**, 1169–1178.
6. Streilein, J. W. and Grammer, S. W. (1989) In vitro evidence that Langerhans cells can adopt two functionally distinct forms capable of antigen presentation to T lymphocytes. *J. Immunol.* **143**, 3925–3933.
7. Witmer-Pack, M. D., Olivier, W., Valinsky, M. D., Schuler, G., and Steinman, R. M. (1987) Granulocyte/macrophage colony-stimulating factor is essential for the viability and function of cultured murine epidermal Langerhans cells. *J. Exp. Med.* **166**, 1484–1498.
8. Heufler, C., Koch, F., and Schuler, G. (1988) Granulocyte/macrophage colony-stimulating factor and interleukin 1 mediate the maturation of murine epidermal Langerhans cells into potent immunostimulatory dendritic cells. *J. Exp. Med.* **167**, 700–705.
9. Steinman, R. M. (1991) The dendritic cell system and its role in immunogenicity. *Annu. Rev. Immunol.* **9**, 271–296.
10. Aiba, S. and Katz, S. I. (1990) Phenotypic and functional characteristics of in vivo-activated Langerhans cells. *J. Immunol.* **145**, 2791–2796.
11. Kripke, M. L., Munn, C. G., Jeevan, A., Tang, J. M., and Bucana, C. (1990) Evidence that cutaneous antigen-presenting cells migrate to regional lymph nodes during contact sensitization. *J. Immunol.* **145**, 2833–2838.
12. Romani, N., Lenz, A., Glassel, H., Stössel, H., Stanzl, U., Majdic, O., Fritsch, P., and Schuler, G. (1989) Cultured human Langerhans cells resemble lymphoid dendritic cells in phenotype and function. *J. Invest. Dermatol.* **93**, 600–609.
13. Teunissen, M. B. M., Wormmeester, J., Krieg, S. R., Peters, P. J., Vogels, I. M. C., Kapsenberg, M. L., and Bos, J. D. (1990) Human epidermal Langerhans cells undergo profound morphologic and phenotypical changes during in vitro culture. *J. Invest. Dermatol.* **94**, 166–173.

14. Teunissen, M. B. M., Wormmeester, J., Rongen, H. A. H., Kapsenberg, M. L., and Bos, J. D. (1991) Conversion of human epidermal Langerhans cells into interdigitating cells in vitro is not associated with functional maturation. *Eur. J. Dermatol.* **1**, 45–54.
15. Symington, F. W., Brady, W., and Linsley, P. S. (1993) Expression and function of B7 on human epidermal Langerhans cells. *J. Immunol.* **150**, 1286–1295.
16. Moulon, C., Péguet-Navarro, J., Courtellemont, P., Redziniak, G., and Schmitt, D. (1993) In vitro sensitization and restimulation of hapten-specific T cells by fresh and cultured human epidermal Langerhans cells. *Immunology* **80**, 373–379.
17. Péguet-Navarro, J., Moulon, C., Caux, C., Dalbiez-Gauthier, C., Banchereau, J., and Schmitt, D. (1994) Interleukin 10 inhibits the primary allogeneic T cell response to human epidermal Langerhans cells. *Eur. J. Immunol.* **24**, 884–891.
18. Péguet-Navarro, J., Dalbiez-Gauthier, C., Dezutter-Dambuyant, C., and Schmitt, D. (1993) Dissection of human Langerhans cells' allostimulatory function: the need for an activation step for full development of accessory function. *Eur. J. Immunol.* **23**, 376–382.
19. Péguet-Navarro, J., Dalbiez-Gauthier, C., Moulon, C., and Schmitt, D. (1993) Enhancement of human Langerhans cell allostimulatory function upon a short in vitro culture is not mediated by GM-CSF. *J. Invest. Dermatol.* **100**, 467.
20. Wood, G. S., Kosek, J., Butcher, E. C., and Morhenn, V. B. (1985) Enrichment of murine and human Langerhans cells with solid phase immunoabsorption using pan-leucocyte monoclonal antibodies. *J. Invest. Dermatol.* **84**, 37–40.
21. Gommans, J. M., Van Erp, P. E. J., Forster, S., Boezeman, J., and Mier, P. D. (1985) Isolation and preliminary biochemical characterization of the human epidermal Langerhans cell. *J. Invest. Dermatol.* **85**, 191–193.
22. Hanau, D., Schmitt, D. A., Fabre, M., and Cazenave, J. P. (1988) A method for the rapid isolation of human epidermal Langerhans cells using immunomagnetic microspheres. *J. Invest. Dermatol.* **91**, 274–279.
23. Teunissen, M. B. M., Wormmeester, J., Kapsenberg, M. L., and Bos, J. D. (1988) Enrichment of unlabeled human Langerhans cells from epidermal cell suspensions by discontinuous density gradient centrifugation. *J. Invest. Dermatol.* **91**, 358–362.
24. Morris, J., Alaibac, M., Jia, M. H., and Chu, T. (1992) Purification of functional active epidermal Langerhans cells: a simple and efficient new technique. *J. Invest. Dermatol.* **99**, 237–240.

5

Growth and Differentiation of Human Adipose Stromal Cells in Culture

Brenda Strutt, Wahid Khalil, and Donald Killinger

1. Introduction

Human adipose stromal cells provide an excellent model for studying a variety of metabolic processes in an in vitro system. These are normal cells derived from subcutaneous or omental adipose tissue. Under specific culture conditions, they will differentiate without replication into cells resembling mature adipocytes or will replicate, become confluent, and grow in subculture. The initial observation that the stromal vascular fraction of human omental adipose tissue contained a fibroblast-like cell that was a possible adipocyte precursor was made by Poznanski et al. *(1)*. Subsequent studies demonstrated enzymological and morphological properties that developed during differentiation *(2–4)*. The primary interest has focused on the conditions required to stimulate these cells to differentiate into adipocytes, which can accumulate lipid and possess the enzymes involved in lipogenesis, including glycerol-3-phosphate dehydrogenase (GPDH) and lipoprotein lipase (LPL).

Extensive studies have been carried out on sublines of 3T3 cells derived from mouse embryo. These cells resemble fibroblasts and will differentiate into mature adipocytes under appropriate conditions *(5)*. Adipose stromal cells derived from mouse cell lines have also provided models to study factors controlling differentiation *(6,7)*. These models, however, differ from adipose stromal cells derived from human tissue such that conditions used to induce differentiation in these model systems were unable to induce differentiation in human cells consistently. Hauner et al. *(8)* published conditions in which 20–80% of stromal cells obtained from human subcutaneous adipose tissue will differentiate into adipocytes containing lipogenic enzymes. These conditions provide an excellent basis for studying the factors involved in differentiation.

From: *Methods in Molecular Medicine: Human Cell Culture Protocols*
Edited by: G. E. Jones Humana Press Inc., Totowa, NJ

Cells grown under conditions that favor replication rather than differentiation have been of interest for the study of steroid metabolism. Adipose stromal cells are believed to be a major site for the conversion of adrenal androgens to estrogens in the postmenopausal female *(9)*. They have also been shown to contain additional steroidogenic enzymes capable of converting adrenal precursors to androgens, estrogens, and 7-hydroxylated metabolites, which may influence the immune system *(10–12)*.

This chapter focuses only on the culture of human adipose stromal cells, and the conditions required to study factors influencing cell growth or differentiation. Reviews of the factors regulating adipocyte differentiation were published recently by Ailhaud and his colleagues *(13–15)*.

2. Materials

2.1. Isolation and Maintenance of Cells

1. Hank's balanced salt solution (HBSS): Ca^{2+}- and Mg^{2+}-free, no phenol red, containing 50 U/mL penicillin G and 50 µg/mL streptomycin sulfate (Sigma, St. Louis, MO).
2. Collagenase: Type II collagenase (Sigma) solution of 1 mg/mL in HBSS.
3. Red blood cell lysis buffer: $0.154 M$ NH_4Cl, 10 mM $KHCO_3$, 0.1 mM EDTA in deionized, distilled H_2O.
4. Growth medium (DMEM/10%): Dulbecco's modified Eagle's medium (1000 mg/L glucose, no phenol red) (*see* Note 4) + 10% heat-inactivated fetal bovine serum, supplemented with 44 mM $NaHCO_3$, 50 U/mL penicillin G, and 50 µg/mL streptomycin sulfate (Sigma).
5. Nylon mesh: 250- and 20–30-µm pore size (B & SH Thompson, Scarborough, Ontario, Canada).
6. Trypsin/EDTA: 0.05% trypsin (Sigma), 0.53 mM EDTA (Sigma) in HBSS.

All solutions are sterilized by filtration through a 0.22-µm Millipore filter.

2.2. Differentiation Media

1. Differentiation media (DMEM/F12+): DMEM/F12 + 15 mM $NaHCO_3$, 15 mM HEPES, 33 µM biotin, 17 µM pantothenic acid, 0.2 nM triiodothyronine, 0.5 µM insulin, 0.1 µM dexamethasone, 50 U/mL penicillin G, 50 µg/mL streptomycin sulfate (all from Sigma).
2. DMEM/F12: 1:1 mixture of DMEM and Ham's nutrient mixture F-12 supplemented with 15 mM HEPES, 15 mM $NaHCO_3$, 50 U/mL penicillin G, and 50 µg/mL streptomycin sulfate (Sigma).
3. Biotin: Prepare stock of 100X final concentration (3.3 mM) in distilled H_2O, and store in 1-mL aliquots at $-20°C$. Add 1.0 mL/100 mL of DMEM/F12.
4. Pantothenic acid: Prepare stock of 1000X final concentration (17 mM) in distilled H_2O and store in 0.5-mL aliquots at $-20°C$. Add 100 µL/100 mL of DMEM/F12.

5. Triiodothyronine: Prepare stock of 0.2 mM in distilled H_2O ($10^6\times$ final concentration) and store at –20°C. Dilute 1/1000 in DMEM/F12, and then add 100 µL/100 mL of DMEM/F12.
6. Insulin: Commercial stock of 100 U/mL is equal to 0.67 mM. This is 1340× the final concentration required. Add 75 µL/100 mL of DMEM/F12.
7. Dexamethasone: Prepare a stock of 5 mM in EtOH and store at –20°C. Dilute to $10^{-5}M$ in DMEM/F12 and add 1 mL/100 mL of DMEM/F12. Sterilize all solutions by filtration through a 0.22-µm Millipore filter.

2.3. GPDH Assay

1. STM buffer: 0.25M Sucrose, 10 mM Tris-HCl, 1 mM 2-mercaptoethanol, pH 7.4.
2. STME buffer: STM buffer + 1 mM EDTA, pH 7.4.
3. 1M Triethanolamine, pH 7.7.
4. 2.5 mM EDTA.
5. 5M 2-mercaptoethanol.
6. 4 mM Dihydroxyacetone phosphate (DHAP).
   All above solutions can be stored at 4°C, except for the DHAP, which should be stored in 1-mL aliquots at –20°C.
7. GPDH cocktail: To a 1-mg vial of β-NADH add the following in the order shown: 666 µL 1M triethanolamine, 64 µL 5M 2-mercaptoethanol, 64 µL 2.5 mM EDTA, 1126 µL H_2O. The cocktail should be kept on ice and wrapped in tin foil since the β-NADH is light-sensitive.
8. Refer to Notes 9 and 10 for calculations involved in determining GPDH activity.

2.4. Protein Determination

1. Bio-Rad (Mississauga, Ontario, Canada) Protein Assay Dye Reagent Concentrate: store at 4°C.
2. Bovine serum albumin (BSA): Prepare a 100 µg/mL stock solution in HBSS, and store in 1-mL aliquots at –20°C.

2.5. Oil Red O Staining for Lipid

1. 10% Formalin.
2. Oil Red O.
3. Absolute isopropanol.
4. Whatman #1 filter paper.
5. Mayer's hematoxylin stain, 0.1%.
6. Glycerol gelatin.

3. Methods
3.1. Isolation of Adipose Stromal Cells

1. Adipose tissue obtained at time of surgery is put into sterile HBSS. It is best to begin dissection immediately after receiving the tissue.
2. Dissect yellow fat as free of white fibrous tissue and blood vessels as possible, and mince finely with scissors.

3. Transfer tissue to a wide-mouth sterile bottle, and add collagenase solution (4 vol collagenase:1 vol tissue).
4. Incubate at 37°C for 30 min with constant shaking.
5. Pour resulting slurry through sterilized 250-µm nylon mesh, and dispense filtrate into 50-mL centrifuge tubes.
6. Centrifuge for 10 min at 800g at room temperature.
7. Aspirate floating fat cake and supernatant, leaving only a small amount (0.5 mL) of liquid on top of each cell pellet.
8. Resuspend pellets with Pasteur pipet and pool in one or two clean tubes (pellets from five or six tubes can be combined into one tube).
9. Add approx 20 mL of red blood cell lysis buffer to each tube, and allow to sit for 10 min at room temperature.
10. Filter suspension through sterilized 20–30-µm nylon mesh into a clean sterile centrifuge tube, and centrifuge for 10 min at 300g. This step removes contaminating endothelial cells.
11. Aspirate supernatants, and wash cell pellets with 20 mL of HBSS.
12. Resuspend final cell pellets in DMEM/10% and plate cells into appropriate cell-culture dishes or flasks. Fifty grams of minced tissue yield enough cells for approx six 6-well plates, twelve 25-cm^2 flasks, or four 75-cm^2 flasks (refer to Notes 1 and 6).
13. Cultures are incubated at 37°C in 5% CO_2 and 100% humidity.

3.2. Maintenance of Cells Under Replication Conditions

1. Twenty-four hours after cell isolation, aspirate medium, wash adherent cells twice with HBSS to remove nonadherent cells and cell debris, and replace with fresh DMEM/10%.
2. Replace with fresh medium every 2 or 3 d.
3. Depending on cell density at time of plating, it will take 10–15 d before confluency is reached. Adipose stromal cells have a doubling time of approx 3 d in culture (*see* Notes 2 and 3).
4. Cells can be detached from the plastic by washing monolayer with HBSS and treating with trypsin/EDTA at 37°C for 5 min.
5. Transfer cells by diluting 1:5 into fresh DMEM/10%.
6. For studies of cell-surface receptors or membrane characteristics that trypsin would interfere with, cells can be removed from the plastic by scraping with a rubber policeman (cell scraper).

3.3. Induction of Differentiation

1. Within 20 h postisolation, culture medium must be switched to serum-free differentiation medium. Exposure of the human preadipocyte to serum for more than 24 h will interfere with the differentiation process.
2. Wash the adherent cells three times with HBSS to ensure total removal of all serum, and add the defined differentiation media (DMEM/F12+).
3. It takes 7–9 d in this serum-free medium before cell differentiation is visible by microscopic examination. By 15–21 d, maximum differentiation has been

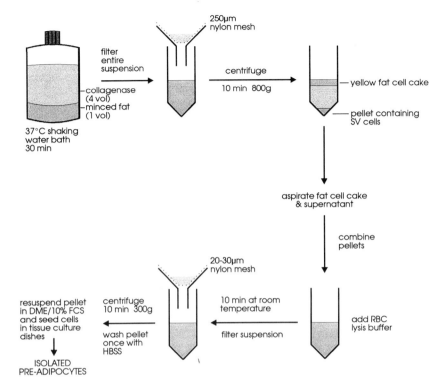

Fig. 1. Protocol for the isolation of stromal-vascular (SV) cells (preadipocytes) from human adipose tissue.

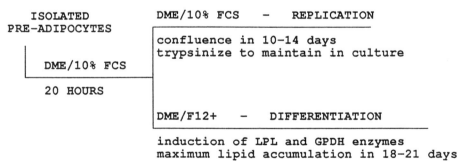

Fig. 2. Schematic illustrating the two culture conditions for adipose stromal cells.

achieved. Fresh differentiation medium should be replaced every 2–3 d throughout this time.

A flowchart of the isolation procedure and a schematic representation of the two conditions for culturing adipose stromal cells are shown in Figs. 1 and 2.

3.4. GPDH Assay

3.4.1. Preparation of Cell Homogenate

1. Keep all buffers and cells on ice during the following procedure.
2. Once the cells have reached the desired state of differentiation, gently wash the monolayer twice with cold HBSS.
3. Gently wash monolayer twice with cold STM buffer.
4. Add cold STME buffer (500 µL/25-cm^2 flask; 300 µL/well in 6-well plate), and let sit for a few minutes.
5. Scrape cells and transfer suspension to microcentrifuge tube.
6. Add another 400 µL STME to 25-cm^2 flask (or 200 µL/well), scrape, and transfer this washing to the same tube.
7. Cell suspensions can be frozen at –80°C at this point.
8. Sonicate cell suspensions using a Fisher 300 series Sonicator set at 35%. Pulse the cells four times, 3–4 s/pulse, moving the sonicator tip through the suspension. Keep the sample on ice or work in a cold room.
9. Centrifuge samples in microfuge for 4 min at 4°C.
10. Transfer supernatant to clean microfuge tube and freeze at –80°C until ready to assay, or remove aliquots and assay immediately.

3.4.2. Assay

1. In a borosilicate glass test tube, mix 150 µL cell homogenate, 90 µL GPDH cocktail, and 60 µL DHAP. The final reaction mixture contains 104 mM triethanolamine, 50 mM 2-mercaptoethanol, 0.025 mM EDTA, 0.220 mM NADH, and 0.8 mM DHAP.
2. Transfer mixture to a microcuvet, and read at 340 nm in a Beckman DU 640 Spectrophotometer. Record measurements for 3 min at room temperature. Reaction is linear over this time period.
3. GPDH converts the substrate dihydroxyacetone phosphate to L-glycerol 3-phosphate. The decrease observed in OD measurement reflects the conversion of NADH to NAD$^+$.

3.5. Protein Determination

1. The protein assay used is the Bio-Rad Protein Assay (a modification of the Bradford method). For smaller amounts of protein (<25 µg/mL), the microassay procedure is followed.
2. Prepare dilutions of BSA from 0–25 µg/mL (0, 1.56, 3.13, 6.25, 12.5, 25 µg/mL). These dilutions comprise a standard curve that should be run with every assay.
3. Mix 800 µL of each standard + 200 µL of dye concentrate. Add 200 µL/well in 3 or 4 wells in a 96-well plate.
4. Samples: Mix 100 µL of cell homogenate as prepared for GPDH assay with 300 µL HBSS. Add 100 µL of dye concentrate, vortex well, and transfer 200 µL to 2 wells of the 96-well plate.
5. The assay is read at OD_{295} using a Titertek Multiscan Plus plate reader.

3.6. Oil Red O Staining for Lipid

1. Adipose stromal cells can be grown on clean, sterile glass coverslips that have been placed in the bottom of wells of a 6-well tissue-culture plate at the time of cell isolation. Fixation and staining of the cells can be carried out in the plate.
2. Prepare Oil Red O stain. Mix 0.70 g Oil Red O with 200 mL of absolute isopropanol, shake, and leave overnight at room temperature. Filter through Whatman #1 (Maidstone, UK) filter paper. Dilute 180 mL of the stain solution with 120 mL of distilled H_2O, and leave overnight at 4°C. Filter solution, let stand for 30 min, and filter once more. This Oil Red O solution is stable for 6–8 mo at room temperature.
3. When ready to stain cells for lipid, aspirate medium and wash cells twice with HBSS. Fix cells by adding 1 mL of 10% formalin, and leave overnight at room temperature.
4. Remove formalin and rinse with 60% isopropanol for 30 s.
5. Stain with Oil Red O for 1 h.
6. Remove stain and rinse in 60% isopropanol for about 5 s while shaking gently.
7. Wash with tap water for 2–3 min.
8. Add hematoxylin for 2–3 min.
9. Wash with tap water for 3 min.
10. Mount coverslip on slide in a drop of glycerol gelatin. Ring coverslip with glycerol gelatin for permanency.
11. The lipid droplets in the cells will be stained a bright orange-red, and the cell nuclei will be stained blue. This double staining allows for quantitation of the number of cells that have undergone differentiation.

3.7. Summary

Cells grown in the presence of fetal bovine serum will replicate in culture. Although they may accumulate small lipid droplets as they become confluent, they do not develop adipogenic enzymes, and maintain a fibroblast-like appearance as seen in Fig. 3A. These cells are very active in the metabolism of adrenal precursors to metabolic products, which include estrogens, androgens, and 7α-hydroxylated compounds (*see* Note 7). This metabolism changes as the cells grow in culture and become confluent *(16)* (Fig. 4).

Human adipose stromal cells grown under the serum-free conditions described begin to accumulate lipogenic enzymes by approximately d 10 and have maximum enzyme activity by d 21. Different lipogenic enzymes appear at different stages of cell differentiation *(14)* (*see* Note 8). There is a progressive accumulation of lipid droplets during differentiation, although under these culture conditions, the characteristic signet ring morphology is not achieved (Fig. 3B).

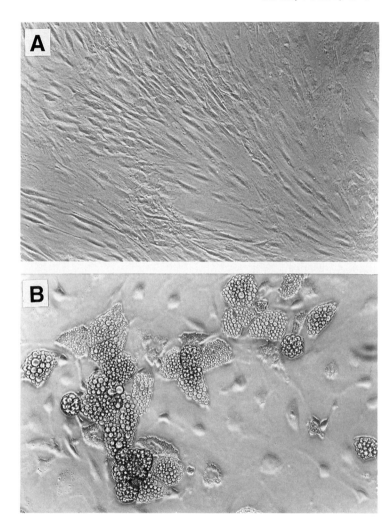

Fig. 3. Inverted-phase photographs of human adipose stromal cells in culture. **(A)** Confluent cells in replicating conditions, DMEM/10% (magnification 100×). **(B)** Cells in differentiating conditions, DMEM/F12+, for 20 d (magnification 200×).

4. Notes

1. The adipose stromal cell is approx 15–20 μm in diameter and assumes a fibroblastic morphology when grown in culture. As is the case with most human cell cultures, large variability is observed in the growth rate of these cells from different subjects. The plating density of these cells can be quite important to their growth rate. If the preadipocytes are seeded too sparsely, they tend to grow very slowly. A good seeding density is approx 5×10^4 cells/well in a 6-well plate

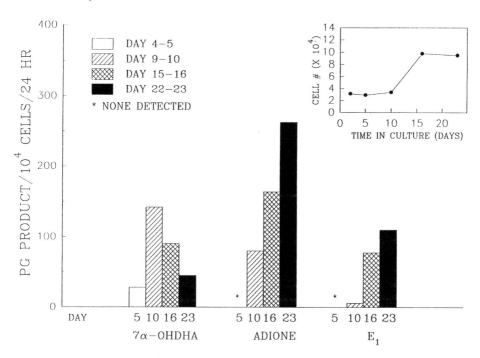

Fig. 4. Growth curve of human adipose stromal cells in primary culture under replicating conditions. Study shows the time-course of metabolism of dehydroepiandrosterone to three major metabolites: 7α-hydroxydehydroepiandrosterone (7α-OHDHA), androstenedione (A-dione), and estrone (E_1).

or 2×10^5 cells/25-cm^2 flask. Also note that the plating efficiency of adipose stromal cells is about 50%, so take this into account when plating cells. On average preadipocytes have a population doubling time of 40–60 h.
2. Ten percent fetal calf serum is optimum for promoting replication of human adipose stromal cells. The preadipocyte growth rate is about 30–50% slower when growth media is supplemented with 5% fetal calf serum, but their androgen metabolism is not altered. At 1% serum, the cells are kept viable, but they do not replicate.
3. Adipose stromal cells can be passaged in tissue culture by trypsinization for up to 6–7 passages. The cells maintain their fibroblastic appearance and their ability to metabolize androgens, although activity decreases with higher passage number.
4. The phenol red indicator present in most tissue-culture media contains some estrogenic properties. Therefore, for androgen/estrogen metabolism studies, media without phenol red should be used.
5. Primary human preadipocytes that have been exposed to serum in tissue culture for more than 24 h cannot be induced to differentiate.

6. For maximum differentiation, preadipocytes should be plated as densely as possible, about 4–5 times the density of cells in regular serum-containing growth media. The percentage of cells that differentiate and accumulate lipid droplets is quite variable from subject to subject, ranging from 20–80%.
7. Stromal cells derived from adipose tissue from different body sites of the same individual will replicate in culture at similar rates. The metabolic characteristics of the cells may, however, differ depending on the site of origin of the adipose tissue. For example, lower body fat is more active in androgen-to-estrogen conversion than upper body fat *(17)*. Hormonal control of lipolysis also differs in fat tissue from different body sites *(18–20)*.
8. Adipose cell differentiation in vitro is characterized by the accumulation of lipid droplets as well as the induction of specific lipogenic enzymes, LPL and GPDH. Levels of both enzymes are undetectable in preadipocytes at d 0. LPL activity is high by d 4 and reaches maximum levels by d 8–12 (~17 mU/mg protein), whereas GPDH activity begins to rise by d 6–8 and reaches maximum levels by d 16 (~500 mU/mg protein) *(8)*. After 8–10 d in the defined serum-free conditions, small lipid droplets are visible in the cells by microscopic examination. By 16–20 d, the cytoplasm of the differentiated cell is filled with these droplets.
9. The micromolar absorption coefficient of NADH at 340 nm is 6.22×10^{-3} L/µmol.

    GPDH enzyme activity = (Δ absorbance/min) × [$1/(6.22 \times 10^{-3}$L/µmol)] × (total volume in cuvet/sample volume in cuvet)

    which defines GPDH activity in µmol/min · L. This can be converted to µmol/min · mL and divided by the protein concentration of the sample (mg/mL) to relate the enzyme activity to the amount of protein – µmol enzyme activity/mg protein/min.
10. One milliunit of GPDH enzyme activity is equal to the oxidation of 1 nmol NADH/min.

References

1. Poznanski, W., Waheed, I., and Van, R. (1973) Human fat cell precursors. Morphologic and metabolic differentiation in culture. *J. Lab. Invest.* **29,** 570–576.
2. Van, R., Bayliss, C., and Roncari, D. (1976) Cytological and enzymological characterization of human adipocyte precursors in culture. *J. Clin. Invest.* **58,** 699–704.
3. Van, R. and Roncari, D. (1978) Complete differentiation of adipocyte precursors. A culture system for studying the cellular nature of adipose tissue. *Cell. Tiss. Res.* **181,** 197–203.
4. Green, H. and Kehinde, O. (1975) An established preadipose cell line and its differentiation in culture. *Cell* **5,** 19–27.
5. Green, H. and Kehinde O. (1975) Sublines of mouse 3T3 cells that accumulate lipid. *Cell* **1,** 113–116.
6. Négrel, R., Grimaldi, P., and Ailhaud, G. (1978) Establishment of a preadipocyte clonal line from epididymal fat pad of ob/ob mouse that responds to insulin and lipolytic hormones. *Proc. Natl. Acad. Sci. USA* **75,** 6064–6068.

7. Grimaldi, P., Djian, P., Négrel, R., and Ailhaud, G. (1982) Differentiation of Ob 17 preadipocytes to adipocytes: requirement of adipose conversion factor(s) for fat cell cluster formation. *EMBO J.* **1**, 687–692.
8. Hauner, H., Entenmann, G., Wabitsch, M., Gaillard, D., Ailhaud, G., Négrel, R., and Pfeiffer, E. F. (1989) Promoting effects of glucocorticoids on the differentiation of human adipocyte precursor cells cultured in a chemically defined medium. *J. Clin. Invest.* **84**, 1663–1670.
9. Grodin, J. M., Siiteri, P. K., and MacDonald, P. C. (1973) Source of estrogen production in postmenopausal women. *J. Clin. Endocrinol. Metab.* **36**, 207–211.
10. Perel, E., Daniilescu, D., Kindler, S., Kharlip, L., and Killinger, D. W. (1986) The formation of 5α-reduced androgens in stromal cells from human breast adipose tissue. *J. Clin. Endocrinol. Metab.* **62**, 314–318.
11. Simpson, E. R., Cleland, W. H., Smith, M. E., and Mendelson, C. R. (1981) Estrogen formation in stromal cells of adipose tissue of women: induction by glucocorticoids. *Proc. Natl. Acad. Sci. USA* **78**, 5690–5695.
12. Khalil, M. W., Strutt, B., Vachon, D., and Killinger, D. W. (1993) Metabolism of dehydroepiandrosterone by cultured human adipose stromal cells: identification of 7α-hydroxydehydroepiandrosterone as a major metabolite using high performance liquid chromatography and mass spectroscopy. *J. Steroid. Biochem.* **46**, 585–594.
13. Ailhaud, G., Grimaldi, P., and Négrel, R. (1994) Hormonal regulation of adipose differentiation. *Topics Endocrinol. Metab.* **5**, 132–136.
14. Ailhaud, G., Armi, E.-Z., Bardon, S., Barcellini-Couget, S., Bertrand, B., Catalioto, R., et al. (1990) The adipocyte: relationships between proliferation and adipose cell differentiation. *Am. Rev. Respir. Dis.* **142**, S57–S59.
15. Négrel, R., Gallard, D., and Ailhaud, G. (1989) Prostacyclin as a potent effector of adipose cell differentiation. *Biochem. J.* **257**, 399–405.
16. Killinger, D. W., Strutt, B. J., Roncari, D. A., and Khalil, M. W. (1995) Estrone formation from dehydroepiandrosterone in cultured human breast adipose stromal cells. *J. Steroid. Biochem. Molec. Biol.* **52**, 195–201.
17. Killinger, D., Perel, E., Daniilescu, D., Kharlip, L., and Lindsay, W. R. N. (1987) The relationship between aromatase activity and body fat distribution. *Steroids* **50**, 1–3.
18. Rebuffé-Scrive, M., Morin, P., and Bjorntorp, P. (1991) The effect of testosterone on abdominal adipose tissue in men. *Int. J. Obesity* **15**, 791–795.
19. Rebuffé-Scrive, M., Enk, L., Crona, N., Lönnroth, P., Abrahamsson, L., Smith, U., and Bjorntorp, P. (1985) Fat cell metabolism in different regions in women. *J Clin Invest* **75**, 1973–1976.
20. Rebuffé-Scrive, M., Lönnroth, P., Morin, P., Wesslau, Ch., Bjorntorp, P., and Smith U. (1987) Regional adipose tissue metabolism in men and postmenopausal women. *Int. J. Obesity* **11**, 347–355.

6

Human Fetal Brain Cell Culture

Mark P. Mattson

1. Introduction

The human brain is arguably the most complex organ system, consisting of more than 10^{11} nerve cells and at least three times as many glial cells. Because of the cellular complexity of even the simplest nervous systems, the development of in vitro technologies that allow isolation and study of nerve and glial cells under conditions in which their environment can be precisely manipulated has proven invaluable *(1,2)*. However, several factors have contributed to the relative dearth of information concerning the mechanisms responsible for the development and proper function of the human brain compared to the knowledge base in lower mammalian and invertebrate species. A major impediment has been the ethical considerations surrounding the use of fetal human tissue obtained from elective abortions and the lack of an alternative source of viable normal human brain cells. Although research that employs human fetal tissue has been limited, what has been done has made a major impact in the development of prophylactics and therapeutics for several important human diseases. For example, the use of fetal tissue from elective abortions was key to development of the polio vaccine *(3)*.

In addition to ethical considerations, technical hurdles related to the ability to maintain long-term cultures and store cells in a cryopreserved state have greatly limited the type and number of experiments that can be performed. The present chapter describes methods developed in my laboratory for procurement of human fetal brain tissue, and the preparation, maintenance, and long-term culture of neurons and glia from different brain regions. In addition, a protocol for cryopreserving fetal human brain cells is included, which provides a means to establish cell stocks that allow performance of experiments over an

From: *Methods in Molecular Medicine: Human Cell Culture Protocols*
Edited by: G. E. Jones Humana Press Inc., Totowa, NJ

extended time period. Finally, examples of applications of such cultured cells to studies of human brain development and disease are briefly discussed.

2. Materials

1. Mg^{2+}- and Ca^{2+}-free Hank's balanced salt solution (HBSS) buffered with 10 mM HEPES (pH 7.2).
2. Serum-containing maintenance medium (SCMM): Eagle's minimum essential medium with Earle's salts (Gibco) supplemented with 20 mM KCl, 1 mM sodium pyruvate, 20 mM glucose, 1 mM L-glutamine, 2 mg/mL gentamicin sulfate, 10% fetal bovine serum (v/v), and 27 mM sodium bicarbonate (pH 7.2) (all supplements from Sigma, St. Louis, MO).
3. Defined maintenance medium (DMM): Neurobasal medium with B27 supplements (Gibco, Grand Island, NY; see Note 4).
4. Cryopreservation medium: SCMM to which dimethyl sulfoxide is added to a final concentration of 8% (v/v).
5. Culture dishes: 35 mm (Corning, Oneonta, NY; #25000) and 60 mm (Costar #3060) dishes, glass-bottom 35-mm dishes from Mat-Tek Inc. (Ashland, MA; #P35G-14-0-C-gm), and 96-well plates (Corning #25860).
6. Polyethyleneimine (50% solution; Sigma) is diluted 1:1000 in borate buffer, which consists of 3.1 g/L boric acid and 4.75 g/L borax in glass-distilled water, pH 8.4.

The nature and sources of other materials discussed can be found in the cited references.

3. Methods

3.1. Tissue Acquisition

Establishing a source of tissue requires correspondence with the directors of local women's surgical clinics, who are generally receptive to contributing to the scientific advancements that result from studies of human fetal tissue. Although surgery is most commonly performed during the first trimester of pregnancy, tissue from 12–15 wk fetuses have been utilized for several reasons, including that brain regions of interest can be identified more readily than at earlier gestational ages, and neuronal viability in cultures established from these gestational windows is much greater than in cultures established from older fetuses. Estimates of quantities of tissue required for particular experiments can be made based on numbers of viable cells obtained/fetus (see Section 3.2.).

1. Ethical considerations. Because of the delicate nature of the issue of use of human fetal tissue for biomedical research, guidelines have been established by various governmental and institutional organizations in order to ensure that proper ethical principles are practiced (4). The major source of human fetal brain tissue is

elective abortions. The guidelines that were followed were developed in conjunction with the University of Kentucky Internal Review Board and include the following key points:
   a. Informed consent for tissue donation is obtained after the patient has consented, in writing, to the abortion.
   b. Individuals involved in the research are not involved in the abortion decision, and have no role in the timing of the abortion or the surgical procedures.
   c. Tissue is donated without any form of compensation to the patient.
   d. Confidentiality of the patient's identity and of the specific use of the tissue is maintained.
2. Handling and storage of brain tissue. Communication between the clinic staff and the investigator allows procurement of the tissue as soon as possible following the surgery. Stocks of sterile refrigerated HBSS are kept on hand at the clinic. A member of the clinic staff obtains the brain tissue and places it in the HBSS. From this point on, handling of tissue and cells is performed using sterile conditions under a laminar flow hood. Brain tissue can be held in cold HBSS for at least 6 h without reducing cell viability. The human tissue should be handled as biohazardous material, since it may harbor infectious agents, notably, HIV and hepatitis. The investigator may choose to have a sample of tissue screened for such infectious agents, which would allow the opportunity to dispose of infected tissue. Alternatively, a sample of the patient's blood may be procured for HIV and hepatitis testing (with appropriate informed consent).

3.2. Cell Dissociation, Cryopreservation, and Culture

1. The most common method of elective abortion involves aspiration of the fetus. This method usually results in compression and fragmentation of the brain tissue, and so hinders identification of specific brain structures. The cerebral hemispheres remain relatively intact and can be recognized by their characteristic gyri. In many cases, other brain regions can also be identified, including the hippocampus and mesencephalon, which I and others *(5,6)* have used in cell-culture studies. Once the brain region of interest is identified and isolated, the tissue is placed in cold HBSS and minced into pieces of approx 1 mm^3 using a scalpel (#10 blade).
2. Tissue pieces are transferred to a 15-mL conical-bottom culture tube and allowed to settle to the bottom. Typically, the amount of tissue per tube will be equivalent to approximately one-half of a cerebral hemisphere or 2 hippocampi/tube. HBSS is removed, replaced with 4 mL of HBSS containing 2 mg/mL trypsin, and incubated for 20–30 min at room temperature. Tissue is rinsed twice with 5 mL HBSS, incubated for 5 min in HBSS containing 1 mg/mL soybean trypsin inhibitor, and rinsed twice with 5 mL HBSS (*see* Note 1). Ten milliliters of either SCMM or cryopreservation medium are then added to the tube, and cells are dissociated by trituration using a fire-polished pasteur pipet (the pipet is fire-polished by holding vertically, tip down, in a Bunsen burner flame until the diameter of the tip is reduced by approx 50%). The resulting cell suspension typically contains 1–5 mil-

lion viable cells/mL (*see* Note 2). Cells are then seeded into culture dishes in volumes calculated to achieve the desired plating cell density.
3. Cells dissociated in cryopreservation medium are aliquoted (0.5 mL) into 1-mL cryovials (with screw-tops). Cells should be frozen relatively slowly (approx –1°C/min) to –80°C, which can be accomplished using a controlled-cooling freezer. Alternatively, cryovials can be placed in freezing containers (simply sandwich the vials between two, 3–4 cm thick, pieces of styrofoam) and are then placed in a –80°C freezer for 12–24 h (it is important that the vials be completely insulated from the freezer air, so that they do not cool too rapidly). Frozen vials are then transferred to liquid nitrogen for long-term storage. In order to establish cultures from the cryopreserved cell stocks, the vials are rapidly thawed by immersion in a water bath at 37°C. Cells have been stored for 3 yr using these methods with cell losses of only 20–40%.
4. I routinely use polyethyleneimine as a growth substrate, and apply it to the surface of the culture dish as follows: Polyethyleneimine is dissolved in borate buffer (1:1000 dilution [v/v]). The surface of the culture dish is covered with this solution and allowed to incubate at room temperature for 3–24 h.
5. The dishes are then washed in sterile deionized distilled water (3 × 2 mL), allowed to dry under a laminar flow hood, and then exposed for 10 min to UV light to ensure sterility. SCMM is then added to the dishes (1.5 mL/35-mm dish; 2.5 mL/60-mm dish; and 100 µL/well in 96-well plates), and they are placed in a humidified CO_2 incubator (6% CO_2/94% room air) at 37°C until the time of cell plating (typically 1–4 d following preparation of dishes).
6. Following cell plating, cultures are left in the incubator for 16–24 h to allow cells to attach to the growth substrate (*see* Note 3). The medium is then removed and replaced with a lesser volume (0.8 mL/35-mm dish; 1.5 mL/60-mm dish; 100 µL/well in 96-well plates), which improves long-term neuronal survival. Cultures can be maintained without medium change for up to 3 wk. To maintain cultures for longer time periods, 50% of the medium can be replaced on a weekly basis.
7. Cultures maintained as just described will contain both neurons and astrocytes (Fig. 1); the astrocytes will progressively proliferate until they form a monolayer on the culture surface. Such cultures are valuable for a variety of studies. However, for many applications, it is desirable to utilize cultures that are essentially pure populations of either neurons or astrocytes. For example, in order to establish whether neurons are directly responsive to a neurotrophic factor, one must rule out the possibility of an indirect action mediated by glial cells *(7)*. Nearly pure neuronal cultures can be obtained by maintaining the cells in DMM (serum-free medium) (*see* Note 4). Alternatively, cells can be maintained in SCMM to which the mitotic inhibitor cytosine arabinoside (10 µM) is added on culture d 2. Essentially pure astrocyte cultures can be obtained by plating the cells on uncoated plastic dishes to which very few neurons will adhere. The death of any remaining neurons can be induced by doubling the medium volume and changing the medium twice per week.

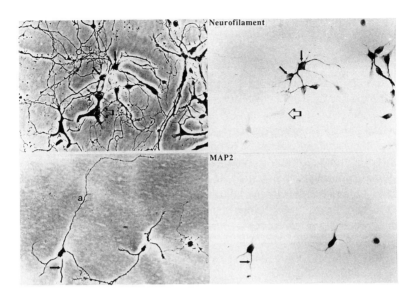

Fig. 1. Cell types present in human fetal brain cell cultures. (upper panels) Phase-contrast (left) and bright-field (right) micrographs of neurons and (e.g., solid arrows) and astrocytes (e.g., open arrow) in a neocortical culture immunostained with antibody to the medium-sized (160-kDa) neurofilament protein. (lower panels) Phase-contrast (left) and bright-field (right) micrographs of neurons in a hippocampal cell culture immunostained with antibody to microtubule-associated protein-2 (MAP2). Note immunoreactivity of cell bodies and dendrites (e.g., arrow), but not axons (a).

3.3. Applications

1. Fetal human brain cell cultures contain cells in various stages of development. Neurons in such cultures will readily elaborate neurites and form synaptic connections. Although relatively little work has been done to characterize the development of human fetal brain neurons in culture, the information that has been obtained has provided interesting insight into the similarities and differences between development of human neurons and their rodent counterparts. Human fetal cortical neurons grow more slowly and exhibit a longer "life span" in culture than do fetal rat cortical neurons *(8)*. Rates of axon elongation were up to 10 times slower in cultured cortical neurons from fetal human brain compared to rat brain. The human fetal neurons also exhibited a protracted development of sensitivity to the excitatory neurotransmitter glutamate *(9)*. Studies of rat brain development had shown that expression of non-NMDA receptors (AMPA/kainate receptors) precedes the appearance of NMDA receptors; the expression of these receptors occurred over a period of approx 4–10 d in cell culture *(10)*. In addition, rat neocortical and hippocampal neurons became vulnerable to excitotoxicity within 1–2 d of the appearance of glutamate receptors. In contrast, human

fetal neocortical neurons expressed non-NMDA and NMDA glutamate receptors over an extended 2–4 wk culture period. Moreover, the human neurons exhibited calcium responses to glutamate for several weeks prior to their being vulnerable to excitotoxicity *(9)*.

An important rationale for studying human fetal brain cells is that very little information is available concerning cellular and molecular mechanisms operative in human brain development vis-à-vis commonly studied laboratory animals. For example, a rapidly growing body of literature is available concerning the roles of neurotrophic factors in brain development in rodents. In contrast, parallel information in the human brain is lacking. Human fetal brain cell culture provides an opportunity to establish the similarities and differences in neurotrophic factor signaling in developing human and rodent brains. Basic fibroblast growth factor (bFGF) promotes neuronal survival in human neocortical cell cultures *(5)*. Immunostaining studies indicated that both neurons and astrocytes possess bFGF, which may be an endogenous source of trophic support. Regulation of neuronal growth cones, the motile distal tips of growing axons and dendrites, can be reliably studied in human fetal brain cell cultures (e.g., Fig. 2) using approaches developed in nonhuman systems *(11,12)*. In the example shown in Fig. 2, intracellular free calcium levels were imaged in an axonal growth cone using the calcium indicator dye fluo-3.

2. Both acute (e.g., stroke and traumatic brain injury) and chronic (e.g., Alzheimer's, Parkinson's, and Huntington's diseases) neurodegenerative disorders are major concerns in our society; taken together, they afflict tens of millions of Americans. An important impediment to conquering these disorders is that in most cases, they can only be studied at the cellular and molecular levels postmortem. The ability to study living human brain cells under controlled conditions allows one to test hypotheses and establish cause–effect-type relationships. For example, in Alzheimer's disease, a 40–42-kDa peptide called amyloid β-peptide (Aβ) forms insoluble deposits (plaques) in the brain (*see* ref. *13* for review). The Aβ arises from a much larger membrane-associated β-amyloid precursor protein (βAPP); mutations of βAPP have been causally linked to a small percentage of cases of Alzheimer's disease. However, the roles of the βAPP mutations and Aβ deposition in the pathogenesis of Alzheimer's disease are not well understood. Recent

---

Fig. 2. *(opposite page)* Changes in intracellular free calcium levels ([Ca^{2+}]$_i$) in an axonal growth cone of a cultured human fetal neocortical neuron following exposure to a calcium ionophore. The calcium indicator dye fluo-3 was introduced into the neuronal cytoplasm (using the membrane-permeant acetoxymethylester form), and images of [Ca^{2+}]$_i$ were acquired using a confocal laser scanning microscope (Molecular Dynamics). Images were acquired immediately prior to (upper panel) and 30 s and 4 min following exposure to 1 μM of calcium ionophore A23187. The [Ca^{2+}]$_i$ is represented on a gray scale (brighter pixel intensity indicates higher [Ca^{2+}]$_i$). Prior to exposure to A23187, the [Ca^{2+}]$_i$ is low throughout the growth cone. At 30 s following exposure to A23187, the [Ca^{2+}]$_i$ rises throughout the growth cone, including in

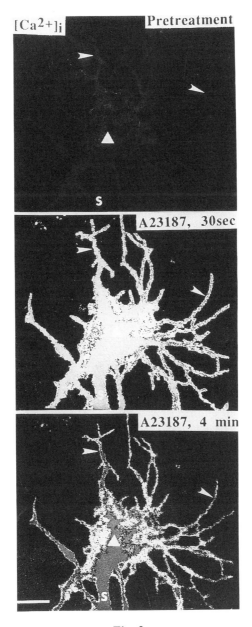

Fig. 2.

filopodia (e.g., arrowheads), the base of the growth cone (triangle), and in the shaft of the axon (s). The $[Ca^{2+}]_i$ subsequently recovers toward baseline levels, as can be seen in the image acquired 4 min following exposure to A23187. Scale bar, 5 µm.

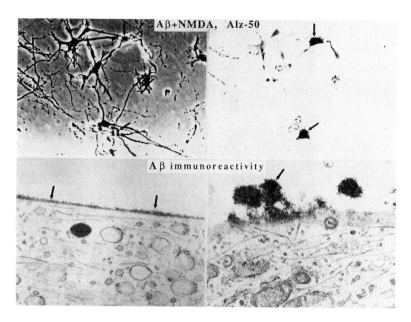

Fig. 3. Use of human fetal brain cell culture to study mechanisms of neuronal degeneration relevant to the pathogenesis of Alzheimer's disease. (upper panels) Phase-contrast (left) and bright-field (right) micrographs of a field of cultured human fetal neocortical neurons (30 d in culture) shown 24 h following exposure to 40 μM amyloid β-peptide (Aβ; amino acids 1–38) plus 50 μM NMDA. The cultures were immunostained with an antibody (Alz-50) that recognizes an altered form of the microtubule-associated protein τ present in neurofibrillary tangles of Alzheimer's disease. Aβ increased neuronal vulnerability to NMDA excitotoxicity and induced the accumulation of Alz-50 immunoreactivity in the cell bodies of some neurons (solid arrows), but not others (open arrows). (lower panels) Immunoelectron micrographs (peroxidase-DAB) showing the cell surface of neurons in human fetal neocortical cell cultures that had been exposed to Aβ for 24 h. Aβ immunoreactivity was localized primarily to the surface of the plasma membrane (e.g., arrows).

studies of cultured human fetal neocortical neurons showed that Aβ can cause neurons to be exquisitely sensitive to excitotoxic cell injury and death (Fig. 3; ref. *14*). When added to human fetal neocortical cultures, Aβ accumulated on or in the plasma membrane (Fig. 3). The mechanism of Aβ neurotoxicity involves elevation of intracellular calcium levels *(14)*, apparently resulting from free radical-mediated damage to plasma membrane proteins involved in the regulation of ion homeostasis *(15,16)*.

3. The ability to study living human neurons in culture has also provided insight into mechanisms of ischemic brain injury, a process relevant to hypoxic conditions in the developing brain (e.g., cerebral palsy) and stroke in the adult brain. Human fetal neocortical cultures have been employed to study the mechanisms

whereby neurotrophic factors protect neurons against hypoglycemic and hypoxic insults. bFGF and nerve growth factor (NGF) each protected neocortical neurons against glucose deprivation-induced injury *(7)*. These neurotrophic factors prevented the delayed elevation of intracellular calcium levels otherwise induced by glucose deprivation. These findings suggest that endogenous neurotrophic factors may serve a neuroprotective function when the brain is subjected to ischemic conditions. Moreover, the data suggest that neurotrophic factors may prove useful in reducing neuronal injury in clinical settings, such as following stroke or traumatic brain injury *(17)*.

Finally, a variety of substances of abuse (e.g., cocaine, alcohol, phencyclidine) are known to adversely affect brain development. Human fetal brain cell cultures can provide valuable information concerning the direct actions of substances of abuse on developing brain cells. Studies of cultured human fetal neocortical neurons have shown that phencyclidine, at concentrations believed to be reached *in utero*, can be neurotoxic *(18)*. The findings were consistent with a mechanism involving blockade of potassium channels by phencyclidine, resulting in membrane depolarization and elevated levels of intracellular calcium. Studies of the effects of ethanol and cocaine *(19)* on human fetal neurons remain to be performed. Further studies of cultured human fetal brain cells are needed and are likely to improve our understanding of human neurodegenerative disorders greatly, as well as mechanisms of brain damage resulting from substance abuse.

4. Notes

1. Bacterial and fungal contamination can be a problem. Its incidence can be reduced by washing the brain tissue several times in HBSS containing relatively high levels of antibacterial and antifungal agents (e.g., 10 mg/mL gentamicin and 5 mg/mL amphotericin B) prior to cell dissociation.
2. Deoxyribonuclease can be used to reduce cell clumping during the dissociation process. A concentration of 10 µg/mL deoxyribonuclease (Sigma) was used.
3. Fetal human neurons attach more slowly to the growth substrate than their rodent counterparts. In order to improve plating efficiency, it is therefore important not to change the culture medium for 16–24 h following plating.
4. DMM (Gibco "Neurobasal" plus B27 supplements) is an excellent serum-free medium for promoting long-term neuronal survival. However, this medium contains antioxidants, which act to reduce neuronal vulnerability to a variety of conditions, including exposure to excitotoxins, ischemia-like conditions, and amyloid toxicity.

References

1. Beadle, D. J., Lee, G., and Kater, S. B. (1988) *Cell Culture Approaches to Invertebrate Neuroscience*. Academic, London.
2. Mattson, M. P., Barger, S. W., Begley, J. G., and Mark, R. J. (1994) Calcium, free radicals, and excitotoxic neuronal death in primary cell culture. *Methods Cell Biol.* **46**, 187–216.
3. Hayflick, L. (1992) Fetal tissue banned…and used. *Science* **257**, 1027.

4. Vawter, D. E., Kearney, W., Gervais, K. G., Caplan, A. L., Garry, D., and Tauer, C. (1990) *The Use of Human Fetal Tissue: Scientific, Ethical and Policy Concerns.* University of Minnesota Press, Minneapolis.
5. Mattson, M. P. and Rychlik, B. (1990) Cell culture of cryopreserved human fetal cerebral cortical and hippocampal neurons: neuronal development and responses to trophic factors. *Brain Res.* **522,** 204–214.
6. Redmond, D. E., Naftolin, F., Collier, T. J., Leranth, C., Tobbins, R. J., Sladek, C. D., Roth, R. H., and Sladek, J. R. (1988) Cryopreservation, culture, and transplantation of human fetal mesencephalic tissue into monkeys. *Science* **242,** 768–771.
7. Cheng, B. and Mattson, M. P. (1991) NGF and bFGF protect rat and human central neurons against hypoglycemic damage by stabilizing calcium homeostasis. *Neuron* **7,** 1031–1041.
8. Mattson, M. P. and Rychlik, B. (1991) Comparison of rates of neuronal development and survival in human and rat cerebral cortical cell cultures. *Mech. Aging Dev.* **60,** 171–187.
9. Mattson, M. P., Rychlik, B., You, J.-S., and Sisken, J. E. (1991) Sensitivity of cultured human embryonic cerebral cortical neurons to excitatory amino acid-induced calcium influx and neurotoxicity. *Brain Res.* **542,** 97–106.
10. Mattson, M. P., Kumar, K., Cheng, B., Wang, H., and Michaelis, E. K. (1993) Basic FGF regulates the expression of a functional 71 kDa NMDA receptor protein that mediates calcium influx and neurotoxicity in cultured hippocampal neurons. *J. Neurosci.* **13,** 4575–4588.
11. Kater, S. B., Mattson, M. P., Cohan, C. S., and Connor, J. A. (1988) Calcium regulation of the neuronal growth cone. *Trends Neurosci.* **11,** 315–321.
12. Mattson, M. P. (1994) Secreted forms of β-amyloid precursor protein modulate dendrite outgrowth and calcium responses to glutamate in cultured embryonic hippocampal neurons. *J. Neurobiol.* **25,** 439–450.
13. Selkoe, D. J. (1993) Physiological production of the β-amyloid protein and the mechanism of Alzheimer's disease. *Trends Neurosci.* **16,** 403–409.
14. Mattson, M. P., Cheng, B., Davis, D., Bryant, K., Lieberburg, I., and Rydel, R. E. (1992) β-amyloid peptides destabilize calcium homeostasis and render human cortical neurons vulnerable to excitotoxicity. *J. Neurosci.* **12,** 376–389.
15. Goodman, Y. and Mattson, M. P. (1994) Secreted forms of β-amyloid precursor protein protect hippocampal neurons against amyloid β-peptide toxicity and oxidative injury. *Exp. Neurol.* **128,** 1–12.
16. Hensley, K., Carney, J. M., Mattson, M. P., Aksenova, M., Harris, M., Wu, J. F., Floyd, R., and Butterfield, D. A. (1994) A model for β-amyloid aggregation and neurotoxicity based on free radical generation by the peptide: relevance to Alzheimer's disease. *Proc. Natl. Acad. Sci. USA* **91,** 3270–3274.
17. Mattson, M. P. and Scheff, S. W. (1994) Endogenous neuroprotection factors and traumatic brain injury: mechanisms of action and implications for therapies. *J. Neurotrauma* **11,** 3–33.
18. Mattson, M. P. and Rychlik, B. (1992) Degenerative and axon outgrowth-altering effects of phencyclidine in human fetal cerebral cortical cells. *Neuropharmacology* **31,** 279–291.
19. Moroney, J. T. and Allen, M. H. (1994) Cocaine and alcohol use in pregnancy. *Adv. Neurol.* **64,** 231–242.

7

Culture of Human Normal Brain and Malignant Brain Tumors for Cellular, Molecular, and Pharmacological Studies

Francis Ali-Osman

1. Introduction

Human brain neoplasms comprise a highly heterogeneous and biologically diverse group of tumors, the most common and most malignant of which are those of neuroepithelial origin *(1)*. Despite intensive research, little is still understood about the cellular and molecular processes involved in the genesis, progression, and response to therapy of these tumors. Much of the progress made to date, however, has resulted, in part from advances in the ability to culture and propagate cells of both normal and neoplastic brain tissue in vitro *(2,3)*. For example, in vitro cultures have contributed significantly to the development of techniques, such as bromodeoxyuridine labeling, that are used to estimate the cell-growth kinetics of gliomas in patients *(4)*. Normal brain and brain tumor cultures have also played a central role in research directed at a better understanding of the complex interplay between the cellular components of the brain, such as that between various glial cells, neurons, and endothelial cells *(5,6)*. In studies of the molecular and cellular mechanisms involved in brain tumor resistance to therapy, in vitro cultures of human glioma cells have played a significant role in the identification of O^6-alkylguanine DNA alkyltransferase *(7)*, glutathione, and glutathione S-transferases *(8)*, as critical factors in human brain tumor alkylator resistance, findings that are providing the basis for novel therapies for human gliomas. Neurobiology and neuro-oncology research will therefore continue to be critically dependent on appropriate in vitro models of normal and neoplastic brain cells. In this chapter, techniques for in vitro culture of normal glial cells and of malignant gliomas are described. Methods for characterizing the cultured cells and for glioma

From: *Methods in Molecular Medicine: Human Cell Culture Protocols*
Edited by: G. E. Jones Humana Press Inc., Totowa, NJ

stem cell cloning are presented with notes on their use for pharmacological and molecular studies.

2. Materials
2.1. Primary Cell Cultures and Cell Lines

1. Ca^{2+}/Mg^{2+}-free Hank's balanced salt solution (CMF-HBSS).
2. Dulbecco's minimum essential medium (DMEM).
3. Fetal calf serum (FCS) heat-inactivated at 65°C for 1 h.
4. 10X Stock enzyme cocktail for tissue disaggregation:
DNase Type I (2000 U/mg)	100 mg
Neutral protease (1 U/mg)	250 mg
Collagenase Type Ia (125 U/mg)	200 mg
CMF-HBSS	add to 50 mL

   The mixture is magnet-stirred to dissolve and filter sterilized through a 0.2-µm filter.
5. Trypsinizing solution for dissociating monolayer cultures into single cells:
Trypsin	125 mg
Na_2EDTA	20 mg
CMF-HBSS	add to 100 mL
6. Sterile scalpels, blades, 60-µm mesh nylon sieve.

2.2. Cytomorphology, Immunocytochemistry, and In Situ Hybridization

1. Four-well tissue-culture-grade chamber slides (Lab-Tek, Nunc, Napperville, CA).
2. Fixatives: 4% paraformaldehyde, absolute methanol, absolute ethanol, acetone.
3. Stains: Harris hematoxylin, 0.5% eosin in ethanol.
4. Scott's Tap Water.
5. Xylene, aqueous and nonaqueous mounting media (e.g., Permount and Aquamount).
6. Primary antibodies (e.g., against glial fibrillary acidic protein, GFAP).
7. H_2O_2, diaminobenzidine.
8. 100 mM Tris-HCl buffers (pH 7.2 and 7.5).
9. Avidin-linked horseradish peroxidase.
10. 20X SSC: 3M NaCl, 0.3M Na_3 citrate; pH 7.0 adjusted with 1M HCl.
11. Blocking solution: 0.1M Na maleate, 0.1% N-lauroylsarcosine, 0.02% Na dodecyl sulfate.
12. Prehybridization buffer: 5X SSC, 1% blocking solution.
13. cDNA probes labeled with digoxigenin-11-dUTP (*see* Section 3.3.3. for labeling procedure).
14. Hexanucleotide mixture: Random hexanucleotides, 0.5M Tris-HCl, 0.1M $MgCl_2$, 1 mM dithioerythritol, 2 mg/mL bovine serum albumin.
15. dNTP mixture: 1 mM each of dCTP, dGTP, and dATP, 0.65 mM dCTP, and 0.35 mM digoxigenin dUTP; pH 6.5.
16. 2 U/mL of DNA polymerase I (Klenow enzyme).

2.3. Tumor Cell Cloning

1. Enriched DMEM cloning medium: 200 µM L-glutamine, 150 µg/mL Na pyruvate, 1 µg/mL gentamycin, 20% heat-inactivated FCS (pretested for ability to support tumor clonogenic cell growth).
2. Borosilicate glass capillary tubes: It is critical that the capillary tubes are of the highest quality. To allow adequate gas exchange during the incubation period, it is necessary for the tube dimensions to be within a specific range. The author uses tubes that are 1.38 mm in diameter and 105 mm in length. Slight deviations from these should not affect colony formation. Commercially available tubes are often much longer than this, in which case, they may be cut down to size, e.g., with a regular ampule file. Before use in cloning, and unless they have been cleaned to tissue-culture grade by the manufacturer, the tubes are soaked overnight in double-distilled water containing laboratory detergent, such as Liquinox, and further cleaned in an ultrasonic bath for approx 6 h. The tubes are allowed to stand under running water until all the detergent has been washed out. They are then rinsed with five to six changes of double-distilled water, making sure that during the rinses, the water actually fills up the capillary tubes. The cleaned and rinsed tubes are dry-heat-sterilized and cooled to room temperature before use for cell cloning.

3. Methods
3.1. Single-Cell Preparation and Culture Initiation

All materials and solutions used in these procedures should be sterile, and aseptic techniques should be used throughout. The following are the standard procedures used in our laboratory to obtain single-cell suspensions from brain tumors and normal brain specimens:

1. Place the tumor biopsy or normal brain specimen (collected aseptically) in a sterile 100-mm Petri dish. Using a scalpel and forceps, clean the specimen of blood clots, gross necrosis, and other visibly contaminating tissue.
2. Finely mince the tissue with "crossed scalpels." It is not necessary to add medium to the specimen during this process. Transfer the minced tissue aseptically into a trypsizing flask. Dilute the 10X stock enzyme cocktail 1:10 with CMF-HBSS, and add 10 mL/g of tissue. Magnet-stir the mixture at low speed at 37°C until all visible solid tissue has disappeared, usually within 60–90 min.
3. Dilute the cell suspension 1:5 with DMEM, and pass it through a 60-µm nylon mesh placed in a fluted funnel with a sterile culture tube under it to collect the filtered cell suspension. Centrifuge the resulting cell suspension at 300g in a precooled (4°C) centrifuge for 5 min and resuspend the cell pellet in DMEM containing 15% FCS (5 mL/g of tissue). At this stage, cell suspensions of normal brain and low-grade astrocytomas will usually contain a significant amount of fine noncellular debris. This material can interfere with culture establishment by covering the culture surface and thus preventing effective cell attachment. The

debris is also phagocytized by the growing astrocytes, which become vacuolated as a result and subsequently die. To reduce the debris and enhance the chances of successful culture of these specimens, allow the cell suspension to stand for 10–15 min at room temperature and remove approx 75% of the top portion of the suspension. Although this portion usually consists of debris with very few cells, it is advisable to monitor an aliquot of it in a Petri dish under phase-contrast microscopy before discarding the entire suspension. Dilute the remaining bottom 25% of the suspension 1:4 with fresh medium, and repeat the process twice or until most of the debris has been eliminated.

4. Perform a viable (Trypan Blue) cell count, and adjust the cell density to 1×10^5 cells/mL with DMEM containing 15% FCS. Transfer 7.5 mL of the suspension to a T75 tissue-culture flask (or 5 mL to a T25 flask). Simultaneously, pipet 500 µL of a 1:4 dilution of the cell suspension into each well of four-well tissue-culture-grade chamber slides. These slides will be used to characterize the cultures later. Incubate both flasks and slides at 37°C, 5% CO_2 in air and 100% humidity.

5. After 3–4 d (earlier if there is a substantial amount of cellular debris still remaining in the initial culture), carefully remove the culture medium and replace it with fresh prewarmed (37°C) DMEM containing 15% FCS. Examine the cultures once a week until they attain confluence. Cells of normal brain and low-grade gliomas generally grow more slowly in primary culture and may require 3–4 wk to achieve confluence. To enhance the chances of successful culture, it is recommended that normal brain and low-grade astrocytomas be seeded at an initial density of $2.5–5 \times 10^5$ cells/mL, instead of the 1×10^5 cells/mL recommended for malignant gliomas. Once confluence has been achieved, primary cultures may be used for a variety of studies or expanded to yield more cells by passaging them, as is described below.

3.2. Establishment of Cell lines

Primary cultures of gliomas are often initiated with the ultimate goal of establishing permanent cell lines from them. These cell lines are particularly useful as models for drug screening, and as reproducible tools for biochemical and molecular studies. The protocol described has been used successfully in our laboratory to establish cell lines from both normal brain and brain tumors.

1. Set up primary cultures as described in Section 3.1. After the cultures have attained approx 75–80% confluence, rinse them twice with CMF-HBSS, and add trypsinizing solution at 4.5 mL/T75 flask, or 2 mL/T25 flask. Reincubate the flasks at 37°C for 1–2 min, and then agitate them vigorously to dislodge the cells. With the help of a pasteur pipet, aspirate the suspension several times to disaggregate the monolayers into single cells.

2. Add 5 mL of DMEM containing 15% FCS to the suspension; antitrypsin in the FCS will inactivate the trypsin. Dilute the suspension 1:4 with fresh medium, and pellet the cells at 300g in a precooled (4°C) centrifuge for 5 min. Discard the super-

natant, and resuspend the cells in fresh DMEM containing 15% FCS. Perform a cell count, adjust the cell density to 1×10^5 cells/mL (2.5×10^5 cells/mL for slow-growing cultures), and transfer 7.5 mL to a T75 flask or 5 mL to a T25 flask. Sometimes instead of passaging cells based on cell count, the cell suspension from one flask is seeded (or split) into a specific number of flasks. Brain tumor cultures are usually passaged at 1:4 to 1:8 splits, depending on the population growth rate of the culture. Normal brain cultures should be initially split at 1:2 to 1:4.

3. Incubate the flasks at 5% CO_2 in air in a fully humidified environment at 37°C, and examine them once a week until they attain confluence, at which time they are repassaged, as described in steps 1 and 2. Generally, for glial cultures, once the cells have successfully undergone at least five passages (approx 3 mo), they will continue on to become permanent cell lines, which can be maintained indefinitely.

3.3. Culture Characterization

An important step in the culture of normal brain and malignant brain tumors is the characterization of the cells of the resultant cultures. This is usually achieved using a combination of cytomorphology and immunocytochemistry. Other methods, such as *in situ* hybridization for specific gene transcripts, enzyme cytochemistry, and electron microscopy, may also be used. This section describes techniques for characterizing glial cells in culture by cytomorphology, immunocytochemistry, immunofluorescence, and *in situ* hybridization. Cells in culture may also be characterized by biological parameters, such as their population growth kinetics, as described in Section 4.

3.3.1. Cytomorphology

Cytomorphological characterization of cultures of both normal and neoplastic glial cells is best performed with hematoxylin-eosin-stained cultures grown on tissue-culture-grade microscope slides, as described in Section 3.3.1. In the author's laboratory, consistently reproducible results are obtained with the following protocol.

1. Set up cultures in tissue-culture-grade chamber slides, and allow them to attain approx 75% confluence. Decant the growth medium from the slides, and remove the walls of the chambers by pressing the ends together with a gentle lifting action. Using forceps, carefully remove the basement glue that holds the chambers to the slides, taking precaution not to injure the cell monolayers. Rinse the slides twice in phosphate-buffered saline (PBS), and place them in 95% ethanol for at least 15 min to fix the cells. Fixation can continue for several hours without any deleterious effects on cellular morphology. After ethanol fixation, rehydrate the cells by standing the slides in PBS for 20 s.
2. Dip the fixed slides in Harris hematoxylin for 2 min, rinse in tap water (5 dips) and then "blue" in Scott's Tap Water (10 dips). After rinsing (5 dips) in tap water,

dehydrate the slides in increasing ethanol concentration (10 dips each in 70, 95, and 100%), and stain in a 0.5% eosin in ethanol solution for 1 min. Rinse in two exchanges of absolute ethanol, and clear by two changes of xylene for 2 min each. After the xylenes, mount the slides in Permount and examine microscopically. Figure 1A–H shows typical cytomorphologies of cultures of normal astrocytes and brain tumors of different histologies that have been stained as described.

3.3.2. Immunocytochemistry

Immunocytochemical staining is a powerful tool for characterizing cultured glial cells and complements other methods, such as cytomorphology. Markers available for immunological characterization of cultures of human normal brain and malignant brain tumors *(9,10)* include: glial fibrillary acidic protein, GFAP (normal and malignant astrocytes and ependymal cells), S-100 (all cells of glial origin), and galactocerebroside (normal and malignant oligodendroglial cells). The following procedure is used in the author's laboratory for routine immunocytochemical staining for GFAP and other cellular proteins.

1. Prepare cell cultures as described in Section 3.1., and fix in absolute methanol for approx 10 min. Staining intensity for cytoplasmic proteins can, in some cases, be enhanced by placing the slides for 60 s in cold (–20°C) acetone prior to methanol fixation.
2. Rehydrate the slides in PBS (1 min). Using a dropper bottle or a pipet, apply with normal rabbit serum (3 drops in 10 mL of 100 mM Tris buffer, pH 7.2) to the cells to block nonspecific binding sites and incubate them for 30 min at room temperature. Rinse the slides with three changes of cold PBS. Apply a 1:500 dilution (in PBS) of the primary antibody, e.g., GFAP, to the cells, and incubate in a humid atmosphere and 37°C or room temperature for 1–2 h (or with a 1:1000 dilution overnight). It is critical that the cells are not allowed to dry up during this incubation. Wash the slides three times with cold PBS. It is recommended that during washings, the slides are placed on a shaking or rotating platform set at a slow speed.
3. Apply a 1:200 dilution of a biotinylated second antibody that has been raised against the source of the primary antibody, and after 1 h at room temperature, wash the cells three times with cold PBS, and apply a solution of avidin-linked horseradish peroxidase, and incubate for 15 min. Rinse three times with PBS, and treat the slides with a 0.01% H_2O_2 containing 0.05% diaminobenzidine in 50 mM Tris-HCl, pH 7.5. After 30 s, rinse the slides under tap water, dehydrate in increasing ethanol concentrations (70, 95, and 100%), clear in xylene, and mount in Permount. Typical GFAP immunocytochemical staining of cultured malignant glioma cells is shown in Fig. 2A,B.
4. For indirect immunofluorescence staining, the procedure is similar to that described for immunocytochemistry, except that the second antibody is conjugated with an appropriate fluorochrome, such as fluoroscein or rhodamine, and the peroxidase visualization steps are omitted. After the second antibody treat-

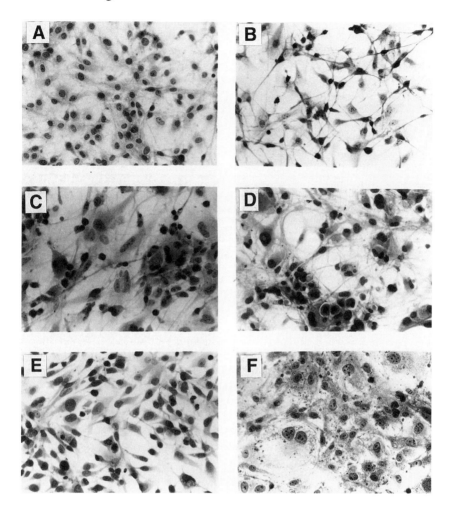

Fig. 1. Typical cytomorphological features of human normal astrocytes and brain tumors in primary culture after staining with hematoxylin-eosin. **(A)** Normal astrocytes, **(B)** low-grade astrocytoma, **(C)** anaplastic astrocytoma, **(D)** glioblastoma multiform, **(E)** malignant oligodendroglioma, and **(F)** meningioma. Note the increasing pleomorphic nature of the cells with increasing malignancy of the astrocytes, the virtual lack of processes in oligodendroglioma cells, and the multiple prominent nucleoli of the meningioma cells.

ment, the cells are rinsed three times with PBS and mounted in an appropriate aqueous mounting solution. For direct immunofluorescence staining, the fluorochrome is conjugated directly to the primary antibody, and as such, the second antibody step is omitted. Figure 2C shows an immunofluoresence staining of

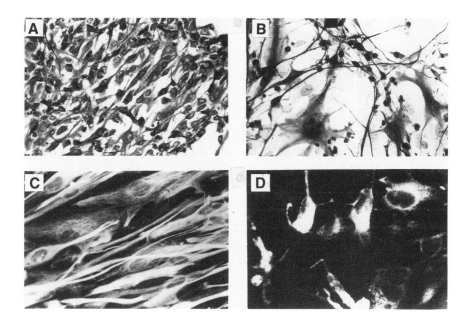

Fig. 2. Immunocytochemical staining of primary cultures of **(A)** an anaplastic astrocytoma and **(B)** a glioblastoma multiform for GFAP, and immunofluorescence of cells of **(C)** an astrocytoma stained for GFAP and **(D)** oligodendroglioma stained live, i.e., without prior fixation, for galactocerebroside.

cultured malignant astrocytomas for GFAP, and Fig. 2D shows oligodentroglioma cells stained live for galactocerebroside.

3.3.3. In Situ *Hybridization*

In situ hybridization is a excellent technique for studying the expression of genes. It has a particular advantage over other methods, such as Northern blotting, because it allows the examination of gene transcripts in intact cells. Both radioactive and nonradioactive procedures are available for *in situ* hybridization. As probes, small cDNAs and, particularly, oligodeoxynucleotides are ideal for *in situ* hybridization, since they penetrate cells easily and provide a greater design flexibility, i.e., can be made complementary to different regions of a given mRNA. For cells cultured from brain tissue and brain tumors, the author has successfully used the following nonradioactive method.

1. Grow cultures to approx 75% confluence in tissue-culture chamber slides as described in Section 3.1. Rinse in cold PBS, and fix in 4% paraformaldehyde for 20 min (room temperature). Denature the DNA by placing the slides in $2N$ HCl for 20 min, followed by 2X SSC buffer for 15 min at 70°C.

2. Postfix the denatured slides in 4% paraformaldehyde for 5 min, wash in PBS for another 5 min, and treat with 0.25% acetic anhydride for 10 min. Place the slides in a prehybridization buffer for 1 h. Apply a solution of the labeled cDNA probe in hybridization buffer to the cells, and incubate the slides overnight at 65°C. In the standard nonradioactive method, cDNA probes are labeled with digoxigenin-11-dUTP by random priming, following the manufacturer's protocol, with a few modifications as follows:
    a. Boil the cDNA probe for 10 min to denature it. Chill quickly on dry ice/ethanol for 90 s and add 1 µg DNA to the following reaction mix (total volume 20 µL): 2 µL hexanucleotide mixture, 2 µL dNTP mixture, 1 µL of a 2 U/mL of DNA polymerase I (Klenow enzyme).
    b. Incubate the mixture at 37°C overnight, and terminate the reaction with 2 µL 200 mM Na$_2$-EDTA, pH 8. Add 1 µL of 20 mg/mL glycogen, followed by 2.3 µL LiCl and 75 µL cold (–20°C) ethanol.
    c. After thorough mixing, precipitate the DNA by standing the solutions at –85°C for 1 h. Thaw at room temperature, and pellet the DNA by centrifugation at 12,000g for 5 min. Wash the pelleted labeled DNA twice with 100 µL 70% ethanol; recentrifuge between washes. After the last wash, dry the pellet, and redissolve it in 50 µL of 10 mM Tris-HCl containing 0.2M Na$_2$-EDTA and 0.1% sodium dodecyl sulfate (SDS).
3. Wash the hybridized slides in an SSC gradation series, and in Tris/NaCl (pH 7.5) for 1 min, followed by 2% normal serum in the same buffer for 1 h. Apply antidigoxigenin antibody conjugated with alkaline phosphatase, and incubate at room temperature for 3 h. Wash the slides in the Tris/NaCl buffer, followed with a buffer (pH 9.5) containing 100 mM Tris-HCl, 100 mM NaCl, and 50 mM MgCl$_2$. Generate hybridization signals by treating the slides with a 0.1% nitroblue tetrazolium solution for 12 h; stop signal development by placing the slides in a buffer (pH 8.0) of 100 mM Tris-HCl/10 mM EDTA. Dehydrate the slides in an ethanol series (70, 95, and 100%), clear in xylene, and mount as described for the immunocytochemical staining.

Figure 3C shows an *in situ* hybridization for glutathione *S*-transferase-pi gene transcripts in cells of a human glioblastoma multiforme cell line in culture.

3.4. Cloning of Malignant Glioma Cells

Human glioma clonogenic cell assays are based on the concept that human tumors contain subpopulations of cells, referred to as tumor stem cells or clonogenic cells, that are characterized by a high proliferative capacity and yield clones of cells of the same progeny in appropriate in vitro assays. Theoretically and conceptually, tumor stem cells are responsible for the growth and progression of the tumor, and the success of any anticancer therapy is dependent on its ability to eradicate these cells. Consequently, much effort has been devoted to developing assays that allow reproducible growth and quantification of tumor stem cells in various human neoplasms. These assays have

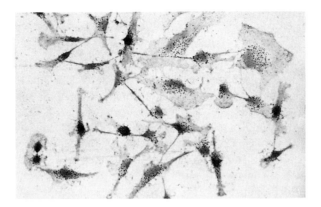

Fig. 3. Nonradioactive *in situ* hybridization for glutathione S-transferase-pi gene transcripts in primary human glioblastoma cells in culture. Cells were hybridized with a full-length cDNA probe labeled with digoxigenin. Visualization was as described in the text.

become powerful tools in the search for new therapies, in predicting patient response to therapy, and in evaluating the action of enzyme inhibitors, growth factors, antisense oligonucleotides, and other biological and molecular regulators of tumor cell growth and differentiation. For human gliomas, a number of techniques for clonogenic cell growth have been described. Anchorage-independent clonogenic assays are, in general, more specific for tumor cells, and of these, the capillary tumor clonogenic cell assay *(11)* yields the highest plating efficiency and the highest rate of cloning success. An attractive feature of the capillary assay is that it requires up to 10-fold less cells and materials than other clonogenic assays. The capillary tumor stem cell assay routinely used in the author's laboratory for malignant gliomas is described.

1. Prepare a single-cell suspension from primary specimens or cell cultures, as described in Section 3.1. Dilute the cell suspension with enriched cloning medium, and adjust the density to 5×10^5 cells/mL.
2. Set up a cloning mixture (total volume 300 µL) in enriched cloning medium containing 5×10^4 cells/mL, 0.2% low-melting point agarose (Seaplaque, Marine Colloids, Rockville, ME), and test agents or drugs at the desired final concentrations.
3. Mix (vortex) the cloning mixture thoroughly, and introduce 50 µL in triplicate into sterile round glass microcapillary tubes (*see* Notes 4–6).
4. Place the capillary tubes on a cold surface, e.g., a glass plate on crushed ice, for a few minutes for the agarose to gel, and then place them in a special holder (can be custom-made in most institutional mechanical shops) or in a large Petri dish. Place a small Petri dish containing double-distilled water in the large dish to ensure a humid atmosphere throughout the incubation period.

5. Incubate the dishes for approx 2–3 wk. Monitor a set of control capillary tubes (specifically set aside for this purpose) once weekly for colony growth.
6. When colonies have attained an average size of approx 60 μm in diameter, the incubation is ended and the colonies are counted. The simplest method with which to count colonies is to flush the contents of the capillary tube on to a glass slide and enumerate the colonies under low power (40–100x) inverted-phase microscopy or indirect illumination. Alternatively, the capillary tubes can be directly viewed under a phase-contrast microscope with direct or indirect illumination.

3.4.1. Determination of Cell Growth Kinetics

Growth kinetics are among the most important biological parameters often used to characterize cell cultures. When cells are first placed in culture, they undergo a lag phase, during which they attach to the culture surface, initially through charge interactions with the attachment surface and later through the production of an extracellular matrix. The lag phase can be as long as 24 h for primary cultures, but often shortens with subsequent passage. Following the lag phase, the cells enter the exponential growth phase, during which the cell number increases logarithmically with time. DNA and protein synthesis usually proceeds at maximum rates during this phase of the culture. It should be noted that usually only a proportion, termed the growth fraction, of the cells are proliferating and, as such, contributing to the increase in cell number of the cell population in culture. Under optimum conditions, cells will continue in the logarithmic phase until the culture attains confluence. Confluent normal glial cell cultures consist typically of a single monolayer of cells, and thereafter, the cells enter a stationary phase in which they stop dividing, a phenomenon referred to as contact inhibition. Tumor cells, on the other hand, continue to grow after they reach confluence and pile up on each as they do so. Eventually, however, tumor cells also enter a stationary phase, not as a result of contact inhibition, but of the low pH, decrease in nutrients, and an accumulation of metabolic waste products in the culture medium. For primary cultures or established cell lines, population growth kinetic data can be obtained using the following procedure:

1. Prepare a single-cell suspension from the tumor or cell line, as described in Section 3.1. Adjust the cell density to exactly 1×10^4 cells/mL (or 2.5×10^4 cells/mL for primary cultures). Mix well with a sterile pipet, and transfer exactly 5 mL of the suspension into seven sets of triplicate 30-mm Petri dishes or T25 tissue-culture flasks. The number of culture sets prepared is determined by the number of data points, i.e., the duration over which the kinetic data are to be determined. The present protocol is designed to collect data over 1 wk (7 d).
2. Incubate the dishes or flasks at 37°C and 5% CO_2 in 95% air in a humidified atmosphere. Every 24 h for 7 d, remove a set of three dishes or flasks, rinse the

cell monolayers with CMF-HBSS, and trypsinize them, as described in Section 3.2. Wash the trypsinized cells once with CMF-HBSS, and resuspend them in 1 mL of culture medium. Perform three independent cell counts (i.e., using different aliquots), and compute the mean cell count for the cell suspension.

3. The number of cells is plotted against time to obtain a growth curve for the cell population. The exponential growth phase of the culture can be described by the following growth kinetic parameters:

$$k = dN/dt \quad (1)$$

$$N = N_o e^{kt} \quad (2)$$

where N_o is the initial cell number, N is the cell number after a given time, t, and k is the regression constant or the growth rate. Integration of Eq. (2) yields:

$$\ln N = \ln N_o + kt \quad (3)$$

The population doubling time, t_d, i.e., time required for the cell number to increase from N_o to $2N_o$, is given by:

$$t_d = \ln 2/k = 0.693/k \quad (4)$$

The number of population doublings, n, i.e., the number of times the initial number of cells has doubled, is given by:

$$n = \text{Log } [N/N_o]/\log 2 = 3.32 \text{ Log } [N/N_o] \quad (5)$$

The computations assume that all the cells in culture are actively proliferating, i.e., the growth fraction is 1. This is an oversimplification, since in reality, only a proportion of the cells in the population are actively dividing and the growth fraction is <1. To be accurate, therefore, N and N_o should be corrected to take into account the actual growth fraction. For most actively proliferating tumor cell cultures, however, the above equations give a usable approximation.

4. Notes

4.1. Initiation, Establishment, and Characterization of Cultures

1. Storage of specimens: If a specimen is not processed immediately, it should be cleaned of blood clots, macroscopic necrotic material, and nontumor tissue, and then cut into 1-mm cube pieces and placed in culture medium, e.g., DMEM containing 10% FCS, in a sterile screw-cap tube. In this condition, the specimen can be stored at 4°C for up to 24 h without significantly impairing the chances of a successful culture. It is, however, recommended that specimens be processed as soon as possible after being obtained in order to ensure that the culture is as representative of the original specimen as possible.

2. Cellular alterations during in vitro culture: For both normal astrocytes and malignant gliomas, establishment of cell lines is relatively simple once a primary culture has been successfully initiated. It is important to note, however, that the high degree of pleomorphism and heterogeneity characteristic of malignant gliomas is

often lost with increasing in vitro passage of the cells *(12)*. Furthermore, over a long period of time and over several in vitro passages, there will be a selection of subpopulations of cells that have the highest growth rates or those that adapt most readily to in vitro growth conditions. Finally, extended in vitro culture often results in alteration in the levels and types of genes that are expressed in the cells. Despite these considerations, if used appropriately, established cultures are powerful tools in neurobiology research and are excellent models for answering a variety of important fundamental questions. A good and useful practice is to cryopreserve aliquots of cells of the primary cell suspension and after every 5 or 10 in vitro passages for up to 30–50 passages. These can be used later to examine any changes that may have occurred in any given parameter, as a result of in vitro passage.
3. Cytomorphological features of cultured cells: Some of the cytomorphological features used to characterize cultures of malignant gliomas include cellular pleomorphism, the extent of process formation, cytoplasmic and nuclear staining characteristics, and the size, shape, and number of nuclei in the cells. In most cases, primary cultures of the brain or brain tumors exhibit some of the characteristic features of cells of the tissue of origin.

4.2. Clonogenic Assay

4. A common problem faced by new users of the capillary cloning technique is that of introducing exactly 50 µL (or some other specified volume) of the cloning mixture into a capillary tube. This is best accomplished with the help of an Eppendorf pipet (200 µL). The pipet is adjusted to the desired volume, a pipet tip is attached to it, and the capillary tube is slid over the tip to achieve a snug airtight fit. The plunger of the pipet is depressed, the free end of the capillary tube is placed in the cloning mixture, and the preset volume is drawn up by releasing the plunger. The capillary tube is then held at an angle to allow the culture to slide to the middle of the tube. The exterior of the tube end that was in contact with the culture mixture is wiped with 70% ethanol to avoid microbial growth.
5. It is critical that at the start of a clonogenic cell assay, the tumor cells are present as single cells and not as cell clumps. For primary specimens, it is often difficult to obtain perfect single-cell suspensions. Therefore, after the assay has been set up, a baseline cell clump count should always be performed and subtracted from the final colony counts. In case of excessive clumping (>25% of final cell counts), it may be necessary to invalidate the experiment.
6. Most brain tumors, particularly highly malignant gliomas and medulloblastomas, yield a significant number of colonies under the conditions described in this chapter. However, some tumors either do not form colonies or else yield low numbers of colonies. For poorly clonogenic tumors, conditioned medium from lethally irradiated cells of a highly clonogenic glioblastoma cell line can enhance colony formation significantly (Table 1). To prepare conditioned medium, cells of the highly clonogenic cell line are grown to approx 80% confluence, the culture medium is replaced with fresh medium (6 mL/T75 flask), and the cells are irradi-

**Table 1
Effect of Conditioned Medium from a Highly Clonogenic Glioblastoma Multiforme Cell Line on Colony Formation by Cells of Primary Malignant Gliomas[a]**

Tumor type	Number of colonies/capillary tube	
	No conditioned medium	+ 25% conditioned medium
Glioblastoma multiforme	67 ± 11	61 ± 6.3
Glioblastoma multiforme	5 ± 1.5	18 ± 3
Anaplastic astrocytoma	11 ± 3.7	23 ± 5.1
Glioblastoma multiforme	33 ± 6.3	42 ± 4.6
Glioblastoma multiforme	0	36 ± 4.3
Anaplastic astrocytoma	12 ± 3	16 ± 2.7

[a]The cell line used to prepare the conditioned medium in this study yields in excess of 110 colonies/capillary tube (containing 2500 cells in 50 µL of culture).

ated with 40 gy γ-irradiation. The cells are then incubated for an additional 72 h, and the medium is collected, centrifuged at 30,000g for 20 min, and added to the cloning mixture at a final concentration of 25%.

4.3. Pharmacological Applications

4.3.1. In Vitro Drug Decay Kinetics

A useful application of in vitro brain tumor cultures is in the study of the action of anticancer agents and in the rational development of therapy *(13)*. These studies often require that the drug exposure conditions are clinically relevant. The design of clinically relevant drug concentrations and exposure times for use in such assays, however, requires knowledge of the in vitro decay kinetics (pharmacokinetics) of the drug under the assay conditions, and this information is often not available. The author describes a bioassay *(14)* that combines the culture methodologies described in this chapter to obtain such quantitative in vitro pharmacokinetic data readily. A prerequisite of the technique is that a cell line be available that is sensitive to the drug of interest. This bioassay procedure can be divided into three main parts:

7. Set up a clonogenic cell assay as described in Section 4., using a glioblastoma multiform (or other) cell line that is sensitive to the drug in question. Add the drug at a concentration range that will ensure that at least 90% cell kill at the maximum test concentration is achieved. It may be necessary to perform a prescreen in order to establish this drug concentration range.
8. From the colony counts, compute surviving fractions (SF) using the equation:

$$SF = N_c/N_o \qquad (6)$$

where, N_c is the number of colonies in drug-treated cultures, and N_o is the number of colonies in control untreated cultures. The SFs are plotted against the drug concentrations, and from the resulting dose–response curves, the following parameters are determined:
  a. The ID_{90} of the drug for the cell line, i.e., the drug concentration required to reduce clonogenic survival by 90%.
  b. A conversion factor, Kc, that relates clonogenic survival with drug concentrations. Kc is the slope of the SF vs drug concentration curve and is given by the equation:

$$Kc = d(SF)/dC \quad (7)$$

or, if the relationship is linear, by:

$$Kc = \Delta(SF)/\Delta C \quad (8)$$

where $\Delta(SF)$ is the change in surviving fraction over a concentration change of ΔC.

9. Incubate the drug at its ID_{90} concentration under the experimental conditions. At different time-points, e.g., every 15 min (or less) for an unstable drug and 0.5–3 h (or longer) for a stable drug, remove aliquots of the drug incubate, and add to a cloning mixture (*see* Section 3.) containing the target cells. Add agarose to 0.2%, mix well, draw up 50-µL aliquots into triplicate capillary tubes, and incubate as described in Section 4. After the incubation period, enumerate the colonies, and compute the SFs at the various time-points that the drug incubates were added to the cloning mixture. Convert the SFs into drug concentrations, using the conversion factor, Kc, computed earlier. Plot the log drug concentrations against time to obtain a concentration–time curve. Since in vitro drug exposure represents a one-compartment pharmacokinetic model, the in vitro half-lives ($t_{1/2}$) and area under the curve (AUC) for the drug can be computed as:

$$t_{1/2} = 0.693/k \quad (9)$$

where k is the drug decay rate constant, and

$$AUC\ (total) = C_o/k \quad (10)$$

where C_o is the initial drug concentration.

These in vitro pharmacokinetic constants are then used to establish drug exposure conditions (drug concentrations and exposure times) required to achieve clinically relevant drug AUCs, under the conditions of the in vitro drug exposure (*see* ref. *14* for more details).

4.3.2. Prediction of Patient Response to Therapy

Tumor stem cell assays provide a means of determining pretherapeutically the inherent response of a patient's tumor to a given chemotherapeutic agent. Such information can be used retrospectively to analyze patients' response to therapy or to design therapy prospectively for a given patient. Although the overall response of a patient's tumor to therapy is dependent on many factors not directly related to tumor sensitivity to the drug, patients whose tumors are inherently resistant to the drug in question will not respond to treatment even if all other factors are favorable. Thus, the ability to quantitate the potential response of the tumor to specific anticancer agents, if used appropriately, can contribute significantly to the planning of effective therapy for the individual patient. In a previous study, the author used the glioma capillary clonogenic cell assay described in Section 3. to demonstrate the ability of the technique to predict the response of both adult and pediatric tumors to chemotherapy *(15)*. Three important points should be noted when clonogenic assays are used to predict patient response to therapy:

10. The clonogenic assays should be performed exclusively with primary specimens. The preparation of single-cell suspensions should occur as soon as possible after acquisition of tumor specimens from surgery. The resulting cell suspensions should be used promptly. These steps will minimize the effect of changes in the expression of genes and of various other factors, such as hypoxia, that can affect the response of the cells to the drugs. If it is unavoidable to delay performing the assay until later, then the cells should be resuspended in cloning medium and stored at 4°C for not more than 6–24 h.
11. The drug concentrations to which the cells are exposed must be clinically relevant, both with respect to the peak plasma concentrations and the AUC. This can be determined using the method described in Section 4.3.1. For drugs whose decay rate constants in vivo and in vitro are significantly different, it may be necessary for cells to be exposed to drug for a specific period of time, e.g., 1 or 2 h, before cloning. Readers are referred to refs. *15* and *16* for additional information on in vivo–in vitro drug exposure.
12. For predicting patient response to therapy, a sensitivity index (SI) is defined that is based on the clonogenic cell survival at a fixed "cutoff" concentration of the drug that is equivalent to that achievable in vivo. The SI can be either the SF at the drug "cutoff" concentration, or it can be computed to take into account the shape of the dose–response curve *(16)*. Indices, such as ID_{50} or ID_{90}, although useful in developmental drug screening, are often inappropriate for predicting patient response to therapy, since by definition, these values are not based on the pharmacokinetics of the drugs and, as such, are often not relevant clinically.

Acknowledgments

The author is indebted to several individuals who have been a part of the development of many of the techniques described in this chapter, in particular, Hans Rainer Maurer and Mark L. Rosenblum, my earlier mentors. I also acknowledge my colleague and friend Mitchel Berger for many years of scientific discourse and collaboration. The expert technical assistance of Dolores Dougherty, Jane Giblin, Patricia Beltz, Jane Caughlin, Donna Stein, and

Evangeline Gagucas is gratefully acknowledged. Various portions of this work were supported by grant awards (CA55835, CA55261, CA46410, N34-CM-57829, and N34-CM-57831) to the author and by a Cancer Center Core Grant to the MD Anderson Cancer Center from the National Cancer Institute, the National Institutes of Health of the United States.

References

1. Zulch, K. J. (1986) *Brain Tumors, Their Biology and Pathology.* Springer Verlag, Berlin, Germany.
2. Baumann, N. (ed.) (1985) Brain cultures—a tool in neurobiology, in *Dev. Neurosci.* **7(5–6),** 249–398.
3. Ali-Osman, F. and Schofield, D. (1990) Cellular and molecular studies in brain and nervous system oncology. *Cur. Opin. Oncol.* **2,** 655–665.
4. Hoshino, T. (1991) Cell kinetics of brain tumors, in *Concepts in Neurosurgery. Neurobiology of Brain Tumors* (Salcman, M., ed.), Williams and Wilkins, Baltimore, MD, pp. 163–193.
5. Silbergeld, D. L., Ali-Osman, F., and Winn, H. R. (1991) Induction of transformational changes in normal endothelial cell by cultured human astrocytoma cells. *J. Neurosurg.* **75,** 604–612.
6. Canady, K. S., Ali-Osman, F., and Rubel, E. W. (1990) Extracellular potassium influences DNA and protein syntheses and glial fibrillary acidic protein expression in cultured glial cells. *Glia* **3,** 368–374.
7. Bodell, W. J., Aida, T., Berger, M. S., and Rosenblum, M. L. (1986) Increased repair of 0^6-alkylguanine DNA adducts in glioma-derived human cells resistant to the cytotoxic and cytogenetic effects of 1,3,-bis (2-chloroethyl)-1-nitrosourea. *Carcinogenesis* **7,** 879–993.
8. Ali-Osman, F., Stein, D. E., and Renwick, A. (1990) Glutathione content and glutathione-S-transferase expression in 1,3-bis(2-chloroethyl)-nitrosourea-resistant human malignant astrocytoma cell lines. *Cancer Res.* **50,** 6976–6980.
9. Bigner, D., Bigner, S. H., and Ponten, J. (1981) Heterogeneity of genotypic and phenotypic characteristics of fifteen permanent cell lines derived from human gliomas. *J. Neuropathol. Exp. Neurol.* **40,** 201–229.
10. Eng, L. F. and Bigbee, J. W. (1978) Immunocytochemistry of nervous system specific antigens. *Adv. Neurochem.* **3,** 43–98.
11. Eng, L. F. and deArmand, S. J. (1983) Immunocytochemistry of the glial fibrillary acidic protein. *Prog. Neuropathol.* **5,** 19–39.
12. Ali-Osman, F. and Beltz, P. A. (1988) Optimization and characterization of the capillary human tumor clonogenic cell assay. *Cancer Res.* **48,** 715–724.
13. Friedman, H. S., Bigner, S. H., Schold, S. C., and Bigner, D. D. (1986) The use of experimental models of human medulloblastoma in the design of rational therapy, in *Biology of Brain Tumors* (Walker, M. D. and Thomas, D. G. T., eds.), Martinus, Nijhoff, Boston, pp. 405–409.
14. Ali-Osman, F., Giblin, J., Dougherty, D., and Rosenblum, M. L. (1987) Application of in vivo and in vitro pharmacokinetics for physiologically rel-

evant drug exposure in a human tumor clonogenic cell assay. *Cancer Res.* **47,** 3718–3724.
15. Ali-Osman, F. (1991) Prediction of clinical response to therapy of adult and pediatric brain tumor patients by chemosensitivity testing in the capillary brain tumor clonogenic cell assay, in *New Trends in Pediatric Neuro-Oncology, vol. 3* (Bleyer, A. W., Packer, R., and Pochedly, C., eds.), Harwood Academic, Basel, Switzerland, pp. 220–230.
16. Ali-Osman, F., Maurer, H. R., and Bier, J. (1983) In vitro cytostatic drug sensitivity testing in the human tumor stem cell assay: a modified method for the determination of the sensitivity index. *Tumor Diagn. Ther.* **4,** 1–6.

8

Cell Culture of Human Brain Tumors on Extracellular Matrices

Methodology and Biological Applications

Manfred Westphal, Hildegard Nausch, and Dorothea Zirkel

1. Introduction

Cell culture is one of the major tools of cell biologists. It has also become an integral part of the daily routine of most oncology laboratories for the purpose of karyotyping, chemoresistance testing, or basic research. It provides investigators with an opportunity to investigate many cellular parameters and interactions in an in vitro system in which the experimental conditions can be controlled and repeated. With many tissues, either human or animal, the problems of cell culture are cell attachment and initial survival. Particularly the primary cultures derived from tumor specimens are a problem in many laboratories. Apart from modifications in the composition of tissue-culture plastic materials, other approaches have been used to get around this problem, such as coating of tissue-culture dishes with attachment enhancers, such as polyamino acids *(1)*, fibronectin *(2)*, laminin *(3)*, and collagen *(4)*. Since it was known that endothelial cells are capable of producing a basement membrane even in vitro, bovine corneal endothelial basement membrane was explored by Gospodarowicz et al. for its role in regeneration and nonregeneration of corneal endothelium in different species. This bovine corneal extracellular matrix (bECM) was found useful in the cell culture of a wide range of different cells *(5,6)*, and bECM as well as other ECMs were employed in the cell biology of tumor cells derived from mammary carcinomas *(7)*, urological tumors *(8)*, and different kinds of pituitary adenomas *(9,10)*, as well as CNS tumors *(11)*, which is the topic of this chapter.

From: *Methods in Molecular Medicine: Human Cell Culture Protocols*
Edited by: G. E. Jones Humana Press Inc., Totowa, NJ

Such biologically produced matrices may have a variety of effects, some of which are merely owing to the provision of an anchor for attachment to a given substrate. The molecules inherent to most matrices bind to the cells via cell-surface receptors, many of them members of the integrin family *(12)*, and thus transmit signals to the cells like soluble growth factors would do via their receptors. In addition it has been shown, however, that ECM contains large amounts of growth factors, the heparin binding factors being the most prominent of them *(13)*, and these include acidic fibroblast growth factor (aFGF), basic fibroblast growth factor (bFGF), vascular endothelial growth factor (VEGF), and the heregulins.

Apart from the many advantages that an ECM coating may offer for primary cultures, it has to be borne in mind that any ECM is not an inert substrate and that it may interfere with the cellular biology of normal as well as neoplastic cells *(14,15)*. This problem becomes even more complicated since matrices from different sources may affect cells in a variety of ways *(16)*. In this neuro-oncological context, it should be mentioned that an ECM derived from human arachnoid cells may induce differentiation in a specific glioma cell line *(17)*. Many different cells, including those derived from neoplastic tissues, have the capacity to produce matrices. In this context, this chapter first focuses on the production and applications of bECM in experimental neuro-oncology. Finally, this chapter deals with tumor-produced matrices and along the way provides an outlook on matrix biology as an interesting field of research in its own right.

Since the previous publication of a chapter dealing with this methodology in an earlier volume of this series *(18)* many years have elapsed, and some minor modifications have been introduced simply through the accumulation of further experience.

2. Materials
2.1. ECM from bECM

1. Bovine eyes: Following the instructions from protocols published earlier and elsewhere *(6,19)*, fresh bovine eyes have to be obtained from the local slaughterhouse (*see* Note 1).
2. Isolation of the cornea (*see* Note 2): Use the following (sterile) instruments:
   a. No. 1 cannulae.
   b. Angled microscissors.
   c. Tight-gripping forceps.
   d. A grooved director or disposable plastic spatula.
   e. Gas burner.
   f. Spray bottle with 70% ethanol.
   g. Laminar flow hood.
   h. 10-cm Petri dish or tissue-culture dish.
   i. Phosphate-buffered saline (PBS).

j. A pipet aid.
k. Pipets.
l. Suction.
3. Cell-culture medium: Dulbecco's modified Eagle's medium (DMEM) with high glucose content supplemented with 10% fetal calf serum, 1 mM glutamine, 2 mM pyruvate, 25 µg/mL gentamycin, and 2.5 µg/mL fungizone.
4. ECM production:
    a. Cell-culture medium.
    b. Dextran (mol wt 40,000).
    c. Filtration units, preferably vacuum-connected, disposable sterile reservoirs.
    d. bFGF.
    e. Sterile ammonium hydroxide solution (20 mM).
    f. Sterile PBS.
    g. Sterile PBS containing 25 µg/mL gentamycin.
    h. Suction.
    i. Pipet aid.

2.2. FGF Isolation (20)

1. Four kilograms of bovine brain (Note 3).
2. Heavy-duty Waring blender.
3. Eight liters of 0.15M ammonium sulfate, pH 5.6.
4. 6M HCl.
5. Heavy-duty stirrer.
6. Preparative cooled centrifuge with 3-L capacity rotor.
7. Ammonium sulfate.
8. Dialysis tubing (mol wt cutoff approx 10,000).
9. CM-Sephadex C-50.
10. 3.5-cm Diameter chromatography column.
11. 0.1M Sodium phosphate, pH 6.0.
12. 0.1M Sodium phosphate, 0.15M NaCl, pH 6.0.
13. 0.1M Sodium phosphate, 0.6M NaCl, pH 6.0.
14. UV spectrophotometer.

2.3. Tumor Cell Isolation (see Note 4)

1. Hank's balanced salt solution (HBSS), Ca- and Mg-free.
2. 50-mL Polypropylene centrifuge tubes (sterile).
3. Tissue-culture dishes.
4. Angled microscissors.
5. 5-mL Tissue-culture pipets with smooth openings (best from Falcon, Becton Dickinson, Heidelberg, Germany or Costar, Cambridge, MA).
6. 15-mL Polystyrene centrifuge tubes (sterile).
7. Option (*see* Note 4).
8. Enzyme cocktail: 0.05% Pronase, 0.03% Collagenase 0.01% DNase in HBSS, filter sterilized.

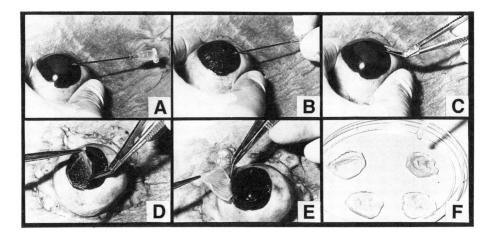

Fig. 1. Different steps in the production of ECM. **(A)** The eye is punctured with a 10-gage needle to drain the anterior eye chamber and relieve the pressure so that the eye wrinkles as seen in **(B)**. **(C)** Starting from the puncture hole, the cornea is excised, being lifted up halfway along the way **(D)** and then cut off after almost complete reflection **(E)**. Four cornea are put face down onto a Petri dish and are covered with PBS **(F)**.

2.4. Immunostaining

1. Round 12-mm diameter glass cover slides.
2. Fine forceps.
3. Antibodies: first antibody against the desired antigen; second antibody with the detection system (i.e., FITC, Rhodamin, or Enzymes).
4. Wet chamber.
5. Citifluor antifading reagent (City University, London).
6. Nail polish.

3. Methods

3.1. Preparation of bECM

1. To prepare the cornea, only eyes that are uninjured and show no signs of inflammation or ulceration are used. They are held in the left hand, rinsed thoroughly with alcohol from a spray bottle, and then held down firmly on a paper towel that is also soaked in alcohol. The eye is then punctured tangentially to the cannula so the fluid from the anterior eye chamber can drain allowing the cornea to shrink after the pressure is taken off the eyeball. Starting from the puncture hole made with the needle, the cornea is carefully cut out with sharp scissors without contaminating it with other tissues from within the eyeball (Fig. 1A–E, Note 2).
2. Once the cornea is removed, it is put external surface down onto a Petri dish on which it will stick owing to adhesive forces. It is then rinsed with a few drops of

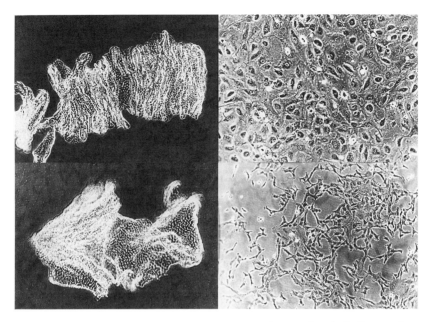

Fig. 2. Appearance of ECM. Two typical sheets of corneal endothelium as it comes off and is seen in the culture dish are shown on the left. Two different confluence stages of endothelial cells are shown on the right.

PBS at room temperature (Fig. 1F). Then the exposed inner surface is gently scraped once with a grooved director or a disposable plastic spatula. Without the exertion of any pressure, this maneuver will yield whole sheets of endothelial cells (Fig. 2). The grooved director is then dipped into a tissue-culture dish already containing the complete cell-culture medium. Use one dish per eye. Process 10 eyes in one session.
3. Transfer dishes containing endothelial cells to the incubator and leave them undisturbed for 5 d. At that point, the first colonies will have formed (Fig. 2). These may be few and far between, but once colonies are there after several days, FGF can be added and proliferation will take off. After the initial culture dishes are about 50% covered with colonies, detach them with trypsin, which may take up to 20 min because the primary cells attach firmly to their support (via an ECM). Put the detached and dispersed cells onto 25 dishes, which have been precoated with gelatin.
4. Dissolve 0.2% gelatin in PBS and autoclave. It is advisable to filter this solution, because it may still contain granular matters which disturb the cell culture. Cover the cell culture dishes (10-cm diameter, Falcon, Costar, or Nunc dishes) with 5 mL of the PBS gelatin, and leave them in the refrigerator overnight. Prior to use, remove the excess PBS gelatin. This gelatin coating facilitates the next passage because the cells are much easier to detach.

5. After the corneal endothelial cells have been expanded on these gelatin-coated dishes, they are best passaged at the point when they are just subconfluent, because then there are fewer intercellular adhesions and the dispersion is more even. The cells from eight to ten 10-cm dishes are combined and added to a 550-mL bottle of complete medium, which also contains 5% dextran (mol wt 40,000, dissolved by heating in a small volume of medium without additives and filter sterilized). Finally, crude bFGF is added (*see* Note 5). This cell suspension is then plated onto any kind of culture dish, which is to be covered with ECM. For example, 6 mL are added to a T25 flask, 25 mL to a T75, and 1 mL/well in a 24-well multiwell tissue-culture dish. At this density and in the presence of bFGF, the cells should reach confluence within 2 to 3 d. After confluence, the cells are left for one more week to 10 d, and then they are lysed.

In any kind of multiwell plates, the cell number needs to be relatively higher than in larger dishes because it should be made certain that the cells become confluent without a tedious second addition of bFGF to hundreds of wells.

6. Lyse the cells by removing the culture medium and replacing it with distilled water containing ammonium hydroxide (20 m*M*). The cells burst rapidly by osmotic shock, and within 3 min the gelatinous cellular debris can be removed and rinsed off. The ECM is left behind and should be washed at least once more with PBS, and finally the ECM-coated dishes can be stored at 4°C while being covered with PBS containing gentamycin (25 µg/mL). At that time, the matrix can be seen with a phase-contrast microscope (Note 6).

3.2. ECM from Human Brain Tumor Cell Lines (tECM)

1. Cells from a large, well-characterized panel of human glioblastoma cell lines *(21)*, most of which are positive for cell-surface fibronectin, were evaluated for ECM production. They all produce different amounts of ECM, which in some cases can be seen in the phase-contrast microscope. More often, the ECM is only detectable by immunofluorescence with the appropriate antibodies. Cells growing in some geometrical arrangement may deposit their matrix in an orderly fashion and impose such order on cells seeded thereon *(18)*.
2. Cell-culture conditions are the same as in bECM production, except that Earle's modified MEM is used as a basal medium.
3. tECM is isolated in the same way as described for bECM in Section 3.1.

3.3. Analysis of ECM by Immunostaining and Protein Determination

1. After the producer cells have been lysed off a glass cover slide that was placed at the bottom of a multiwell culture dish, they are thoroughly rinsed in PBS and then set atop the small pedestals of a staining tray, which can be covered to give a wet chamber. The primary antibody is added, left for 30 min at room temperature, and then rinsed off. The second, fluorescent antibody is then added, left for another 30 min, and again rinsed off. The coverslips are then put face down onto a heavier microscope slide with 20 µL of Citifluor antifading reagent between

them. This "wet chamber" is then sealed of with nail polish. These slides can be stored for many months at 4°C.
2. Several papers have published the lysis of ECM and measurement of protein *(22)*. In our hands, the lysis of the protein with sodium hydroxide worked better than any cocktail of detergents. The cells from which a protein determination is desired are seeded onto six-well multiwell dishes and are left for 8 d after having become confluent. The cells are then lysed with distilled water, and the debris is rinsed off. Thereafter, 600 µL of $0.5M$ NaOH are left for 3 h at 60°C on the ECM. The protein content of the lysate is determined with a commercially available BCA-protein assay (Pierce Biochemical Company). NaOH or BSA dissolved in NaOH is used as the reference point or as standard, respectively.

3.4. Isolation of FGF

The addition of FGF to the dispersed stock cells before plating is of great importance for the production of ECM. This speeds up the rate at which the cultures become confluent and also influences the quality of the matrix. The isolation of bFGF has been described by Gospodarowicz et al. *(20)*, but it is not necessary for the production of ECM to purify bFGF to homogeneity.

1. Homogenize eight frozen brains in 8 L of $0.15M$ ammonium sulfate, pH 5.6.
2. Thereafter, adjust the homogenate to pH 4.5 with $6M$ HCl, and stir for at least 1 h at 4°C.
3. Centrifuge the homogenate at 20,000g, and collect the burgundy-colored supernatants. **Caution: If the color has turned brown because the pH was too low, even if this was only for a short time, the procedure should be stopped and started fresh because bFGF is very intolerant to acidic treatment.**
4. Ammonium sulfate is now added to the extract to give a concentration of 200 g/L, and the suspension is centrifuged again at 20,000g after having been stirred for 60 min.
5. Again the supernatants are collected, and again ammonium sulfate is added to give a final concentration of 450 g/L. After centrifugation, the supernatants are discarded and the pellets collected.
6. The collected pellets are dissolved in a small volume of water (approx 250 mL), adjusted to pH 6.0 with formic acid, and dialyzed against water. (Use dialysis tubing with mol-wt cutoff between 10,000 and 12,000.)
7. The dialysate is then loaded onto a CM Sephadex C50 column (3.5 × 20 cm), which has been pre-equilibrated with $0.1M$ sodium phosphate, pH 6.0. The ionic strength of the dialysate should be equal to or less than that of the running buffer.
8. The material retained on the column is eluted with a stepwise increase of NaCl concentration in the sodium phosphate buffer, first to $0.15M$ and then to $0.6M$.
9. The material eluting at the high salt concentration is collected. It is <0.1% pure, but this is sufficient for use in the production of ECM. It can be stored frozen as well as lyophilized until reconstitution. The concentration at which the extract has to be used should be tested in a bioassay on endothelial cells. It is variable between extractions.

3.5. Isolation and Culture of Brain Tumor Cells

Brain tumors are a very heterogeneous group of tumors originating from various cellular sources. On top of this, tumors within one histological subtype are distinguished by their biological aggressiveness, which is to be reflected by grades given according to various grading systems, of which the WHO grading seems the most widely used. The general distinction should be made between differentiated tumors and anaplastic tumors, and this will decide which method should be taken for the initial dispersion of the tissue specimen.

1. Representative portions of a tumors are obtained at surgery (*see* Note 7). The selected tissue fragment or different fragments from different parts of the tumor are placed immediately into 15 mL of sterile HBSS without calcium and magnesium (CMF). The tissue can be left at room temperature until processing for several hours (*see* Note 8).
2. The tissue is then freed from coagulated blood and minced with angled scissors. The fragments are transferred into a centrifuge tube containing 10 mL of a cocktail of pronase (0.05%), collagenase (0.03%), and DNase (0.01%) in HBSS. After 20 min of incubation at 37°C and shaking of the incubation mixture, the fragments are mechanically disrupted by repeated trituration. The undispersed fibrous fragments should be allowed to settle, and the turbid supernatant transferred to a centrifuge tube.
3. After mincing, the fragments are agitated by repeated trituration until a turbid solution is obtained. Allow large pieces or undispersed fibrous fragments to settle, and remove the supernatant into a centrifuge tube.
4. The cells and small-cell aggregates obtained from both dissociative procedures are pelleted at 80g for 10 min and thereafter redispersed in cell-culture medium containing 10% fetal calf serum (FCS). In the case of enzymatical dispersion, great care should be taken to remove all the enzyme carefully, because otherwise the leftover enzymes will digest the ECM. To be on the safe side, the cells can be washed once in medium.
5. Before seeding this mixture onto whatever culture dishes will be used in the subsequent experiments, a drop of the suspension should be microscopically examined to rule out the possibility that either owing to the nature of the selected tissue or as a consequence of the isolation procedure, it is composed of cellular debris. (A Trypan Blue exclusion assay is optional at this point, but with growing experience, the experimenter will be able to do without it.)
6. The suspension is then aliquoted onto ECM-coated dishes and left to attach for at least 6 h, but preferably overnight.
7. Cells are maintained at 37°C in a humidified atmosphere supplemented with 5–8% CO_2, depending on the medium used or the tissues cultured. We keep meningiomas in DMEM-H21 with 10% FCS, 2 mM glutamine, 1 mM pyruvate, 25 µg/mL gentamycin, and 2.5 µg/mL amphotericin B. Gliomas are maintained in MEM-Earle with the same supplements. Splitting and passaging of the cells are done with a 0.05% trypsin/0.04% EDTA solution in PBS/CMF (STV). All

media and supplements in our laboratory were obtained from Seromed, Biochrom/Berlin, now a division of Nunc, Roskilde, Denmark.
8. Careful inspection of the cells on the next day is mandatory. In extreme cases, the cultures are confluent, and the debris from red blood cells and unattached fragments needs to be rigorously shaken up and rinsed off. In any case, it is advisable to change the medium as early as possible (see Note 9).

4. Notes
4.1. Materials and Methods

1. Eyes can be conveniently transported in a plastic bag and do not need any special prerequisites for transportation. If they are obtained by slaughterhouse workers at inconveniently early hours, they should be kept in a refrigerator until the experimenter will come to pick them up later. Although corneal endothelial cells are rather sturdy, it is advisable to use the eyes within 6–8 h after the animals have been killed.
2. The cornea is a very tough tissue, and the forceps to lift the cornea, after it is excised halfway, should have a precise grip. Also, the scissors should be sharp (Fig. 1A–F).
3. After the brains are obtained from slaughter, they should be frozen rapidly. If they are kept at –20°C, they should be thawed at 4°C the evening before the isolation to change gradually to a waxy consistency. This will take longer when they are kept at –80°C. It is necessary that the brains be at least frozen once before they are used because that facilitates the homogenization owing to denaturing of the tissue. Any attempt to use fresh tissue will already be difficult and messy at the stage of homogenization, and will fail ultimately at the point of the first centrifugation because no firm pellet will be obtained.
4. For gliomas and meningiomas alike, the decision has to be made whether the tissue is to be digested with enzymes or only mechanically dispersed. Enzymatic dispersion has the advantage that a more homogeneous suspension is obtained that can already be seeded for experiments in which it will be advantageous to use aliquots that are more homogeneous. Purely mechanical dispersion has the advantage that only the loosely connected tumor cells will be obtained either as single cells or small aggregates. The blood vessels sometimes are left intact as whole networks, and thus, endothelial contamination is minimized. Even from small fragments, mainly tumor cells will attach and migrate from small original spheroids (Fig. 3).

If the decision is made to use mechanical dispersion by trituration with a tissue-culture pipet, these should have a smooth tip. Some manufacturers leave their cut pipet tips unrefined, resulting in very rough mouths that will tear up the cells. The decision has to be made, however, on an individual basis depending on the softness of the tissue and the experimental goal. Medulloblastomas, pituitary adenomas, pineocytomas, and many of the malignant primary gliomas are usually very soft and will readily disperse by repeated trituration through a 5-mL tissue-culture pipet alone. Acoustic neuromas and other schwannomas, as well as

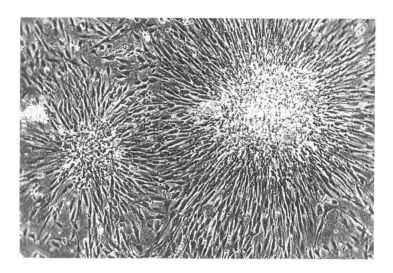

Fig. 3. Radial outgrowth of glial tumor cells from an aggregate found in the primary culture of an anaplastic astrocytoma.

meningiomas, are often very tough and fibrous, so that enzymes help to disrupt the tumor tissue.
5. The addition of FGF is not quantified because it will depend on the respective batch and how much needs to be added. The protein peak eluting from the CM Sephadex C-50 contains both acidic and basic FGF. The shape of the peak will depend on the geometry of the column used and the exact running conditions. Also, the final FGF concentration will depend on the width of the peak and when it is decided to stop the collection. It is best to take an aliquot and test various dilutions to discover the most effective concentration.
6. bECM will have a different appearance from batch to batch. It may look homogeneous and bubbly, but it may also look more like a web or show pronounced crests and ridges, depending on how long the corneal endothelial cells had been propagated in culture, what the initial density was, and how long it took for the cells to become confluent *(18)*. The matrix can be stored for up to 2 yr as in our own experience, or maybe even longer. Likewise, the bovine corneal endothelial cells can be frozen in log–growth phase in a mixture of 4.5 parts medium, 4.5 parts FCS, and 1 part DMSO. They have been stored in liquid nitrogen or at −80°C for up to 8 yr without losing their biological properties.

The higher the passage numbers get for the endothelial cells, the "tougher" they appear to become toward osmotic lysis. It is therefore important to inspect random dishes very closely after lysis to hunt for leftover cell islands. We have observed revitalization of endothelial cells after presumed lysis and storage at 4°C for several days. In questionable cultures, a chromosome analysis should be helpful (Fig. 4).

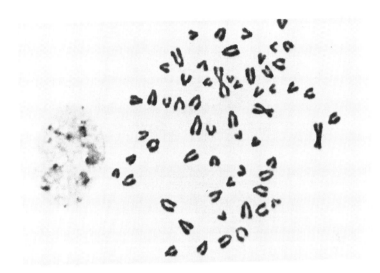

Fig. 4. Simple metaphase spread of bovine endothelial cells stained with Orcein. The normal chromosome set contains 60 chromosomes, which are mostly acrocentric and thus characteristically different from normal human chromosomes.

7. It is advisable to obtain a preliminary histological diagnosis by frozen section. Viable tissue can only be selected on the basis of surgical experience. In the ideal situation, the neurosurgeon is the principal investigator. If this is not the case, a close collaboration between surgeon and scientist should be established with sufficient feedback to improve the sampling technique. In any case, frozen sections from the immediate vicinity of the removed specimen can be helpful.
8. In HBSS CMF solution, the material can be maintained at least for 4 h at room temperature. A yellow discoloration of the phenol red indicator in the CMF around the tissue fragment at the bottom of the tube confirms its metabolic activity and thereby its viability, thus making it unlikely that the selected specimen is part of necrosis (Fig. 5). Usually 1 cm^3 is plenty of material and any temptation to use more should be avoided. When the tissue is derived from an anaplastic tumor, the tissue is mostly soft, and almost liquefied, and can be easily dispersed mechanically. If the specimen was obtained from a differentiated tumor with a lot of fibrous components, such as a fibrillary astrocytoma or a tough meningioma, an enzymatic dispersion can be tried.
9. After the overnight attachment period, attention should be paid to the ECM, which may be intact or can be completely lysed. Such slow lysis of the ECM has been noted in several cases of malignant gliomas in many hundreds of cases. Such lysis is owing to release of proteolytic enzymes after cell death or to the many proteases that are known to be produced by glioma cells and tumor cells in general for the process of invasion *(23)*.

Fig. 5. The arrow points toward the area in which a yellow discoloration of the tissue-culture medium around the tumor specimen can be seen after the tube has been standing for 30 min.

4.2. Applications of ECM

10. Meningiomas are usually benign epithelial lesions that are regularly homogeneous and of rather tough consistency. They occur in a variety of histological subtypes, and despite many efforts, there is no reliable grading system. These tumors usually attach very well even without bECM. There are few cases, however, that do not attach to plastic, but to ECM. This seems to be rather a feature of atypical meningiomas, like those in patients with von Recklinghausen's disease, malignant meningiomas, or those secondary to cranial irradiation. In general, there will be very little difference in the cellular morphology between the coated and uncoated dishes in the primary culture of meningiomas. Because of a superior initial plating efficiency, the cultures on ECM become confluent earlier and can be expanded more rapidly, which may be advantageous for some applications. One useful application of ECM in meningioma cultures may be the rapid availability of large cell numbers for laboratories concerned with cytogenetic investigations. This allows very early chromosome analysis, reducing the

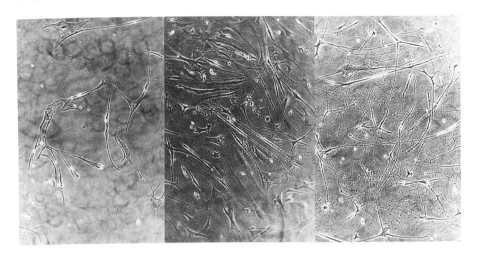

Fig. 6. Three different primary culture conditions are compared in which the advantages of ECM coating over plastic (left) can be seen, as well as the high density of cells despite serum-free conditions (right). The cells were obtained from a small cell anaplastic astrocytoma.

concerns about additional tissue-culture artifacts. Second, bECM allows the rapid reduction of serum concentrations, helps with serum-free conditions, and therefore, facilitates experiments evaluating growth factor requirements for tumor cells.

11. Gliomas have to be divided into low-grade and high grade lesions. In addition, the primary tumors are to be distinguished from those that are operated as recurrent tumors, which may have recurred after irradiation or other treatment. Gliomas, especially the more malignant lesions, are often very heterogeneous with numerous different cell lines within one tumor *(24)*. It has to be assumed that this heterogeneity is the prime basis for the therapeutic difficulties that are intrinsic to glioma management. Only the understanding of this heterogeneity and the investigation of the different growth control mechanisms present in one glioma will lead to an advance in specific therapies. It appears as if the use of ECM as an attachment substrate helps to maintain the heterogeneity of these tumors in early cultures.

12. In contrast to meningiomas, the behavior of primary glioma cultures on plastic and bECM is difficult to predict. In general, the initial attachment already allows speculation about the biological aggressiveness of the tumor, because the well-differentiated gliomas show only poor attachment to plastic and adhere well to ECM, both in medium containing serum and serum-free cultures to which conditioned serum-free medium from an established glioma cell line has been added, but also cultures from anaplastic tumors adhere better to ECM than to untreated culture dishes (Fig. 6).

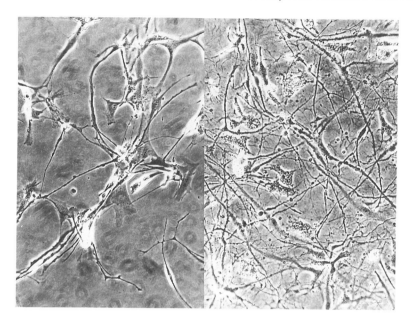

Fig. 7. Two phase-contrast photomicrographs comparing the primary cultures from a recurrent oligodendroglioma grade WHO 2 on plastic and on ECM. In this case, the cellular interaction/differentiation as seen by the extent of glial process formation is much more pronounced on ECM.

Among the most common experimental situations in which efficient primary culture is desirable are chemoresistance analysis, cytogenetics, and differentiation biology using immunocytochemical markers. The advantages of using ECM in the first two instances are given by the complexity of the culture composition that will be obtained. This will result in a more realistic reflection of the behavior of the tumor toward chemotherapy or will result in a more authentic picture of the cytogenetic composition of the tumor. If cellular differentiation and interaction are the target of biological investigation, it can be seen already from phase-contrast microscopy of cultures after 24 h, that the cellular interaction reflected by the intricate network of glial processes is much more elaborate in cultures kept on ECM (Fig. 7).

13. The use of ECM in established cell lines is limited. Mainly there are two applications. In experiments in which it is desirable to establish cell lines in serum-free conditions and maintain them that way, this will need the help of some adhesive substrate after each passage because the cells will not readily attach without serum, stay afloat, and die. Over a long period, cell lines initiated from serum-free cultures and from serum containing medium show a very different growth rate and morphology (Fig. 8). The cells maintained in serum-free medium will have reached at most half the passage numbers of the parallel line maintained in

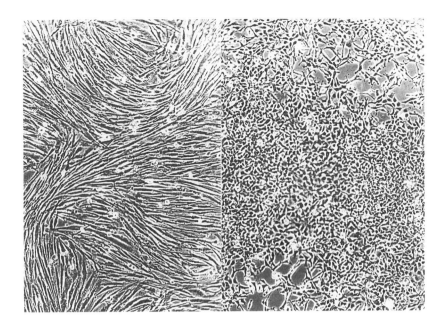

Fig. 8. Phase-contrast photomicrographs from two cell lines established from the same initial specimen obtained from a glioblastoma. The cell line depicted at the top is in passage 68 and was maintained on plastic after initiation on ECM. It was also maintained in medium containing 10% FCS. The cell line shown at the bottom is only in passage 30, but was continuously maintained in serum-free medium on ECM. Both cell lines are distinguished by their different morphologies.

the presence of serum. In part, this could be the result of the differentiating effect of ECM as well, but this is difficult to verify.

14. The most interesting application of ECM as a research area in its own right is the analysis of cell migration. ECM components are an important determinant of cell migration *(25)*, since most glioma patients will eventually die from an unresectable recurrence of their tumor, which has arisen far away from the original site owing to the migratory activity of the tumor cells.

The production of an ECM by normal cells as well as tumor cells in culture is a common phenomenon, although it can be visualized in most cases only by means of immunofluorescence (Fig. 9). The developmental biologists are just beginning to explore this field of ECM in the CNS and its role in regional cell differentiation to determine the fate of migratory grafts for therapeutic purposes *(26)*. From our previous work, it is clear that ECMs influence migration as well as differentiation or the response to growth factors *(18,25)*. These effects can be highly diverse, depending very much on the sources of the ECMs (Westphal et al., unpublished results).

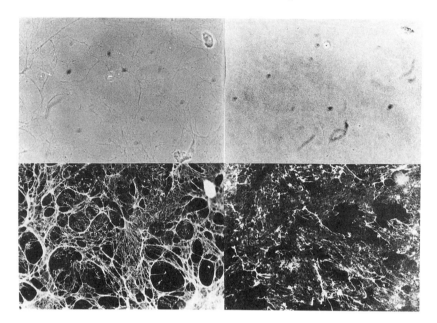

Fig. 9. Immunostaining of glass cover slides covered with bECM (left) and tECM from NCE-G140, glioblastoma (right). The corresponding phase-contrast photomicrographs (top) show only barely visible fine fibrous structures. Below, the bECM is stained with an antiserum against fibronection, and the tECM is stained with antitenascin.

15. Usually, these tumors are soft and can be dispersed mechanically. Even with mechanical dispersion, however, fibroblast overgrowth will be a problem after 3–6 wk of continuous culture, because pituitary adenoma cells do not proliferate, but fibroblasts do. The adenoma cells usually attach within a few hours and can then be used in experiments studying the pharmacological regulation of hormone secretion. The test substances can be added, and the medium can be completely removed to be assayed without the cells coming off because they are firmly attached. Note the exception: Cell cultures initiated from ACTH-secreting adenomas from patients with Cushing's disease attach much more slowly and sometimes need 2 d to flatten. The time it takes for the cells to attach is correlated to the degree of cortisol excess in vivo before surgery *(9)*. In addition to the much improved attachment over plastic, the most striking effects of bECM are the rapid and obvious changes in cellular morphology after exposure to hypothalamic releasing factors, such as GRF for HGH-secreting cells and CRF or vasopressin for ACTH-secreting cells *(9,10)*.
16. Acoustic neuromas as well as other benign schwannomas are difficult to culture even on ECM. If they are soft and cellular, they also attach well to untreated culture dishes. The same holds true for malignant schwannomas to which ECM does not appear to offer any advantage.

17. Pineal tumors are derived from a very heterogeneous group of tissue sources. Those that belong to the glial cell lineage behave like the gliomas. Other derivatives, such as germinomas, pineocytomas, or pineoblastomas, attach well to bECM in comparison to plastic. In general, these tumors behave much like pituitary adenomas, especially since they do not appear to proliferate without any specific additives.
18. Medulloblastomas: This particular type of tumor is very difficult to maintain in culture, and bECM has not provided a major breakthrough. In addition to providing an attachment substrate, the addition of specific additives or conditioned media seems to be the crucial point in a successful culture of this type of early glial tumor (numerous personal communications).
19. Neurocytoma: Neurocytomas are rare neuroglial precursor tumors located at the foramen of Monroi in the lateral ventricles. They need a support for culture, and only the use of bECM has allowed the study of the differentiation potential of these tumor cells *(27)*.

4.3. General

20. Corneal endothelial cells can be frozen and stored in liquid nitrogen. They can be cultured for many in vitro passages and will produce matrix. It has to be noted, however, that the bECM gets thinner and less homogeneous with increasing age of the corneal endothelial cells. ECM can be stored at 4°C for up to 2 yr. It can also be dried after all salt has been washed off.
21. ECMs are different in their chemical composition, depending on the tissue they are derived from. This relates to the matrix proteins as well as to growth factors that are sequestered into the ECM by the producer cells. This aspect has to be taken into account when investigating autocrine mechanisms of cell proliferation *(13,14,28)*. These factors are slowly released and act as intrinsic mitogens. Since heparin binding growth factors will also bind to ECM, any binding studies with such factors performed on ECM may become difficult to interpret.
22. The producer cells for any ECM should be regularly tested for mycoplasma, since otherwise contaminations may spread. It is best to use the PCR-based kits. Although mycoplasma is unlikely to survive the hypo-osmotic lysis and the ammonia, it is safer to dispose of infected producer cells.

Acknowledgments

The authors are grateful to D. Gospodarowicz in whose laboratory the introduction to ECM biology and the methodological training were obtained. We express our thanks to the departmental operating room staff for their collaboration and appreciate the photographical work of S. Freist. The karyotype analyses were performed by M. Haensel. Throughout, work in the Laboratory for Brain Tumor Biology was alternately supported by the Deutsche Forschungsgemeinschaft, the Deutsche Krebshilfe, and the Heinrich-Bauer Stiftung für Hirntumorforschung. The continuous support of the department chairman, Hans-Dietrich Herrmann is vital to the ongoing work in the laboratory.

References

1. Bottenstein, J. E. and Sato, G. H. (1980) Fibronectin and polylysine requirement for proliferation of neuroblastoma cells in defined medium. *Exp. Cell Res.* **129**, 361–366.
2. Terranova, V. P., Aumalley, M., Sultan, L. H., Martin, G. R., and Kleinman, H. K. (1986) Regulation of cell attachment and cell number by fibronectin and laminin. *J. Cell. Physiol.* **127**, 473–479.
3. Couchmann, J. R., Hook, M., Rees, D., and Timpl, R. (1983) Adhesion, growth and matrix production by fibroblasts on laminin substrates. *J. Cell. Biol.* **96**, 177–183.
4. Varani, J., Carey, T. E., Fligiel, S. E. G., McKeever, P. E., and Dixit, V. (1987) Tumor type-specific differences in cell-substrate adhesion among human tumor cell lines. *Int. J. Cancer* **39**, 397–403.
5. Gospodarowicz, D., Vlodavsky, I., and Savion, N. (1981) The role of fibroblast growth factor and the extracellular matrix in the control of proliferation and differentiation of corneal endothelial cells. *Vison Res.* **21**, 87–103.
6. Gospodarowicz, D., Cohen, D., and Fujii, D. K. (1982) Regulation of cell growth by the basal lamina and plasma factors: relevance to embryonic control of cell proliferation and differentiation, in *Cold Spring Harbor Conference on Cell Proliferation*, vol. 9. Cold Spring Harbor Laboratory Press, Cold Spring Harbor, NY, pp. 95–124.
7. Lichtner, R. B., Belloni, P. N., and Nicolson, G. L. (1989) Differential adhesion of metastatic rat mammary carcinoma cells to organ-derived microvessel endothelial cells and subendothelial matrix. *Exp. Cell Biol.* **57**, 146–152.
8. Pavelic, K., Bulbul, M. A., Slocum, H. K., Pavelic, Z. P., Rustum, Y. M., Niedbala, M. J., and Bernacki, R. J. (1986) Growth of human urological tumors on extracellular matrix as a model for the in vitro cultivation of primary tumor explants. *Cancer Res.* **46**, 3653–3662.
9. Westphal, M., Jaquet, P., and Wilson, C. B. (1986) Long-term culture of human corticotropin-secreting adenomas on extracellular matrix and evaluation of serum free conditions. *Acta Neuropathol.* **71**, 142–149.
10. Westphal, M., Hahn, H., and Lüdecke, D. K. (1987) Culture of dispersed cells from human pituitary adenomas from acromegalic patients on extracellular matrix, in *Growth Hormone, Growth Factors and Acromegaly* (Lüdecke, D. K. and Tolis, G., eds.), Raven, New York, pp. 125–133.
11. Westphal, M., Hänsel, M., Brunken, M., König, A., Köppen, J. A., and Herrman, H. D. (1987) Initiation of primary cell cultures from human intracranial tumors on extracellular matrix from bovine corneal endothelial cells. *Exp. Cell Biol.* **55**, 152–163.
12. Hynes, R. O. (1992) Integrins: versatility, modulation and signalling in cell adhesion. *Cell* **69**, 11–25.
13. Alanko, T., Tienari, J., Lehtonen, E., and Saksela, O. (1994) Development of FGF-dependency in human embryonic carcinoma cells after retinoic acid-induced differentiation. *Dev. Biol.* **161**, 141–153.
14. Adams, J. C. and Watt, F. M. (1993) Regulation of development and differentiation by extracellular matrix. *Development* **117**, 1183–1198.

15. Vlodavsky, I., Lui, G. M., and Gospodarowicz, D. J. (1980) Morphological appearance, growth behaviour and migratory activity of human tumor cells maintained on extracellular matrix versus plastic. *Cell* **19**, 607–616.
16. Payne, H. R. and Lemmon, V. (1993) Glial cells of the O-2A lineage bind preferentially to N-cadherin and develop distinct morphologies. *Dev. Biol.* **159**, 595–607.
17. Rutka, J. T. (1986) Effects of extracellular matrix proteins on the growth and differentiation of an anaplastic glioma cell line. *Can. J. Neurol. Sci.* **13**, 301–306.
18. Westphal, M., Hänsel, M., Nausch, H., Rohde, E., and Herrmann, H. D. (1990) Culture of human brain tumors on an extracellular matrix derived from bovine corneal endothelial cells and cultured human glioma cells, in *Methods in Molecular Biology, vol. 5, Animal Cell Culture* (Pollard, J. W. and Walker, J. M., eds.), Humana Press, Clifton, NJ, pp. 113–131.
19. Weiner, R. I., Bethea, C. L., Jaquet, P., Ramsdell, J. S., and Gospodarowicz, D. J. (1983) Culture of dispersed anterior pituitary cells on extracellular matrix. *Methods in Enzymol.* **103**, 287–294.
20. Gospodarowicz, D., Cheng, J., Lui, G. M., Baird, A., and Böhlen, P. (1984) Isolation of brain fibroblast growth factor by heparin sepharose affinity chromatography: identity with pituitary fibroblast growth factor. *Proc. Natl. Acad. Sci. USA* **81**, 6963–6967.
21. Westphal, M., Hänsel, M., Hamel, W., Kunzmann, R., and Hölzel, F. (1994) Karyotype analysis of 20 human glioma cell lines. *Acta Neurochir.* **126**, 17–26.
22. Cardwell, M. C. and Rome, L. H. (1988) Evidence that an RGD-dependent receptor mediates the binding of oligodendrocytes to a novel ligand in a glial-derived matrix. *J. Cell Biol.* **107**, 1541–1549.
23. Liotta, L. A. and Stetler-Stevenson, W. G. (1991) Tumor invasion and metastasis: an imbalance of positive and negative regulation. *Cancer Res.* **51**, 5054–5059.
24. Shapiro, J. R. (1986) Biology of gliomas: chromosomes, growth factors and oncogenes. *Sem. Oncol.* **13**, 4–15.
25. Giese, A., Rief, M., and Berens, M. (1994) Determinants of human glioma cell migration. *Cancer Res.* **54**, 3887–3904.
26. Mucke, L. and Rockenstein, E. M. (1993) Prolonged delivery of transgene products to specific brain regions by migratory astrocyte grafts. *Transgene* **1**, 3–9.
27. Westphal, M., Stavrou, D., Nausch, H., Valdueza, J. M., and Herrmann, H.-D. (1994) Human neurocytoma cells in culture show characteristics of astroglial differentiation. *J. Neurosci. Res.* **38**, 698–704.
28. Vlodavsky, I., Folkman, J., Sullivan, R., Fridman, R., Ishai-Michaeli, R., Sasse, J., and Klagsbrun, M. (1987) Endothelial cell derived basic fibroblast growth factor: synthesis and deposition into subendothelial extracellular matrix. *Proc. Natl. Acad. Sci. USA* **84**, 2292–2296.

9

Isolation and Culture of Human Umbilical Vein Endothelial Cells

David M. L. Morgan

1. Introduction

The vascular endothelium was, in the past, considered to be relatively passive, merely acting as a filter between the blood and the vessel wall. It is now clear that endothelial cells actively contribute to the maintenance of vascular homeostasis. Endothelial cells synthesize and secrete activators as well as inhibitors of both the coagulation system and the fibrinolysis system, and mediators that influence the adhesion and aggregation of blood platelets. They also release molecules that control cell proliferation and modulate vessel wall tone *(1–3)*. Many of these, and other, processes can be studied in vitro using cultured cells, and umbilical veins are the most readily available source of human vascular endothelial cells. The method described here is based on that of Jaffe et al. *(4,5)* and yields primary cultures that contain 95–98% endothelial cells and >98% after the first passage. Contaminating cells in primary cultures are not smooth muscle (if the procedure has been followed correctly), as can be demonstrated by examining the level of staining with antibody to smooth muscle α-actin, but may include monocytes/macrophages at a very low level.

All human tissue should be treated as potentially hazardous, and it is strongly recommended that those engaged in isolating human umbilical vein endothelial cells should be immunized against hepatitis B. To ensure sterility and operator safety, all work should be carried out in a Class II microbiological safety cabinet conforming to BS5726-1992 (Britain), NSF 49 (USA), DIN 12950 (Germany), NF X44 201 (France), AS 2252.2 (Australia), or local equivalent.

From: *Methods in Molecular Medicine: Human Cell Culture Protocols*
Edited by: G. E. Jones Humana Press Inc., Totowa, NJ

2. Materials

Items 35–42 should be prepared under sterile conditions.

1. 250-mL Sterile plastic specimen containers (e.g., Greiner [Frickenhausen, Germany], cat. no. 229170).
2. 7 mL Sterile plastic Bijou containers.
3. 30 mL Sterile plastic Universal containers.
4. 25 and 75 cm^2 tissue-culture flasks.
5. Sterile 5- and 10-mL pipets.
6. Sterile 10- and 20-mL syringes.
7. Sterile 50-mL centrifuge tubes.
8. Sterile 0.2-µm filters (Sartorius [Göttingen, Germany] Minisart NML or equivalent).
9. 0.5-µm Prefilters (Sartorius Minisart GF or equivalent).
10. Scalpel, scissors, forceps, and so forth. Scrub after use, wrap in aluminum foil, and sterilize in an autoclave for 20 min at 121°C.
11. Strong thread (e.g., Barbour [Lissburn, Northern Ireland] no. 40 flax suture).
12. Bench centrifuge, able to take 50-mL tubes.
13. Cling film (50-µm PVC sealing film).
14. Aluminum foil.
15. Cannulae (Bio-Rad [Hercules, CA] 731-8223 Female Luer fitting, 3.2-mm barb to female Luer, or similar). Clean thoroughly after use, wrap in aluminum foil, and sterilize in an autoclave for 20 min at 121°C.
16. Crocodile clips (e.g., Farnell Electronics [Leeds, UK] 495-177 miniature crocodile clips). Scrub after use, wrap in aluminum foil and heat sterilize in an autoclave for 20 minutes at 121°C.
17. 15 mL McCartney bottles, screw-capped.
18. Glass bottles, screw-capped, for prepared medium, and so on, 500 and 200 mL (e.g., medical flats). Cover aperture with aluminum foil and sterilize at 160°C for 90 min.
19. Plastic bottle caps. Wrap in aluminum foil, and sterilize in an autoclave for 20 min at 121°C.
20. Incubator trays.
21. Hank's balanced salt solution (HBSS).
22. Phosphate-buffered saline Dulbecco's "A", (PBS).
23. Medium 199 (without glutamine or bicarbonate). If an adequate supply of tissue-culture-grade water is available, considerable savings can be made by purchasing this and PBS as 10X concentrates.
24. Endothelial cell growth factor (ECGF) (either purchased; Sigma [St. Louis, MO] E2759 or E9640 [100X solution]; Advanced Protein Products, Brierley Hill, UK, GF-100-36; or prepared, see Note 5).
25. Bicarbonate (4.4%; 0.52M)–phenol red (0.03%) solution: Dissolve 44 g sodium hydrogen carbonate plus 30 mg phenol red in 1000-mL tissue-culture-grade deionized water, and sterilize in an autoclave for 10 min at 115°C.

Store as 15-mL aliquots in glass screw-capped McCartney bottles at 4°C. Shelf life ~6 mo; any change in color indicates a loose bottle cap, with the possibility of contamination.

26. 1M HEPES solution: dissolve 47.6 g of HEPES (Ultrol-grade, Calbiochem, San Diego, CA) in 200-mL tissue-culture-grade deionized water, sterilize by filtration through a 0.2-µm filter, and store as 5-mL aliquots in sterile plastic bijous at −20°C. Shelf life ~6 mo.
27. Penicillin and streptomycin solution (~4000 U/mL): Dissolve 479 mg penicillin (penicillin G, sodium salt, e.g., Sigma P3032, ~1650 U/mg) and 1.05 g streptomycin sulfate in 200-mL tissue-culture-grade deionized water. Filter this solution through a 0.5-µm prefilter and a 0.22-µm sterile filter connected in series; dispense into 5-mL aliquots and store at −20°C. Shelf life ~6 mo.
28. Gentamycin solution (0.75%): Dissolve 750 mg gentamycin sulfate (e.g., Sigma G3632, ~600 µg/mg) in 100-mL tissue-culture-grade deionized water, filter through a 0.2-µm filter, and store as 5-mL aliquots in sterile plastic bijous at −20°C. Shelf life ~6 mo.
29. Gelatin solution (1%): Dissolve 1 g gelatin (from bovine skin, e.g., Sigma G9382) in tissue-culture-grade deionized water, and sterilize in an autoclave for 20 min at 121°C. Store at 4°C for up to 6 mo.
30. ECGF (1 mg/mL)–heparin (4.5 mg/mL): Dissolve 80 mg (or other suitable amount, *see* Note 5) ECGF and 360 mg heparin (Sigma H3393, from porcine mucosa, sodium salt) in 80 mL of serum-free medium; then filter this solution through a 0.5-µm prefilter and a 0.22-µm sterile filter connected in series; dispense into 4-mL aliquots and store at −20°C. Shelf life ~3 mo.
31. Trypsin solution (2.5%): Dissolve 2.5 g trypsin (from porcine pancreas, e.g., Sigma T0646) in 100 mL PBS "A" and sterilize by filtration through a 0.22-µm filter. Store at −20°C in 10 mL aliquots. Shelf life ~6 mo.
32. EDTA solution (1%): Dissolve 500 mg of EDTA sodium salt in 50 mL of tissue-culture-grade deionized water, sterilize by filtration through a 0.22-µm filter, and store at room temperature in 5-mL aliquots. Shelf life ~6 mo.
33. Trypsin (0.1%)–EDTA (0.02%; 0.5 mM): Add 10 mL trypsin and 5 mL EDTA solution to 250 mL PBS "A"; store at 4°C. Shelf life ~2 mo.
34. Glutamine solution (200 mM): Dissolve 5.84 g L-glutamine (Sigma tissue-culture-grade) in 200 mL of tissue-culture-grade deionized water. Filter this solution through a 0.5-µm prefilter and a 0.22-µm sterile filter connected in series; dispense into 5-mL aliquots, and store at −20°C. Shelf life ~6 mo.
35. Cord collection medium: To 400 mL of HBSS, add 15 mL bicarbonate–phenol red solution, 10 mL of HEPES, and 5 mL of gentamycin. Divide the resulting solution between five specimen containers and store at 4°C. Use within 1 wk.
36. Fetal calf serum (FCS) (*see* Notes 2 and 3): Dispense into 20-mL aliquots in sterile plastic Universal containers and store at −20°C. Shelf life ~1 mo. Serum that has been kept frozen can be stored for 6 mo at −20°C and 1–2 yr at −70°C.
37. Newborn calf serum: Dispense into 20-mL aliquots in sterile plastic Universal containers and store at −20°C. Shelf life ~1 mo.

38. Serum-free medium: To 400 mL of medium 199, add 5 mL of penicillin–streptomycin solution, 20 mL of bicarbonate–phenol red solution, and 5 mL glutamine (final concentration in serum-containing medium will be ~100 U/mL of each antibiotic, 21 mM bicarbonate, 2 mM glutamine). Store at 4°C and use within 2 wk.
39. Culture medium: To 160 mL of serum-free medium, add 20 mL FCS and 20 mL newborn calf serum (final concentration is 20% serum). Store at 4°C and use within 1 wk. Human umbilical vein endothelial cells require 20% serum for growth; partial substitution with newborn calf serum results in a reduction of costs without affecting cell growth.
40. Growth factor-containing medium: Add 4 mL of ECGF–heparin to 200 mL of serum-containing culture medium (final concentration 20 µg/mL ECGF and 90 µg/mL heparin). Store at 4°C, and use within 1 wk.
41. Gelatin coating of tissue-culture flasks: Place 3 mL of the gelatin solution in a 25-cm^2 tissue-culture flask (5–8 mL in a 75-cm^2 flask) and tilt to ensure complete wetting of the culture surface; at this stage, prepared flasks can be stored for 1–2 wk at 4°C. Before use, remove the gelatin solution, and rinse the flask with serum-free medium.
42. Collagenase solution (*see* Notes 2 and 4): Make up just prior to use in the proportion of 10 mL/20 cm of cord. Dissolve sufficient collagenase (Sigma Type II or Boehringer Mannheim [Mannheim, Germany] collagenase A, from *Clostridium histolyticum*) in a small volume of serum-free medium 199, filter through a 0.5-µm prefilter and a 0.22-µm sterile filter connected in series, and then dilute with serum-free medium to a final concentration of 0.5 mg/mL.

3. Methods
3.1. Collection of Umbilical Cords

As soon as possible after delivery, the umbilical cord should be separated from the placenta, placed in cold collection medium and stored at 4°C. It is important to explain to the staff of the Delivery Suite that it is the *whole* cord that is required and not just a specimen. Umbilical cords collected and stored in this way can be used for endothelial cell isolation up to 48 h after delivery.

A small refrigerator should be available in the Delivery Suite to house both specimen jars containing cord collection medium, into which freshly delivered cords should be placed, and cords awaiting collection by the laboratory.

3.2. Isolation and Culture of Human Umbilical Vein Endothelial Cells
3.2.1. Isolation and Primary Culture

1. Warm all solutions to 37°C.
2. Make up sufficient collagenase solution for the number and length of cords to be processed.

3. Remove the cord from the specimen jar, and blot with a paper towel to remove blood and collection medium.
4. Make a clean transverse cut across one end of the cord to expose the two umbilical arteries and the umbilical vein. The latter can be identified by its thinner wall and larger, stretchable lumen.
5. Insert a cannula into the vein, and remove tissue and arteries to expose about 1 cm of vein. It is important to remove a short section of the adjacent arteries; otherwise it may be difficult to tie the ligature tightly enough, and the cannula may slip from the vein during cell isolation.
6. Secure the cannula by tying the vein, above the barb, with strong thread.
7. Repeat the process at the other end of the cord; at this point, very long cords can be divided to give two shorter lengths of 20–25 cm.
8. Fill a 20-mL syringe with warmed PBS "A," attach it to the cannula at one end, and then use this solution to flush the vein free from blood, clots, and so forth, collecting the eluate in a beaker.
9. Repeat the procedure from the other end (gentle massage may facilitate the removal of clots). Before expelling all the wash solution, occlude the cord at one end, and check for leaks from perforations made when blood samples were taken. Any holes found can be closed using crocodile clips.
10. Attach an empty syringe to one cannula, and then use another syringe to fill the vein with sufficient prewarmed collagenase solution to ensure that the lumen is distended (approx 10 mL/20 cm of cord).
11. Cover the cord in cling-film and, with the syringes still attached, place the cord in an incubator at 37°C for 10 min.
12. Remove the cord from the incubator, and massage it gently to assist the detachment of cells from the vessel wall.
13. Apply suction with one syringe while exerting finger pressure along the cord in the direction of flow to draw out as much of the cell suspension as possible.
14. Place the cell suspension in a sterile 50-mL centrifuge tube.
15. Leaving one syringe still attached to the cord, take up 20 mL of prewarmed PBS "A" into a syringe, and use this to flush free any further loosened cells, working the solution backward and forward between the two syringes a few times. Draw off as much of this second cell suspension as possible, and add to the tube containing the collagenase eluate.
16. Pellet the cells by centrifugation at 1000 rpm (~150g) in a benchtop centrifuge for 5 min.
17. Decant the supernatant fluid, and resuspend the cells in 4 mL of serum-containing culture medium (without growth factor).
18. Transfer the cell suspension to a gelatin-coated 25-cm^2 tissue-culture flask, and place in an incubator at 37°C in an atmosphere of 5% CO_2/95% air. At this point, the flask will contain large numbers of erythrocytes in addition to endothelial cells.
19. The next day, remove the medium from the culture flask, rinse the cells with prewarmed serum-free medium to remove any residual red cells, and add 4 mL of fresh culture medium.

20. Every second day, feed the cells by removing and discarding half the medium in the flask and replacing it with fresh serum-containing culture medium. The cells should reach confluence in 7 d (or less) and are then ready for use, or they can be subcultured to obtain larger numbers of cells.

3.2.2. Subculture of Endothelial Cells

1. Warm all solutions to 37°C.
2. Remove the medium from a confluent primary monolayer of cells, and rinse the cells with serum-free medium (serum contains a trypsin inhibitor).
3. Add 1 mL of trypsin–EDTA solution to a 25-cm^2 flask, and tilt the flask to thoroughly wet the cell monolayer. Return the flask to the incubator.
4. After 5 min, remove the flask from the incubator, and examine under the microscope. The cells should have rounded up and begun to detach from the surface. Smartly tap the side of the flask two to three times to aid cell detachment. Prolonged exposure to trypsin can damage endothelial cells and reduce viability.
5. Add an equal volume of growth factor-containing medium, and draw the cell suspension up and down three to four times through a narrow-bore pipet to break up any cell clumps. Place the cell suspension in a gelatin-coated 75-cm^2 flask (the split ratio will be 1 to 3), and add sufficient growth factor-containing medium to bring the volume to 10 mL.
6. Return the cells to the incubator, and feed every second day with growth factor-containing medium.

The subcultured cells, which are now at passage 1, should again become confluent within 7 d. If required, passage 1 cells can again be subcultured by dividing the contents of one 75-cm^2 flask between three new flasks (split ratio again 1 to 3), using 2.5 mL trypsin–EDTA for cell detachment. These passage 2 cells should again reach confluence by 7 d. Further subculture is not advisable, since the cell growth rate begins to decline.

3.3. Identification of Endothelial Cells

Vascular endothelial cells can be identified by the typical "cobblestone" morphology exhibited by confluent monolayers *(5,6)*; positive staining for von Willebrand factor *(7)*, and the ability to take up acetylated low-density lipoprotein *(8)*. Human umbilical vein endothelial cells also stain positively and specifically with several commercially available MAbs, e.g., EN-4 *(9)* and Pal-E *(10)* (to undefined antigens), and END-10 (to CD34) *(11)*, plus antibodies to PECAM-1 (CD31) *(12)* and P-selectin (CD62P) *(13)*. Cells pretreated with interleukin-1, tumor necrosis factor-α, or bacterial lipopolysaccharide also selectively express E-selectin (CD62E) *(14)* and VCAM-1 (CD106) *(15)*. The original papers should be consulted for methodological details. In the UK, sources of suitable antibodies include: Biogenisis, Stinisford Road, Poole, BH17 7NF; Cambridge BioScience, Newmarket Road, Cambridge, CB5 8LA;

R & D Systems, Barton Lane, Abingdon, OX14 3YS; Serotec, Kidlington, Oxford, OX5 1JE; and TCS Biologicals, Botolph Claydon, Buckingham, MK18 2LR.

4. Notes

1. Useful hints: Cell cultures left confluent for more than 2–3 d will start to detach from the substratum and cease to be viable. Primary cultures that start to grow, but fail to become confluent in 7 d can sometimes be stimulated by detaching the cells with trypsin, adding fresh medium, and then leaving them to readhere in the same flask.
2. Testing of reagents: Not all batches of FCS will support the growth of human umbilical vein endothelial cells; also, commercial preparations of collagenase usually contain varying proportions of other proteases that are harmful to endothelial cells and whose presence results in reduced yields of viable cells. Hence, samples of both materials should be obtained and tested before bulk purchases are made.
3. FCS: The growth-promoting potential of different batches of serum can be assessed by measuring their effect on the incorporation of [^{3}H]leucine into protein by human umbilical vein endothelial cells (assessing growth by measuring changes in cell protein may not be sufficiently sensitive). The procedure is described by Morgan et al. *(6)*.
    a. Confluent first-passage cells are trypsinized, resuspended in growth factor-containing medium at a concentration of 2.5×10^4 cells/mL, and 200 µL dispensed into 60 wells of a 96-well flat-bottomed tissue-culture plate, the outer 36 wells of which each containing 200 µL of PBS (for some reason, cell growth is less reproducible in the outer wells).
    b. After incubation for 24 h, the medium in each row is removed from the wells and replaced by medium containing 20% of one of the serum samples under test (without growth factor); 10 replicate wells are used for each serum sample, one of the rows being reserved for the current batch of serum as a control.
    c. [^{3}H]Leucine, 1 µCi (1 µL), is then added to each well.
    d. After a further 24 h, the medium is removed, the cell monolayer in each well is rinsed twice with PBS "A" (200 µL), exposed to 5% trichloroacetic acid (200 µL) for 5 min, and then rinsed with methanol (200 µL).
    e. Finally, the cells are digested by the addition of 200 µL of $25M$ formic acid. Each formic acid digest is transferred, with one rinse of PBS "A," to a plastic β-scintillation vial; 3.5 mL of scintillation fluid (e.g., Pharmacia [Uppsala, Sweden] "Optisafe") are added, and the amount of radioactivity incorporated into the trichloroacetic acid insoluble protein is determined in a scintillation counter.

    The higher the incorporated counts, the better that serum supports the growth of endothelial cells.
4. The suitability of collagenase samples is determined by the yield of attached cells after 24 h and by the time taken to reach confluence in primary cultures isolated using different batches of the enzyme.

5. Preparation of ECGF: This material is expensive and is not always available from commercial sources as a separate item. If extensive culture of endothelial cells is to be undertaken, then consideration should be given to the possibility of preparing it "in house." The method of isolation is relatively simple and, if the necessary equipment is available, an amount sufficient for a 1–2 yr supply can be obtained from a single preparation. The method described here is based on that of Maciag et al. *(16,17)*.

All steps are carried out at 4°C.
   a. Collect one or two porcine brains from the slaughterhouse, and transport on ice to the laboratory. (The original procedure called for bovine brain, but porcine brain has also been found to yield a satisfactory product, thus avoiding any possible risk from handling material that may be contaminated by bovine spongiform encephalitis.)
   b. Rinse each brain with distilled water (3 L/kg of tissue) to remove blood and other fluids, and blot the tissue dry using paper towels.
   c. Homogenize the brain tissue in $0.15M$ NaCl (1.25 L/kg of tissue) in batches in a Waring blender for no longer than 3 min.
   d. Extract the homogenates by mechanical agitation on an orbital shaker for 2 h.
   e. Separate the insoluble material by centrifugation (13,000g, 40 min).
   f. Collect the supernatant fluid with as little as possible of the lipid-rich buoyant layer.
   g. Adjust the pH (if necessary) to between 6.5 and 7.4.
   h. Dissolve streptomycin sulfate (7.5 g/kg tissue) in distilled water (50 mL/ 7.5 g solid). Adjust the pH of this solution to 8.0 using $1M$ NaOH, and then cool to 4°C.
   i. Add the streptomycin solution dropwise with stirring to the brain extract. The final pH should be 6.5–7.0.
   j. Allow to stand overnight at 4°C.
   k. Separate the precipitated material by centrifugation (27,000g; 30 min), and collect the supernatant.
   l. Lyophilize the extract as soon as possible after separation, and store desiccated at –20°C. Growth factor prepared in this way can be stored frozen for up to 2 yr.

Preparations should be tested for activity by adding to heparin-containing growth medium at concentrations of 5–50 µg/mL and examining their effect on endothelial cell growth by the [^{3}H]leucine incorporation method. Suitable preparations should give maximum incorporation at 10–20 µg/mL; concentrations >30 µg/mL may be inhibitory.

References

1. Fajardo, L. F. (1989) The complexity of endothelial cells. *Am. J. Clin. Pathol.* **92**, 241–250.
2. Pearson, J. D. (1991) Endothelial cell biology. *Radiology* **179**, 9–14.

3. Moncada, S., Palmer, R. M. J., and Higgs, E. A. (1991) Nitric oxide: physiology, pathophysiology, and pharmacology. *Pharmacol. Rev.* **43**, 109–142.
4. Jaffe, E. A., Nachman, R. L., Becker, C. G., and Minick, C. R. (1973) Culture of human endothelial cells derived from umbilical veins: identification by morphologic and immunologic criteria. *J. Clin. Invest.* **52**, 2745–2756.
5. Jaffe, E. A. (1984) Culture and identification of large vessel endothelial cells, in *Biology of Endothelial Cells* (Jaffe, E. A., ed.), Martinus Nijhoff, Boston, pp. 1–13.
6. Morgan, D. M. L., Clover, J., and Pearson, J. D. (1988) Effects of synthetic polycations on leucine incorporation, lactate dehydrogenase release, and morphology of human umbilical vein endothelial cells. *J. Cell. Sci.* **91**, 231–238.
7. Hoyer, L. M., de los Santos, R. P., and Hoyer, J. R. (1973) Antihaemophilic factor antigen. Localisation in endothelial cells by immunofluorescent microscopy. *J. Clin. Invest.* **52**, 2737–2744.
8. Voyta, J. C., Via, D. P., Butterfield, C. E., and Zetter, B. R. (1984) Identification and isolation of endothelial cells based on their increased uptake of acetylated low-density lipoprotein. *J. Cell Biol.* **99**, 2034–2040.
9. Cui, Y. C., Tai, P.-C., Gatter, K. C., Mason, D. Y., and Spry, C. J. F. (1983) A vascular endothelial cell antigen with restricted distribution in human foetal, adult and malignant tissues. *Immunology* **49**, 183–189.
10. Schlingemann, R. O., Dingjan, G. M., Emeis, J. J., Blok, J., Warnaar, S. O., and Ruiter, D. J. (1985) Monoclonal antibody PAL-E specific for endothelium. *Lab. Invest.* **52**, 71–76.
11. Schlingemann, R. O., Rietveld, F. J. R., de Waal, R. M. W., Bradley, N. J., Skene, A. I., Davies, A. J. A., Greaves, M. F., Denekamp, J., and Ruiter, D. J. (1990) Leukocyte antigen CD34 is expressed by a subset of cultured endothelial cells and on endothelial abluminal microprocesses in the tumor stroma. *Lab. Invest.* **62**, 690–696.
12. Newman, P. J., Berndt, M. C., Gorski, J., White, G. C., Lyman, S., Paddock, C., and Muller, W. A. (1990) PECAM-1 (CD31) cloning and relation to adhesion molecules of the immunoglobulin gene superfamily. *Science (Wash. DC)* **247**, 1219–1222.
13. Johnston, G. I., Cook, R. G., and McEver, R. P. (1989) Cloning of GMP-140, a granule membrane protein of platelets and endothelium: sequence similarity to proteins involved in cell adhesion and inflammation. *Cell* **56**, 1033–1044.
14. Bevilaqua, M. P., Stengelin, S., Gimbrone, M. A., and Seed, B. (1989) Endothelial leukocyte adhesion molecule 1: an inducible receptor for neutrophils related to complement regulatory proteins and lectins. *Science (Wash. DC)* **243**, 1160–1165.
15. Osborn, L., Hession, C., Tizard, R., Vassallo, C., Luhowskyj, S., Chi-Rosso, G., and Lobb, R. (1989) Direct cloning of vascular cell adhesion molecule 1, a cytokine induced endothelial protein that binds to lymphocytes. *Cell* **59**, 1203–1211.
16. Maciag, T., Cerundolo, J., Ilsley, S., Kelley, P. R., and Forand, R. (1979) An endothelial cell growth factor from bovine hypothalamus: identification and partial characterisation. *Proc. Natl. Acad. Sci. USA* **76**, 5674–5678.
17. Maciag, T., Hoover, G. A., Stevenson, M. B., and Weinstein, R. (1984) Factors which stimulate the growth of human umbilical vein endothelial cells *in vitro*, in *Biology of Endothelial Cells* (Jaffe, E. A., ed.), Martinus Nijhoff, Boston, pp. 87–96.

10

Human Thymic Epithelial Cell Cultures

Anne H. M. Galy

1. Introduction

The thymus is a very complex organ that regulates T-cell production. Thymocytes (immature T-cells) constitute by far the largest cellular population in the organ (several billions of thymocytes in a child's thymus), but small numbers of other hematopoietic cells are found in the intrathymic microenvironment, such as thymic monocytes, macrophages, interdigitating dendritic cells, and B-cells. Stromal cells of nonhematopoietic origin comprise thymic epithelial cells (TEC) organized in a network throughout the organ, thymic fibroblasts of the capsule and interlobular septae, and endothelial cells of the thymic vasculature. The proportion of these thymic cellular components varies with age *(1)*. Fat infiltration becomes significant at puberty and increases throughout adulthood. Intrathymic T-cell maturation is supported in part by TEC, which express cell-surface molecules interacting with counterreceptors on the maturing thymocytes *(2–4)*. Importantly, TEC induce positive selection and major histocompatibility complex restriction of T-cells *(5)*. In vitro studies have shown that TEC produce numerous cytokines *(4,6,7)*, which may directly and/or indirectly contribute to T-cell maturation. Reciprocally, TEC functions are affected by interactions with T-cells *(4,8)*. Monolayer cultures of TEC provide in vitro systems to study the biology of TEC *(2–7)*. However, investigators should be aware that monolayer cultures may not be representative of a three-dimensionally structured TEC network, and the function of isolated cells may differ from that of cells in the midst of a complex environment. One method to obtain highly purified monolayer cultures of human TEC is described here. Nontransformed human TEC can be propagated and passaged up to six times before cells become senescent. Careful "budgeting" of the cell stocks by freezing early passages can provide a long-lasting supply of purified TEC strains with determined purity.

From: *Methods in Molecular Medicine: Human Cell Culture Protocols*
Edited by: G. E. Jones Humana Press Inc., Totowa, NJ

1.1. Principles of the Method

The method to isolate and grow TEC relies on cell density, cell-adhesion properties, and modulation of cell growth by appropriate factors (Note 1). Stromal cells need to be released from the tissue by mechanical or enzymatic disruption to establish a TEC culture. A large number of thymocytes are removed by density sedimentation prior to culturing. Stromal cells (mostly thymic epithelial cells and thymic fibroblasts) are separated from the lesser or nonadherent hematopoietic cells (mostly thymocytes, thymic monocytes, and interdigitating dendritic cells) by culture. The cell-culture medium favors TEC expansion, particularly because of its high serum concentration and the presence of epidermal cell growth factor (EGF), a mitogen for TEC *(9,10)*. Hydrocortisone (OHC) and cholera toxin (CT) further improve TEC proliferation, and OHC promotes apoptosis of residual cortical thymocytes (Note 2). Thymic stromal cell cultures tend to be overgrown by highly proliferating thymic fibroblasts, and special treatment of the cultures with EDTA is required to remove fibroblasts *(11)* (Note 3). The purity of TEC cultures can be measured by immunofluorescent staining for intracellular keratin, which is expressed exclusively in epithelial cells, but not in fibroblasts or hematopoietic cells *(12)*.

2. Materials

1. Tissues: Human thymi are obtained ethically in compliance with the appropriate regulations. Two common sources of thymic specimens are fetal thymi collected from electively aborted fetuses and postnatal thymi obtained from patients (generally children) undergoing cardiac surgery during which a portion of the thymus is removed. Thymic specimens are aseptically placed in a tube containing culture medium with antibiotics and transported to the laboratory, preferably within 24 h of collection. Human tissues are biohazardous material and should be handled with precaution. In particular, the investigator should wear a labcoat and gloves at all times, dispose of waste in appropriate containers, and manipulate the tissue in a certified biological safety cabinet (Biosafety Level 2). Hepatitis B vaccination is recommended.
2. RPMI 1640.
3. Phosphate-buffered saline (PBS) without calcium and magnesium.
4. Dimethyl sulfoxide (DMSO).
5. Heat-inactivated (56°C for 30 min) fetal bovine serum (FBS), preferably triple 0.1 μm filtered to ensure that it is mycoplasma-free. Once inactivated, aliquot in 30-mL fractions in 50-mL conical tubes, and store frozen at $-20°C$.
6. Versene solution (phosphate-buffered salt solution containing 0.02% EDTA).
7. Collagenase type IAS (Sigma, St. Louis, MO). Reconstitute in RPMI 1640 at 5 mg/mL (or 2000 U/mL). Aliquot and store at $-20°C$.
8. Deoxyribonuclease (DNase) type II-S (Sigma): Reconstitute in RPMI 1640 at 1 mg/mL and store at $-20°C$.

Thymic Epithelial Cell Cultures

9. Trypsin 10X or 0.5% (Gibco, Grand Island, NY) mycoplasma and parvovirus screened, frozen. Thaw the content of the bottle, and aliquot the 10X trypsin in 10-mL fractions that will be frozen again at –20°C. When needed, an aliquot of 10X trypsin will be thawed and diluted in Versene. Thawed 10X trypsin aliquots can be kept at 4°C for 1 wk.
10. Medium: Prepared by mixing equal volumes of DMEM (with 4.5 g/L glucose, L-glutamine, and 10 mM HEPES) and HAM F12 (with L-glutamine and 25 mM HEPES) supplemented with 15% FBS, 50 U/mL and 50 µg/mL, respectively of penicillin and streptomycin antibiotics (P/S), 1 mM sodium pyruvate, 4 mM additional L-glutamine, 0.4 µg/mL OHC, 12.5 ng/mL sterile, endotoxin-tested tissue-culture-grade EGF from mouse submaxillary glands, and 10 ng/mL cholera toxin (CT). The DMEM, HAMF12, and pyruvate solution can be mixed and stored for months at 4°C. Antibiotics, L-glutamine, FBS, OHC, EGF, and CT are stored in small aliquots at –20°C and added freshly to TEC medium, which is subsequently kept no more than 3 wk at 4°C. A frozen stock of OHC-EGF-CT supplement can be prepared as follows: To 16.5 mL of FBS, add 0.7 mL of an OHC solution at 5 mg/mL in ethanol, add 100 µg of EGF, and add 75 µL of CT solution prepared at 1 mg/mL in endotoxin-free bovine serum albumin (BSA) (BSA at 5 mg/mL in water). Aliquot this sterile supplement in 0.5-mL fractions, and store frozen at –20°C for several months to a year. Use 0.5 mL of supplement for 230 mL of medium (i.e., 200 mL of DMEM-HAMF12 medium with 30 mL of FBS).
11. Eight-well Labtek chamber slide (acetone-resistant Labtek chambers, Nunc, Naperville, IL).
12. Acetone.
13. Tween-80 (Sigma).
14. Mouse monoclonal antibody (MAb) antikeratin KL1 (Amac, Westbrook, ME).
15. Irrelevant mouse IgG1 isotype-matched negative control (MOPC 21, Sigma).
16. Fluorescein-conjugated polyclonal goat antimouse immunoglobulin (GAM-FITC) (TAGO, Burlingame, CA).
17. Coplin jars.
18. Evans Blue 0.5% solution in PBS (Sigma).
19. Fluorescence mounting medium (Dako, Carpinteria, CA).

3. Methods

Unless otherwise indicated, all procedures are performed under sterile conditions (Biosafety level 2). Volumes of medium and times of incubation are given for a medium-sized thymus specimen of approx 1 × 2 × 2 cm.

3.1. Dissection and Digestion of Thymic Tissue

1. With forceps, extract thymus from the transport tube, and place into a 100 × 20-mm Petri dish with 5 mL of cold RPMI 1640 medium containing 50 µg/mL DNase.

2. Carefully peel off and cut away the conjunctive capsule with fine stainless-steel forceps and scissors. Hemorrhagic areas, blood vessels, and nonthymic tissues are also carefully excluded (Note 4).
3. Cut the loosened thymic lobular structure in small cubes (approx 1 × 1 mm or smaller) using 4-in.-long curved microscissors. This takes about 5 min of repeated cutting and is necessary to facilitate the enzymatic dissociation.
4. Add RPMI-DNase if necessary.
5. Pipet the resulting cell mixture into a 50-mL conical tube using RPMI-DNase to transfer cells remaining in the Petri dish.
6. Fill tube up to 45 mL with RPMI-DNase.
7. Decant the cell suspension at 1g for 5 min at room temperature. Fragments should settle to the bottom of the tube.
8. Discard thymocytes remaining in the upper 40 mL.
9. Repeat the procedure at least three times or until the supernatant becomes clearer (complete elimination of thymocytes is not possible) (Note 5).
10. Resuspend the washed fragments in 20 mL of RPMI-DNase.
11. Transfer fragments to a sterilized 100-mL glass bottle containing a magnetic stirring bar.
12. Add collagenase to the final concentration of 200 U/mL (or 500 µg/mL).
13. Place the bottle into a 37°C water bath, and gently stir its contents for 20–30 min. This can be achieved simply by placing the bottle into a water-containing glass beaker placed on a magnetic hot plate. Temperature is carefully equilibrated and controlled with a thermometer.
14. Pipet the released cells (in suspension or in the form of very small aggregates), and transfer to a 50-mL conical tube containing cold RPMI and 50% FBS kept on ice.
15. Add fresh collagenase and DNase to the rest of the tissue, and pursue digestion for another 20 min.
16. Repeat the procedure once. At the end of the third digestion, most of the tissue should be reduced to cell suspension.
17. Collect cells in RPMI + FBS and spin down gently (5 min, 200g at 4°C). Cells in the pellet still contain thymocytes, but this cell preparation can be placed in culture at this point. Further enrichment in stromal cells is achieved by layering the cells over an FBS cushion.
18. Resuspend cells in 20 mL of cold RPMI + DNase, and layer 2 mL of this mixture onto 3 mL of cold FBS in 15-mL conical tubes that are kept undisturbed at 4°C for 40 min.
19. After decantation, carefully discard the top half fraction and collect the bottom fraction. Staining the cells for CD45 antigen expression (a marker of leukocytes that is not expressed on nonhematopoietic stroma [3]) shows significant enrichment in CD45$^-$ stromal cells in the bottom fraction (Galy, unpublished observations).

3.2. Culture and Purification of TEC

1. Resuspend isolated cells in TEC medium and plate into three 75-cm^2 tissue-culture flasks incubated in a humidified atmosphere of 95% air, 5% CO$_2$ at 37°C for

3 d without disturbing the flasks. The resulting culture should be relatively dense and ideally would consist of plating $0.5–1 \times 10^5$ stromal cells/75 cm^2.
2. After 3 d, wash nonadherent cells away by removing the medium, and gently rinsing the flask with 10 mL of RPMI before adding fresh TEC medium. Examination of the flasks under a phase-contrast microscope shows a growing stromal monolayer and probably very small adherent explants crowned by emerging TEC.
3. Eliminate the few remaining thymocytes by further washing and medium change, twice a week.
4. Examine the culture. After about 1–2 wk, large stromal areas are visible in the culture. TEC are recognized by their polygonal morphology and fine intracytoplasmic perinuclear granulations (Fig. 1A). As seen in Fig. 1B, thymic fibroblasts are spindle-shaped cells often oriented similarly to adjacent cells (Note 5). Thymic monocytes and macrophages are generally round cells with a central nucleus and an irregular outer membrane projecting multiple pseudopodes. Thymocytes are small lymphoid cells that appear refractive since they are mostly nonadherent, but can also be bound to stromal cells. Thymic fibroblasts, thymic monocytes, and thymocytes are less adherent to the culture flask than TEC and are therefore preferentially detached on treatment of the cultures with an EDTA solution (Versene).
5. Treat cultures with EDTA as soon as cell growth is well established and areas of thymic fibroblasts develop. The treatment will be repeated as often as needed to obtain a pure TEC culture.
6. To treat the cultures, remove the medium, and gently wash the flasks twice with 5 mL of PBS and 5 mL of Versene.
7. Flush fibroblasts by forcefully projecting 5 mL of Versene (at room temperature) at a 45° angle on the cells with a cotton-plugged glass Pasteur pipet and a hand-operated bulb.
8. Repeat this procedure 5–10 times to flush the entire surface of the flask. Concentrate on areas where fibroblasts were seen growing. Collect and discard the detached cells. Repeat the procedure with another 5 mL of fresh Versene. This drastic procedure will remove a lot of cells, including some TEC. Therefore, the cultures will be significantly depleted. However, areas of highly adherent pure TEC should be spared and will grow back.
9. Wash the flask gently with 5 mL of TEC medium. Discard. Add 15 mL of TEC medium and place the flask back in the incubator.
10. Observe cultures under the microscope for appearance of a confluent TEC monolayer within 2 wk.
11. Repeat the EDTA washing procedure if thymic fibroblasts grow again.

3.3. Passage of TEC Cultures

Confluent monolayers of TEC are passaged with trypsin EDTA.

1. Remove medium from the flask, and wash cells twice with approx 10 mL of Versene gently spread over the entire surface of culture. Make sure all serum-containing medium is removed, since it will inhibit the action of trypsin.

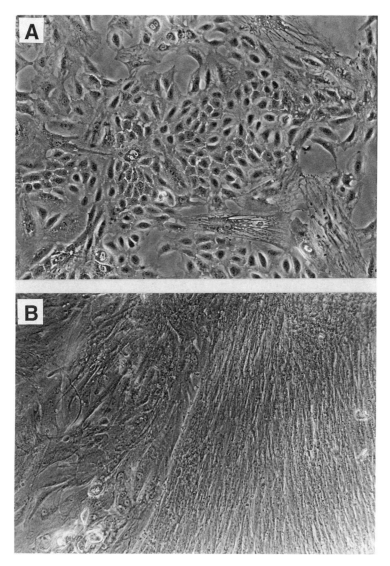

Fig. 1. (A) Phase-contrast photomicrograph (objective 10×) of a culture showing a highly pure TEC area. Note the proliferative cobblestone-shaped small TEC and more senescent large flattened TEC. (B) Phase-contrast photomicrograph (objective 10×) of a culture largely overgrown by thymic fibroblasts that are spindle-shaped cells organized unidirectionally. An area of TEC is seen on the left side of the photograph.

2. Add 10 mL of a 1X trypsin solution (freshly prepared by 1:10 dilution of the 10X stock into Versene) onto the cell monolayer to cover the entire surface for about 30 s.

3. Remove trypsin, and repeat the procedure once. Remove excess trypsin again leaving only a liquid film over the cells.
4. Return flask to 37°C for 5–7 min of incubation, or until all cells loosen from plastic.
5. Bang flask once on its large side in order to detach all cells.
6. Harvest TEC with 10–20 mL of cold TEC medium, and spin cells down (7 min at 400g at 4 °C).
7. Plate the resulting pellet at approx $0.5–1 \times 10^6$ cells/75 cm^2 flask.
8. Carefully record the number of passages, since TEC will lose their proliferative potential after the 5th or 6th passage.

3.4. Freezing TEC Cultures

The supply of TEC can be extended by freezing aliquots of passaged cells and expanding only a fraction of the cultured cells.

1. Detach cells as indicated in Section 3.3.
2. Resuspend TEC at 2×10^6 cells/mL in cold TEC medium.
3. Add slowly an equal volume of cold DMSO solution (20% DMSO in TEC medium) to obtain a final cell suspension of 1×10^6 TEC/mL in 10% DMSO solution.
4. Dispense 1 mL of the cold cell–DMSO mixture in cold cryovials.
5. Freeze in controlled-freeze apparatus or more simply in styrofoam racks placed inside a styrofoam box at –80°C overnight. When frozen, the vials are transferred and stored indefinitely in liquid nitrogen.

3.5. Immunofluorescence Measurement of Keratin-Positive Cells

The proportion of cells expressing intracytoplasmic cytokeratin filaments can be measured by immunostaining and serves to identify TEC (Note 6).

1. Detach cells as indicated in Section 3.3.
2. Plate 2×10^4 TEC in 200 µL of TEC medium/well of an acetone-resistant eight-well Labtek chamber slide. Distribute cells in at least two wells (one for negative isotype control, the other for keratin staining).
3. Place chamber slides in incubator at 37°C for adherence overnight.
4. Remove medium. At this point, the procedure does not need to be carried out under sterile conditions.
5. Wash cells in the Labtek chamber three times with warm (37°C) PBS, which is gently pipeted in and out of the chamber well.
6. Blot excess liquid after the last wash.
7. Add acetone (approx 400 µL) in each well to fix the cells and permeabilize the cellular membranes for 10 min.
8. Remove acetone, and discarded in waste containers for volatile solvents.
9. Air-dry chambers for 30 min, and then store at –20°C for no more than 3 wk.
10. For immunostaining, rehydrate slide with approx 500 µL of PBS/well for 30 min.

11. Discard PBS and add 250 µL of the antikeratin antibody or of the negative control each at the concentration of 20 µg/mL in PBS with 0.04% tween for 40-min incubation in a humidified chamber at room temperature.
12. Discard antibodies, and detach the upper part of the chamber.
13. Wash the slide in a coplin jar filled with PBS-tween.
14. Repeat procedure twice for 5-min-long washes each time.
15. Incubate slides in the dark for 40 min with approx 250 µL/well of the secondary goat antimouse FITC reagent diluted 1:40 in PBS-tween, and carefully distributed over the entire surface of the chamber.
16. Wash slides twice as described in step 14, and a third time in Evans Blue counterstaining solution (0.01% solution in PBS-tween). Protect slides from light.
17. Blot excess liquid, and mount slides with fluorescence mounting medium.
18. Examine slides under epifluorescence microscopy.

4. Notes

1. There seems to be ontogeny-related differences in the proliferative potential of thymic stromal cells. Fetal thymi or young postnatal specimen (less than a year old) will grow better than older thymi.
2. An improved medium formulation can be obtained by adding 1% of serum supplement LPSR-1 (Sigma) to TEC medium. However, this product is no longer available for import in the United States. A bovine pituitary extract (Collaborative Research, Lexington, MA) is mitogenic for TEC at 50 µg/mL.
3. Thymic fibroblasts are thought to contribute significantly to early T-cell development *(13)*. Thymic fibroblast cultures can be easily obtained after digestion of the capsule and culture in TEC medium. The nature of the culture can be assessed by its keratin negativity and lack of CD45 expression.
4. Careful initial dissection of the thymic capsule is very important to reduce the chance of overgrowth by thymic fibroblasts.
5. Fetal thymi can be placed in culture after dissection into very small fragments and thymocyte wash without enzymatic digestion. In that case, explants will attach and TEC will grow out of the explants after 3–5 d, and the monolayer will extend rapidly to cover the surface of the dish by d 8 (plate one fetal thymus of 1.5 × 1 × 1 cm into one 75-cm² flask). Subsequently, the culture will be treated as described in Section 3.2., steps 5–11 to remove fibroblasts if needed.
6. It is possible to analyze the intracellular keratin content of cultured TEC by flow cytometry. Trypsinized cell suspensions are washed in PBS, and the loosened pellet is fixed in 1 mL of cold methanol for 10 min. About 10 mL of PBS are added to dilute the methanol, and cells are spun down and washed in PBS twice. Washed cells are immunostained with an antikeratin antibody (CAM-5.2) specific for keratin 8,18 (Becton Dickinson, Mountain View, CA), which gives better results on the fluorescence-activated cell scanner (FACscan) than KL1. The antibody is diluted at 10 µg/mL in PBS with 0.2% BSA. Cells are then washed and incubated with the secondary antibody (GAM-FITC 1:40 in PBS-BSA) and washed twice. Cells are analyzed on the FACscan, adjusting setting for large cells.

References

1. Nakahama, M., Mohri, N., Mori, S., Shindo, G., Yokoi, Y., and Machinami, R. (1990) Immunohistochemical and histometrical studies of the human thymus with special emphasis on age-related changes in medullary epithelial and dendritic cells. *Virchows Arch. B Cell Pathol.* **58,** 245–251.
2. Singer, K. H., Denning, S. M., Whichard, L. P., and Haynes, B. F. (1990) Thymocyte LFA-1 and thymic epithelial cell ICAM-1 molecules mediate binding of activated human thymocytes to thymic epithelial cells. *J. Immunol.* **144,** 2931.
3. Galy, A. H. M. and Spits, H. (1992) CD40 is functionally expressed on human thymic epithelial cells. *J. Immunol.* **149,** 775–782.
4. Galy, A. H. M. and Spits, H. (1991) IL-1, IL-4 and IFN-γ differentially regulate cytokine production and cell surface molecule expression in cultured human thymic epithelial cells. *J. Immunol.* **147,** 3823–3830.
5. Martin Fonchecka, A., Schuurman, H. J., and Zapata, A. (1994) Role of thymic stromal cells in thymocyte education: a comparative analysis of different models. *Thymus* **22,** 201–213.
6. Galy, A. H. M., de Waal Malefyt, R., Barcena, A., Mohan-Peterson, S., and Spits, H. (1993) Untransfected and SV40-transfected fetal and postnatal human thymic stromal cells: analysis of phenotype, cytokine gene expression and cytokine production. *Thymus* **22,** 13–33.
7. Galy, A. H. M., Spits, H., and Hamilton, J. A. (1993) Regulation of M-CSF production by cultured human thymic epithelial cells. *Lymphokine Cytokine Res.* **12,** 265–270.
8. Surh, C. D., Ernst, B., and Sprent, J. (1992) Growth of epithelial cells in the thymic medulla is under the control of mature T cells. *J. Exp. Med.* **176,** 611–616.
9. Sun, L. Serrero, G., Piltch, A., and Hayashi, J. (1987) EGF receptors on TEA3A1 endocrine thymic epithelial cells. *Biochem. Biophys. Res. Commun.* **148,** 603–608.
10. Hadden, J. W., Galy, A., Chen, H., and Hadden, E. M. (1989) A pituitary factor induces thymic epithelial cell proliferation in vitro. *Brain Behav. Immun.* **3,** 149–159.
11. Singer, K. H., Harden, E. A., Robertson, A. L., Lobach, D. F., and Haynes, B. F. (1985) In vitro growth and phenotypic characterization of mesodermal-derived and epithelial components of normal and abnormal thymus. *Hum. Immunol.* **13,** 161–176.
12. Moll, R., Franke, W. W., Schiller, D. L., Geiger, B., and Krepler R. (1982) The catalog of human cytokeratins: patterns of expression in normal epithelia, tumors and cultured cells. *Cell* **31,** 11.
13. Anderson, G., Jenkinson, E. J., Moore, N. C., and Owen, J. J. T. (1993) MHC class II positive epithelium and mesenchyme cells are both required for T-cell development in the thymus. *Nature* **362,** 70–73.

11

Generation and Cloning of Antigen-Specific Human T-Cells

Hans Yssel

1. Introduction

The ability to grow antigen-specific human T-cell clones in vitro has been instrumental in understanding T-cell function. A major breakthrough in T-cell culture in vitro was the discovery of the T-cell growth-inducing properties of interleukin-2 (IL-2), originally called T-cell growth factor or TCGF, by Morgan et al. *(1)*, who for the first time were able to grow human T-lymphocytes, isolated from human bone marrow. Shortly thereafter, Gillis and Smith *(2)* reported the cloning and long-term culture of mouse cytotoxic T-cells, using TCGF. In spite of the success of growing human T-cells in vitro, the cloning of these cells, however, turned out to be more difficult: in contrast to mouse cell lines that could be maintained in culture in the presence of TCGF-containing supernatant only, long-term cultures of cloned human allo-antigen-specific T-cell lines *(3,4)* needed repetitive stimulation with specific allo-antigen for their growth *(3)*. This led to the general assumption that antigen-specific IL-2-dependent T-cell lines and T-cell clones would lose their antigen responsiveness when propagated with IL-2 in the absence of specific antigen.

In view of the presumed requirements for antigenic stimulation, a culture system was devised for the generation and expansion of stable allo-antigen-specific cloned human T-cell lines. At the beginning of each culture, T-cells are stimulated with a feeder cell mixture, consisting of irradiated peripheral blood mononuclear cells (PBMC), an Epstein–Barr virus-transformed lymphoblastoid cell line (EBV-LCL), expressing the specific allo-antigen, and phytohemagglutinin (PHA). The use of this feeder cell mixture was found to improve cloning efficiencies and growth rates of allo-antigen-specific T-cell clones dramatically *(5–7)*. Between repetitive restimulations with feeder cells, T-cell clones were expanded in medium supplemented with TCGF and later, when it became available, recombinant (r)IL-2. Interestingly, it was found that T-cell clones, specific for various other antigens, could be cultured and propagated in this feeder cell mixture as well, in the absence of specific antigen. Using this polyclonal stimulation protocol with feeder cells, it was possible to generate successfully stable T-cell clones specific for tumor antigens *(8,9)*, and recall antigens, such as tetanus toxoid *(10)*, bacterial antigens *(11,12)*, and allergens *(13)* successfully, as well as T-cell receptor (TCR) γδ$^+$ T-cell clones *(14)*. Therefore, in our culture system, the presence of antigen does not seem to be a prerequisite to maintain antigen specificity and growth properties of antigen-specific T-cells, and in this respect, our methodology to grow antigen-specific T-cell clones differs from methods reported by others who favor the presence of specific antigen *(15,16)*.

The mechanism by which the feeder cell mixture exerts its growth-promoting effects is not clear. It is noteworthy to mention that the addition of rIL-2 alone only induces a transient activation of the T-cells, indicating that this growth factor is not able to promote long-term growth and proliferation by itself. This is most likely because of the inability of IL-2 to induce a sustained expression of a high-affinity IL-2R. Apparently, PHA can replace the requirement of antigen-mediated triggering of the TCR for T-cell activation by inducing the expression of the IL-2Rα chain (CD25) on the T-cells. PBMC, as well as EBV-LCL, are likely to function as accessory cells, the presence of which is needed for this mitogen to be effective. In addition, PBMC will produce growth and costimulatory factors, following activation with PHA or EBV-LCL, which will further enhance the proliferation of the T-cell clones. Each stimulation of T-cell clones with feeder cells is followed by a growth factor-dependent expansion phase, during which the growing T-cells are expanded with exogenous IL-2 or other growth factors. Since this culture method is effective in expanding T-cell clones of various origins and specificities, one does not have to worry about the availability of the relevant antigen-presenting cells (APC).

Generally, foreign antigens are processed by APC and presented on their cell surface as small linear protein fragments, where they are recognized by the TCR in association with autologous major histocompatibility complex (MHC) molecules to which they are bound. Whereas autologous MHC molecules function as a restriction element in the recognition of antigen, foreign (mismatched) MHC itself can function as allo-antigen for T-cells as well. However, there are no differences in the way the TCR and additional accessory molecules on the T-cells, such as CD4 and CD8, interact with MHC molecules, whether or not they carry foreign antigen: CD4$^+$ T-cells recognize antigen in the context of an MHC class II molecule, whereas the recognition of antigen by CD8$^+$ T-cells is restricted by MHC class I molecules. In the same vein, allospecific CD4$^+$ and CD8$^+$ T-cells directly recognize MHC Class II and class I molecules, respectively.

In the following sections, the generation of allo-antigen- and (soluble) antigen-specific T-cell clones are described. There are no differences in methodology for the generation of each type of cell. As mentioned, depending on the interaction with, or recognition of MHC class II or class I antigens, allo-antigen-specific cells are CD4$^+$ or CD8$^+$, respectively. Antigen-specific T-cell clones can be identified based on their capacity to either kill specific target cells, to proliferate, or to secrete cytokines, following antigen-specific stimulation. Once antigen-specific T-cell clones have been selected they can be cultured for extended periods of time, while maintaining their antigen-specific properties, using the polyclonal stimulation protocol, described in this chapter (*see also* Fig. 1).

2. Materials

2.1. Culture Medium

All cell cultures are grown in Yssel's medium *(17)*, which was originally described as a serum-free medium and is based on a modification of Iscove's Modified Dulbecco's Medium (IMDM, containing L-glutamine, 25 mM HEPES, but no α-thiogycerol, Gibco-BRL, Grand Island, NY; cat. no. 21056-015) (*see* Note 1). To prepare Yssel's medium, dissolve freshly in the following order in IMDM:

1. Bovine serum albumin (BSA) (Sigma, St Louis, MO; A2153), at a final concentration of 0.25% (w/v).
2. 2-Amino ethanol (Sigma; E0135), at a final concentration of 1.8 µg/L.
3. Transferrin (Holo form; Boehringer, Indianapolis, IN; cat. no. 1317 423), at a final concentration of 40 mg/L. An excellent source is 30% saturated transferrin (652-202) in liquid form from Boehringer, to be diluted according to the manufacturer's instructions (*see* Note 2).
4. Insulin (Sigma; I5500) at a final concentration of 5 mg/L (make stock solution of 1 mg/mL in 0.01M HCl and add to medium).
5. Linoleic acid (Sigma; L1376) and oleic acid (Sigma; O3879), both at a final concentration of 2 mg/L. Make stock solution in ethanol (20 mg/mL) and add to medium (*see* Note 3).
6. Palmitic acid (Sigma; P5917), at a final concentration of 2 mg/L (stock solution [20 mg/mL] in ethanol can be stored at 4°C for mo).
7. Penicillin/streptomycin (Gibco-BRL; cat. no. 59-60277P).

Filtrate through 0.2-µm filter, aliquot, and store at –20 or –80°C.

2.2. Cells and Reagents

1. PBMC, isolated from the peripheral blood of healthy volunteer donors by centrifugation over fycoll hypaque *(18)*, to be used as feeder cells for the culture of T-cell clones. Keep cells at 4°C on ice prior to use.

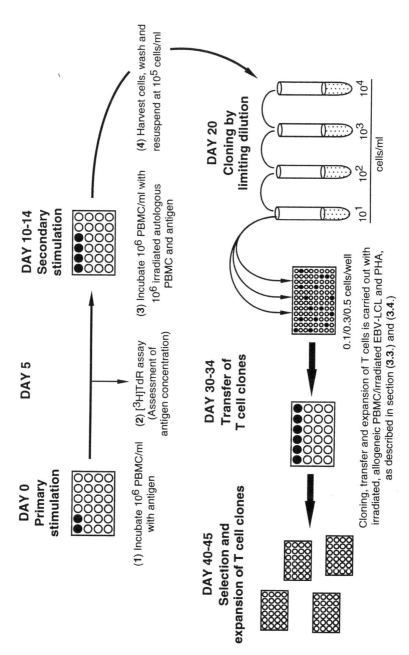

Fig. 1. Generation and cloning of antigen-specific T-cells.

2. Peripheral blood or tissue samples from healthy donors or patients to be used for the generation of T-cell lines.
3. EBV-LCL, generated by infection of B-cells with EBV (*see* Section 3.1.).
4. EBV-producing marmoset cell line B98.5. (EBV-containing supernatant is produced by growing B98.5 cells until medium is completely exhausted. After centrifugation for 10 min at 2000 rpm to remove residual B98.5 cells and filtration through a 45-µm filter, virus-containing supernatant can be stored at –80°C for years.)
5. Antigen (*see* Note 4).
6. PHA (Wellcome, Beckenham, UK; HA-16 or HA-17).
7. Phosphate-buffered saline (PBS), supplemented with 1% human serum (HS).
8. Tissue-culture plates (24-well flat-bottom and 96-well round- and flat-bottom Linbro plates; Flow Lab, McLean, VA).
9. Recombinant IL-2 (IL-4, IL-7, IL-12, IL-15).
10. Cyclosporin-A® (Sandoz, East Hanover, NJ).
11. Phorbol ester (PMA or TPA, Calbiochem, La Jolla, CA; cat. no. 524400).
12. Calcium ionophore (Calbiochem, cat. no. 100105).
13. Anti-CD3 and CD28 monoclonal antibodies (MAbs).
14. Neutralizing anti-HLA-DR, DQ, and DP MAbs.
15. Glutaraldehyde grade I (Sigma; G5882).
16. Dimethyl sulfoxide (DMSO) (Sigma; D5879).
17. [^{3}H] TdR (Amersham, Arlington Heights, IL).
18. ^{51}Cr (Amersham).

3. Methods
3.1. EBV-Transformed B-Cell Lines

1. Isolate PBMC by centrifugation over ficoll hypaque.
2. Incubate 2×10^6–10^7 PBMC/mL in a 15-mL centrifuge tube for 2 h at 37°C in 1 or 2 mL of EBV-containing culture supernatant.
3. Wash the cells twice with PBS/1% HS, and incubate in culture medium at a concentration of 10^5 cells/mL in the presence of 1 µg/mL Cyclosporin® in a 96-well tissue-culture plate at a final volume of 100 µL.
4. Add 100 µL of fresh culture medium after 5 d.
5. After 10–14 d, growing cultures of blastoid cells can be observed, which can be split if necessary and transferred to 24-well plates or tissue-culture flasks. Cells should be maintained at concentrations between 2×10^5 and 5×10^5 cells/mL, and cultures should be split with fresh culture medium when they reach concentrations >10^6 to 2×10^6 cell/mL.

3.2. Generation of Allo-Antigen-Specific T-Cells

1. Stimulate 10^6 PBMC with 2×10^5 irradiated (5000 rad) HLA-mismatched EBV-LCL in a total volume of 1 mL in a 24-well Linbro plate (*see* Note 5).
2. Culture the cells for 10 d at 37°C, 5% CO_2 without addition of medium or growth factors.

3. To evaluate the proliferative activity of the responding T-cells, transfer 100 µL of cells, in duplicate or triplicate, into a 96-well plate at d 5 of culture, and pulse with 1 µCi (=37 kBq) [^{3}H]TdR for 4 h. Harvest cells onto a glass fiber filter, and determine amount of incorporated [^{3}H]TdR by liquid scintillation counting.
4. At d 10 of culture, collect the T-cells, wash once with medium, and restimulate 10^6 T-cells with 2×10^5 irradiated (5000 rad) EBV-LCL, used in the first stimulation, in a final volume of 1 mL in a 24-well tissue-culture plate.
5. Clone the cells by limiting dilution between d 5 and 7 after restimulation (*see* Sections 3.4. and 4.2. and Note 6).

3.3. Generation of (Soluble) Antigen-Specific T-Cells

1. Incubate freshly isolated PBMC from immunized donors or patients at a concentration of 10^6 cells/mL in a 24-well Linbro plate, and add soluble antigen at different concentrations (i.e., 0.1, 1, and 10 µg/mL) (*see* Notes 7 and 8).
2. Culture for 10 d at 37°C, 5% CO_2 without addition of medium or growth factors.
3. Evaluate the proliferative activity of the responding T-cells by transferring 100 mL of cells into a 96-well plate at d 5 of culture and pulsing with 1 µCi (=37 kBq) [^{3}H]TdR for 4 h. Harvest cells onto a glass fiber filter, and determine amount of incorporated [^{3}H]TdR by liquid scintillation counting.
4. Collect the T-cells at d 10 of culture, wash once with medium, and restimulate 10^6 T-cells with an optimal (based on the proliferation results of 3.2.3) concentration of antigen, in the presence of 10^6 irradiated (4000 rad) autologous PBMC, in a final volume of 1 mL in a 24-well Linbro plate (*see* Notes 9 and 10).
5. Clone the cells by limiting dilution between d 5 and 7 after restimulation (*see* Sections 3.4. and 4.2. and Note 6).

3.4. Cloning by Limiting Dilution

1. Prepare a cloning feeder cell mixture by adding together in a 50-mL centrifuge tube 5×10^5 cells/mL of irradiated (4000 rad) PBMC from any healthy donor, 5×10^4 cells/mL of an irradiated (5000 rad) EBV-LCL, and 50 ng/mL PHA.
2. Bring the cloning feeder cell mixture at 37°C, and use immediately.
3. Collect T-cells (obtained from Sections 3.2. or 3.3.), wash once, and resuspend in culture medium at a concentration of 10^5 cells/mL.
4. Make serial dilutions of 10^4, 10^3, 10^2, and 10 T-cells/mL with the feeder cell mixture as diluent.
5. Starting from 10 cells/mL, make a series of higher dilutions, i.e., 5, 3, and 1 cells/mL.
6. Transfer 100 µL of the latter cell suspensions into 96-well round-bottom plates. The number of wells to be filled depends on the expected cloning frequencies (*see* Note 11).
7. Add 100 µL of culture medium containing rIL-2 (20 U/mL) after 5 d of culture.
8. After identification of growing T-cell clones, usually after 10–14 d of culture, transfer the cells in 200 µL to a 24-well Linbro plate, add 300 µL of culture medium, and add 0.5 mL of a 2X feeder cell mixture to expand the cells (*see* Section 3.6.).

3.5. Selection of Allo-Antigen-Specific T-Cell Lines

3.5.1. Assay for Cytotoxic Activity

1. Wash 10^6 EBV-LCL to be used as target cells in the assay with PBS, and remove supernatant.
2. Incubate pellet with 100 µCi ^{51}Cr in a water bath at 37°C for 45 min.
3. Wash the cells three times with PBS, and resuspend in culture medium.
4. Seed 10^3 to 2×10^3 target cells/well of a 96-well round-bottom plate in a volume of 100 µL.
5. Add 100 µL of effector T-cells in triplicate wells at different effector/target cell ratios.
6. Keep separate wells with target cells that have been incubated with medium alone (spontaneous ^{51}Cr release) or with 1% Triton-X100 (maximal ^{51}Cr release).
7. Centrifuge the plates for 5 min at 800 rpm.
8. Incubate for 4–8 h at 37°C, 5% CO_2.
9. Harvest 100 µL of culture supernatant from each well, and count amount of released ^{51}Cr in a γ-counter (Packard Instrument Corp., Downers Grove, IL).
10. Calculate the percentage of specific ^{51}Cr release with the formula:

$$(\text{Release of sample} - \text{spontaneous release})/(\text{maximal release} - \text{spontaneous release}) \times 100\% \quad (1)$$

and select responding T-cell clones

11. HLA-restriction elements can be determined by carrying out the cytotoxicity assay in the presence of anti-HLA-DR, DQ, or DP MAb, and/or by using various HLA-typed EBV-LCL as target cells.

3.5.2. Assay for Antigen-Specific Proliferation

1. When using autologous EBV-LCL as APC, incubate 2×10^4 T-cell with 4×10^4 irradiated (5000 rad) EBV-LCL, in a 96-well round-bottom plate in a final volume of 200 µL in the presence or absence of antigen. When autologous PBMC are used as APC, incubate 10^5 T-cells with 10^5 irradiated (4000 rad) PBMC, in a 96-well flat-bottom plate, in the presence or absence of antigen in a final volume of 200 µL. Optimal concentration of antigen has been determined in Section 3.3., step 3.
2. Incubate for 72 h at 37°C, pulse with 1 µCi (=37 kBq) [^{3}H]TdR, and incubate for another 4 h.
3. Harvest cells onto a glass fiber filter, determine amount of incorporated [^{3}H]TdR by liquid scintillation counting, and select responding T-cell clones.
4. HLA-restriction elements can be determined by carrying out the proliferation assay in the presence of anti-HLA-DR, DQ, or DP MAb, and/or by using various HLA-typed EBV-LCL as APC.

3.5.3. Stimulation of T-Cell Clones for Antigen-Specific Cytokine Production Assay

1. Incubate 10^6 T-cells/mL with 2×10^6 autologous EBV-LCL or 10^6 autologous PBMC as APC, in the presence or absence of antigen in either a 24 (final volume 1 mL), 48- (500 µL), or 96-well (200 µL) plate.

2. For polyclonal stimulation, use combinations of PMA (1 ng/mL) and A23187 (500 ng/mL); PMA and soluble anti-CD3 MAb (1 µg/mL); anti-CD28 MAb (1 µg/mL) and coated anti-CD3 MAb (to coat plates with anti-CD3 MAb, incubate a 96-well flat-bottom plate with 10 µg/mL anti-CD3 MAb diluted in PBS overnight at 4°C, and wash twice with 100 µL of culture medium).
3. Culture the cells at 37°C, 5% CO_2 for 48–72 h, and harvest the supernatants, which can be frozen once, prior to cytokine analysis.
4. Determine cytokine production levels by cytokine-specific ELISA (*see* Note 12).

When EBV-LCL and peptides are used, in Sections 3.5.2. and 3.5.3., EBV-LCL can be fixed with glutaraldehyde.

1. Preincubate EBV-LCL in culture medium with 1–5 µg/mL peptide for 2–24 h.
2. Wash the cells once in PBS, and resuspend between 10^6 and 10^7 cells in 1 mL PBS.
3. Add 1 mL of 0.05% stock glutaraldehyde, and incubate for 30 s at room temperature.
4. Add 2 mL of PBS, and incubate for an additional 10 min at room temperature.
5. Wash twice with culture medium, and use in the assays, described in Sections 3.5.2. and 3.5.3.

3.6. Expansion of Cloned T-Cell Lines

1. Prepare a 2X concentrated feeder cell mixture, consisting of 2×10^6 cells/mL irradiated (4000 rad) PBMC from any healthy donor, 2×10^5 cells/mL irradiated (5000 rad) EBV-LCL, and 100 ng/mL PHA, and keep at 4°C until use (*see* Notes 13–15).
2. Collect the T-cell clones, wash with medium or with PBS, supplemented with 2% HS or BSA, and bring the cells to a concentration of 4×10^5 cells/mL.
3. Transfer 0.5 mL of the T-cell clone suspension into a 24-well plate, add 0.5 mL of the 2X concentrated feeder cell mixture, and incubate at 37°C and 5% CO_2 (final concentrations of feeder cells per well are 10^6 PBMC and 10^5 EBV-LCL, and concentration of PHA is 50 ng).
4. T-cell clones should be split with fresh culture medium, supplemented with growth factors (rIL-2 20 U/mL ± rIL-4 10 ng/mL), usually between d 3 and 4 after restimulation with the feeder cells (*see* Notes 16–18).
5. Continue to expand the cultures by adding medium and growth factor(s) over the next 7–10 d.
6. When the T-cells become smaller and round off, usually between d 10 and 14 of culture, repeat the culture cycle starting with step 1 (*see also* Fig. 2 and Notes 19 and 20).

3.7. Cryopreservation of T-Cell Clones

3.7.1. Cell Freezing

1. Centrifuge cells in a 15-mL centrifuge tube, and resuspend the pellet in RPMI-1640, supplemented with 10% HS, fetal calf serum (FCS) or 1% BSA, at concentrations between 10^6 and 10^8 cells/ml and put on ice (*see* Note 21).

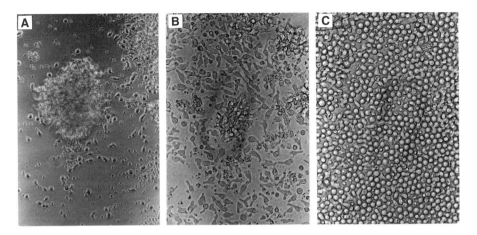

Fig. 2. Different stages in the culture of cloned T-cell lines. (**A**) Culture aspect 24 h after addition of the feeder cell mixture. Mostly, PBMC and EBV-LCL can be seen. (**B**) Growing T-cells between 3 and 6 d after restimulation. The cells are growing in aggregates and have a blastoid appearance. (**C**) T-cells that have been cultured for 10–14 d in the presence of growth factors. The cells have stopped proliferating and have become small and round.

2. Make a fresh solution of 20% DMSO in RPMI-1640, supplemented with HS, FCS, or BSA, and put on ice (add DMSO to medium and not vice versa) (*see* Note 22).
3. Add dropwise the same volume of the DMSO solution to the cell suspension while gently shaking for 2–3 min.
4. Transfer the cells to the freezing vials, and freeze using a temperature-controlled freezing apparatus.

3.7.2. Cell Thawing

1. Thaw vial quickly in 37°C water bath, transfer the cells to 15-mL tube, and put on ice.
2. Add dropwise 2 mL of cold PBS, at first very slowly, while gently shaking, over a time period of about 2–3 min.
3. Using a 2-mL pipet, put 2 mL of HS or FCS under the cell suspension.
4. Centrifuge cells for 5 min at 1000 rpm and remove supernatant.
5. Wash cells once with culture medium.

4. Notes
4.1. Culture Medium

1. Yssel's medium is suitable for the culture of T-cell clones in the absence of added HS or FCS. The addition of 1% human AB[+] serum will, however, optimize

growth conditions. The BSA preparation listed in Section 2.1. contains a significant amount of endotoxin, which, however, does not affect T-cell growth. If endotoxin-free conditions are required for subsequent experiments with these cells, the preparation can be replaced with endotoxin-free BSA (Sigma A3675). Yssel's medium, supplemented with 1% human AB$^+$ serum, can be purchased from Gemini Bioproducts Inc. (Calabasas, CA; cat. no. 400-113).
2. Both the Holo form and the Apo form of transferrin can be used. However, individual batches may have inhibitory, rather than growth-promoting, effects on T-cell growth and should therefore be tested prior to use. Furthermore, transferrin preparations may contain human Ig in detectable amounts, which may cause background problems in antibody production assays, notably those for IgE determinations. For such experiments, replace the transferrin source with Ig-free transferrin (Pierce, Rockford, IL; cat. no. 31152).
3. Stock linoleic and oleic acid should be stored, as free acids, at $-20°$ under nitrogen to prevent oxidation of the unsaturated bond.

4.2. Generation, Cloning, and Selection of Antigen-Specific T-Cell Lines

4. Commonly used antigens are tetanus toxoid (Lederle, Pearl River, NY), purified protein derivative (PPD) (Staten Serum Institute, Copenhagen, Denmark), house dust mite (*Dermatophagoides* spp), and grass pollen (*Lolium* spp) proteins (ALK, Copenhagen, Denmark). Most genes coding for these protein have been cloned, and recombinant or synthetic peptides are available.
5. For the generation of allo-antigen-specific T-cells, stimulate PBMC with EBV-LCL at concentrations of about 25%, since higher concentrations will likely result in the generation of T-cells with nonspecific, NK-like activity *(3)*.
6. It is recommended to include some lower dilutions of T-cells (10, 10^2, and 10^3 cells/well) in cloning experiments. The growth of these cultures will give an early indication that culture conditions for the cloning experiment were optimal.
7. For the generation of CD4$^+$ allo-antigen-specific cells, use HLA class II-mismatched EBV-LCL, whereas for the generation of CD8$^+$ allo-antigen-specific cells, HLA class I-mismatched EBV-LCL should be used. The protocol, described in Section 3.3., has been used successfully for the generation of both types of allo-antigen-specific T-cell clones *(3,5,19,20)*.
8. For the generation of antigen-specific Th0 cells, PBMC from donors who have been previously immunized with tetanus toxoid can be used successfully. Ideally, donors should have received a booster injection about 1 mo, but not earlier, prior to the use of their PBMC. Th1 cells, reactive with PPD, can be generated using PBMC from donors who have previously been vaccinated with BCG or from those who have a positive Mantoux reaction. Human Th2 cells can be generated from the peripheral blood of atopic donors by stimulation with soluble allergens, such as house dust mite allergen *Dermatophagoides* spp or grass pollen *Lolium perenne*. Suitable patients can be selected, based on their allergen-specific serum IgE levels and RAST scores. Although there is not always a

positive correlation between RAST score and success rate in the generation of allergen-specific Th2 clones, donors with a RAST of +++ or more should be chosen preferentially. Since the frequency of allergen-specific T-cell clones in peripheral blood is rather low *(21)*, better results can be obtained using skin biopsies from patients with atopic dermatitis.

9. Although EBV-LCL are efficient APC when peptides are used, their capacity to present native antigens, such as tetanus toxoid or PPD, is often limited. Therefore, preferentially autologous or HLA-matched PBMC should be used in proliferation or cytokine production assays with native antigen.
10. EBV-LCL function better as APC at higher R/S ratios (3:1 or 5:1). However, high concentrations of EBV-LCL give high background readings in proliferation assays. This problem is avoided by using glutaraldehyde-fixed cells.
11. Cloning efficiencies can be calculated most reliably from serial dilutions. For the calculation of cloning frequencies, plot the fraction of negative wells (y-axis) against the number of seeded wells (x-axis). The cell dose containing one clonogenic cell is determined by locating on the x-axis the dose that corresponds to 0.37 cells/culture.

    Alternatively, cloning efficiencies can be calculated from one single dilution with the formula:

$$-\ln[(\text{number of negative wells/number of total wells})/ \text{number of cells seeded/well}] \times 100\% \qquad (2)$$

12. Detailed protocols for double-sandwich cytokine-specific ELISA have been described in refs. *22* and *23*. Most ELISAs can be purchased commercially. A list of MAbs and polyclonal antibodies available for the detection of cytokines can be found in ref. *23*.

4.3. T-Cell Culture

13. Since culturing of T-cell clones requires large amounts of PBMC from different donors, donor serum should be tested for the presence of antibodies against hepatitis B, CMV, and HIV antigen prior to processing of the cells, whereas all cell cultures containing human PBMC should be handled according to the biosafety guidelines of the NIH *(24)*. Freshly isolated, nonirradiated, PBMC can be stored in culture medium at 4°C for at least 2 d prior to use.
14. To remove the abundance of platelets present in peripheral blood after isolation of the PBMC, wash at least four times with PBS, supplemented with 2% HS or BSA, by centrifugation at 800 rpm for 10 min in 50-mL centrifuge tube.
15. After irradiation of cells with a cesium source, wash the cells at least once with culture medium to remove oxygen (O^-) radicals that may have formed owing to the irradiation.
16. Using the conditions described in Section 3.6., it is generally not necessary to add any medium and growth factors to the T-cell cultures for the first 3–4 d, unless higher numbers of T-cell clones are present in the feeder cell mixture. For optimal growth, when culturing the T-cell clones in tissue-culture plates or flasks,

it is important to keep growing T-cells in log phase by allowing the cells to grow at rather high density (10^6–2×10^6 cells/mL) and by frequently splitting the cultures with fresh culture medium, supplemented with growth factors.

17. As the primary growth factor for the expansion of T-cell clones, rIL-2 should be used, whereas rIL-4 might be added when Th2 cell clones are grown. IL-12 has been found to have T-cell growth-promoting effects. However, it strongly induces the production of IFN-γ in peripheral blood *(25)* and even in established Th2 clones *(26)*. Although it has yet been impossible to evaluate the amount of IL-12 in the culture system described in this chapter, the use of PBMC and EBV-LCL, which have been described to be producers of IL-12, may bias cultured human Th2 clones toward the Th0 phenotype and may explain the frequently observed intrinsic capacity of many of Th2 clones to produce IFN-γ.

18. Proliferating Th0 and Th1 cell clones will usually remove dead feeder cells and cell debris within the first 4 d of culture (*see* Fig. 2). Although the mechanism for these scavenger properties is unknown, it may be related to the cytotoxic potential of these cells. Indeed, Th2 cells, which have no lytic activity, grow less well under our culture conditions, and often much debris remains in the wells, which seems to have inhibitory effects on the growth of the cells. It is recommended to collect the cells and wash them several times with medium or PBS supplemented with 2% HS or BSA by centrifugation at 800 rpm for 10 min, after which they should be transferred to new wells at a concentration of 5×10^5 cells, in culture medium containing growth factor(s). This procedure can be carried out repeatedly.

19. Keep track of the age of the T-cell clones, and make frozen stocks of the cells, especially at an early stage. The cells generated just after cloning are the most valuable and should be kept as a backup storage. Although some long-term cultured T-cells seem to surpass the 30–50 cell divisions, known as the Hayflick limit, in the author's experience, the growth of the T-cell clones usually slows down after about 15–20 stimulations with the feeder cell mixture, and it becomes more and more difficult to expand the cells.

20. Cells contaminated with mycoplasma show decreased growth and altered functional capacities. Check cell cultures, especially those that are frequently handled and maintained in culture for extended periods of time, for the presence of mycoplasma at least every 14 d. Cells that are contaminated with mycoplasma should be discarded immediately. The success of T-cell cloning and culture is greatly affected by the presence of mycoplasma.

21. T-cell clones can be frozen at any time and, after thawing, can be expanded immediately with the feeder cell mixture. Cells frozen between d 4 and 6 after restimulation with the feeder cells, which are at that time still highly proliferative, can be cultured for several days in medium with growth factors on thawing. Cells frozen at later stages should be cultured with feeder cells prior to expansion with growth factors.

22. Do not use IMDM, Yssel's medium, or any other medium containing HEPES when freezing T-cells, since the permeabilization of the cell membrane by DMSO results in the penetration of HEPES into the cells, which may lead to cell death.

Acknowledgment

DNAX Research Institute for Molecular and Cellular Biology is supported by Schering-Plough Corporation.

References

1. Morgan, D. A., Ruscetti, F. W., and Gallo, R. (1976) Selective in vitro growth of T lymphocytes from normal human bone marrows. *Science* **193,** 1007,1008.
2. Gillis, S. and Smith, K. A. (1977) Long term culture of tumour-specific cytotoxic T cells. *Nature* **268,** 154–156.
3. Spits, H., de Vries, J. E., and Terhorst, C. (1981) A permanent human cytotoxic T-cell line with high killing capacity against a lymphoblastoid B cell line shows preference for HLA A, B target antigens and lacks spontaneous cytotoxic activity. *Cell. Immunol.* **59,** 435–441.
4. Krensky, A. M., Reiss, C. S., Mier, J. W., Strominger, J. L., and Burakoff, S. J. (1982) Long-term cytolytic T-cell lines allospecific for HLA-DRw6 antigen are OKT-4$^+$. *Proc. Natl. Acad. Sci. USA* **79,** 2365–2369.
5. Spits, H., Yssel, H., Terhorst, C., and de Vries, J. E. (1982) Establishment of human T lymphocyte clones highly cytotoxic for an EBV transformed B cell line in serum-free medium: isolation of clones that differ in phenotype and specificity. *J. Immunol.* **128,** 95–99.
6. Malissen, B., Kristensen, T., Goridis, C., Madsen, M., and Mawas, C. (1981) Clones of human cytotoxic T lymphocytes derived from an allo sensitized individual: HLA specificity and cell surface markers. *Scand. J. Immunol.* **14,** 213–224.
7. Spits, H., Borst, J., Terhorst, C., and de Vries, J. E. (1982) The role of T cell differentiation markers in antigen-specific and lectin-dependent cellular cytotoxicity mediated by T8$^+$ and T4$^+$ cytotoxic T cell clones directed at class I and class II MHC antigens. *J. Immunol.* **129,** 1563–1569.
8. de Vries, J. E. and Sptis, H. (1984) Cloned human cytotoxic T lymphocyte (CTL) lines reactive with autologous melanoma cells. I. In vitro generation, isolation and analysis to phenotype and specificity. *J. Immunol.* **132,** 510–519.
9. Yssel, H., Spits, H., and de Vries, J. E. (1984) A cloned human T cell line cytotoxic for autologous and allogeneic B lymphoma cells. *J. Exp. Med.* **134,** 239–254.
10. Yssel, H., Blanchard, D., Boylston, A., de Vries, J. E., and Spits, H. (1986) T cell clones which share T cell receptor epitopes differ in phenotype, function and specificity. *Eur. J. Immunol.* **16,** 1187–1193.
11. Yssel, H., Shanafelt, M.-C., Soderberg, C., Schneider, P. V., Anzola, J., and Peltz, G. (1991) *B. burgdorfi* activates T cells to produce a selective pattern of lymphokines in Lyme arthritis. *J. Exp. Med.* **174,** 593–601.
12. Haanen, J. B., de Waal Malefijt, R., Res, P. C., Kraakman, E. M., Ottenhoff, T. H., de Vries, R. R., and Spits, H. (1991) Selection of a human T helper type 1-like T cell subset by mycobacteria. *J. Exp. Med.* **174,** 583–592.
13. Yssel, H., Johnson, K. E., Schneider, P. V., Wideman, J., Terr, A. R. K., and de Vries, J. E. (1992) T cell activation inducing epitopes of the house dust

mite allergen *Der p* I. Induction of a restricted cytokine production profile of *Der p* I-specific T cell clones upon antigen-specific activation. *J. Immunol.* **148,** 738–745.
14. Alarcon, B., de Vries, J. E., Pettey, C., Boylston, A., Yssel, H., Terhorst, C., and Spits, H. (1987) The T cell receptor γ/CD3 complex implication in the cytotoxic activity of a $CD3^+$ $CD4^-$ $CD8^-$ human natural killer clone. *Proc. Natl. Acad. Sci. USA* **84,** 3861–3865.
15. Sredni, B., Volkman, D., Schwartz, R. H., and Fauci, A. S. (1981) Antigen-specific human T cell clones: development of clones requiring HLA-DR compatible presenting cells for stimulation in the presence of antigen. *Proc. Natl. Acad. Sci. USA* **78,** 1858–1862.
16. Eckels, D. S., Lamb, J. R., Lake, P., Woody, J. N., Johnson, A. H., and Hartzman, R. J. (1982) Antigen-specific human T lymphocyte clones. Genetic restriction of influenza virus-specific responses to HLA-D region genes in man. *Hum. Immunol.* **4,** 313–324.
17. Yssel, H., de Vries, J. E., Koken, M., Van Blitterswijk, W., and Spits, H. (1984) Serum-free medium for the generation and propagation of functional human cytotoxic and helper T cell clones. *J. Immunol. Methods* **72,** 219–227.
18. Boyum, A. (1968) Separation of leukocytes from blood and bone marrow isolation of mononuclear cells and granulocytes from human blood isolation of mononuclear cells by 1 inst centrifugation and of granulocytes by combining inst centrifugation and inst sedimentation at 1 g. *Scand. J. Clin. Lab. Invest.*, **21(Suppl. 97),** 77.
19. Spits, H., Borst, J., Giphart, M., Colligan, J., Terhorst, C., and de Vries, J. E. (1984) HLA-DC antigens can serve as recognition elements for human cytotoxic lymphocytes. *Eur. J. Immunol.* **4,** 299–304.
20. Spits, H., Yssel, H., Thompson, A., and de Vries, J. E. (1983) Human $T4^+$ and $T8^+$ cytotoxic T lymphocyte clones directed at products of different class II major histocompatibility complex loci. *J. Immunol.* **131,** 678–683.
21. Sager, N., Feldmann, A., Schilling, G., Kreitsch, P., and Neumann, C. (1992) House dust-mite-specific T cell in the skin of subjects with atopic dermatitis: frequency and lymphokine profile in the allergen patch test. *J. Allergy Clin. Immunol.* **89,** 801–810.
22. Abrams, J. S., Roncarolo, M.-G., Yssel, H., Andersson, U., Gleich, G. J., and Silver, J. (1992) Strategies and practice of anti-cytokine monoclonal antibody development: immunoassay of IL-10 and IL-5 in clinical samples. *Immunol. Rev.* **127,** 5–24.
23. Abrams, J. S. (1995) Immunoenzymatric assays of mouse and human cytokines, using NIP-linked anti-cytokine antibodies, in *Current Protocols in Immunology* (Coligan, J. E., Kruisbeek, A. M., Margulies, D. H., Shevach, E. M., and Strober, W., eds.), Wiley, New York, pp. 6.20.1–6.20.15.
24. Centers for Disease Control and Prevention and National Institutes of Health (1993) *Biosafety in Microbiological and Medical Laboratories.* HHS publication no. CDC 93-8395, US Government Printing Office, Washington, DC.

25. Chan, S. H., Perussia, B., Gupta, J. W., Kobayashi, M., Pospisil, M., Young, H. A., Wolf, S. F., Young, D., Clark, S. C., and Trinchieri, G. (1991) Induction of interferon γ production by natural killer cell stimulatory factor: characterization of the responder cells and synergy with other inducers. *J. Exp. Med.* **173,** 869–879.
26. Yssel, H., Fasler, S., de Vries, J. E., and de Waal Malefyt, R. (1994) Interleukin 12 (IL-12) transiently induces IFN-γ transcription and protein synthesis in human $CD4^+$ allergen-specific helper type 2 (Th2) T cell clones. *Int. Immunol.* **6,** 1091–1096.

12

Protection of Purified Human Hematopoietic Progenitor Cells by Interleukin-1β or Tumor Necrosis Factor-α

Robert J. Colinas

1. Introduction

Hematopoiesis is the process of blood cell formation. In the adult human, the bone marrow (BM) is the primary hematopoietic organ. Each day, the BM produces billions of leukocytes, erythrocytes, and thrombocytes, which enter the circulation. Production of such enormous numbers of mature blood cells results from the exponential expansion and differentiation of pluripotent hematopoietic stem cells through a series of increasingly more differentiated hematopoietic progenitor cells (HPCs). Regulation of hematopoiesis is accomplished by the transduction of signals that follow interactions of the HPCs and developing blood cells with BM stromal cells, cytokines, and the extracellular matrix.

Functional assessment of the hematopoietic potential of human HPCs is accomplished using in vitro colony-forming cell (CFC) and long-term culture (LTC) methodologies that were initially developed using rodent bone marrow *(1,2)*. Of the techniques available, the CFC assay is the most commonly employed and has been adapted for use with multiple sources of human HPCs, including BM, fetal liver, umbilical cord blood, and peripheral blood. Colony formation by HPCs in the CFC assay relies on the addition of exogenous cytokines to cells suspended in semisolid medium. Since no colonies will form in the absence of added cytokines, the investigator has control over which hematopoietic lineage is being assessed.

It has been shown that the oxygen partial pressure (pO_2) in human BM is approx 2–5% *(3)*. This is considerably lower than the near-ambient pO_2 (19.3%), which exists in the typical cell-culture incubator maintained at 7%

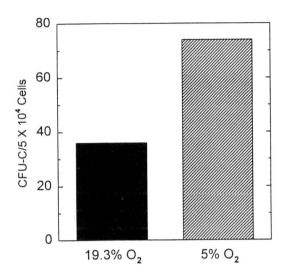

Fig. 1. Effects of near-ambient (19.3%) and physiological (5%) pO_2 on colony formation by human HPCs in vitro. Normal human BMMNCs were plated in semisolid medium containing 2 ng/mL rHuGM-CSF. The cultures were incubated at 37°C in either air/7% CO_2 or 5% O_2, 7% CO_2 and 88% N_2, and colonies of 250 cells were counted after 14 d. The data shown are from one representative experiment.

CO_2. Numerous studies have demonstrated that, even in the presence of exogenously added reducing agent, both CFC and LTC assays perform nearly twofold better when conducted at physiological pO_2 *(4,5)* (Colinas, unpublished) (Fig. 1). Furthermore, in experiments using either mouse or human BM, it has been shown that the inhibitory effects of several oxidative stress-inducing hematotoxic agents and near-ambient pO_2 on granulocyte/macrophage-colony-stimulating factor (GM-CSF)-induced colony formation are additive *(6)*. Thus, it is important to conduct studies on human HPCs at physiological pO_2.

As stated, hematopoiesis is largely controlled by the interactions of HPCs with cytokines. Interleukin-1β (IL-1β) and tumor necrosis factor-α (TNFα) are pleiotropic inflammatory cytokines that are believed to play a central role in hematopoiesis *(7,8)*. Cellular sources of IL-1β and TNFα include macrophages, lymphocytes, granulocytes, fibroblasts, and endothelial cells. Using unfractionated BM cells or purified HPCs (pHPCs), IL-1β synergizes in vitro with other cytokines, such as IL-3, GM-CSF, IL-6, or c-kit ligand (KL), resulting in both increased colony size and numbers *(9,10)*. In contrast, when unfractionated bone marrow mononuclear cells (BMMNC) are treated with TNFα, colony formation is inhibited *(11–13)*. However, investigations using pHPCs have shown that TNFα stimulates more primitive HPCs, while inhibit-

Table 1
Studies Demonstrating Protection by IL-1α, IL-1β, or TNFα

Experimental system	Cytokine	Treatment[a]	References
Mouse, in vivo, in vitro	IL-1, TNFα	γ-Radiation	*(17–19)*
Mouse, in vivo	IL-1α, TNF-α	Endotoxin	*(20,21)*
Mouse, in vivo	IL-1, TNF-α	Ischemia/reperfusion	*(22,23)*
Mouse, in vivo	TNF	5-FU, MTX, VB	*(24)*
Mouse, in vivo	IL-1α	γ-Radiation with DX or DDP	*(25)*
Mouse, in vivo	IL-1α	DX	*(26–28)*
Mouse, in vivo	IL-1, TNF-α	hyperoxia	*(29,30)*
Human, in vitro	IL-1	γ-radiation	*(31)*
Human, in vitro	IL-1, TNF-α	4HC	*(32,33)*
Human, in vitro	IL-1, TNF-α	DX	*(34)*
Human, in vitro	TNF-α	HQ	*(34)*

[a]The specific treatment protocols employed and the cell types protected in these studies are available in the corresponding references. Abbreviations are as follows: 5-FU, 5-fluorouracil; MTX, methotrexate; VB, vinblastine; DX, doxorubicin; DDP, cis-dichlorodiammine platinum; 4HC, 4-hydroperoxycyclophosphamide; HQ, hydroquinone.

ing those progenitors committed to the erythroid and myeloid lineages *(14–16)*. Thus, it appears that the net effect of TNFα on hematopoiesis depends on the differentiation state of the HPC and the specific lineage to which it is committed.

IL-1β or TNFα treatment also has been shown to protect HPCs as well as other cell types against the effects of numerous inhibitory agents (Table 1). Experimental evidence suggests that the inhibitory effects of these agents result from the production of oxidative stress through mechanisms that generate reactive oxygen intermediates (ROIs) or deplete glutathione (GSH). Thus, it is likely that IL-1β and TNFα produce a state within the HPC that makes it more resistant to oxidative stress. This state is most likely achieved by the induction of gene expression, the products of which are involved in antioxidant and/or detoxification responses (Fig. 2). Consistent with this hypothesis, IL-1β and TNFα have been shown to induce manganous superoxide dismutase (MnSOD) expression in both normal and tumor cell lines *(35–37)*. In addition, TNFα induces aldehyde dehydrogenase (ALDH) expression in human HPCs. ALDH is an enzyme that detoxifies the reactive 4-hydroperoxycyclophosphamide (4HC) metabolite aldophosphamide *(38)*. Moreover, TNFα also induces an unidentified response that protects human HPCs from phenylketophosphamide toxicity *(39)*.

The following method describes an experimental protocol that has been used to determine whether IL-1β or TNFα could protect human HPCs from the inhibitory effects of doxorubicin (DX) at physiological pO_2 and temperature

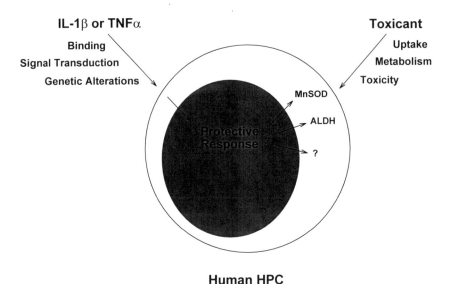

Fig. 2. A model for protection of human HPCs by IL-1β or TNFα. It is believed that these cytokines induce genetic changes in the HPC, which are capable of protecting against a wide range of toxic agents. Some of these genetic changes are known (MnSOD and ALDH), and others are unknown (?).

(34). A summary of this procedure is represented schematically in Fig. 3. Briefly, pHPCs are isolated from normal human BM (Section 3.2.), and pretreated with IL-1β or TNFα in the absence or presence of DX (Section 3.3.). The cells are washed (Section 3.4.) and plated in semisolid medium containing colony-stimulating cytokines (Section 3.5.). After a 14-d incubation, colony numbers and types are delineated (Section 3.6.). IL-1β or TNFα is considered protective if two criteria are met. First, pretreatment with either cytokine alone does not significantly affect CFU-C frequency or type relative to the medium-pretreated control HPCs. Second, the addition of either cytokine to DX-exposed pHPCs results in greater numbers of CFU-C than in the samples exposed to DX alone. Although the protocol that follows is specific for normal $CD34^+$ BM cells pretreated with combinations of IL-1β, TNFα, and DX, it could be easily adapted to determine whether any identifiable HPC subpopulation can be protected by a given bioactive molecule against the inhibitory or toxic effects of any modifier (*see* Note 8).

2. Materials
2.1. Human BM Source

An approved source of normal human HPCs must be available. In our studies, aspirates from the posterior iliac crest of consenting hematologically

HPC Protection Assay

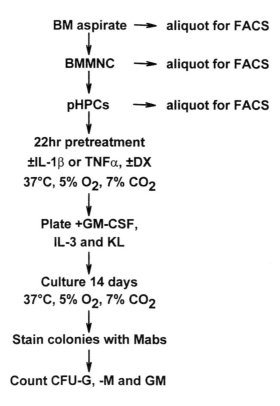

Fig. 3. Schematic representation of the HPC protection assay.

normal 18–40-yr-old donors have been used in an Institutional Review Board-approved outpatient procedure. The aspiration procedure is conducted by a licensed hematologist using a Sternal/Iliac BM aspiration needle (Baxter Scientific, Valencia, CA; cat. no. DIN1515X).

2.2. Special Equipment

1. A laminar flow biological safety cabinet (TC hood) (Baker Co., Sanford, ME).
2. A sealed glove box with an air lock (PlasLabs, Lansing, MI, model no. 830 ABC-SPEC). The glove box chamber is fitted with an HEPA outlet filter, a vacuum line controlled from the inside, and an inlet valve controlled from the outside. The airlock has both externally controlled inlet and vacuum lines. The inlet valve is connected to a cylinder of compressed 5% O_2, 7% CO_2, and 88% N_2 fitted with an in-line 0.2-µm gas filter.
3. A flow cytometer, such as a FACScan (Becton Dickinson, San Jose, CA).

4. A water-jacketed CO_2 incubator (Forma Scientific, Marietta, OH, model 3158) retrofitted with an OxyReducer™ (Reming Bioinstruments, Redfield, NY, model 311) (*see* Note 3) or an incubator designed to regulate both pO_2 and pCO_2 (Hareus Inc. Equipment Group, South Plainfield, NJ).
5. A fluorescence microscope, such as a Zeiss Axioskop (Zeiss, Thornwood, NY).

2.3. CD34+ Immunoselection Materials

There are several commercially available systems for immunoselecting human HPCs. All operate with similar effficiencies (approx 65–80% CD34+, with ≤15% expressing cell-surface antigens found on mature blood cells, in the recovered population, and a 50% yield) and are based on monoclonal antibodies (MAbs) to the CD34 cell-surface glycoprotein that has been shown to be an HPC marker *(40)*. The CellPro biotinylated antiCD34 MAb–avidin column-based system (Bothell, WA; cat. no. LC34) has been found to be reasonably economical, relatively quick, and can be used in the glove box.

2.4. Cytokines

In our studies, researchers have utilized IL-1β, TNFα, IL-3, GM-CSF, and KL. There are many commercial sources of recombinant human (rHu) cytokines, and significant differences between the same cytokines obtained from different manufacturers have not been found. However, it is much more economical, though somewhat slower and more laborious, to obtain the needed cytokines from generous researchers that produce them in large quantities in their own laboratories. The Biological Response Modifiers Program of the National Cancer Institute (Frederick, MD) is another economical source for cytokines. Cytokines should be prepared as 100X stock solutions in PBS/0.1% BSA, aliquoted and stored at $-70°C$.

2.5. Disposable Tissue-Culture Supplies

Sterility of tissue cultureware is crucial, since the cells will be cultured for over 2 wk. Nearly all of the supplies used are generic and come presterilized. Vacutainer brand evacuated test tubes without anticoagulant can be obtained from Fischer Scientific (Pittsburgh, PA; cat. no. 02-685 A). Obtain nonsterile 1.9-mL polypropylene microcentrifuge tubes from Baxter Diagnostics Inc. (McGaw, IL; cat. no. C3520-1) and autoclave for 20 min at 121°C, then dry at 60°C before use.

2.6. Media, Fetal Bovine Serum, and Reagents

1. Low endotoxin Iscove's modified Dulbecco's medium (IMDM) is available from Biowhittaker (Walkersville, MD) and is stored at 4°C (*see* Note 4).
2. Defined fetal bovine serum (FBS) is purchased from Hyclone (Logan, UT), thawed, aliquoted in 5- and 10-mL volumes, and stored at $-20°C$.

3. Phosphate-buffered saline (PBS) and 0.9% NaCl (saline) are available from Biowhittaker.
4. Ficoll-Hypaque density gradient medium is from Pharmacia (Uppsala, Sweden) and is maintained at room temperature in the dark.
5. A 1000 U/mL solution of preservative-free sodium heparin (PF heparin) can be obtained from Lyphomed (Deerfield, IL).
6. DX can be purchased as a sterile 2 mg/mL solution from Chiron Therapeutics (Emeryville, CA) and should be stored at 4°C in the dark.
7. Powdered IMDM is available from Gibco-BRL (Gaithersburg, MD; cat. no. 12200-036). To make 2X IMDM, follow the manufacturer's directions for preparation, except bring the final volume to 500 mL instead of 1 L. Filter the medium through a 0.22-µm filter, and store at 4°C for no more than 1 mo.
8. Low-melting-point (LMP) Seaplaque agarose is available from the FMC Corp. (Rockland, ME; cat. no. 50102). Stocks of 0.9% LMP agarose are made in 250-mL glass bottles by adding 1.8 g of agarose to 200 mL chilled pyrogen-free double-distilled deionized H_2O, mixing well, and autoclaving at 121°C for 25 min. After the autoclave has returned to ambient pressure, but the agarose is still molten, tighten the caps and mix well. Store agarose at room temperature.

2.7. Immunophenotyping Reagents

1. A phycoerythrin (PE)-conjugated MAb against the HPC antigen CD34 (HPCA-2) can be obtained from Becton Dickinson.
2. The PE-conjugated monocyte-specific anti-CD14 and the fluorescein isothiocyanate (FITC)-conjugated granulocyte-specific anti-CD66b MAbs are available from Becton Dickinson and Immunotech (Westbrook, ME), respectively. Since fluorochrome-conjugated MAbs are light- and heat-sensitive, these reagents should be stored at 4°C in the dark.
3. Paraformaldehyde can be obtained from Fisher Scientific Co. Make PBS/2% paraformaldehyde in a chemical fume hood by stirring 2 g of paraformaldehyde into 100 mL of 70–80°C PBS. When the solution clears, cool, pass through a 0.22-µm filter, and store at 4°C.
4. Prepare PBS, 0.1% (w/v) NaN_3 by dissolving 1 g NaN_3 in 1 L PBS. NaN_3 is a hazardous substance. Follow the manufacturer's directions for safe handling and disposal. Make PBS, 0.1% NaN_3, 0.1% BSA, pH 8.0, by adding 0.4 g BSA to 400 mL PBS, 0.1% NaN_3 and adjusting the pH to 8.0 with $10N$ NaOH. Add 1.25 mg mouse IgG (Sigma, St. Louis, MO; cat. no. 15381) to a 25-mL aliquot of this solution. Store at 4°C.

3. Methods

3.1. Preliminary Preparations

1. Two hours before beginning this procedure, clean the interior of the glove box and airlock with disinfectant and allow to dry.
2. Stock the glove box with the necessary sterile supplies (i.e., pipeters, serological pipets, pipet tips, a waste container, several 15- and 50-mL conical polypropy-

lene centrifuge tubes, tube racks, sterile scissors, and the CellPro column stand) and seal the inner airlock door.
3. With the HEPA filter completely open, flush the air out of the glove box at 10 psi for 10 min with the 5% O_2, 7% CO_2 88% N_2 gas mixture, and then decrease the flow rate to 1 psi to maintain a slight positive pressure. Partially close the HEPA filter (*see* Note 5).
4. While the glove box is being gassed, prepare 50 mL of PBS, 10 U/mL PF heparin, two 15-mL cushions of Ficoll-Hypaque in 50-mL conical polystyrene (clear) centrifuge tubes, 150 mL PBS, 5% FBS, 50 mL PBS, 1% BSA (made from CellPro 5% BSA), and 40 mL IMDM, 20% FBS, 4 m*M* L-glutamine/20 µg/mL Gentamicin sulfate (complete IMDM, 20% FBS).
5. With the caps loosened slightly, pass the reagents prepared in step 5, a small box of ice, the CellPro biotinylated anti-CD34 MAb, 5% BSA, and the avidin column into the glove box chamber by evacuating the airlock to ≤15 in Hg, and then flushing with 5% O_2, 7% CO_2 88% N_2 back to ambient atmospheric pressure three times. From this point on everything entering the glove box must be cycled through the airlock in this manner (*see* Note 6).

3.2. HPC Isolation, Phenotyping, and Yield

1. In order to prevent coagulation, 4–6 mL of BM are aspirated into a syringe containing 500 U of PF heparin.
2. Following aspiration, a ≤20-gage hypodermic needle is immediately attached to the syringe, air is forced out, the aspirate is injected into a Vacutainer evacuated test tube without anticoagulant, and the specimen is transported to the laboratory.
3. On receiving the BM in the laboratory, disinfect the exterior of the tube and introduce it into the glove box. Except for centrifugations, overnight storage, 37°C incubations, or unless otherwise noted, all further manipulations of the BM cells are conducted in the glove box.
4. Transfer the aspirate to a 50-mL tube, and dilute to 50 mL with PBS/10 U/mL PF heparin. It is important to obtain a cell count at this point in order to calculate HPC yield. In general, undiluted BM aspirates with low peripheral blood contamination have nucleated cell counts of $35–100 \times 10^6$/mL (*see* Note 7a).
5. Remove 3×10^6 cells for later flow cytometric phenotyping, and place on ice.
6. Carefully layer 25 mL of the diluted aspirate onto the top of each Ficoll-Hypaque cushion, seal the tubes tightly, and centrifuge at room temperature for 10 min at 1000*g*.
7. Carefully remove the BM mononuclear cells (BMMNCs) banded at the interface, and pool them in a new 50-mL tube.
8. Dilute the BMMNCs to 50 mL with PBS, 5% FBS, and pellet at 400*g* for 10 min at room temperature.
9. Discard the supernatant, resuspend the pellet in 15 mL PBS, 5% FBS, transfer to a 15-mL tube, and pellet at 400*g* at 4°C. From this point on, conduct all centrifugations at 400*g* at 4°C.
10. Wash the BMMNCs once more with 15 mL PBS/1% BSA, remove the supernatant, and place the cells on ice.

Table 2
CD34+ Cell Content of BM Subpopulations

Donor	BM fraction	Total cell number, ×10⁻⁶ᵃ	Percent CD34⁺ᵇ	Total CD34⁺ ×10⁻⁶ᶜ	Percent yieldᵈ
1	BM aspirate	252	2.1	5.3	100
	BMMNC	185	1.0	1.9	35
	CD34⁺	2.0	72	1.4	27
2	BM aspirate	413	2.4	9.9	100
	BMMNC	249	0.5	1.3	13
	CD34⁺	1.3	65	0.8	8

^{a}Total nucleated cell counts in unfractionated 4–6 mL BM aspirates and BMMNC and CD34$^+$ cell fractions from two normal donors were determined using a Coulter electronic particle counter.
^{b}The percentage of CD34$^+$ cells in each fraction was determined by flow cytometry.
^{c}The total number of CD34$^+$ cells was calculated from the percent CD34$^+$ cells and the total cell number in each BM fraction.
^{d}The yield of CD34$^+$ cells in each BM fraction represents the percentage of the total number of CD34$^+$ cells in the unfractionated BM aspirate.

11. The BMMNC recovery should be approx 50–75% of the original total cell count (Table 2). Therefore, resuspend the cells in a volume of PBS/1% BSA that will give 100–200 × 10^6 cells/mL.
12. Obtain a cell count again, remove 3 × 10^6 cells again for later flow cytometric analysis, and store on ice.
13. Add 40 µL of the CellPro biotinylated anti-CD34 MAb/mL of BMMNCs, mix well, and incubate on ice for 30 min, mixing again after the first 15 min. While the cells are incubating with the MAb, prepare the avidin column as instructed by the CellPro protocol.
14. Dilute the labeled BMMNCs to 15 mL with PBS/1% BSA, and pellet at 400g for 10 min at 4°C.
15. Wash once with 15 mL of PBS/1% BSA, and resuspend the BMMNC pellet in CellPro 5% BSA at 100–200 × 10^6 cells/mL.
16. Follow the CellPro instructions to apply the BMMNCs to the column, wash away the unbound cells (CD34$^-$) into one 15-mL tube, and mechanically release and elute the bound CD34$^+$ (pHPC) subpopulation from the column into a new 15-mL tube.
17. At this point, pellet the BM aspirate and the BMMNC fractions reserved for FACS analysis and the isolated CD34$^+$ cells, remove the supernatant, and resuspend each in 1 mL of complete IMDM, 20% FBS. Add 20 U of PF heparin to the aliquot of unfractionated BM aspirate to prevent coagulation.
18. Obtain a cell count of the pHPCs, and remove 50 × 10^3 for flow cytometric analysis. In general, approx 0.5–1% of the BMMNCs are isolated as pHPCs.
19. Store the pHPCs in sealed tubes overnight in an ice bath in the refrigerator. Storage of the cells in this manner does not result in any loss of HPC viability or colony-forming ability.

20. To block nonspecific MAb binding, add an equal volume of PBS, 0.1% NaN_3, 1% BSA, 50 µg/mL mouse IgG to the BM fractions reserved for flow cytometric analysis.
21. After blocking for at least 4 h at 4°C and without washing the cells, stain the BM cell fractions with anti-CD34-PE or fluorochrome-matched isotype control MAbs and analyze by flow cytometry using standard methods. Typical pHPC frequency in the various BM cell populations and yield are shown in Table 2.

3.3. Pretreatment of pHPCs

1. The following morning, disinfect, restock, and flush the glove box with 5% O_2, 7% CO_2, and 88% N_2 again.
2. Plan the following treatment groups in triplicate: untreated control, 50 ng/mL IL-1β, 25 ng/mL TNFα, 100 nM DX, IL-1/DX, TNF/DX, and IL-1/TNF/DX.
3. Thaw IL-1β and TNFα on ice, and prepare the following in the TC hood: 10 µM DX stock by adding 43.5 µL 2 mg/mL DX to 14.95 mL chilled saline, 10 mL complete IMDM/5% FBS, and 21 capped, sterile 1.9-mL polypropylene microcentrifuge tubes.
4. Introduce the items prepared in step 3 into the glove box, and equilibrate for 1.5 h.
5. Pellet the pHPCs, resuspend them at 16.7×10^3/mL in the complete IMDM, 5% FBS, and aliquot 0.3 mL into each 1.9-mL tube.
6. To the indicated tubes first add the IL-1β and TNFα, and then the DX to final concentrations of 50 ng/mL, 25 ng/mL, and 100 nM, respectively, mixing well after each addition.
7. Cap the tubes and place them in a 37°C incubator maintained at 5% O_2, 7% CO_2, and 88% N_2 for 22 h.

3.4. Removal of the Cytokines and Toxicants

Since the cells are kept ice-cold, and the cytokines and DX are immediately diluted and then washed away, Sections 3.4. and 3.5. can be performed in the TC hood.

1. Remove the tubes from the incubator, and place them on ice in the TC hood.
2. Add 1.5-mL ice-cold complete IMDM, 5% FBS to each tube, recap, mix by inversion, and pellet the cells.
3. To prevent cell loss during the washing procedure, aspirate all but the last 100 µL of the supernatant.
4. Wash the cells twice with 1.5 mL ice-cold complete IMDM, 5% FBS/tube, and store on ice until being plated in semisolid medium.

3.5. CFU-C Assay

1. Loosen the bottle cap, and melt the stock 0.9% LMP agarose in a boiling water bath. Then place it in a 40°C water bath to cool.
2. While the agarose is cooling, label 35-mm TC dishes, thaw the aliquots of GM-CSF, IL-3, and KL on ice and, in a 50-mL tube, combine 14 mL 2X IMDM, 7 mL FBS, and 17.5 µL 40 mg/mL gentamicin sulfate, cap it tightly, and warm it to 40°C.

3. When both the agarose and the medium reach 40°C, remove both from the water bath, and in the TC hood, quickly add 14 mL 0.9% agarose to the medium, mix, and then add GM-CSF, IL-3, and KL to 2, 5, and 20 ng/mL, respectively.
4. To each 35-mm TC dish, first add 1.5 mL of medium, and then the resuspended cells from one tube, minimizing bubble production (*see* Notes 7c and d).
5. Mix well by swirling the dishes in both directions, and place the dishes on an ice-cold leveled metal tray to solidify. Following this procedure, plating can be done conveniently in sets of 8–10 dishes (*see* Note 7e).
6. When all the samples are plated and have solidified, place the dishes in a well-humidified incubator containing 5% O_2, 7% CO_2, and 88% N_2 for 14 d (*see* Note 7f).
7. At this point, count colonies (CFU-C) of >50 cells using a dissecting microscope at 40X with oblique backlighting.

3.6. In Situ *Fixation and Staining of CFU-C*

1. After counting CFU-C, fix the cells by adding 2 mL of 2% paraformaldehyde in PBS to the center of each dish, and rock gently for 15 min at room temperature.
2. Carefully aspirate the fixative without disturbing the agarose, rinse once with 3 mL PBS, and then wash the dishes twice for 30 min by adding 2.5 mL of PBS, rocking gently at room temperature.
3. Add 1 mL PBS, 0.1% BSA, 0.1% NaN_3, 50 µg/mL of mouse IgG to each dish, and incubate for ≥4 h at 4°C.
4. For each experimental group of three dishes, add fluorochrome- and isotype-matched control MAbs to one control dish, and anti-CD66b-FITC and anti-CD14-PE to the two remaining replicate dishes.
5. Rock the dishes for 24 h at 4°C in the dark.
6. Remove unbound MAbs by rinsing once and then washing 4 times for 8–12 h each with 2.5 mL cold PBS, 0.1% BSA, 0.1% NaN_3, pH 8.0, at 4°C in the dark.
7. Enumerate CFU-G, CFU-M, and CFU-GM using a fluorescent microscope at 100X magnification. Relative to the isotype controls, CFU-G are $CD66b^+/CD14^{neg/lo}$, CFU-M are $CD14^+/CD66b^-$, and CFU-GM contain both $CD66b^+$ and $CD14^+$ cells.

4. Notes

1. The inhibitory effects of ambient pO_2 exposure and oxidative stress-inducing treatments on both mouse and human in vitro colony-forming assays have been shown to be additive (*6*). Therefore, in studies designed to assess the effects of potentially oxidizing toxicants, either alone or in the presence of protective cytokines, it is critical that these studies be conducted at physiological pO_2.
2. Additional HPC sources may be used, such as BM harvested under general anesthesia for transplantation purposes, umbilical cord blood, peripheral blood, or fetal liver. However, it is important to remember that several subtle differences may exist between HPCs derived from different sources.
3. The OxyReducer consists of an O_2 sensor, which is placed in the incubator, linked to an electronic controller that maintains a preset pO_2 by flushing the incubator

with N_2. A liquid N_2 (LN_2) tank capable of gas takeoff at pressures ≥20 psi (Taylor Wharton, Theodore, AL; model XL-45) is an economical primary N_2 source. As a backup, connect a cylinder of compressed N_2 gas into the N_2 line between the LN_2 tank and the OxyReducer, with the regulator outlet pressure set 10 psi lower than that of the primary N_2 source. The higher back-pressure from the primary N_2 source will prevent gas use from the compressed N_2 tank until the LN_2 tank is empty.

4. Endotoxin contamination is a persistent concern in experiments assessing the biological effects of cytokines. Thus, all reagents and supplies should be certified pyrogen-free. Rinse glassware well with pyrogen-free, double-distilled deionized H_2O, dry, and bake for 2 h at 180°C. If any doubts about endotoxin remain, determine levels using the Limulus Amoebocyte Lysate Assay kit from Biowhittaker.
5. The glove box chamber cannot withstand a pressure differential greater than a few psi. Watch carefully to avoid pressure damage.
6. All solutions should be equilibrated in the low-O_2 environment for at least 1.5 h before being used.
7. Troubleshooting:
    a. Low total BM cell count: On occasion, a BM aspirate will have a very low cell count (<100 × 10^6 total nucleated cells). In these cases, it should be determined whether there are enough $CD34^+$ cells to perform the experiment. To make this determination, stain an aliquot of BMMNC with HPCA-2-PE, and analyze by flow cytometry. BMMNC containing less than three times the absolute number of CD34+ cells needed should not be used.
    b. High variability between CFU-C counts in replicate dishes: The most likely source for this sort of variability is from the loss of cells during cytokine/toxicant removal. Since there are only 5000 cells/sample, it is very difficult to detect cell loss. Therefore, to prevent loss of cells, follow the meniscus down as the supernatant is aspirated, and leave the last 100 µL in each tube.
    c. Semisolid medium solidifies before completion of plating: To prevent the semisolid medium used in the CFU assay from solidifying before all the samples are plated out, set it in a small 37°C water bath inside the TC hood.
    d. Microscopic observation of the CFU-C is distorted: Although distortion cannot be avoided around the edge, bubbles produced during the plating procedure will also cause optical distortions. Keep bubbling to a minimum.
    e. All the CFU-C have developed on the bottom of the dish: This results from the cells settling to the bottom of the dish before the agarose has solidified. Thoroughly resuspend the contents of the dishes immediately before setting them on the chilled metal tray.
    f. The dishes dry out during the 14-d culture period: It is very important to maintain high humidity in the incubator. This is most easily accomplished by bubbling the N_2 in the incubator through a beaker of sterile water and placing several evaporation pans in the incubator.

8. This protocol can be easily adapted to assess the effects of any cytokine or inhibitory treatment on any identifiable HPC. However, it is important first to determine appropriate cytokine and toxicant concentrations, as well as the kinetics of their effects. For example, the inhibitory effects of DX are manifest more slowly than the IL-1β- or TNFα-mediated protective effects. As a result, both cytokines and toxicants can be added to the pHPCs at the time of culture initiation. In contrast, the inhibitory effects of agents, such as 4HC and γ-radiation, occur very quickly. Therefore, the HPCs must be pretreated with IL-1β or TNFα for at least 14 h before being exposed to either 4HC or γ-radiation.

Acknowledgments

The research from which this protocol was derived was done in the laboratory of David A. Lawrence and in the Molecular Immunology Core Facility at the Wadsworth Center of the New York State Department of Health. Funding for this research was provided by the NIH National Research Service Award ES05538 (R. J. Colinas) and NIH grant ES05020 (D. A. Lawrence) from the National Institute of Environmental Health Sciences.

References

1. Bradley, T. R. and Metcalf, D. (1966) The growth of mouse bone marrow cells *in vitro. J. Exp. Biol. Med.* **44**, 287–300.
2. Dexter, T. M., Allen, T. D., and Lajtha, L. G. (1977) Conditions controlling the proliferation of haemopoietic stem cells *in vitro. J. Cell. Physiol.* **91**, 335–344.
3. Cater, D. B. and Silver, I. A. (1960) Quantitative measurements of oxygen tension in normal tissues and in tumors of patients after radiotherapy. *Acta Radiol.* **53**, 233–241.
4. Rich, I. N. (1986) A role for the macrophage in normal hemopoiesis. II. Effect of varying physiological oxygen tensions on the release of hemopoietic growth factors from bone-marrow-derived-macrophages *in vitro. Exp. Hematol.* **14**, 746–751.
5. Smith, S. and Broxmeyer, H. E. (1986) The influence of oxygen tension on the long term growth *in vitro* of haematopoietic progenitor cells from human cord blood. *Br. J. Haematol.* **63**, 29–34.
6. Colinas, R. J., Burkart, P. T., and Lawrence, D. A. (1994) *In vitro* effects of hydroquinone, benzoquinone, and doxorubicin on mouse and human bone marrow cells at physiological oxygen partial pressure. *Toxicol. Appl. Pharmacol.* **129**, 95–102.
7. Fibbe, W. E. and Willemze, R. (1991) The role of interleukin-1 in hematopoiesis. *Acta Haematologica* **86**, 148–154.
8. Trinchieri, G. (1992) Effects of TNF and lymphotoxin on the hematopoietic system. *Immunol. Ser.* **56**, 289–313.
9. Muench, M. O., Scheider, J. G., and Moore, M. A. S. (1992) Interactions among colony-stimulating factors, IL-1β, IL-6, *kit*-ligand in the regulation of primative murine hematopoietic cells. *Exp. Hematol.* **20**, 339–349.

10. Kobayashi, M., Imamura, M., Gotohda, Y., Maeda, S., Iwasaki, H., Sakurada, K., Kasai, M., Hapel, A. J., and Miyazaki, T. (1991) Synergistic effects of interleukin-1β and interleukin-3 on the expansion of human hematopoietic progenitor cells in liquid cultures. *Blood* **78,** 1947–1953.
11. Khoury, E., Lemoine, F. M., Baillou, C., Kobari, L., Deloux, J., Guigon, M., and Najman, A. (1992) Tumor necrosis factor alpha in human long-term bone marrow cultures: distinct effects on nonadherent and adherent progenitors. *Exp. Hematol.* **20,** 991–997.
12. Beran, M., O'Brian, S., Gutterman, J. U., and McCredie, K. B. (1988) Tumor necrosis factor and human hematopoiesis. I. Kinetics and diversity of human bone marrow cell response to recombinant tumor necrosis factor alpha in short-term suspension cultures in vitro. *Hematol. Pathol.* **2,** 31–42.
13. Akahane, K., Hosoi, T., Urabe, A., Kawakami, M., and Takaku, F. (1987) Effects of recombinant tumor necrosis factor (rhTNF) on normal human and mouse hematopoietic progenitor cells. *Int. J. Cell Cloning* **5,** 16–26.
14. Caux, C., Saeland, S., Favre, C., Duvert, V., Mannoni, P., and Banchereau, J. (1990) TNF alpha strongly potentiates IL-3 and GM-CSF induced proliferation of human CD34$^+$ hematopoieitic progenitor cells. *Blood* **75,** 2292–2302.
15. Backx, B., Broeders, L., Bot, F. J., and Lowenburg, B. (1991) Positive and negative effects of tumor necrosis factor on colony growth from highly purified normal marrow progenitors. *Leukemia* **5,** 66–70.
16. Caux, C., Favre, C., Saeland, S., Duvert, V., Durand, I., Mannoni, P., and Banchereau, J. (1991) Potentiation of early hematopoiesis by tumor necrosis factor-α is followed by inhibition of granulopoietic differentiation and proliferation. *Blood* **78,** 635–644.
17. Neta, R., Douches, S. D., and Oppenheim, J. J. (1986) Interleukin 1 is a radioprotector. *J. Immunol.* **136,** 2483–2485.
18. Constine, L. S., Harwell, S., Keng, P., Lee, F., Rubin, P., and Siemann, D. (1991) Interleukin 1 alpha stimulates hemopoiesis but not tumor cell proliferation and protects mice from lethal total body irradiation. *Int. J. Radiation Oncology Biol. Phys.* **20,** 447–456.
19. Zucali, J. R., Moreb, J., Gibbons, W., Alderman, J., Suresh, A., Zhang, Y., and Shelby, B. (1994) Radioprotection of hematopoietic stem cells by interleukin-1. *Exp. Hematol.* **22,** 130–135.
20. Sheppard, B. C., Fraker, D. L., and Norton, J. A. (1989) Prevention and treatment of endotoxin and sepsis lethality with recombinant human tumor necrosis factor. *Surgery* **106,** 156–162.
21. Vogel, S. N., Kaufman, E. N., Tate, M. D., and Neta, R. (1988) Recombinant interleukin-1α and recombinant tumor necrosis factor α synergize in vivo to induce early endotoxin tolerance and associated hematopoietic changes. *Infect. Immun.* **56,** 2650–2657.
22. Eddy, L. J., Goeddel, D. V., and Wong, G. H. W. (1992) Tumor necrosis factor-α pretreatment is protective in a rat model of myocardial ischemia-reperfusion injury. *Biochem. Biophys. Res. Commun.* **184,** 1056–1059.

23. Brown, J. M., White, C. W., Terada, L. S., Grosso, M. A., Shanley, P. F., Mulvin, D. W., Banerjee, A., Whitman, G. J. R., Harken, A. H., and Repine, J. E. (1990) Interleukin 1 pretreatment decreases ischemia/reperfusion injury. *Proc. Natl. Acad. Sci. USA* **87,** 5026–5030.
24. Slordal, L., Warren, D. J., and Moore, M. A. S. (1990) Protective effects of tumor necrosis factor on murine hematopoiesis during cycle-specific cytotoxic chemotherapy. *Cancer Res.* **50,** 4216–4220.
25. Evans, M. J., Kovacs, C. J., Gooya, J. M., and Harrell, J. P. (1991) Interleukin-1α protects against the toxicity associated with combined radiation and drug therapy. *Int. J. Radiat. Oncol. Biol. Phys.* **20,** 303–306.
26. Eppstein, D. A., Kurahara, C. G., Bruno, N. A., and Terrell, T. G. (1989) Prevention of doxorubicin-induced hematotoxicity in mice by interleukin 1. *Cancer Res.* **49,** 3955–3960.
27. Lynch, D. H., Rubin, A. S., Miller, R. E., and Williams, D. E. (1993) Protective effects of recombinant interleukin-1α in doxorubicin-treated normal and tumor-bearing mice. *Cancer Res.* **53,** 1565–1570.
28. Damia, G., Komschlies, K. L., Futami, H., Back, T., Gruys, M. E., Longo, D. L., Keller, J. R., Ruscetti, F. W., and Wiltrout, R. H. (1992) Prevention of acute chemotherapy-induced death in mice by recombinant human interleukin 1: protection from hematological and nonhematological toxicities. *Cancer Res.* **52,** 4082–4089.
29. Tsan, M. F., White, J. E., Santana, T. A., and Lee, C. Y. (1990) Tracheal insufflation of tumor necrosis factor protects rats against oxygen toxicity. *J. Appl. Physiol.* **68,** 1211–1219.
30. White, C. W., Ghezzi, P., Dinarello, C. A., Caldwell, S. A., McMurtry, I. F., and Repine, J. E. (1987) Recombinant tumor necrosis factor/cachectin and interleukin 1 pretreatment decreases lung oxidized glutathione accumulation, lung injury, and mortality in rats exposed to hyperoxia. *J. Clin. Invest.* **79,** 1868–1873.
31. Moore, M. A. S., Muench, M. O., Warren, D. J., and Laver, J. (1990) Cytokine networks involved in the regulation of haemopoietic stem cell proliferation and differentiation, in *Molecular Control of Haemopoiesis* (Bock, G. and Marsh, J., eds.), Wiley, New York, pp. 43–61.
32. Moreb, J., Zucali, J. R., Gross, M. A., and Weiner, R. S. (1989) Protective effects of IL-1 on human hematopoietic progenitor cells treated *in vitro* with 4-hydroperoxycyclophosphamide. *J. Immunol.* **142,** 1937–1942.
33. Moreb, J., Zucali, J. R., and Rueth, S. (1990) The effects of tumor necrosis factor-α on early human hematopoietic progenitor cells treated with 4-hydroperoxycyclophosphamide. *Blood* **76,** 681–689.
34. Colinas, R. J., Burkart, P. T., and Lawrence, D. A. (1995) The effects of interleukin-1β and tumor necrosis factor-α on *in vitro* colony formation by human hematopoietic progenitor cells exposed to doxorubicin or hydroquinone. *Exp. Hematol.* in press.
35. Eastgate, J., Moreb, J., Nick, H. S., Suzuki, K., Taniguchi, N., and Zucali, J. R. (1993) A role for manganese superoxide dismutase in radioprotection of hematopoietic stem cells by interleukin-1. *Blood* **81,** 639–646.

36. Wong, G. H. W. and Goeddel, D. V. (1988) Induction of manganous superoxide dismutase by tumor necrosis factor: possible protective mechanism. *Science* **242,** 941–944.
37. Kizaki, M., Sakashita, A., Karmaker, A., Lin, C. W., and Koeffler, H. P. (1993) Regulation of manganese superoxide dismutase and other antioxidant genes in normal and leukemic hematopoietic cells and their relationship to cytotoxicity by tumor necrosis factor. *Blood* **82,** 1142–1150.
38. Manthey, C. L., Landkamer, G. J., and Sladek, N. E. (1990) Identification of the mouse aldehyde dehydrogenases important in aldophosphamide detoxification. *Cancer Res.* **50,** 4991–5002.
39. Moreb, J., Zucali, J. R., Zhang, Y., Colvin, M. O., and Gross, M. A. (1992) Role of aldehyde dehydrogenase in the protection of hematopoictic progenitor cells from 4-hydroperoxycyclophosphamide by interleukin 1β and tumor necrosis factor. *Cancer Res.* **52,** 1770–1774.
40. Civin, C. I., Strauss, L. C., Brovall, C., Fackler, M. J., Schwartz, J. F., and Shaper, J. H. (1984) Antigenic analysis of hematpoiesis III. A hematopoietic progenitor cell surface antigen defined by a monoclonal antibody raised against KG-1a cells. *J. Immunol.* **133,** 157–165.

13

Human Mononuclear Phagocytes in Tissue Culture

Yona Keisari

1. Introduction

Peripheral blood human monocytes (HuMo) are the major source for human mononuclear phagocytes. Such monocytes, when cultured, differentiate into monocyte-derived macrophages (MoDM), and undergo various structural, biochemical, and functional changes.

The most common method used for the separation of mononuclear cells (MNC) from the blood is Ficoll-Hypaque density gradient centrifugation, essentially described by Boyum *(1)*. Ficoll-Hypaque at a density of 1.077 g/L is used to separate the denser granulocytes and erythrocytes from the lighter lymphocytes, monocytes, and thrombocytes. The mononuclear cells stay at the top of the Ficoll-Hypaque layer, whereas the denser cells sink to the bottom of the centrifuge tube.

Peripheral blood monocytes are purified from the mononuclear fraction by adherence to plastic. Adherence can be carried out either directly onto tissue-culture plates in which they will be grown further (24- or 96-well plates), or onto tissue-culture flasks from which they will be removed and recultured in the required plates or chambers *(2)*.

When an enriched monocyte cell suspension is required, MNC harvested on Ficoll-Hypaque gradients can be further separated on Percoll gradients into lymphocytes and monocytes. The method initially described by Ulmer and Flad *(3)* was modified by Orlandi et al. *(4)*, which used only one Percoll concentration. After separation on Percoll, the monocytes that are less dense than lymphocytes stay on top of the Percoll layer, whereas the lymphocytes go through to the bottom of the tube.

From: *Methods in Molecular Medicine: Human Cell Culture Protocols*
Edited by: G. E. Jones Humana Press Inc., Totowa, NJ

Long-term incubation of monocytes under tissue-culture conditions results in the differentiation of the cultured cells, and in the appearance of monocyte-derived macrophages *(5)*. Long-term incubation of the cells in culture results in a substantial loss of cells. In various studies, it was found that granulocyte-macrophage colony-stimulating factor (GM-CSF) or IL-3 *(6–10)*, as well as PKC activators/tumor promoters *(11,12)* can be used to facilitate the survival of MoDM.

The adherent human monocytes bind firmly to plastic substrata, and it is very difficult to remove the cells for quantitative measurements. There are several methods to enumerate or quantitate adherent monocytes/macrophages, of which four are described.

2. Materials

1. Earle's balanced salt solution (EBSS).
2. Dulbecco's PBS without Ca^{2+} and Mg^{2+}.
3. Ficoll-Hypaque (density 1.077 g/L).
4. RPMI-1640, supplemented with 100 µg/mL streptomycin 100 U/mL penicillin, 300 µg/mL (2 mM) L-glutamine.
5. Newborn bovine serum (NBS) or pooled human AB serum (HABS), heat-inactivated (56°C, 30 min).
6. Percoll (Pharmacia LKB Biotechnology AB, Uppsala, Sweden).
7. 10X EBSS.
8. Human recombinant GM-CSF and IL-3.
9. Phorbol retinoyl acetate (PRA).
10. 12-*O*-Tetradecanoyl-13-phorbol acetate (TPA).
11. Mezerein (MEZ).
12. Bio-Rad Protein Assay, cat. no. 500-0006, Bio-Rad Laboratories (Munich, Germany). Dilute 1:5 in double-distilled H_2O, and filter before use. Prepare fresh for each assay.
13. Hemacolor® color reagents (Merck, Darmstadt, Germany), or Diff-Quik® reagents (Harleco, Gibbstown, NJ).
14. 0.5% SDS dissolved in double-distilled H_2O.
15. Methanol.
16. Isopropanol (2-propanol) for analysis.
17. MTT (3-[4,5-dimethyl-thiazol-2-yl]-2,5-diphenyl tetrazolium bromide) (5 mg/mL in PBS).
18. Limulus Amebocyte Lysate (LAL) reagent (Pyrogent, Whittaker, M.A., Bioproducts Inc., Walkersville, MD) (*see* Note 1).

3. Methods

3.1. The Isolation of MNC *(see Note 2)*

Peripheral blood human monocytes can be separated from heparinized blood samples or from the buffy coats of normal blood bank donors with sodium

Mononuclear Phagocytes 155

citrate as an anticoagulant. Blood bank buffy coats are obtained after centrifugation of blood bank bags (original volume of 400 mL) and removal of the upper layer of the content. The separated fraction (25–40 mL) contains more than 90% of the leukocytes of the blood donation, 10% of the erythrocytes, and 5% of the plasma.

1. Dilute buffy coats 1:4 (v/v) or heparinized blood samples 1:2 (v/v) with Dulbecco's PBS without Ca^{2+} and Mg^{2+}.
2. Add 40 mL of the diluted blood to 50-mL conical polypropylene tubes.
3. Add 10 mL of Ficoll-Hypaque to the bottom of each tube. To prevent mixing, insert a sterile Pasteur pipet into the diluted blood down to the bottom of the tube, and inject the Ficoll-Hypaque into the pipet by a 10-mL syringe (0.8 × 40 mm needle).
4. Centrifuge the tubes at 700g for 30 min at room temperature with brakes off.
5. Remove the cellular fraction that is on top of the Ficoll-Hypaque layer, and transfer to a new 50-mL conical tube. For blood bank buffy coats from the same donor, pool cells from four tubes into one.
6. Wash three times with 50 mL cold Dulbecco's PBS without Ca^{2+} and Mg^{2+} under the following conditions: (a) once at 350g for 10 min, and (b) twice at 230g for 7 min.
7. Resuspend the cells in supplemented RPMI-1640 containing 10% serum.
8. Place the tube for 16 h at 4°C to allow for separation of the MNC (cell pellet) from the thrombocytes (fluid suspension).
9. Discard the fluid gently by low-rate aspiration, and resuspend the cells in 10 mL cold supplemented RPMI-1640 + 2% serum.
10. For cell counts, dilute first 1:10 in supplemented RPMI-1640, and then 1:2 with 0.1% Trypan Blue solution, and count the cells with a hemocytometer. The average yield is $540 \pm 130 \times 10^6$ MNC (range 400–770)/original 400 mL blood.

3.2. The Isolation of Mononuclear Phagocytes

3.2.1. Separation of Monocytes by Adherence in Microtiter Plates

1. Prepare MNC at $2-3 \times 10^7$ cells/mL in supplemented RPMI-1640 + 2% serum.
2. Add 0.1 mL of the cell suspension to each well ($2-3 \times 10^6$ cells/well) of a 96-well tissue-culture plate.
3. Incubate for 30 min at 37°C and 7.5% CO_2.
4. Remove nonadherent cells by washing the wells three times with warm (37°C) EBSS. For this purpose, use a 5- or 10-mL syringe (without a needle).
5. Add 0.2 mL of supplemented RPMI-1640 + 10% serum. The resulting monolayers contains 2.43 ± 0.22 to $3.04 \pm 0.4 \times 10^5$ cells/well.

3.2.2. Separation of Monocytes by Adherence in Tissue-Culture Flasks

1. Resuspend MNC at 5×10^6/mL in supplemented RPMI-1640 + 2% serum.
2. Add 5 mL of the cell suspension to 25-cm^2 or 20 mL to 75-cm^2 tissue-culture flasks.
3. Incubate for 30 min at 37°C and 7.5% CO_2.
4. Remove nonadherent cells by washing the flasks extensively three times with warm (37°C) EBSS.

5. Add 5 or 25 mL of supplemented RPMI-1640 + 10% serum to the 25- or 75-cm^2 tissue-culture flasks, respectively.
6. Incubate for 16 h at 37°C and 7.5% CO_2.
7. Shake the flasks firmly, and remove the medium that contains the detached monocytes. To remove still adherent cells, wash the flasks with cold EBSS using a 10-mL syringe and a 0.8 × 40 mm needle.
8. Centrifuge the cells at 350g for 10 min, and resuspend the cells in supplemented RPMI-1640 + 10% serum.
9. Culture 2–3 × 10^5 cells/well/0.2 mL in 96-well plates, or 5 × 10^5 cells/well/ 0.5 mL in 24-well plates.

3.2.3. Separation of Monocytes on an Iso-Osmotic Percoll Gradient

1. Prepare iso-osmotic supplemented RPMI-1640 + 10% serum by adjusting the osmolality to 285 mosM.
2. Prepare iso-osmotic Percoll solution by mixing Percoll with EBSS (10X) 93:7 (v/v), and adjust to 285 mosM.
3. Prepare a 46% solution of iso-osmotic Percoll with iso-osmotic RPMI-1640.
4. Resuspend the mononuclear fraction obtained after Ficoll-Hypaque separation in iso-osmotic RPMI-1640 at 5–10 × 10^6 cells/mL.
5. Add 5 mL of the cell suspension to 10–12 mL tubes.
6. Add 5 mL 46% Percoll solution to each tube using a Pasteur pipet as described in Section 3.1. for Ficoll-Hypaque.
7. Centrifuge at 600g for 30 min at room temperature with brakes off.
8. Remove interface with a sterile pipet, and wash the cells twice with supplemented RPMI-1640.
9. Count cell number.
10. The recovery is 11.3 ± 3% (range 8–18%) of the MNC fraction, and 86 ± 6% of the cells are monocytes (range 77–95%).

3.3. Long-Term Cultures of MoDM

3.3.1. MoDM Cultured in the Presence of Colony-Stimulating Factors

1. Prepare cultured monocytes in 96- or 24-well plates as described (*see* Note 3).
2. Add 50–200 U/mL GM-CSF or IL-3 to the monocyte cultures (*see* Note 4).
3. Incubate without changing the medium for at least 10 d to obtain human monocyte-derived macrophages (HuMoDM).
4. If extended incubation periods are required, add the indicated amount of CSF every 2 wk by replacing half of the volume of the culture medium.

3.3.2. MoDM Cultured in the Presence of PKC Activators/Tumor Promoters (see Note 5)

1. Prepare cultured monocytes in 96- or 24-well plates as described (*see* Note 3).
2. Add TPA, MEZ, or PRA to monocyte cultures at a final concentration of 2–5 nM (*see* Note 6).
3. Incubate, without changing the medium, for at least 10 d to obtain HuMoDM.

4. If extended incubation periods are required, add the indicated amount of PKC activators every 3 wk by changing half of the volume of the culture medium.

3.4. General Methods for the Quantitation of Cultured Adherent Mononuclear Phagocytes

3.4.1. Cell Count

1. Remove the culture medium from the cultured monocytes.
2. Add ice-cold PBS (0.2 mL) without Ca^{2+} and Mg^{2+}.
3. Scrape gently the adherent cells, and count with a hemocytometer.

3.4.2. Protein Determination (see Note 8)

1. Wash the cultured cells extensively with EBSS to remove medium and serum.
2. Lyse the cells with 200 µL of 0.1% Triton X-100 for 30 min at 37°C.
3. Add 20-µL samples of the cell lysates to 200 µL of Bio-Rad reagent in 96-well plates.
4. Incubate the samples for 15 min, and read at 600 nm in an automated spectrophotometer. Protein concentration is determined using a standard curve of BSA (5–1000 µg/mL).

3.4.3. Hemacolor Colorimetric Microtiter Assay (see Note 9)

1. Remove the supernatant from cell monolayers cultured in 96-well plates, and dry the cells quickly in the air.
2. Fix the monolayers with methanol (50 µL/well) for 30 s (do not rinse the wells between steps 2 and 4).
3. Add 80–100 µL/well of Hemacolor Reagent 2 for 60 s.
4. Add 80–100 µL/well of Hemacolor Reagent 3 for another 60 s.
5. Rinse the plates three times with tap water.
6. Fill again with water and decolorize for 5 min.
7. Remove the water, and dry the plates extensively. (Following this step, the stained cultures can be kept for several weeks in the dark.)
8. For stain extraction, add 0.2 mL/well of SDS (0.5%) dissolved in double-distilled H_2O for at least 90 min.
9. Measure OD at 600 or 630 nm with an automated microplate reader.

3.4.4. MTT Assay (see Note 10)

1. Remove the culture medium from cell monolayers cultured in 96-well plates.
2. Reconstitute each well with 0.2 mL supplemented RPMI-1640 containing 1 mg/mL MTT.
3. Incubate the cultures for 2–4 h at 37°C.
4. Remove the supernatants from the wells.
5. Add 0.2 mL/well of lysing reagent containing $0.04N$ HCl in isopropanol.
6. Mix the wells thoroughly.
7. Read the plates in an automated microplate spectrophotometer at 570 and 630 nm as reference.

3.4.5. Alternative MTT Method

1. If the cultured cells are not tightly adherent and might be removed with the supernatant, it is recommended to remove only 0.1 mL of the culture medium from the cells cultured in 96-well plates.
2. Add to each well 0.025 mL of 5 mg/mL MTT in PBS.
3. Incubate the cultures for 2–4 h at 37°C.
4. Add 0.1 mL/well of lysing reagent containing $0.04N$ HCl in isopropanol.
5. Mix the wells thoroughly.
6. Read the plates in an automated microplate spectrophotometer at 570 and 630 nm as reference.

4. Notes

1. All the media and buffers used should be assayed for the presence of bacterial endotoxin by the Gel-clot technique *(13)* using the Limulus Amebocyte Lysate (LAL) reagent. Reagents should be used only if no detectable Lipopolysaccharide (LPS) is found (sensitivity—0.064 Endotoxin U/mL).
2. Steps 1–5 of this procedure are carried out at room temperature (T_r), and all the reagents should be at T_r. The use of cold reagents should be avoided at this stage of the separation.
3. For extended incubation periods (more than 4 d), it is recommended not to culture cells in the wells at the periphery of the plates. These wells should be filled with sterile water to the top to reduce evaporation of liquid from the cultures.
4. The optimal effect of CSF was achieved when added on the first day of culture. After 6 d in culture, the cells did not respond to the addition of CSF, and they behaved as nontreated cells. Microscopic observations of MoDM obtained in the presence of CSF revealed a homogenous population of large spread-out cells, whereas nontreated cultures were more heterogeneous in their appearance and some small round cells were also apparent *(6,7)*.
5. Adherent HuMo cultured in 96-well plates showed a substantial loss (51%) of adherent cells in nontreated monocyte cultures after 2 wk of incubation. In comparison, HuMoDM cultures treated with various PKC activators/tumor promoters lost only 0–26% of the cells after incubation for 2 wk *(11,12)*.
6. Prepare stock solutions of PRA, TPA, and MEZ in DMSO at 10 µM, and store in the dark at –20°C. When diluting the reagents in culture media before adding to the cells, the final concentration of DMSO should not exceed 0.1%.
7. In our laboratory, we maintained MoDM cultures for 4 mo by adding TPA (2 nM) every 3 wk.
8. Determination of protein concentration of cells cultured in 96-well plates is according to the Bradford method *(14)*.
9. The Hemacolor colorimetric microtiter assay *(15)* uses reagents generally used to stain blood cells and cells in tissue cultures. The staining kit holds three solutions: Solution 1 contains methanol for fixation, solution 2 contains a xanthene dye (orange color), and solution 3 is a Thiazine solution containing a mixture of

azure I dyes and methylene blue (blue reagent). A spectrophotometric analysis of a mixture of solutions 2 and 3 in SDS 0.5% revealed a peak of absorption at 517 nm caused by the xanthene dye, and a second peak at 634 nm caused by the thiazine solution. Measurements of stained cells are carried out in an automated microplate reader using 630- or 600-nm filters. Hemacolor reagents can be substituted by Diff-Quik reagents that serve a similar purpose.

10. The MTT assay is based on the observation that tetrazolium salts are reduced to formazan by cellular respiratory enzymes. This activity is performed only by viable cells, and thus the method may indicate the amount of viable cells present in culture. The method was initially described by Mosmann *(16)* for MTT, but other tetrazolium reagents may also be used *(17)*.

References

1. Boyum, A. (1968) Isolation of mononuclear cells and granulocytes from human blood. *Scand. J. Clin. Lab. Invest.* **21(Suppl. 97),** 77–89.
2. Treves, A. J., Yagoda, D., Haimovitz, A., Ramu, N., Rachmilewitz, D., and Fuks, Z. (1980) The isolation and furification of human peripheral blood monocytes in cell suspension. *J. Immunol. Methods* **39,** 71–80.
3. Ulmer, A. J. and Flad, H.-D. (1979) Discontinuous density gradient separation of human mononuclear leukocytes using Percoll as gradient medium. *J. Immunol. Methods* **30,** 1–10.
4. Orlandi, M., Bartolini, G., Chiricolo M., Minghetti, L., Franceschi, C., and Tomasi, V. (1985) Prostaglandin and thromboxane biosynthesis in isolated platelet-free human monocytes. I. A modified procedure for the characterization of the prostaglandin spectrum produced by resting and activated monocytes. *Prostaglandins, Leukotrienes Med.* **18,** 205–216.
5. Zuckerman, S. H., Ackerman, S. K., and Douglas, S. D. (1979) Long-term human peripheral blood monocyte cultures: establishment, metabolism and morphology of primary human monocyte-macrophage cell cultures. *Immunology* **38,** 401–411.
6. Robin, G., Markovich, S., Athamna, A., and Keisari, Y. (1991) Human recombinant granulocyte-macrophage colony stimulating factor augments the viability and cytotoxic activities of human monocyte derived macrophages in long term cultures. *Lymphokine and Cytokine Res.* **10,** 257–263.
7. Dimri, R., Nissimov, N., and Keisari, Y. (1994) Effect of human recombinant granulocyte-macrophage colony stimulating factor and IL-3 on the expression of surface markers of human monocyte derived macrophages in long term cultures. *Lymphokine and Cytokine Res.* **14,** 237–243.
8. Elliot, M. J., Vadas, M. A., Eglinton, J. M., Park, L. S., Bik To, L., Cleland, L. G., Clark, S. C., and Lopez. A. F. (1989) Recombinant human interleukin-3 and granulocyte-macrophage colony-stimulating factor show common biological effects and binding characteristics on human monocytes. *Blood* **74,** 2349–2359.
9. Markowicz, S. and Engleman, E. G. (1990) Granulocyte-macrophage colony-stimulating factor promotes differentiation and survival of human peripheral blood dendritic cells in vitro. *J. Clin. Invest.* **85,** 955–961.

10. Eischen, A., Vincent, F., Bergerat, J. P., Louis, B., Faradji, A., Bohbot, A., and Oberling, F. (1991) Long term cultures of human monocytes in vitro. Impact of GM-CSF on survival and differentiation. *J. Immunol. Methods* **143,** 209–221.
11. Keisari, Y., Bucana, C., Markovich, S., and Campbell, D. E. (1990) The interaction between human peripheral blood monocytes and tumor promoters: Effect on in vitro growth, differentiation and function. *J. Biol. Response Modif.* **9,** 401–410.
12. Markovich, S., Kosashvilli, D., Raanani, E., Athamna, A., O'Brian, C. A., and Keisari, Y. (1994) Tumor promoters/protein kinase C activators augment human peripheral blood monocyte maturation in vitro. *Scand. J. Immunol.* **39,** 39–44.
13. Yin, E. T., Galanes, C., Kinsky, S., Bradshaw, R., Wessler, S., and Luderitz, O. (1972) Picogram-sensitive assay for endotoxin: Gelation of limulus polyuphemus blood cell lysate induced by purified lipopolysaccharide and lipid A from gramnegative bacteria. *Biochim. Biophys. Acta* **261,** 284–289.
14. Bradford, M. M. (1976) A rapid and sensitive method for the quantitation of microgram quantities of protein utilizing the principle of protein-dye binding. *Anal. Biochem.* **72,** 248–254.
15. Keisari, Y. (1992) A colorimetric microtiter assay for the quantitation of cytokine activity on adherent tissue culture cells. *J. Immunol. Methods* **146,** 155–161.
16. Mosmann, T. (1983) Rapid colorimetric assay for cellular growth and survival: Application to proliferation and cytotoxicity assays. *J. Immunol. Methods* **65,** 55–63.
17. Alley, M. C., Scudiero, D. A., Monks, A., Hursey, M. L., Czerwinski, M. J., Fine, D. L., Abbott, B. J., Mayo, J. G., Shoemaker, R. H., and Boyd, M. R. (1988) Feasibility of drug screening with panels of human tumor cell lines using a microculture tetrazolium assay. *Cancer Res.* **48,** 589–601.

14

Purification of Peripheral Blood Natural Killer Cells

Ian M. Bennett and Bice Perussia

1. Introduction

The ability to perform biological studies on Natural Killer (NK) cells requires effective methods for their isolation from hematopoietic cells of other lineages. NK cells are a discrete lymphocyte subset distinguishable from B- and T-lymphocytes on the basis of both physical and phenotypic characteristics that can be exploited for their purification. Techniques based on differential cell buoyancy (centrifugation on discontinuous density gradients, such as Percoll *[1]*) have been used to enrich NK cells from mixed lymphocyte populations, but do not allow purification of these cells to homogeneity. The mononuclear cell suspensions obtained, although enriched in NK cells, also contain variable proportions of other cell types (notably monocytes and/or activated T- and B-lymphocytes) *(2)* and subsets of NK cells of higher density are lost in these preparations.

The most satisfactory purification techniques for NK cells, as well as for other leukocyte subsets, rely on their distinctive phenotype and make use of monoclonal antibodies (MAbs) directed to lineage-specific surface antigens (Ag). Although NK cell-specific surface markers have not been identified yet, lack of surface expression of T-cell receptor/CD3 complex and surface Ig, and expression of CD16 (low-affinity receptor for the Fc portion of IgG, FcγRIIIA) *(3)* and CD56 (and N-CAM isoform) *(4)* serve to identify NK cells within mononuclear cell populations. MAbs to both antigens are available, and cells sensitized with them can then be detected with a variety of secondary reagents to permit their identification and physical separation. Using this approach, homogeneous preparations of NK cells are isolated from mixed mononuclear cell populations following either of two schemes: direct isolation of NK cells using MAbs to surface Ag expressed on these cells (positive selection) and depletion of all cells other than NK using a mixture of MAbs directed to Ag expressed on the former, but absent from the latter population (negative selection). The advantage of positive selection is the rapid and specific isolation of NK cells. However, Ab binding to antigens capable of signal transduction, like CD16, leads to modulation of NK cells' biological functions *(5,6)*, making them unusable for certain applications. Negative selection techniques, instead, yield cells in their least altered state that may be most suitable for most functional studies. Choosing between the two systems depends on the specific experimental requirements.

Here we describe in detail the use of a dependable and relatively inexpensive method of isolating NK cells (indirect antiglobulin resetting) that is suitable for both positive and negative selection, results in good yields, and allows easy and rapid manipulation of large numbers of cells. Indirect antiglobulin resetting is based on the use of erythrocytes (E) coated with anti-mouse Ig Ab as a secondary reagent to detect cells that have murine MAb bound at their surface which recognizes lineage-specific Ag, leading to the formation of rosettes. The subsequent physical separation of Ag^+ (rosetted) and Ag^- (nonrosetted) cells is obtained by simple centrifugation on density gradients. Other reliable techniques exist that use secondary reagents coupled to different detection systems but, unlike indirect antiglobulin resetting, may not be practical for all investigators owing to unavailability of specialized equipment, low yields, or prohibitive costs. For example, fluorochrome-labeled secondary reagents and fluorescence-activated cell sorting *(7)* are used to purify highly homogeneous NK cell populations: This requires availability of a flow cytometer, is time and money consuming, and has the disadvantage of allowing recovery of relatively low numbers of cells; methods using magnetic beads *(8)*, although fast and efficient are extremely expensive; panning the Ab-sensitized cells on dishes coated with antimouse Ig *(7)* is efficient, fast, and economical but may become impractical when large numbers of cells need to be processed; complement (C)-dependent lysis *(7)*, which is practical and efficient, can be performed only with C-fixing MAb and may result in nonspecific toxicity and, consequently, the need for screening numerous batches of sera for optimal use. These methods (described accurately in the references provided) may however be used efficiently instead of indirect antiglobulin resetting when specific needs make them appropriate and practical. The general approach to cell isolation discussed can also be applied to a variety of additional needs such as subfractionation of NK cell subsets (e.g., $CD8^+$ and $CD8^-$ cells) substituting appropriate MAb in purification steps following the isolation of NK cells by negative selection.

Since the number of NK cells that can be obtained directly from peripheral blood is low and may not be sufficient for some studies, a protocol is also provided to increase the number of NK cells in short-term cultures in vitro. These cultures can be used as a starting population to separate numbers of NK cells larger than those that would he obtained from equivalent volumes of fresh peripheral blood. NK cells prepared in this way, however, have some characteristics of activated NK cells *(9)*, and it is advisable that results of studies using these cells be confirmed with primary resting NK cells.

2. Materials

1. Culture medium: RPMI-1640, supplemented with 10% heat-inactivated (45 min, 56°C) fetal bovine serum (FBS), 2 mM glutamine, and, if desired, antibiotics (0.5 U/mL penicillin, 0.5 µg/mL streptomycin) (complete medium).
2. 1.077 g/mL Ficoll-Na Metrizoate density gradient, such as Ficoll-Hypaque (F/H) (Pharmacia, Uppsala, Sweden); it is stored at 4°C in the dark.
3. Phosphate-buffered saline (PBS): 12 mM NaH$_2$PO$_4$, 12 mM Na$_2$HPO$_4$, pH 7.2, 0.15M NaCl.
4. 0.15M NaCl.
5. CrCl$_3$ Solution: 0.1% CrCl$_3$ · 7H$_2$O in 0.15M NaCl, pH 4.5 (stock solution). This must be prepared in advance and aged at least 1 mo before use. The stock solution must be stored at room temperature in a glass container protected from exposure to light; the shelf life for this solution is at least 1 yr. During the first week after preparing the solution, its pH needs to be checked every other day and adjusted to 4.5, if needed. Repeat the same once per week for the following 3 wk. This solution is used to couple antimouse antibodies to erythrocytes (E). Because CrCl$_3$ causes E agglutination by linking membrane proteins, each new batch of CrCl$_3$ solution must be titrated to determine the subagglutinating dilution to he used. For this, sheep E (25 µL of a 2% suspension in 0.15M NaCl containing 0.1% bovine serum albumin [BSA]) are incubated (1:1 [v:v]) in round-bottom 96-well plates with serial 1:2 dilutions of the CrCl$_3$ stock solution in 0.15M NaCl. The lowest dilution not causing E agglutination is determined after a 30-min incubation at room temperature.
6. 1 mg/mL Goat antimouse Ig (GaMIg) in 0.15M NaCl, adsorbed on human Ig and affinity-purified. Affinity-purified GaMIg is commercially available, but may contain human Ig-crossreactive Ab that, if present, may bind Ig-bearing cells (e.g., B-lymphocytes or opsonized monocytes) and lead to contamination of the NK cell preparation with these cell types. Their depletion can be easily obtained by passing the preparation over a human IgG-CNBr Sepharose 4B column. Phosphates need to be removed from the preparations by dialysis against 0.15M NaCl (four changes are usually sufficient). After dialysis, the preparation is filter sterilized (minimum concentration 1 mg/mL) and stored in 1–2-mL aliquots at 4°C for years without loss of titer. It is used to prepare the SE detection system.
7. MAbs reacting with leukocyte subsets: anti-T-cells: CD3, CD4, CD5; anti-monocytes: CD14, CD32, CD64: anti-B-cells: CD19; anti-NK cells: CD16, CD56. If needed (*see* Notes 1 and 2): antihuman E (antiglycophorin A) and anti-PMN (CD15). The murine B-cell hybrids producing these MAbs are all available

from the American Type Culture Collection (ATCC); culture supernatants or ascites can be used; alternatively. MAbs can he purchased from commercial sources.
8. Sheep erythrocytes (SE): These can be obtained from several commercial sources and are stored in Alsever's solution at 4°C for approx 1 mo. These cells are coated with the affinity-purified GaMIg using the procedure described in Section 3.
9. B-Lymphoblastoid cell lines to be used as feeder cells: RPMI-8866, Daudi, or possibly other B-cell lines.

3. Methods
3.1. Preparation of the Starting Lymphocyte Populations
3.1.1. Peripheral Blood Lymphocytes (PBL)

1. Peripheral blood mononuclear cells (PBMC) are first prepared by density gradient centrifugation. The expected cell yield is approx 1×10^6 cells/mL of blood from healthy donors (range $0.5-2 \times 10^6$). Place 15 mL F/H in a 50-mL conical centrifuge tube and overlay 30 mL of blood (anticoagulated with heparin) slowly on top of this solution. For optimal recovery, care has to be taken not to disrupt the surface tension of the density gradient material.
2. Centrifuge at 800g, 15–20°C, for 20–30 min. Make sure that the centrifuge brake has been turned off to obtain a sharp PBMC band at the gradient's interface.
3. Carefully collect the mononuclear cell band at the interface using a 10-mL pipet, and transfer it to a new 50-mL tube. Remove all cells in this band, trying to take as little of the density gradient as possible. Mix the cell suspension with PBS (1:1 [v:v]) to dilute any F/H carried over.
4. Centrifuge the cells at 350–400g, for 5 min at room temperature. Decant the cell-free supernatant (which is turbid owing to the presence of platelets) and rap the tube against a solid surface to resuspend the pellet; if a large number of E are present in the PBMC band, which sometimes happens because of variability in E density, it may be necessary to use vacuum aspiration to remove the cell-free supernatant and avoid cell loss (*see* Notes 1–3).
5. Resuspend the cells in PBS, and centrifuge at 150g for 7 min. Repeat additional washes in the same conditions twice, and finally resuspend the cells in complete medium for counting. The low-speed centrifugations are needed to reduce platelet contamination of the final cell suspension.
6. PBL are prepared from the PBMC isolated above using an adherence step to remove the majority of monocytes. For this, PBMC in complete medium are plated in tissue-culture-treated Petri dishes. The number of cells and volume of the cell suspension that will allow an even settling of cells depend on the size of the dish used. As an example, 50×10^6 cells in 5 mL medium form an evenly distributed monolayer when placed in 100-mm^2 dishes; proportionally lower numbers of cells (~5×10^6 cells/10-mm^2 surface) are placed in smaller dishes, but in this case, a relatively larger volume of medium may he needed to cover the plate evenly. Be careful to avoid adding bubbles to the plates since they will prevent the cells from evenly contacting the bottom of the dishes.

7. Incubate the cells at 37°C for 30 min in a 5% CO_2 atmosphere.
8. Collect nonadherent cells without detaching adherent monocytes: for this, add some PBS to each plate (not directly onto the cells, but onto the wall of the Petri dish), and swirl/rock the solution back and forth to resuspend nonadherent cells. Transfer this supernatant to a tube.
9. Add PBS to one of the plates, and repeat the washing step. Transfer the cell suspension to the next dish, and continue until all plates have been washed. Repeat this step again, continuing until no significant number of nonadherent cells can be seen under an inverted microscope (usually 4–5 washes). All cells collected in the washes are pooled with those collected in step 8.
10. Centrifuge the cells, resuspend them in compete medium, and count.

3.1.2. Short-Term PBL-B-Lymphoblastoid Cell Line Cocultures (9)

1. Grow the feeder cells in culture as needed; 2×10^5 feeder cells are needed for each 1×10^6 PBL that will be put into culture. The ability of B-lymphoblastoid cell lines to act in vitro as feeders to sustain preferential proliferation of NK cells from PBL depends on the quality of the feeder cells before they are added to the cultures (see Note 4). Exponentially growing, viable cells are essential for successful cultures.
2. Irradiate these cells with 30 gy. RPMI-8866 cells should be irradiated the day before they are needed and kept in a 37°C incubator until use. Daudi cells can be irradiated and placed into culture on the same day.
3. Immediately before use, centrifuge the cells ($200g$, 5 min) and resuspend them in fresh complete medium (potentially inhibitory cytokines produced by these cells during the overnight incubation are removed in this way).
4. Mix the feeder cells with the PBL, prepared as in the previous section, at a 1:5 feeder cells:PBL ratio, and a final PBL concentration of 2.5×10^5/mL complete medium.
5. Add 2 mL of the cell suspension to each well of a 24-well tissue-culture plate, and place in an incubator (37°C, humidified 5% CO_2 atmosphere). Cultures can be set up in flasks, but the yield of total cells, and of NK cells in particular, is lower.
6. On d 6 of culture, aspirate approximately half of the medium from the wells, and replace it with fresh medium (see Note 5).
7. On d 10, collect the cultures (see Note 6). The proportion of NK cells present can be determined by surface phenotyping (indirect immunofluorescence is the simplest method) the day before. On average, a fivefold increase in total cell number is achieved in the cultures at this time (e.g., $\sim 50 \times 10^6$ cells are recovered from cultures started from 10×10^6 PBL). Typically, NK ($CD16^+$/$CD56^-$) cells represent ~70–80 and 50% of the cell population when RPMI-8866 and Daudi cells are used as feeders, respectively (this represents a ~20-fold increase in the total number of NK cells compared to the starting PBL population). The remainder cells are $CD3^+$ T-cells. B-cells and monocyte/macrophages are not detectable at the end of the culture. If active proliferation is observed before d 10, the cells can be collected earlier.

3.2. Indirect Antiglobulin Rosetting for Positive and Negative Selection of NK Cells

3.2.1. Preparation of CrCl₃ SE Coated with GaMIg (10)

1. Wash SE three times with $0.15M$ NaCl (never use PBS: phosphates inhibit the CrCl$_3$-dependent coupling of proteins to cell membranes). Spin the cells ($800g$, 7 min) and aspirate the cell-free supernatant (decanting the supernatant may result in loss of erythrocytes, if the cells were in a loose pellet).
2. Dilute the CrCl$_3$ stock solution in $0.15M$ NaCl to the appropriate subagglutinating concentration previously determined for that batch (final optimal concentration is usually 0.01–0.005%). Filter-sterilize the solution using a 0.45-µm filter.
3. In a 50-mL conical centrifuge tube, mix the following in the given order to prepare 50 mL of a 4% suspension of E-CrCl$_3$-GaMIg: 28 mL $0.15M$ NaCl, 2 mL packed SE, 2 mL GaMIg (the Ig are dissolved in NaCl, $0.15M$ usually at 1 mg/mL; depending on the batch, lower concentrations can be used), 8 mL CrCl$_3$ solution. Smaller volumes can be prepared, depending on the need, modifying the volumes of the different reagents, but maintaining their relative proportions (*see* Note 8).
4. Incubate the suspension for 15 min at room temperature with occasional mixing.
5. Add PBS to stop the reaction and centrifuge ($800g$, 5 min).
6. Wash the E-CrCl$_3$-GaMIg twice with PBS (800 g/7 min), and resuspend in 50 mL complete medium. The suspension is stored at 4°C and can be used up to 1 mo. Each time before use wash the cell suspension once with PBS to remove membranes of lysed E or free Ig, which may have come off the cells and, if present, can complete with the intact SE for binding to the MAb-sensitized cells. Resuspend E at 4% in fresh complete medium.

3.2.2. Lymphocyte Sensitization with MAb

1. Resuspend the PBL preparations from which NK cells are to be purified (20×10^6/mL complete medium) in an appropriately sized centrifuge tube.
2. Dilute the desired MAb in PBS to the concentration previously determined to be optimal for rosette formation with cells known to express the Ag of interest, and mix (1:1 [v:v]) with the PBL. Culture supernatants, ascites, or purified Ig (or their F[ab']$_2$ fragments) are appropriate for use. In general, culture supernatants work best at a 1:2–1:4 dilution, ascites at a 10^{-3}–10^{-4} dilution, and purified Ig at 0.5–1 µg/mL. The optimal concentration to be used, however, has to be determined experimentally for each batch of Ab preparation (*see* Note 7). For negative selection of NK cells from PBL, use a mixture of anti-T (CD3, CD5), anti-B (anti-HLA-DR, CD19), and antimonocyte (CD14) MAb. If NK cells are to be purified from the PBL-B-lymphoblastoid cells cocultures, a mixture of anti-T and antimonocyte MAb is sufficient, since B-cells are not detectable in the cultures. For positive selection, use a mixture of anti-NK cell MAb (CD16, CD56) in all cases.
3. Incubate the cell suspension on ice for 30 min. Prechill the centrifuge and the tube carriers at this time (5–10°C).

Peripheral Blood NK Cells

4. Wash the excess unbound Ab with ice-cold PBS (5 min centrifugation, 400g, in the cold).
5. Decant the supernatant, and wash twice more as in step 4.
6. Resuspend the cells in 10 mL ice-cold complete medium, and place on ice. For optimal rosette formation, a maximum of 200×10^6 cells can he placed in a 50-mL tube; up to 50×10^6 cells are instead placed in a 15-mL round-bottom culture tube.

3.2.3. Rosette Formation with E-CrCl$_3$-GaMIg

1. Mix 2.5 mL of the 4% suspension of E-CrCl$_3$-GaMIg, in ice-cold fresh complete medium, with 200×10^6 PBL presensitized, as in Section 3.2.2., with the desired combination of murine MAb. Volumes of E suspension are proportionally modified to treat different numbers of cells.
2. Centrifuge in the chilled carriers/centrifuge (400g, 7 min).
3. Incubate the pelleted cells on ice for 30 min.
4. Resuspend the cells with a Pasteur pipet until all clumps have been disaggregated (rosettes do not break apart). Place a drop of the cell suspension on a slide with coverslip to check for percent rosettes (optical microscopy, count at least 200 cells) (*see* Note 9).

3.2.4. Enrichment/Purification of NK Cells

1. After resuspending the pellet (Section 3.2.3., step 4), underlay F/H, carefully displacing the lymphocyte–SE mixture upward (13 or 5 mL F/H solution are underlayed in 50- and 15-mL tubes, respectively).
2. Being careful not to jar the tubes, centrifuge them at 800g for 15 min.
3. After centrifugation, the cells expressing the antigens recognized by the MAb used are in the pellet (rosetted), and those not expressing them are at the interface of the gradient. If MAb-negative cells are to be obtained (negative selection), carefully transfer the nonrosetted cells from the interface of the gradient to a new tube, and add 50% by volume of sterile PBS (*see* Notes 10–12).
4. Wash the cells twice with PBS, and resuspend them in complete culture medium.
5. To recover the Ab-positive cells (positive selection), aspirate the F/H, resuspend the pellet in a small volume of PBS, transfer the cells to a new tube (in order to avoid contaminating these cells with Ab-negative cells, which may have adhered to the wall of the tube), fill it with PBS, and centrifuge. After decanting the PBS, resuspend the pellet by rapping the tube against a solid surface, and wash twice more.
6. In order to achieve the highest degree of purification (>98%), it is necessary to repeat the rosetting step, without adding new MAb, on the cells obtained in step 4 or 5 (be sure to keep the cells at 4°C). For this, the cells collected at the interface (negative selection, step 4) are pelleted with additional E-CrCl$_3$-GaMIg as before (steps 1–3 in Section 3.2.1.), and those collected from the pellet (positive selection, step 5) are resuspended in medium and pelleted again without adding more E. After 30 min of incubation on ice, the pellets are resuspended, F/H is

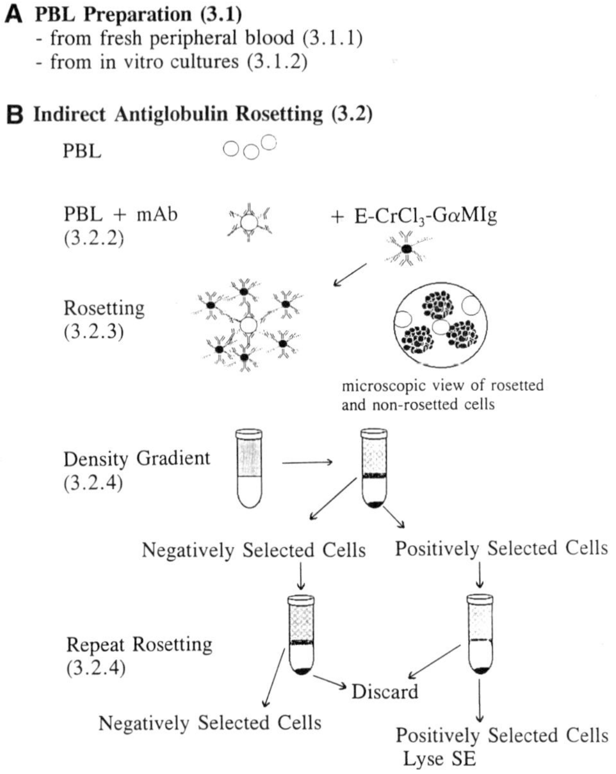

A PBL Preparation (3.1)
  - from fresh peripheral blood (3.1.1)
  - from in vitro cultures (3.1.2)

B Indirect Antiglobulin Rosetting (3.2)

PBL

PBL + mAb (3.2.2)   + E-CrCl$_3$-GαMIg

Rosetting (3.2.3)

microscopic view of rosetted and non-rosetted cells

Density Gradient (3.2.4)

Negatively Selected Cells   Positively Selected Cells

Repeat Rosetting (3.2.4)

→ Discard

Negatively Selected Cells   Positively Selected Cells
                              Lyse SE

Fig. 1. Schematic outline of the indirect antiglobulin rosetting method of NK cell purification.

underlayed, and the tubes are centrifuged as before (steps 1 and 2). NK cells are collected from the interface of the second F/H gradient performed with cells negatively selected (the pellet is discarded) or from the pellet of the second F/H gradient performed with positively selected cells (cells at the interface are discarded). In this latter case, SE can be lysed by adding 0.5 mL H$_2$O and pipeting carefully for 1 min, after which the tube is quickly filled up with PBS and the cells are washed twice.

Following this protocol, homogeneous NK cell preparations are reproducibly obtained (>98% CD16$^+$/CD56$^+$/CD3$^-$ cells, as determined by indirect or direct immunofluorescence [9] using a panel of MAb). The actual yield is, on average, 60% (range 50–75%) of the theoretical one expected on the basis of the proportion of NK cells in the starting population: ~8 × 10^6 and 30 × 10^6 NK cells can thus be obtained from 100 × 10^6 PBL or cultured lymphocytes, respectively, where NK cells are ~15 and 80% (*see* Notes 13 and 14). A scheme of the steps involved in the separation procedure is shown in Fig. 1.

4. Notes

4.1. Possible Problems in Preparing the Starting Lymphocyte Populations from Peripheral Blood

1. Occasionally, significant numbers of erythrocytes may contaminate the PBL preparations. These cells have an unusual buoyancy, will be carried over in each step, and will actually be enriched during the purification procedure. Depending on the application, it may be required that the E be lysed. For this, use H_2O (*see* Section 3.2.4., step 6), or $0.15M$ NH_4Cl, $0.01M$ $KHCO_3$, 0.1 mM EDTA, pH 7.2, buffer. To lyse E with this lysis buffer, the PBL suspension is incubated for 10 min on ice with the buffer (3 mL buffer/each 100×10^6 cells), and washed once with PBS. Alternatively, antiglycophorin A MAb can be added to the mixture of MAb used for negative selection.
2. PMN usually are not found at the interface of the F/H when using peripheral blood from healthy donors. However, in rare donors and some patients, PMN with altered densities are present that may not pellet through the density gradient, and variable proportions of them may contaminate the PBL preparation. They can be eliminated adding an anti-CD15 MAb to the mixture used for negative selection.
3. Low yield of PBL: This is usually owing to loss of cells following clumping during the isolation procedure. There are three primary causes of this:
   a. Insufficient or too high speed washes of the PBMC: in this case a large number of platelets that are subsequently activated (especially in the monocyte adherence step) and aggregate with other cell types may be carried over;
   b. Cell clumps are easily formed when the cells are either spun at too high a speed, or resuspended inaccurately after washing; and
   c. Release of DNA from dying cells, which acts as a tenacious adhesive in vitro: addition of 50 μg/mL DNase I, to these cells and incubation at 37°C for 2–3 min will clear the clumps (store aliquots of sterile DNase I, 5 mg/mL medium without FBS, for easy use). Continuing the purification without eliminating these clumps results in major cell loss.

4.2. Limited NK Cell Proliferation in the In Vitro PBL-Lymphoblastoid Cells Cocultures

4. Individual laboratory clones of the RPMI-8866 and Daudi cell lines may vary in their ability to support NK cell proliferation from PBL; moreover, this property decreases after they are kept in culture for long periods of time (4–6 mo). Defrost a new batch of low-passage cells when the NK cell yields start declining (usually this happens gradually).
5. The precise feeding and harvesting schedule for a particular feeder cell or isolate may need to be optimized following the kinetics of NK cell proliferation. Maximum NK cell proliferation may actually occur one or more days earlier or later than that described.
6. Although proliferation of NK cells in the cocultures is observed with PBL from most healthy donors, donor variability may account for occasional unsuccessful cultures.

4.3. Lack of Rosette Formation

7. Incorrect dilutions of the MAb used for PBL sensitization: To solve the problem, the working dilution for each MAb preparation used has to be determined. For this, serial dilutions of the MAb are incubated with lymphocytes in round-bottom 96-well plates under the conditions described in Sections 3.2.2. and 3.2.3. Rosette formation is assessed microscopically on an aliquot of the cells. Choose the dilution resulting in the maximum percentage of rosettes, corresponding to that of cells detected with the MAb by indirect immunofluorescence.
8. Insufficient amount of GaMIg coating the SE.
    a. Each new batch of GaMIg has to be titrated to determine the correct concentration of Ig to be coupled to E. This is achieved by testing for rosette formation SE suspensions prepared with serial dilutions of the GaMIg (2–0.25 mg/mL).
    b. Phosphates are present in one of the reagents used for $CrCl_3$ coupling. Because even trace amounts of phosphate inhibit the $CrCl_3$-dependent protein coupling to cell membranes, it is essential that E are washed with saline, and that all reagents used are prepared and diluted in the same solution (avoid PBS at any time).
    c. On rare occasions, coupling of GaMIg to E is unsuccessful for no apparent reason. Just start again. Each new batch of E-$CrCl_3$-GaMIg has to be prepared and tested in advance.
9. Low numbers of rosettes compared to the proportion of Ab^+ cells. If the reasons in Notes 1 and 2 can be excluded, the most likely explanation for this is that the temperature was not kept low during all steps. This may result in capping and down-modulation of the antigen from the cell surface, thus preventing efficient rosette formation.

4.4. Contamination of the Final NK Cell Preparation with Other PBL Subsets

10. Make sure that all procedures are performed in the cold and that rosetting is performed twice.
11. In the case of negative selection, the most likely explanation is a short time of centrifugation of the F/H gradient: increase the time to 20 min and/or modify the centrifugation speed.
12. In the case of positive selection, inaccurate resuspension of the pellets after rosette formation will result in trapping rosette Ab^- PBL in the clumps. These will sediment in the pellet of the F/H gradient, thus contaminating the rosette Ab^+ cells.

4.5. Low Recovery of NK Cells

13. In the case of negative selection the problem is, for the most part, owing to inaccurate resuspension of the pellets after rosette formation before the separation on F/H. Rosette-negative PBL trapped in the clumps reach the pellet, resulting in loss of cells.
14. When the recovery of positively selected cells is low, loss of cells probably occurred during the E lysis step. Careful resuspension of the cells during the lysis and use of the NH_4Cl buffer instead of H_2O should solve the problem.

References

1. Timonen, T., Ortaldo J. R., and Herberman, R. B. (1981) Characteristics of human large granular lymphocytes and relationship to Natural Killer cells. *J. Exp. Med.* **153**, 569–582.
2. Perussia, B., Fanning, V., and Trinchieri, G. (1985) A leukocyte subset bearing HLA-DR antigens is responsible for in vitro alpha interferon production in response to viruses. *Natl. Immun. Cell Growth Regul.* **4**, 120–137.
3. Perussia, B., Starr, S., Abraham, S., Fanning, V., and Trinchieri, G. (1983) Human Natural Killer cells analyzed by B73.1, a monoclonal antibody blocking Fc receptor functions. I. Characterization of the lymphocyte subset reactive with B73.1. *J. Immunol.* **130**, 2133–2141.
4. Lanier, L. L., Chang, C., Azuma, M., Ruitenberg, J., Hemperly, J., and Phillips, J. (1991) Molecular and functional analysis of Natural Killer cell-associated Neural Cell Adhesion molecule (N-CAM/CD56). *J. Immunol.* **146**, 4421–4426.
5. Perussia, B., Acuto, O., Terhorst, C., Faust, J., Lazarus, R., Fanning, V., and Trinchieri, G. (1983) Human Natural Killer cells analyzed by B73.1, a monoclonal antibody blocking Fc receptor functions. II. Studies of the B73.1 antibody antigen interaction at the lymphocyte membrane. *J. Immunol.* **130**, 2142–2148.
6. Anegon, I., Cuturi, M. C., Trinchieri, G., and Perussia, B. (1988) Interaction of Fc receptor (CD16) with ligands induces transcription of interleukin 2 receptor (CD25) and lymphokine genes and expression of their products in human Natural Killer cells. *J. Exp. Med.* **167**, 452–472.
7. Coligan, J., Kruisbeek, A., Marguilees, D., Shevach, E., and Strober, W. (eds.) (1991) *Current Protocols in Immunology*, Wiley, New York, Sections 3 and 5.
8. Naume, B., Nonstad, U., Steingker, B., Funderud, S., Smeland, E., and Espevic, E. (1991) Immunomagnetic isolation of NK and LAK Cells. *J. Immunol.* **148**, 2429–2436.
9. Perussia, B., Ramoni, C., Anegon, I., Cuturi, M. C., Faust, J., and Trinchieri, G. (1987) Preferential proliferation of Natural Killer cells among peripheral blood mononuclear cells cocultured with B lymphoblastoid cell lines. *Natl. Cell Growth Regul.* **6**, 171–188.
10. Goding, J. W. (1976) The chromium chloride method of coupling antigens to erythrocytes: definition of some important parameters. *J. Immunol. Methods* **10**, 61–66.

15

Cystic Fibrosis Airway Epithelial Cell Culture

Manuel A. Vega

1. Introduction
1.1. Cystic Fibrosis

Cystic fibrosis (CF) is the most frequent (incidence around 1/2500 live births) genetic cause of death among Caucasians. It is an autosomal recessive disorder compromising the secretory epithelia. Clinically, CF is a polymorphic disease showing abnormal functioning of the airways, the digestive apparatus (pancreas and intestine), the reproductive tract, and the sweat glands, leading to respiratory insufficiency, malnutrition, male sterility, and production of salty sweat. The average life-span of CF patients falls around 25–30 yr of age in the United States and Europe, and around 10 yr of age in Latin America *(1,2)*. Respiratory infections are the cause of death of more than 90% of CF patients. No curative treatments are as yet available for CF.

Chloride transport is the primary function affected in CF epithelial cells. The Cl⁻ transporter molecule involved in CF cells is a cAMP-dependent, apical membrane protein called cystic fibrosis transmembrane conductance regulator (CFTR). CFTR mutant cells have lost their ability to move chloride ions in response to cAMP *(3,4)*.

CFTR, a 1480 amino acid polypeptide, is encoded by the CF gene. The CF gene spans over 250 kb and contains 27 exons. At the time this chapter was written, more than 500 different mutations had been reported on CF alleles found in CF patients. The so-called ΔF508 mutation (an in-frame deletion of 3 bases causing the loss of phenylalanine at position 508 on the CFTR protein) is the mutation most frequently found in all populations so far tested; its relative frequency varies from 30–80% among different ethnic groups. The ΔF508 mutation affects the processing of CFTR along the endoplasmic reticulum and the Golgi apparatus leading to the absence of CFTR on the plasma membrane *(4–9)*.

CFTR expression has been reported in secretory epithelial cells (airways, intestine, pancreas, and epididymus). Moreover, there is some expression of CFTR in nonepithelial cells like lymphocytes and fibroblasts. Recently, expression of CFTR-mRNA in human ejaculated sperm cells (*10,11*, Vega et al., unpublished results) has been found.

In the airways of CF patients, a deficient CFTR-mediated Cl^- transport together with the observed decreased reabsorption of Na^+, leads to the alteration of the ionic and osmotic properties of the luminal mucus overlaying the epithelium. Deprived of salt ions, the mucus becomes dehydrated and more viscous, interfering with normal withdrawal and cillia-mediated cleaning up of the epithelium. The effect on the normal secretory functions of the airway epithelium thus leads to accumulation of mucus, bacterial colonization, inflammation, and final alteration of the bronchiolar and alveolar histological structure.

1.2. Airway Epithelial Cells

As far as CF is concerned, detailed knowledge on airway epithelial cells plays a pivotal role in two key ways: in studies aimed at getting insight into the basic phenomena involved in the molecular pathogenesis of the disease and in the development of gene transfer with therapeutic purposes into the epithelium in vivo.

Human airways extend from the nose inward, into trachea, bronchi, bronchioles, and alveoli. The epithelium covering the respiratory tract includes a mixture of different cell types with specialized functions, all laying on a basement membrane. The relative distribution of the different cell types varies along the different zones of the airways. CFTR expression level is, as well, not uniform along the different cell types found in human airway epithelia: it is highest in submucosal gland cells, found all along airway epithelium. Alterations caused by mutant CFTR molecules are restricted, however, to the bronchial and bronchiolar tubes *(12,13)*.

The major cell types found in the bronchial pseudostratified epithelium are basal cells (pyramidal cells not reaching the lumen of the epithelium), ciliated cells, undifferentiated cells (presumably the airway epithelium stem cells), goblet cells, and serous cells. Branching from the airway, the epithelium forms submucosal glands covered by goblet and serous cells, resting on a membrane contiguous to the airway epithelium basement membrane. Goblet and serous cells, either from the submucosal glands or from the airway epithelium, produce mucus that fills the submucosal glands and forms "islands" over the pericellular fluid lining the airway surface *(14,15)*.

The epithelium of the bronchiole is mainly constituted by two cell types: ciliated cells and Clara cells, forming a columnar epithelium that lays on a basement membrane and is lined by a mucus layer *(14,15)*.

A difference in cell type composition of bronchial, but not nasal, epithelia, between CF and normal patients is observed: relatively less ciliated cells and more basal, undifferentiated and secretory cells are found in CF bronchial epithelium than in normal bronchial epithelium *(16)*.

1.3. Epithelial Cell Culture

When cultured in vitro, epithelial cells show the characteristic epithelial morphology: cells are near isodiametric; cells are compacted in well-defined colonies, with precise limits and do not spread out of the colonies. Cells of some CF airway epithelial cell lines are, however, fibroblast-shaped, and aggregate in colonies with no precise limits and with cells spreading out of the colonies.

Epithelial cells express characteristic cytokeratins and form cell-to-cell junctions typical of epithelia. Monolayers formed by in vitro growth are real epithelial sheets that generate measurable potential differences on both sides of the monolayer.

Cell lines are grown on plastic. Colonies can be grown up from as little as 1 cell/well (1 mL of culture media/well). There is no difference in the mean size of the colonies grown from either 1 cell/well (1 mL) or 100–1000 cells/well (1 mL). No feeding is then necessary for growing individual isolated clones. Division rate is about 24 h.

Primary cells, however, are grown on collagen-coated dishes and usually on feeder layers. Division rate is around 100 h, and cells support only three to five passages before they stop dividing.

1.4. Mycoplasma Decontamination

Airway epithelial cells survive antimycoplasma treatment, consisting of culturing in normal medium containing 1% mycoplasm removal agent (ICN Flow, Costa Mesa, CA; cat. no. 30-500-44) for 2 wk. Growth rate and morphology are not affected by the treatment.

1.5. Sensitivity of Airway Epithelial Cells to G-418

As expected, different airway epithelial cell lines show different sensitivity to the neomycin derivative G-418. At day 7 of culture in the presence of G-418, survival of cells varies from cell line to cell line between 10 and 80–100% (50 μg G-418/mL) and between 0 and 50% (100 μg G-418/mL).

An interesting observation, however, has been made in the laboratory concerning the natural resistance of airway cell lines to G-418, suggesting that CF cell lines might be more sensitive to G-418 than normal cell lines. This difference allows for a selective recovery of the more resistant (normal) cell line in the presence of G-418, from mixtures of two different (normal and CF) cell lines with differential sensitivity to G-418 (unpublished data).

1.6. Differential Sensitivity of CF and Normal Cells to Epinephrine

It has recently been reported that CF airway epithelial cells, either primary or cells lines, show a higher sensitivity to epinephrine-induced toxicity than equivalent normal airway epithelium cells *(17)*.

Treatment with either epinephrine or forskolin rapidly kills CF cells, whereas normal cells are mainly unaffected. Killing of CFTR⁻ cells by epinephrine is quite rapid: 16–24 h. CF cells transfected with CFTR expression vectors become epinephrine-resistant (Vega et al., unpublished results). In fact, both CF (CFTR⁻) and normal (CFTR⁺) cells are sensitive to epinephrine, but CF cells respond to lower concentrations and shorter times of treatment. Sensitivity to epinephrine is dependent on cell density. CFTR⁺ cells can be recovered out of mixtures of CFTR⁻ and CFTR⁺ cells by epinephrine treatment. The epinephrine-based selection method is rapid and easy to perform. However, it demands a skilled operator as long as it depends on a relatively narrow window given by the differential sensitivity of CFTR⁻ and CFTR⁺ cells.

1.7. Sensitivity of CF Cells to Temperature

As mentioned, the ΔF508 mutation causes the CFTR to be retained at the Golgi apparatus and to not to reach the plasma membrane *(9,18)*. However, it has been reported that when growing (ΔF508) CF cells at temperatures lower than usual (e.g., 20°C), the (ΔF508) CFTR can be detected on the plasma membrane, and there it accounts for the partially recovered cAMP-dependent Cl⁻ transport. Therefore, ΔF508/ΔF508 CF cells behave more like "normal" cells when cultured at 20°C.

2. Materials

1. Culture media: For cell lines (either normal or CF cells) use Dulbecco's modified Eagle's medium (DMEM):Ham's F12 (1:1), 10% heat-inactivated fetal calf serum (mycoplasma free), 2 mM L-glutamine, 50 µg/mL penicillin-G, and 50 µg/mL streptomycin.

    Filter sterilize either the individual components or the final mixture, and store at 4°C for no longer than 2 wk. Antibiotics and L-glutamine are prepared in water, filter-sterilized, aliquoted in 5 mL stocks, and stored at –20°C. Fetal calf serum is heat-inactivated (30 min, 56°C), aliquoted in 50 mL, stored at –20°C, and filter sterilized just before adding to the culture medium.

    For primary cells (either normal or CF cells), use DMEM:Ham's F12 (1:1), 10% heat-inactivated fetal calf serum (mycoplasma free), 2 mM L-glutamine, 50 µg/mL penicillin-G, 50 µg/mL streptomycin, 5 µg/mL transferrin, 10 µg/mL insulin, 100 ng/mL hydrocortisone, and 25 ng/mL epidermal growth factor (EGF).

    Stock solutions (100–1000X) are prepared as follows: transferrin (human, iron-free) in DMEM:Ham's F12, insulin (bovine pancreas) in 0.001M HCl in

water, EGF (mouse) in water, and hydrocortisone (first dissolved in 100% ethanol to a concentration of 5 mg/mL) in HEPES-buffered Earle's salts (HBES) + 5% fetal calf serum. Stocks are filter sterilized, aliquoted, and stored at $-20°C$ (insulin solution can be stored at $4°C$ for a couple of weeks).
2. Coating with collagen: Dissolve collagen (Human placenta, type IV, Sigma, St. Louis, MO) in 0.1% acetic acid by stirring for 48 h at $4°C$. Filter the collagen solution through 2–3 sterile gauze layers. Dilute with water (1/20). Pour the collagen solution into plastic dishes: 1 mg/mL, 10–50 µg/cm². Air dry for 1–2 d (temperature not higher than $37°C$). Store coated dishes at $4°C$ in and humid atmosphere (for several months).
3. Storage medium:
   a. Medium 1: 10% dimethylsulfoxide (DMSO), 80% heat-inactivated fetal calf serum, 10% culture medium.
   b. Medium 2: 10% DMSO, 60% heat-inactivated fetal calf serum, 30% culture medium.
   Store at $-20°C$ for months. Add the DMSO component immediately before use.
4. Epinephrine solution: Prepare a 125-mM stock solution of epinephrine (Sigma) in water. The stock solution (1000X) can be stored at $-20°C$ for months and resists repeated cycles of freeze and thawing. Keep thawed epinephrine solutions on ice. Prepare the working dilutions immediately before use.
5. Loading solution: 108 mM NaCl, 4.7 mM KCl, 1 mM CaCl$_2$, 1 mM MgCl$_2$, 20 mM NaHCO$_3$, 0.8 mM Na$_2$HPO$_4$, 0.4 mM NaH$_2$PO$_4$, 10 mM glucose, and 5 mM HEPES, pH 7.2.
6. Immortalization buffer: 10 mM Na$_2$HPO$_4$, 10 mM NaH$_2$PO$_4$, 1 mM MgCl$_2$, and 250 mM sucrose, pH 7.4. Filter sterilize and store at $4°C$.
7. Detachment solution: 0.5 mM EDTA, 0.1% trypsin in (Ca^{2+}-free, Mg^{2+}-free) PBS. Prepare a 10X stock. Filter-sterilize, aliquot, and store at $-20°C$. Dilute in PBS to make a 1X working solution that can be stored at $-20°C$ and resist thawing several times.

3. Methods

3.1. Obtaining and Culturing of Primary Airway Epithelial Cells (see Notes 1–6)

1. Cells can be obtained by epithelial brushing of the airways through fiberoptic bronchoscopy; by scraping off nasal polyps following surgery; by scraping off the epithelium from lung, bronchi, or tracheal necropsy, or biopsy specimens.
2. In the two latter cases (scraping off from surgery pieces), protease treatment (2.5 mg/mL pronase, 60 min, $37°C$, in DMEM medium containing 1% fetal calf serum) is used to liberate cells.
3. Recover cells in culture medium for primary cells.
4. Centrifuge at 800g, $4°C$, for 5 min.
5. Recover in fresh medium, and culture in collagen-coated dishes at $37°C$, under 5% CO$_2$. Feed with fresh medium every 2 d.

3.2. Immortalization of Airway Epithelial Cells (see Notes 7 and 8)

1. Airway epithelial cells can be immortalized with the large T-antigen of SV40 virus.
2. Grow primary epithelial cells as usual.
3. Harvest cells with detachment solution (5–10 min, room temperature), and centrifuge for 10 min at 4°C, 800g.
4. Resuspend cells in 0.5 mL immortalization buffer at a density of 10^6–10^7 cells/mL.
5. Mix 10 µg of a SV40 large T-antigen plasmid expression vector with the 0.5 mL cell suspension.
6. Incubate for 10 min on ice.
7. Perform electroporation as indicated in Chapter 16, Section 3.1.2.
8. Incubate for 10 min at room temperature.
9. Recover cells in culture medium (culture medium for cell lines), and culture at 37°C, 5% CO_2, as usual. Replace the culture medium with fresh medium every 3–5 d.
10. After a couple of weeks, colonies of immortalized cells will be clearly distinguishable.
11. Continue culture of these immortalized clones as usual for cell lines.

3.3. Culture of Airway Epithelial Cells

Cells are grown at 37°C under 5% CO_2 in the respective culture media described above. Duplication time under the conditions described is around 24 h for the cell lines and around 100 h for the primary cultures.

3.4. Detachment of Cells

1. Completely remove the culture medium lying on the cells by aspiration.
2. Immediately overlay with 1–3 mL detachment solution for a small (30-mL) bottle or a 10-cm plate.
3. Incubate at room temperature in the hood for 3–5 min, avoiding direct contact of the plate/bottle with the metallic parts of the hood.
4. Shake gently by hand from time to time, and look by eye against a light or a window until the monolayer detaches. Firmly close the bottle and agitate energically several times to homogenize the cell suspension and disaggregate cell clumps.
5. Once the cell suspension is homogeneous, add 4 vol of culture medium to neutralize trypsin and recover cells by centrifugation at 800g for 10 min at 4°C.

3.5. Cryopreservation of Cells (see Notes 9–12)

3.5.1. Freezing Down Cells

1. Grow cells up to no more than 70% confluence in normal culture medium.
2. Detach cells as previously described.
3. Centrifuge at 4°C, 800g for 10 min.
4. Resuspend cellular pellet in ice-cold storage medium to a density of around 10^6 cells/mL. Always keep the cell suspension on ice and process immediately. Split into ice-cold cryotubes, 1 mL each, and keep on ice.
5. Freeze down using any automatic procedure, and store in liquid nitrogen.

3.5.2. Thawing Cells

1. Thawing of cells has to be performed with care. Take the cryotubes out the liquid nitrogen and immediately put on ice.
2. Thaw by putting the tube in a water bath previously set up at 37°C.
3. Immediately after thawing, put on ice and process.
4. Gently recover the cell suspension and pipet, drop by drop, into 5 mL of ice-cold normal culture medium.
5. Centrifuge at 4°C, 800g for 10 min.
6. Resuspend in an appropriate volume of culture medium, and culture as usual.

3.6. Epinephrine Selection Procedure (see Notes 13–15)

1. Grow cells up to 70–80% confluence as described.
2. Replace culture medium with fresh medium containing the desired concentration of epinephrine. Final epinephrine concentrations that allow for distinction between CFTR$^-$ and CFTR$^+$ cells are from 100–300 mM.
3. Culture as usual.
4. After 12–16 h, follow up CFTR$^-$ cell death under the microscope, observing every 2–3 h.
5. Gently shake the bottle or the dish by hand to remove partially detached cells.
6. To stop the selection, replace medium containing epinephrine with fresh epinephrine-free culture medium, and culture as usually.
7. Continue culture of surviving cells, or analyze the surviving population for the expression of active CFTR.

3.7. Sensitivity of (ΔF508) CF Cells to Temperature (see Note 16)

1. Grow cells up to 50–70% confluence at 37°C under 5% CO_2, as usual.
2. Replace culture medium with fresh medium, and continue culture at 20°C, under 5% CO_2, for 24 h.
3. Assay for activity of the CFTR on the plasma membrane. For instance, using the epinephrine-resistance assay:
   a. Change the old medium with fresh medium containing the desired concentration of epinephrine (100–300 µM final concentration).
   b. Continue culture at 20°C (or at 37°C, control cultures), under 5% CO_2, for another 24- or 48-h period.

3.8. CFTR Activity Assessment: Isotope Efflux Measurement (see Notes 17 and 18)

1. Grow cells in six-well plates until about 90% confluent, under the usual conditions.
2. Replace the culture medium with 0.5 mL of loading solution containing ^{125}I$^-$ (3–5 µCi/mL), and incubate at 37°C for 30 min.
3. Remove the extracellular isotope by washing three times, at room temperature, with 3 mL of loading solution.

4. Measure basal isotope efflux for 3–4 min at 37°C, by addition and replacement every 30 s of 1 mL of loading solution on the cells.
5. Add forskolin (10 µM) to the subsequent 30-s-replacements of loading solution to measure stimulated isotope efflux.
6. Measure $^{125}I^-$ in every (1-mL) isotope basal and stimulated efflux sample.

4. Notes
4.1. Obtaining and Culturing Primary Airway Epithelial Cells

1. The epithelial nature of the cells can be verified by their ability to form characteristic epithelial sheets. (Transepithelial potential differences can be measured across the epithelial sheet.)
2. Plating efficiency of these primary cells is approx 30%. Confluence is reached around 7 d of culture.
3. Under the described conditions, primary epithelial cells proliferate until three to four passages (splitting 1:4). After that, proliferation stops.
4. Airway epithelial primary cells can be grown on a feeder layer of 3T3 fibroblasts (like Todaro's Swiss mouse 3T3, 3T3J2, NIH 3T3, and Balb/c-3T3). For preparation of feeder cells:
   a. Grow 3T3 cells in 10% fetal calf serum in DMEM (they can be maintained as continually growing stocks for 1–2 mo) until confluence.
   b. Prepare confluent cultures of 3T3 cells as feeders by irradiation (30 gy). At this step, cells can be maintained in the incubator in fresh medium for several days before proceeding.
   c. Wash twice with detachment solution, and recover disaggregated cells.
   d. Split 1/3 and replace in 10% fetal calf serum in DMEM.
   e. Within 48 h, replace the DMEM medium by epithelial cell culture medium containing the airway epithelial primary cells, and culture as indicated.
5. Since fibroblasts may be contaminating the airway epithelial cells, they can be selectively eliminated as follows:
   a. Culture the epithelial cells until colonies of 50–200 cells are developed.
   b. Replace the culture medium by 0.02% EDTA in PBS, and incubate for half a minute in the hood.
   c. Pipet up and down the EDTA-PBS solution rather vigorously over all the area of the dish. (By this procedure, fibroblasts—as well as feeder cells—but not the epithelial cells are detached from the dish.
   d. Replace the EDTA-PBS with fresh solution, and repeat the operation until no fibroblasts can be seen under the microscope.
   e. Wash attached cells with serum-free medium, and finally add fresh complete culture medium.
   f. Add new irradiated feeder cells to the dish, and continue culture as usual.
6. Epithelial cells can be characterized by immunocytochemical detection of the cytokeratins typical of epithelial cells, by detection of epithelial cell–cell interaction structures through electron microscopy, or by measurement of potential differences across the epithelial monolayer.

4.2. Immortalization

7. Immortalized epithelial clones can be subsequently characterized by detection of the large T-antigen expression by immunocytochemistry, by Northern analysis, or by their ability to pass over a critical number of passages (necessary for a cell culture to become a cell line).
8. Alternative to the use of SV40 large T-antigen alone as immortalizing factor, hybrid SV40-adenovirus 12 has been used to immortalize airway epithelial cells *(19)*. When the cell lines are to be used for the study of CF gene transfer through (adeno)virus vectors, however, SV40 large T-antigen alone is preferable.

4.3. Cryopreservation

9. Cells stored at lower densities (10^4–10^5 cells/mL) are also viable.
10. Keep cells in DMSO-containing media always on ice. Unless cold, the DMSO will affect the cells.
11. Freezing down cells can be efficiently achieved by the following simple procedure: overnight at $-20°C$, followed by overnight freezing at $-80°C$, followed by final storage in liquid nitrogen. Cells can be stored in liquid nitrogen for years.
12. No higher concentrations of serum in the medium used for recovery and initial culture of the thawed cells is required. Cells usually recover rapidly and grow fast, so it is desirable to dilute the thawed suspension to a convenient cell density in order to avoid the necessity of rapid splitting.

4.4. Epinephrine Selection

13. Cell densities <70–80% increase $CFTR^+$ cell (resistant cells) sensitivity to epinephrine, whereas higher densities increase $CFTR^-$ cell (sensitive cells) resistance to epinephrine. Avoid trying epinephrine selection on very high-density cultures or on cultures containing cell clumps, because most cells will behave as epinephrine-resistant.
14. Selection time, on epinephrine treatment, depends mainly on cell density (apart from epinephrine concentration).
15. Surviving cells recover rapidly, after epinephrine removal from the culture medium.

4.5. Sensitivity to Temperature

16. Culture of cells at $20°C$ and 5% CO_2 can be easily performed by putting the dishes into a closed container like a sealed desiccator. The desiccator is first cooled down to $20°C$. It is then introduced open, with the dishes inside, into a $37°C$, 5% CO_2 cell incubator to equilibrate to 5% CO_2. Finally, the desiccator is quickly closed and put into a $20°C$ incubator (this incubator can be just a simple oven placed inside a $4°C$ cold room). The procedure is repeated three times a day in order to replace the atmosphere above the growing cells.

4.6. CFTR Activity

17. A stimulation in isotope efflux should be measurable following forskolin addition into non-CF epithelial cell cultures.

18. Results are expressed as the fractional efflux per time unit: amount of isotope released by the cells in a given time interval divided by the amount of isotope present in the cells at the beginning of that interval, divided by the length of the time interval.

Acknowledgments

The author thanks Lila N. Drittanti (Argentina), Jorge Gabbarini (Argentina), Claude Besmond (France), Michel Goossens (France), Pascale Fanen (France), and Bruno Costes (France), as well as AFLM (Association Française de Lutte contre la Mucoviscidose) (France), CONICET (National Research Council, Argentina), Hospital Int. Gral. J. Penna (Argentina), and MICROGEN S.A.-Biotechnology (Argentina).

References

1. Boat, T. F., Welsh, M. J., and Beaudet, A. L. (1989) Cystic fibrosis, in *The Metabolic Basis of Inherited Disease*, vol. 6 (Scrivier, C. L., Beaudet, A. L., Sly, W. S., and Valle, D., eds.), McGraw-Hill, New York, pp. 2649–2680.
2. McKusick, V. A. (1990) Cystic fibrosis, in *Mendelian Inheritance in Man. Catalogs of Autosomal Dominant, Autosomal Recessive and X-linked Phenotypes*. John Hopkins University Press, Baltimore, MD, pp. 1120–1126.
3. Quinton, P. M. (1983) Chloride impermeability in cystic fibrosis. *Nature* **301**, 421,422.
4. Kerem, B. S., Rommens, J. M., Buchanan, J. A., Markiewicz, D., Cox, T. K., Chakravarti, A., et al. (1989) Identification of the cystic fibrosis gene: genetic analysis. *Science* **245**, 1073–1080.
5. Riordan, J. K., Rommens, J. M., Kerem, B. S., Alon, N., Rozmahel, R., Grzelczak, Z., et al. (1989) Identification of the cystic fibrosis gene: cloning and characterization of complementary DNA. *Science* **245**, 1066–1073.
6. Rommens, J. M., Jannuzzi, M. C., Kerem, B. S., Drumm, M. L., Melmer, G., Dean, M., et al. (1989) Identification of the cystic fibrosis gene: chromosome walking and jumping. *Science* **245**, 1059–1065.
7. Romeo, G. and Devoto, M. (eds.) (1990) Population analysis of the major mutation in cystic fibrosis. *Hum. Genet.* **85**, 391–445.
8. *Cystic Fibrosis mutation data*. Privileged communication prepared for members of the CF Genetic Analysis Consortium, April 1995.
9. Cheng, S. H., Gregory, R. J., Marshall, J., Paul, S., Souza, G. A., O'Riordan, C. R., and Smith, A. E. (1990) Defective intracellular transport and processing of CFTR is the molecular basis of most cystic fibrosis. *Cell* **63**, 827–834.
10. Crawford, I., Maloney, P. C., Zeitlin, P. L., Guggino, W. B., Hyde, S. C., Turley, H., Gatter, K. C., Harris, A., and Higgins, C. F. (1991) Immunocytochemical localization of the cystic fibrosis gene product CFTR. *Proc. Natl. Acad. Sci USA* **88**, 9262–9266.
11. Denning, G. M., Ostedgaard, L. S., Cheng, S. H., Smith, A. E., and Welsh, M. J. (1992) Localization of cystic fibrosis transmembrane conductance regulator in chloride secretory epithelia. *J. Clin. Invest.* **89**, 339–349.

12. Englehardt, J. F., Yankaskas, J. R., Ernst, S. A., Yang, Y., Marino, C. R., Boucher, R. C., Cohn, J. A., and Wilson, J. M. (1992) Submucosal glands are the predominant site of CFTR expression in the human bronchus. *Nature Genet.* **2,** 240–248.
13. Trapnell, B. C., Chu, C. S., Paakko, P. K., Banks, T. C., Yoshimura, K., Ferrans, V. J., Chernick, M. S., and Crystal, R. G. (1991) Expression of the cystic fibrosis transmembrane conductance regulator in the respiratory tract of normal individuals and individuals with cystic fibrosis. *Proc. Natl. Acad. Sci. USA* **88,** 6565–6569.
14. Jeffery, P. K. (1983) Morphologic features of airway surface epithelial cells and glands. *Am. Rev. Respir. Dis. Suppl.: Compar. Bio. Lung* **128,** S14–S19.
15. Breeze, R. G. and Wheeldon, E. B. (1977) The cells of the pulmonary airways. *Am. Rev. Respir. Dis.* **116,** 705–777.
16. Crystal, R. (1992) *Protocol for Gene Therapy of the Respiratory Manifestations of Cystic Fibrosis Using a Replication Deficient, Recombinant Adenovirus to Transfer the Normal Human CFTR cDNA to the Airway Epithelium.* Recombinant DNA Advisory Committee (USA).
17. Vega, M. A., Goossens, M., and Besmond, C. (1994) A powerful method for in vitro selection of normal versus cystic fibrosis airway epithelial cells. *Gene Ther.* **1,** 59–63.
18. Welsh, M. J. and Smith, A. E. (1993) Molecular mechanisms of CFTR chloride channel dysfunction in cystic fibrosis. *Cell* **73,** 1251–1254.
19. Maarten, K. (1992) *Chloride Transport in Normal and Cystic Fibrosis Epithelial Cells.* Thesis, Erasmus University, Rotterdam, pp. 78–96.

16

Gene Transfer into Cystic Fibrosis Airway Epithelial Cells

Manuel A. Vega

1. Introduction

Gene transfer into airway epithelial cells becomes a particularly motivating goal as far as cystic fibrosis (CF) is concerned. As mentioned in Chapter 15, approx 90% of deaths caused by this devastating disease are the result of infections of the respiratory tract owing to dysfunction of the Cl^- transport in airway epithelial cells. Efficient transfer of the cystic fibrosis transmembrane conductance regulator (CFTR) gene into the airway epithelium of CF patients in vivo is one of the current challenges of gene therapists, and much is being made in that direction in the United States, Europe, and Argentina. The airway epithelium has a very slow turnover, with presumably only 1–2% of cells at division at a given time. Therefore, retroviruses are not the most suitable vector system for gene transfer of the CF gene *(1)*. Adenovirus vectors, adeno-associated virus vectors, or liposomes are being tested in the gene therapy clinical trials currently going on *(2,3)*. In vitro CF gene transfer into airway epithelial cells has relevance in the optimization of gene transfer vectors, of CFTR activity tests, of endogenous as well as vector-produced CFTR-mRNA analysis, and so forth, related to the in vivo application. In the following sections, analysis will restrict itself to in vitro gene transfer approaches.

2. Materials
1. One liter of phosphate-buffered saline (PBS): 8 g NaCl, 0.2 g KCl, 2.14 g $Na_2HPO_4 \cdot 7H_2O$, and 0.2 g KH_2PO_4, pH 7.4. Autoclave and store at 4°C.
2. Electroporation buffer: 10 mM HEPES in 1X PBS (pH 7.5). Filter sterilize and store at 4°C for months.

3. DMEM-1 medium: Mix together: 450 mL DMEM medium, 50 mL fetal calf serum (heat-inactivated), 5 mL L-glutamine (200 mM), 2.5 mL penicillin (100 U/mL), 2.5 mL streptomycin (100 μg/mL). Sterilize by filtration and store at 4°C.
4. CsCl solutions for adenovirus purification: Prepare the three solutions as follows:

Density	CsCl, g	Buffer, mL
1.25	36.16	100
1.34	51.20	100
1.40	62.00	100

Buffer (1 L, 1X): 8.0 g NaCl, 0.38 g KCl, 3.0 g Tris, 0.187 g $Na_2HPO_4 \cdot 7H_2O$, pH 7.4–7.5. Filter and store buffer at room temperature. Autoclave and store CsCl solutions at room temperature.

5. Dialysis buffer for adenovirus purification: 700 mL sterile water, 100 mL 100 mM Tris, pH 7.4, 10 mM $MgCl_2$, 100 mL sterile glycerol. Complete to 1 L with sterile water. Alternatively, sterile physiological saline (0.9% NaCl), made up with 10% glycerol, can be used as dialysis buffer. Cool dialysis buffer to 4°C prior to use.

3. Methods
3.1. Transfection of CF Airway Cells
3.1.1. PREPARATION OF TRANSFECTING DNA (SEE NOTE 1)

1. Obtain plasmid DNA as usual by standard preparation procedures.
2. Wash the DNA pellet, after ethanol precipitation, with 70% ethanol, and dry briefly. Redissolve the DNA in 50–100 μL of sterile water.
3. Take 5–10 μL for restriction analysis to confirm identity and to estimate DNA concentration.
4. Cut the appropriate amount of DNA in the remaining DNA solution with a convenient restriction enzyme to obtain the transfecting DNA in a linear form.
5. Extract DNA with 1/1 (v/v) (chloroform:isoamyl alcohol [24:1])/phenol, precipitate with 2–3 vol of absolute ethanol, wash with 70% ethanol, and dissolve in 500 μL of water.
6. Reprecipitate DNA with absolute ethanol as usual, wash pellet with 70% ethanol, and resuspend in 200–300 μL of cell-culture-quality, sterile water.
7. Repeat the procedure twice in the hood using sterile pipet tips and tubes. Finally, recover DNA pellet in a volume of cell-culture-quality, sterile water to give a final concentration of 2–5 μg DNA/μL.

3.1.2. Electroporation (see Notes 2 and 3)

1. Grow cells up to 70–80% confluence as described.
2. Detach cells with a given volume of detachment solution as described in Chapter 15.
3. Add 4 vol of culture medium.
4. Centrifuge at 800g for 10 min at 4°C.
5. Wash twice with 1X PBS. The first time, recover cells in a volume of PBS that is half of the starting volume. The second time, recover in a volume that is 1/50 of the starting volume.

6. Centrifuge at 800g for 10 min at 4°C.
7. Pipet 15–20 µL of DNA solution (40 µg DNA) into 0.4-cm deep electroporation cuvets and keep on ice.
8. Resuspend cells by gently pipeting with a blue tip in 3.5 mL of ice-cold electroporation buffer.
9. Split cells in electroporation cuvets: 350 µL of cell suspension ($0.5–1.0 \times 10^7$ cells) each cuvet.
10. Mix the DNA solution with the cells by gently pipeting.
11. Keep on ice for 10 min.
12. Remove the rest of the water and ice from the outside of the cuvet with absorbent paper, and gently shake by hand to resuspend cells.
13. Pulse: 250 V, 960 µF.
14. Keep 10 min at room temperature.
15. Recover cells from the cuvet with 2 mL of culture medium by very gently pipeting up and down.
16. Split the 2-mL cell suspension into five 10-cm plates, complete volume with culture medium, and culture for 48 h.

3.1.3. Lipofection

1. Grow cells on plates or bottles up to 70–80% confluence.
2. Mix 10 µg DNA solution with 300 µL Opti-MEM (Gibco BRL).
3. Add 300 µL of (1/5) lipofectine (Gibco-BRL, Gaithersburg, MD)/Opti-MEM, and mix gently.
4. Incubate 15 min at room temperature. The mixture should become somewhat turbid.
5. Aspirate the culture medium on the cells, and wash two times with 2–5 mL of Opti-MEM.
6. Add 2.5 mL to the mixture of lipofectine/DNA/Opti-MEM, and gently mix.
7. Replace the Opti-MEM on the cells with the mixture of lipofectine/DNA/Opti-MEM.
8. Incubate at 37°C, 5% CO_2 for at least 5 h.
9. Aspirate the mixture of lipofection/DNA/Opti-MEM laying on the cells and overlay the usual culture medium.
10. Continue culture under the same conditions for an additional 48 h.

3.2. Transduction of CF Airway Cells

3.2.1. Gene Transfer Using Retroviral Vectors

1. The day before the gene transfer experiment, split airway epithelial cells, previously grown as usual, to 1/10–1/15 of confluence, and replace the old medium on the retroviral vector producer cell line with fresh medium (8 mL/10-cm plate).
2. On the day of the experiment, perform the following steps.
3. Recover the 8-mL medium (supernatant) from the retroviral vector producer cell line using a syringe or a pipet, and replace with fresh medium on the cells.

4. Filter the virus-containing supernatant through a 0.45-μm Millipore filter. The filtered supernatant can eventually be stored at this stage at –80°C.
5. Add 40 μL of a 200X solution of polybrene (final concentration 8 μg/mL) to the 8 mL of virus-containing supernatant.
6. Replace the medium on the airway epithelial cells to be infected by the virus-containing (polybrene containing) supernatant. For a 10-cm plate: 4 mL supernatant; for every well of a six-well plate: 0.5 mL supernatant.
7. Mix gently and incubate 60–90 min at 37°C, with periodic and gentle shaking by hand.
8. Replace the virus-containing supernatant by fresh culture medium, and continue culture.
9. If a second infection round is to be performed, repeat the steps as described for the first infection.
10. Expression of CFTR in infected cells can be easily assessed by either the epinephrine-resistance test (*see* Chapter 15) or by RT-PCR (*see* Chapter 17) as described. Northern analysis can also be performed to detect the virus-produced CFTR-mRNA on the basis of its different size compared to the endogenous CFTR-mRNA.

3.2.2. Gene Transfer Using Adenoviral Vectors

3.2.2.1. OBTAINING CRUDE LYSATES AND PURE PREPARATIONS OF E1-DEFICIENT ADENOVIRUS (*SEE* NOTES 4–10)

1. Plate 293 cells using DMEM-1 medium, and culture at 37°C, 5% CO_2 until 70–80% confluence.
2. Prepare the infection medium by mixing 10 mL of DMEM-1 medium with either 80 μL of crude viral lysate or 1–2 μL of pure virus preparation (equivalent to 1–10 PFU/cell).
3. Aspirate the medium over the 293 cells, and immediately add the appropriate amount (10 mL every 10-cm plate) of fresh infection medium.
4. Incubate the cells at 37°C, 5% CO_2 until the cytopathic effects (CPE) are complete (*see* Note 6).
5. Detach the cells pipeting up and down.
6. Collect the cells in a 50-mL tube and centrifuge at 800g, 10 min, at 4°C.
7. Resuspend the cell pellet in 10 mL of 10 mM Tris-HCl, pH 8.0, or in DMEM-1 medium.
8. Store the cell suspension at –70°C (if resuspended in Tris buffer) or at –20°C (if resuspended in DMEM-1 medium).
9. Release the virus from the cells by freezing on dry ice and thawing at 37°C followed by vortexing 20–30 s. Repeat the freeze–thaw cycle five times. The final lysate is the so-called crude viral lysate.
10. Store as the prelysate cell suspension described in step 8.
11. Centrifuge the crude viral lysate in a 2059 Falcon tube at 9400g, 4°C, for 5 min in a Sorvall HS4 rotor.
12. Recover the supernatant in a fresh 2059 Falcon tube.

13. Overlay the supernatant on a two-phase gradient made of 2.5 mL CsCl (density 1.25) and 2.5 mL CsCl (density 1.40) in an ultraclear SW41 Beckman tube, and complete the tube with DMEM-1 medium.
14. Centrifuge in a SW41 rotor at 150,000g, 20°C, for 1 h.
15. Recover the virus band (lower opalescent band) by gentle aspiration through the tube wall using a 1-mL syringe with a 21-gage needle.
16. Overlay the complete band on 8 mL of CsCl (density 1.33) in an ultraclear SW41 Beckman tube.
17. Centrifuge in a SW41 rotor at 150,000g, 20°C, for 18 h.
18. Recover the opalescent (virus) band as mentioned above, and collect in a sterile Eppendorf tube.
19. Bring the virus suspension to 10% final concentration glycerol, and from now on, keep on ice.
20. Split into two dialysis bags (¾-in. dialysis tubing) and dialyze in the cold room against precooled (cold room) dialysis buffer as follows: 2 × 500 mL, 30 min each, and 3 × 1000 mL, 60 min each.
21. Recover the virus suspension and aliquot (100 µL/tube) in sterile Eppendorf tubes.
22. Store the purified virus at –70°C.

3.2.2.2. CFTR GENE TRANSFER USING ADENOVIRAL VECTORS (SEE NOTES 11 AND 12)

1. Grow airway epithelial cells up to 70–80% confluence as usual.
2. Replace the medium with fresh medium.
3. Infect cells with the adenovirus, either with pure preparations or with crude viral lysates at multiplicities of infection (MOI) ranging from 10–100 depending on the adenoviral vector used.
4. Incubate cells under the usual conditions.
5. After 24 h, expression of CFTR from the adenoviral vector can be easily detected in the epithelial cells by assessment of the epinephrine resistance of the infected culture (*see* Chapter 15) or by detection of the normal CFTR-mRNA by RT-PCR (*see* Chapter 17) as described. As mentioned for retroviral vectors, Northern analysis also allows for differentiation of the virus-produced and the endogenous CFTR-mRNAs.

4. Notes
4.1. Preparation

1. The QIAGEN method (DIAGEN GmbH, Germany) is easy to perform, very rapid and gives good-quality DNA appropriate for transfection of airway epithelial cells.

4.2. Electroporation

2. Voltage has been optimized to 250 V. Total efficiency at both sides of 250 V is much lower: with higher voltages (around 300 V) there is a 10-fold decrease in the number of cells surviving the electric shock, and with lower voltages (around 200 V) there is a two- to fourfold decrease in the transfection efficiency (the number of transfected clones over the number of total cells).

3. Cell aggregates formed by the electroshock can be partially disgregated by gently pipeting cells through blue tips. Remaining aggregates can be separated by a round of low-speed centrifugation. Alternatively, since they do not attach, they will be discarded by replacement of the culture medium once living cells have attached.

4.3. Crude Lysates and Pure Preparations

4. The 293 cells to be infected with adenovirus should not be at densities higher than 70–80%.
5. Do not let 293 cells dry when changing media (they are sensitive to dessication), and do not use cold media, but always use room temperature or prewarmed media (293 cells are sensitive to temperature). When adding culture media on 293 cells, add the medium gently on the corner and not in the middle of the plate. The 293 cells detach easily and are sensitive to pH changes. Do not keep them out of the incubator for prolonged periods of time.
6. The cytopathic effect after adenovirus infection of 293 cells can be detected after 21–24 h and is full at approx 30–40 h, depending on the MOI. Different E1-deficient adenovirus vectors need different times for giving maximal cytopathic effect on 293 cells. Cytopathic effects developed before 24 h may not have been caused by adenovirus infection, and the cultures should be discarded.
7. If cells showing the cytopathic effect are not easily detachable, incubate longer and try again.
8. Prepare the ultraclear SW41 Beckman tubes to be used in the CsCl gradient centrifugation of the adenovirus as follows: soak in 70% ethanol, then in sterile water, and then dry in the hood.
9. Before dialysis, the virus preparation can be stored for 2 d at 4°C.
10. If dilutions have to be made to the final stock virus suspension containing 10% glycerol, dilute with DMEM-1 medium in order to stabilize the virus.

4.4. CFTR Gene Transfer

11. Epinephrine-resistance test: Change the virus-containing medium with fresh medium containing 100–300 mM epinephrine, and incubate for 24 h as usual. Count the percentage of living cells after the 24 h. The percentage varies around 40–50% living cells in CFTR-virus-infected cells compared to 0–5% of survival of either noninfected control cells or non-CFTR adenoviral-vector-infected cells.
12. RT-PCR: As mentioned in Chapter 17, the normal (non-ΔF508) mRNA allele can be easily differentiated from the ΔF508 allele by RT-PCR analysis. Therefore, a virus-produced normal CFTR-mRNA is easily detected in ΔF508/ΔF508 cells. However, detection of either virus-produced normal CFTR-mRNA in ΔF508/normal cells or of virus-produced ΔF508 CFTR-mRNA in ΔF508/ΔF508 cells (used as a control in transfer experiments of the normal allele into ΔF508/ΔF508 cells) is hampered by the fact that the same allele produced by the virus is endogenously produced by the host cell. This problem can be overcome by diluting the cell lysate prior to performing the RT-PCR reaction. Since the level of endog-

enous CFTR-mRNA is (usually) far below that of the virus-produced CFTR-mRNA, this late can be detected from dilutions of the starting lysate at which the band corresponding to the endogenous CFTR-mRNA has disappeared. A series of 10 times dilutions of airway epithelial cell lysates giving up to approx 1 cell/RT-PCR reaction still allow for detection of the CFTR-mRNA. Higher dilutions do not allow for detection of the endogenous CFTR-mRNA, but if (and only if) the cells were infected with a CFTR-expression vector, a CFTR-mRNA band can still be easily detected. This band varies in intensity from vector to vector, but dilution analysis indicates that the vector-produced CFTR-mRNA can be represented as high as 10^2–10^4 times the endogenous CFTR-mRNA.

Acknowledgments

The author thanks Lila N. Drittanti (Argentina), Jorge Gabbarini (Argentina), Claude Besmond (France), Michel Goossens (France), Pascale Fanen (France), and Bruno Costes (France), as well as AFLM (Association Française de Lutte contre la Mucoviscidose) (France), CONICET (National Research Council, Argentina); Hospital Int. Gral. J. Penna (Argentina), and MICROGEN S.A.-Biotechnology (Argentina).

References

1. Drumm, M. L., Pope, H. A., Cliff, W. H., Rommens, J. M., Marvin, S. A., Tsui, L. C., Collins, F. S., Frizzell, R. A., and Wilson, J. M. (1990) Correction of the cystic fibrosis defect in vitro by retrovirus-mediated gene transfer. *Cell* **62,** 1227–1233.
2. Zabner, J., Couture, L. A., Gregory, R. J., Graham, S. M., Smith, A. E., and Welsh, M. J. (1993) Adenovirus-mediated gene transfer transiently corrects the chloride transport defect in nasal epithelia of patients with cystic fibrosis. *Cell* **75,** 207–216.
3. Coutelle, C. (1994) *Approaches to Gene Therapy for Cystic Fibrosis* (Lecocq, J. P., ed.), Conference on Gene Therapy, Strasbourg.

17

Genetic Analysis of Cystic Fibrosis Airway Epithelial Cells

Manuel A. Vega

1. Introduction

More than 500 different mutations have been described to date on the cystic fibrosis (CF) gene *(1)*. The most frequent mutation is the so-called ΔF508 mutation, which accounts for 30–50% of CF chromosomes in southern European countries and for 50–80% of CF chromosomes in the United States, Canada, Argentina, and central and northern European countries.

The presence of the ΔF508 mutation on the cystic fibrosis transmembrane conductance regulator (CFTR)-mRNA can be easily evidenced by simple electrophoresis of the PCR amplification products of exon 10 on 6% polyacrylamide gels. Homoduplex (when both strands are perfectly complementary) DNA, either normal or ΔF508, gives a single band on electrophoresis gels. On PCR amplification of cells carrying only the normal or, alternatively, only the ΔF508 allele, only homoduplex DNA molecules are obtained: either normal/normal or ΔF508/ΔF508 double-stranded DNA, respectively. When the starting material is heterozygous, e.g., normal, and ΔF508 molecules are present, the PCR amplification reaction will give three kinds of molecules: homoduplexes normal/normal, homoduplexes ΔF508/ΔF508, and heteroduplexes normal/ΔF508. The heteroduplex normal/ΔF508 is composed of double-stranded DNA molecules carrying one strand of normal DNA and the other strand of ΔF508 DNA. The heteroduplex normal/ΔF508 can be easily detected on electrophoresis gels on the basis of its different migration compared to the homoduplex (either normal/normal or ΔF508/ΔF508 molecules *(2)*.

Thus, the status of a cell with respect to the presence or absence of the ΔF508 mutation can be readily determined either:

From: *Methods in Molecular Medicine: Human Cell Culture Protocols*
Edited by: G. E. Jones Humana Press Inc., Totowa, NJ

1. At the genomic level (analysis of DNA), it can be:
   a. Homozygous for the normal allele;
   b. Homozygous for the ΔF508 allele; or
   c. Heterozygous for the normal/ΔF508 alleles; or
2. At the expression level (analysis on RNA):
   a. Exogenous expression of the normal allele on a endogenous genomic ΔF508 background, or
   b. Exogenous expression of the ΔF508 allele on a endogenous genomic normal background.

This latter type of analysis (2a) becomes a key feature in the field of CF gene therapy, in order to assess the efficacy of the expression vectors carrying the normal allele on ΔF508/ΔF508 patient cells.

Although homoduplex normal/normal molecules cannot be distinguished from homoduplex ΔF508/ΔF508 molecules on the PAGE system mentioned, each type of molecule can be converted to heteroduplexes when denatured and renatured in the presence of the other, e.g., normal/normal molecules will generate heteroduplexes only when renatured in the presence of ΔF508 molecules and vice versa. Thus, if a PCR amplification product that originally gives a single band shows two bands after renaturation in the presence of ΔF508 molecules, it means that the original product was a homoduplex normal/normal molecule. If, on the contrary, it still gives a single band, that means it was a ΔF508/ΔF508 molecule. An analogous analysis can be made for the renaturation in the presence of normal/normal molecules.

Similar reasoning and procedures for the analysis of CFTR-mRNA are applied to the assessment of the genotypic status (either normal or ΔF508 on exon 10 of the genomic locus) of the airway cells. In this case, the PCR amplification reaction is performed on genomic DNA obtained from the airway cells.

If the interest is centered on CF cells harboring mutations other than the ΔF508, their genotypic status can be determined as well, although more complicated procedures are necessary. The strategy includes three steps:

1. PCR amplification on different exons (one exon per tube or up to three different exons in the same reaction tube);
2. Analysis of the PCR amplification products by denaturing gradient gel electrophoresis (DGGE) in order to detect those exons harboring mutations or polymorphisms; and
3. Sequencing of the selected exons to determine the identity of the genetic change (mutation, polymorphism) *(3)*.

This three-step analysis can also be performed on DNA extracted from dried blood spots, which is usually a more accessible source of genomic DNA for genotypic analysis than airway cells.

2. Materials

1. RT mix: Mix together 7 µL 1M KCl, 5 µL 1M Tris-HCl (pH 8.3), 2 µL 0.25M $MgCl_2$, 2.5 µL dNTPs (20 mM each), 2 µL 700 mM (β-mercaptoethanol), 1 µL RNasin, 1 µL (10 U/µL) AMV reverse transcriptase (RT). Prepare the mix immediately before use, and keep on ice. Stock solutions of KCl, Tris, and MgCl2 are kept at room temperature. dNTPs stocks, RNasin, and AMV RT are stored at –20°C. β-Mercaptoethanol stock is stored at 4°C and diluted immediately before use.
2. Polyacrylamide gels: Six percent polyacrylamide gels are made of: 1 mL 10X TBE, 3 mL 40% acrylamide, 13 µL TEMED, 105 µL 10% ammonium persulfate (APS), and 16 mL water. Prepare the gels immediately before use or the day before, and store at 4°C in Saran wrap™. The 40% acrylamide solution is made of 39 g acrylamide + 1 g bis-acrylamide in 100 mL final water solution, and stored at 4°C for months. APS solutions should be prepared every 2 wk in water, although some stocks keep their activity for much longer periods.
3. PCR buffer: 500 mM KCl, 100 mM Tris (pH 8.4), 15 mM $MgCl_2$, and 0.1% gelatin. Autoclave, aliquot in 0.5 mL stocks, and store at –20°C.
4. IG buffer: 4M guanidinium thiocyanate, 25 mM Na-citrate (pH 7.0), 0.5% sarcosyl, and 0.1M 2-β mercaptoethanol. For preparation, dissolve 250 g guanidinium thiocyanate in 296 mL water. Add 17.6 mL 0.75M Na citrate (pH 7), 26.4 mL 10% sarcosyl (10 g sarcosyl in 100 mL water, dissolved at 65°C). This presolution can be stored at 4°C for about 3 mo. Just before use, add the β-mercaptoethanol (360 µL every 50 mL of presolution).
5. Lysis buffer: 0.32M sucrose, 10 mM Tris-HCl (pH 7.5), 5 mM $MgCl_2$, and 1% Triton X-100. Filter-sterilize and store at –20°C.
6. PCR-2 buffer: 50 mM KCl, 10 mM Tris-HCl (pH 8.4), 2.5 mM $MgCl_2$, 0.1 mg/mL gelatin, 0.45% Nonidet P40, and 0.45% Tween-20. Autoclave and store at –20°C. Just before use, add 1.2 µL proteinase K (10 mg/mL) every 200 µL PCR-2 buffer.

3. Methods
3.1. RT-PCR Analysis of CFTR-mRNA (see Notes 1–6)

1. Grow cells on plates or bottles as described.
2. Suspend cells by trypsin treatment (*see* Chapter 15), neutralize with 4 vol of culture medium, and centrifuge at 4°C and 800*g* for 10 min.
3. Wash twice with 1X PBS.
4. Discard the rest of PBS by inverting the tubes on absorbent paper and by gently aspirating small drops with a pipet.
5. Tubes with the dried cell pellets can be frozen down and stored at –20°C for later processing.
6. Isolate RNA as follows:
   a. Lyse the cells with 2 mL IG buffer every 1×10^7 cells.
   b. Add and mix after addition of every reagent: 0.2 mL 2M Na-acetate (pH 4.0), 2 mL water-saturated phenol, and 0.5 mL chloroform.
   c. Mix vigorously, and incubate on ice for 15 min.
   d. Centrifuge for 15 min at 2500*g* at 4°C.

e. Take the aqueous phase, and put into an ice-cold 15-mL Falcon tube.
   f. Add 1.1 mL isopropanol and incubate for 60 min (or overnight) at –20°C.
   g. Centrifuge for 30 min, 4°C, 9500g in HB4 rotor.
   h. Recover the pellet in 1 mL of IG buffer, and split into three Eppendorf tubes.
   i. Precipitate RNA with 2.5 vol of 100% ethanol + 0.5M NaCl.
   j. Incubate for 30 min at –80°C.
   k. Centrifuge for 15 min at 4°C, at maximum speed in an Eppendorf microfuge.
   l. Wash the RNA pellet with 70% ethanol.
   m. Resuspend RNA in 50–100 µL water.
7. Mix on ice: 1 µL (100 ng) random hexanucleotide mix, 10 µL (10 µg/µL) RNA solution, and 29.5 µL water (DEPC-treated).
8. Incubate for 10 min at 70°C to denature RNA.
9. Put on ice.
10. Add 20.5 µL of RT mix.
11. Incubate for 40 min at 42°C and put on ice.
12. Take 2–5 µL of the RT reaction products for PCR amplification (for a final reaction volume of 50 µL).
13. Mix: 2–5 µL RT reaction products, 1 µL primer 1 (10 pmol/µL), 1 µL primer 2 (10 pmol/µL), 5 µL 10X PCR buffer, 5 µL dNTPs (2 nM each), 33 µL water (or to 50 µL), and 2.5 U Taq polymerase.
14. Perform PCR amplification according to either of the columns in Table 1.
15. Analyze 10 µL of PCR products by PAGE (0.5X TBE buffer, 180–200 V) on 20 mL, 6% gels.

3.2. Sequencing of CFTR-mRNA (see Note 7)

1. Perform RT-PCR on airway epithelial cells RNA as described in Section 3.1., using the appropriate primers.
2. If necessary, run the PCR products on a 6% polyacrylamide gel as described in order to recover a noncontaminated single band. In this case, following electrophoresis, cut off the band corresponding to the fragment of interest and split it in three to four pieces. Extract DNA from the pieces of gel by heating for 20 min at 80°C in a small volume of sterile water.

3.2.1. Sequencing of Single-Stranded DNA

1. Perform two asymmetric PCR reactions (in both directions) on the product of the RT-PCR reaction. Mix:
   a. 5 µL 10X buffer;
   b. 5 µL dNTPs (2 nM each);
   c. 5 µL primer 1 (10 pmol/µL, total 50 pmol); and
   d. 1 µL primer 1 (2 pmol/µL, total 2 µL); and
   e. 5 µL PCR products of the former PCR reaction;
   f. 29 µL water (or extracted DNA solution) (to 50 µL); and
   g. 2.5 U Taq polymerase.

Table 1
PCR Amplification Analysis of the CFTR-mRNA

Primers, exon, 5'->3'	Product on CFTR-mRNA	PCR amplification
AGAACTGGAGCCTTCAGAGGG (10) GTTGGCATGCTTTGATGACGC (10)	158[a] bp (exon 10)	5 min 94°C; 1 min 94°C; 1 min 55°C; 2 min 72°C; 7 min 72°C (35 cycles)
CGGATAACAAGGAGGAACGC (4) GCCTTCCGAGTCAGTTTCAG (7)	565 bp (exons 4–7)	5 min 94°C; 1 min 94°C; 1 min 60°C; 2 min 72°C; 7 min 72°C (35 cycles)
CGGATAACAAGGAGGAACGC (4) TTCTGCACTAAATTGGTCGA (13)	1637 bp (exons 4–13)	5 min 94°C; 1 min 94°C; 1 min 55°C; 2 min 72°C; 7 min 72°C (35 cycles)
CTGCCTTCTGTGGACTTGGTT (6a) TTCTGCACTAAATTGGTCGA (13)	1403 bp (exons 6a–13)	5 min 94°C; 1 min 94°C; 1 min 55°C; 2 min 72°C; 7 min 72°C (35 cycles)
GGGGAATTATTTGAGAAAGC (9) GGAAAACTGAGAACAGAATG (10)	270 bp (exons 9–10)	5 min 94°C; 1 min 94°C; 1 min 55°C; 2 min 72°C; 7 min 72°C (35 cycles)
GTTTTCCTGGATTATGCCTGGC (10) TTCTGCACTAAATTGGTCGA (13)	467 bp (exons 10–13)	5 min 94°C; 1 min 94°C; 1 min 55°C; 2 min 72°C; 7 min 72°C (35 cycles)

[a]The PCR product of 158 bp covers the codon deleted in the ΔF508 mutation. The ΔF508-PCR product can be differentiated from the normal PCR product by PAGE (see Section 3.1., step 15).

   Amplify as follows: 5 min, 94°C; 60 cycles of: 1 min, 94°C, 1 min, 55°C, 2 min, 72°C; and 7 min, 72°C.
2. Analyze on 6% polyacrylamide gels to test for the presence of single-stranded DNA.
3. Sequence asymmetric PCR products (single-stranded DNA) by standard single-stranded DNA sequencing procedures.

3.2.2. Sequencing of Double-Stranded DNA

1. Clone PCR products obtained from the first PCR reaction, after running an electrophoresis gel and extracting the DNA as described in Section 3.2.1. using the TA Cloning kit with the pCR™ vector (Invitrogen Corporation, San Diego, CA).
2. Analyze transforming clones by standard DNA miniprep analysis or by PCR amplification directly on heated (10 min, 80°C) bacterial cells using the same primers as for the first PCR reaction.
3. Sequence plasmid DNA by standard double-stranded DNA sequencing procedures.

3.3. Easy Detection of (ΔF508) CFTR-mRNA

1. Perform a PCR amplification reaction on the RT products obtained as described, under the following conditions:
   a. Primers: 5'-AGAACTGGAGCCTTCAGAGGG-3' and 5'-GTTGGCATGC-TTTGATGACGC-3'.
   b. Amplification reaction: 5 min 94°C; 40 cycles of 0.5 min, 94°C; 0.5 min, 62°C; 1 min, 72°C; and 7 min, 72°C.
2. Analyze on 6% polyacrylamide gels as described.
3. If two bands are observed on the polyacrylamide gel, the starting material was a mixture of normal and ΔF508 molecules (either different alleles on the cellular endogenous genes, or a mixture of cellular endogenous genes and vector-produced exogenous alleles).
4. If a single band is obtained, only either ΔF508 or normal molecules were present in the cell. Further analyze the sample as follows in order to elucidate the identity of the molecules present in the single band.
5. Put 5 µL of the PCR product of the former reaction in each of two separate PCR amplification tubes.
6. Add 5 µL of a DNA solution (at a concentration of DNA similar to that of the PCR amplification products) made of a (homoduplex) DNA known to be either normal/normal or ΔF508/ΔF508 (for instance, genomic DNA or cDNA coming from a normal [homozygous on exon 10] individual or from a ΔF508/ΔF508 CF patient).
7. Incubate for 10 min at 94°C followed by 10 min at 55°C.
8. Analyze on a 6% polyacrylamide gel as described. It two bands are obtained from the mixture with the DNA known to be normal/normal and a single band is obtained from the mixture with the DNA known to be ΔF508/ΔF508, then the unknown sample is ΔF508. Inverse results indicate that the unknown sample is normal/normal.

3.4. DNA Genotypic Analysis (see Note 8)

1. Isolate genomic DNA as follows:
   a. Resuspend cells (2×10^6 cells) in 1 mL of lysis buffer, and put into an Eppendorf tube.
   b. Centrifuge for 20 s at maximum speed in a Eppendorf microfuge, discard supernatant, and resuspend the pellet in 1 mL of lysis buffer.
   c. Repeat the procedure three times.
   d. Resuspend the pellet in PCR-2 buffer, and incubate for 60 min at 56°C.
   e. Inactivate proteinase K by incubating for 10 min at 95°C.
   f. Take 25 µL of lysate (containing approx 1 µg genomic DNA), and perform PCR amplification as usual in a total volume of 100 µL.
2. Determine first if the ΔF508 mutation is present: Perform a PCR amplification reaction on genomic DNA using the primers and conditions described in Section 3.3. for analysis of exon 10 on CTFR-mRNA.

3. If mutations other than the ΔF508 are present, they can be detected and identified as indicated in Section 1. *(3–5)*.

4. Notes
4.1. RT-PCR Analysis

1. RNA isolation and the RT reaction have to be performed using sterile tubes, tips, and solutions; wearing gloves; keeping all solutions on ice, and, when possible, using disposable plasticware. Water (and water for preparation of solutions) should be DEPC-treated.
2. RNA concentration can be precisely determined by spectrophotometry at 260/280 nm. Alternatively, when no precise determinations are necessary, an estimation can be made on an agarose gel by eye comparison with a sample of RNA of known concentration.
3. The hexanucleotide mix used in random priming DNA-labeling procedures can be used for priming the RT reaction.
4. Airway epithelial cells, but not any tissues, contain an amount of CFTR-mRNA high enough to allow for its detection by PCR using volumes of RT reaction products even lower than the 2 μL indicated in Section 3.1.
5. The RT reaction products can be stored at –20°C before proceeding into the PCR amplification reaction.
6. Each of the PCR reactions described (*see* Table 1) gives a single band of the indicated size when airway epithelial cells are used, although additional bands can be obtained when RNA from other tissues is analyzed.

4.2. Sequencing of CFTR-mRNA

7. DNA extraction from gel pieces can be performed in the same PCR amplification tube where the asymmetric PCR reaction will take place. In this case, consider the extraction volume of water for the calculation of the final volume of water to be added to the PCR reaction.

4.3. DNA Genotypic Analysis

8. When setting the PCR reaction on genomic DNA isolated as described in Section 3.4., consider the composition of the lysis PCR-2 buffer in which the DNA is dissolved for the calculation of the volumes of the reagents to be added.

Acknowledgments

The author thanks Lila N. Drittanti (Argentina), Jorge Gabbarini (Argentina), Claude Besmond (France), Michel Goossens (France), Pascale Fanen (France), and Bruno Costes (France), as well as AFLM (Association Française de Lutte contre la Mucoviscidose) (France), CONICET (National Research Council, Argentina), Hospital Int. Gral. J. Penna (Argentina), and MICROGEN S.A.-Biotechnology (Argentina).

References

1. *Cystic Fibrosis mutation data*. Privileged communication prepared for members of the CF Genetic Analysis Consortium, April 1995.
2. Fanen, P., Ghanem, N., Vidaud, M., Besmond, C., Martin, J., Costes, B., Plassa, F., and Goossens, M. (1992) Molecular characterization of cystic fibrosis: 16 novel mutations identified by analysis of the whole cystic fibrosis conductance transmembrane regulator (CFTR) coding regions and splice site junctions. *Genomics* **13**, 770–776.
3. Ghanem, N., Fanen, P., Martin, J., Conteville, P., Yahia-Cherif, Z., Vidaud, M., and Goossens, M. (1992) Exhaustive screening of exon 10 CFTR gene mutations and polymorphisms by denaturing gradient gel electrophoresis: applications to genetic counselling in cystic fibrosis. *Mol. Cell. Probes* **6**, 27–31.
4. Costes, B., Girodon, E., Ghanem, M., Chassignol, M., Thuong, N. T., Dupret, D., and Goossens, M. (1993) Psoralen-modified oligonucleotide primers improve detection of mutations by denaturing gradient gel electrophoresis and provide an alternative to GC-clamping. *Hum. Mol. Genet.* **2**, 393–397.
5. Vidaut, M., Fanen, P., Martin, J., Ghanem, M., Nicoles, S., and Goossens, M. (1990) 3 point mutations in the CFTR gene in french cystic patients: identification by denaturing gradient gel electrophoresis. *Hum. Genet.* **85**, 446–449.

18

Human Tracheal Gland Cells in Primary Culture

Marc D. Merten

1. Introduction

For several years, tracheal gland cells have been cultured from different animal species, such as the cat *(1)*, cow *(2)*, and ferret *(3)*. There are differences, however, in the structure and function of the various animal airways, rendering it difficult to extrapolate to humans. In this chapter, the author describes techniques that facilitate the isolation and culture of tracheal gland cells from humans. These techniques allow high reproducibility, optimal cell isolation, and high phenotypic expression, rendering them appropriate for physiological, pharmacological, and biomedical applications.

1.1. General Considerations

Human bronchotracheal submucosal glands have long been recognized as the major secretory structure in the bronchotracheal tree *(4)*. They are composed of mucous and serous cells (Fig. 1), surrounded by myoepithelial cells, and are connected to the bronchotracheal lumen by collecting ducts *(5)*. The tracheal submucosal tissue is richly innervated, vascularized, and also contains smooth muscle, neuroendocrine cells, and mastocytes. These elements are embedded in a parenchyme, which has numerous fibroblasts. Glands are considered the principal source of the secretion of mucus, which is involved in the defense of the airway. The mucus is a complex mixture composed of various macromolecules. Mucins, probably the most widely known macromolecules of bronchial secretion, originate predominantly from the mucous component of the glands and from the goblet cells in the surface epithelium. The other proteins present in mucus stem from serum exudation and also from the serous component of the glands. Gland serous cells secrete antibacterial proteins, such as lactoferrin, lysozyme, and peroxidase *(6)*. In addition, they secrete an

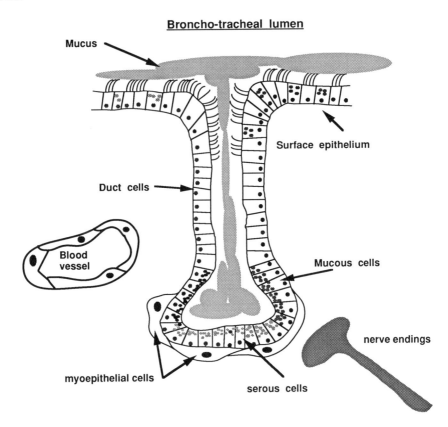

Fig. 1. Schematic representation of an HTG. Both mucous and serous cells contribute to the secretion of mucus, which is then evacuated to the bronchotracheal lumen through collecting ducts. Mucus is then transported by the ciliary beat to the larynx.

antiprotease—the bronchial inhibitor—and are the primary site in the lung for secretory IgA transcytosis. After serous and mucous cells have exocytosed their secretory products in the gland lumen, the myoepithelial cells appear to help in the evacuation of mucus, via collector ducts, into the bronchotracheal tree.

In many bronchopathologies, there is either an overabundance of mucus or a failure in the pulmonary defense system. It is therefore valuable to culture tracheal gland cells in order to provide a cellular, molecular, and pharmacological basis for the comprehension of these diseases. The initial problem that arose when the author began the isolation and culture of human tracheal gland (HTG) cells was the complexity of the bronchotracheal mucosa. This, in turn, prompted the question of how tracheal gland cells could be detected after culture and how a functional homogeneous cell monolayer of this cell type could be obtained.

2. Materials

1. Transport medium for the surgical specimen: RPMI 1640, 0.1% LPSR-I (a serum replacement), 10 g/L glucose, 0.33 g/L sodium pyruvate, 200 µg/mL gentamycin, 5 U/mL amphotericin B (all provided from Sigma, St. Louis, MO). Storage at 4°C for 1 mo.
2. Detachment medium: 110 mM NaCl, 5 mM KCl, 1 mM Na$_2$HPO$_4$, 1 mM KH$_2$PO$_4$, 20 mM N-2-hydroxyethyl-piperazine-N'-2-ethanesulfonic acid (HEPES), 10 mg/mL of fraction V human serum albumin, 4 g/L glucose, 0.11 g/L pyruvate, and 2 mM ethylenediaminetetraacetic acid (EDTA), pH 7.4. Store at –20°C after filter sterilization in 25-mL aliquots, until required.
3. Digestion medium: Identical to the detachment medium but without EDTA, and containing 1 mM MgCl$_2$ and 2 mM CaCl$_2$, in addition to the following enzymes: 200 U/mL type IA collagenase (*see* Note 1), 200 U/mL type I-S hyaluronidase, 0.1 mg/mL type I porcine pancreatic elastase, 200 U/mL type II DNase (all from Sigma). Store at –20°C after sterile-filtration in 25-mL aliquots, until required.
4. Collagen solution: The preparation of collagen that was found to be the most adaptable to HTG cell culture is as follows: Use rat tails, which must first be frozen for at least 1 wk. Then soak them in a 95% ethanol solution for 3 min. Subsequently, break the tails into 1-cm-long segments using strong sterile pliers. Dissect the isolated tendons, and wash them in sterile water. The tendons are gathered and put into a 1 mM acetic acid solution for 24 h at 4°C (100 mL/tail). After centrifugation to remove the nonsolubilized material, the stock collagen solution can be used. Store at 4°C for no more than 2 mo.
5. Culture medium (*see* Note 2): Dulbecco's modified Eagle's medium (DMEM), Ham's F12, 50/50% (v/v) containing 1% LPSR-I (a serum replacement from Sigma), and the following substances made up to the indicated final concentrations: 10 g/L glucose; 0.33 g/L pyruvate; 0.2 g/L leucine, isoleucine, and valine; 0.1 g/L glutamic acid and cysteine; 3 µM epinephrine; and antibiotics, 100 U/mL penicillin G and 100 µg/mL streptomycin. Glucose, pyruvate, and amino acids are added to DMEM/F12 prior to filtration when the media are prepared from powdered mixtures. LPSR-I and antibiotics are prepared together in aliquots that will be added just before each medium change. Epinephrine is also prepared independently (stored at –80°C in a 1 mM HCl solution) and added just before use.

3. Methods
3.1. Isolation of HTG
3.1.1. Dissection of the Tissue

1. Spread out the human bronchial tissue (*see* Note 3) on a dissection board and carefully clean the mucous material present on the organ surface with sterile gauze.

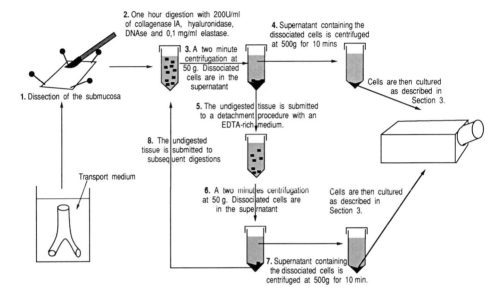

Fig. 2. Isolation of HTG cells. After the mucosa of the bronchial tissue have been dissected, they are submitted to successive enzymatic digestion procedures followed by EDTA treatment. Cells isolated at each step are seeded onto collagen substrate; $30 \pm 10 \; 10^6$ cells/g of dissected fresh tissue are obtained.

2. Dampen the sample for 10 s with ethanol at 95%. This considerably reduces the possibility of bacterial contamination. It also leads to the destruction of the surface epithelium.
3. Sponge up the ethanol, and thoroughly soak the surface with complete cell-culture medium. During the following operations, the surface must be kept continually damp with this solution.
4. The mucosa, submucosa, and particularly the tissue between the cartilagenous rings can now be dissected.
5. Cut this tissue into small pieces (<1 mm^3) with scissors, and put them into digestion medium.

3.1.2. Enzymatic Digestion of the Dissected Tracheal Mucosae

The dissected tissue can now be submitted to successive 1-h digestions, followed by 15 min of cell detachment (Fig. 2). The protocol for each operation was adapted from descriptions by Culp et al. *(1)* and others by Finkbeiner et al. *(2)*, which had been perfected for the isolation of bovine and feline tracheal gland cells.

1. Incubate for 1 h in gently agitating digestion medium at 37°C (25 mL solution/10 cm^2 of bronchial sample).

2. Centrifuge the tissue for 2 min at 50g, and separate the pellet and the supernatant. The supernatant and the pellet are now submitted to different operations.
3. Centrifuge the supernatant at 500g for 10 min. Discard the new supernatant. The existing pellet contains isolated cells.
4. Incubate the first pellet with detachment medium for 15 min at 37°C, and centrifuge at 50g for 2 min to separate the undigested tissue and the detached cells. Remove this supernatant, and centrifuge it at 500g for 10 min to retrieve the cells. Submit the undigested tissue to a further 1-h digestion.
5. Two pellets of centrifuged cells per digestion operation are obtained. After the cells have been counted, seed at an initial density of 25,000–45,000 cells/cm^2 (*see* Note 4). Repeat this operation until all the tissue is totally digested. The number of successive operations (from three to over eight times) varies according to the age and sex of the donor (*see* Note 5).
6. The two first digestions yielded few cells, but a lot of cell debris. The following digestions are much more productive. As a result, after the full digestion procedure is achieved, $30 \pm 12\ 10^6$ cells/g of fresh dissected tissue are actually obtained. This value may greatly vary depending on the time delay between death and collection of the tissue.

3.2. Culture of HTG Cells

3.2.1. Types of Cultures

Like many epithelial cells, HTG cells are able to grow in monolayers on plastic supports. In vivo, however, the epithelia are laid on an extracellular matrix composed of collagen, glycoproteins, and proteoglycans. This matrix participates in the polarity of the cells and plays an important role in their growth and differentiation. HTG cells proved to be very sensitive to the substratum on which they are seeded *(7–9)*. The most convenient substratum was collagen Type 1. This substratum not only promoted cell differentiation, but was the only one to permit an optimal cell growth. There are several different procedures to prepare type I collagen either from calf skin or rat tail, both of which give satisfactory results. Two types of HTG cell cultures can be realized.

3.2.1.1. Monolayer Cultures

In this culture type, cells grow in two dimensions on thin collagen-film-coated surfaces. A 1/100 dilution of collagen in H$_2$O (stock solution) is dropped onto the plastic surfaces (0.2 mL/cm^2) and left overnight. This collagen solution must be removed just before seeding the cells so that the collagen film does not dry out. Under these conditions, about 10 µg/cm^2 of collagen are absorbed.

3.2.1.2. The Collagen Lattice

Bell et al. *(10)* first developed this technique for fibroblast culture, and it has recently been adapted for HTG cells. In this cell-culture type, HTG cells are first cultivated in the aforementioned conditions and are then introduced into a thick collagen gel where they grow three dimensionally. A complete protocol is described in detail elsewhere *(11)*. This protocol was modified to simplify the technique, which is as follows:

1. Prepare a mixture, kept in ice, of 1 mL of complete medium containing 2% LPSR-I, 0.5 mL of a collagen type 1 stock solution (2 mg/mL in 0.1% acetic acid/135 mM NaCl), and 0.25 mL of a 35-mM NaOH/100 mM NaCl.
2. Add to this mixture 0.25 mL of complete medium containing 2% LPSR-I and 10^5 HTG cells.
3. Transfer immediately onto a 35-mm diameter Petri dish, and place the dish in the 5% CO_2 incubator. The collagen polymerizes to form a thick gel matrix in which cells are embedded.
4. After the gel has polymerized, add fresh culture medium. After 3 wk of culture, glandular-like branching structures are noticeable within the gel. In many aspects, these structures resemble the in vivo morphology *(12)*.

3.2.2. Evolution and Serial Passaging

After isolation, microscopic examination indicates the presence of cells of undetermined phenotypes and abundant cell debris. During the first 4–6 d, culture medium must not be changed. At this time, it is extremely important to avoid bacterial contamination (*see* Note 6). Cells of different origin begin to grow (Fig. 3A). Small islets of highly proliferating epithelioid cells can be distinguished *(8)*. When cells reach about 75% confluence, it is necessary to perform a partial trypsinization (Fig. 3B,C) to separate HTG cells from all other contaminating cells (*see* Note 7). These islets will then continue to grow and will reach confluence within 5–7 d (Fig. 3D). One of the most difficult and delicate aspects of HTG cell culturing is the determination of the moment and the conditions of the passage. Since the cells grow in clusters, in the same flask, there are cells at various states of differentiation. The cells inside the clusters are differentiated and actively secrete, but have a low proliferation rate. The cells at the periphery of the clusters proliferate rapidly and are undifferentiated. The passage consists of a trypsin treatment (0.025% trypsin and 0.02% EDTA in a PBS buffer) over a sufficient period of time (3–5 min) to enable the isolation of only the cells growing at the periphery of the clusters. This is possible since the differentiated cells are much more adherent to the substrate. The isolated cells are seeded at 25,000 cells/cm^2, and the original flask can be cultured further—new cells will spread again from the remaining clusters.

The progression of HTG cells in their culture life time is accompanied by a decrease in the proliferation rate. The density of the cell at confluence, the viability, and the capacity to secrete also decrease, and characteristics of apoptosis begin to appear. This limitation of the cell life-span is a necessary

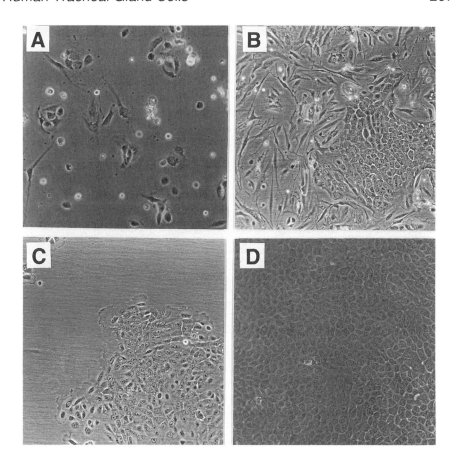

Fig. 3. Phase-contrast micrographs of cultured HTG cells. **(A)** Cells 2 d after isolation. A mixture of cells of a different nature is isolated and will grow on the collagen gel (magnification: ×320). **(B)** Among the cells, clusters of HTG cells are visible 1 wk after isolation (in the middle of the micrograph) (magnification: ×160). **(C)** The partial trypsinization allows the specific selection of the HTG cells, which are strongly adhesive to the substrate collagen (magnification: ×160). **(D)** Multiple partial trypsinizations lead to an homogeneous monolayer of HTG cells (magnification: ×320).

condition of the "normality." The cell-growth parameters (population doubling time, cell density at confluence, interdivision time, the lag time, etc.) are identical during the first three passages and decrease dramatically thereafter. Culturing HTG cells for more than six passages was not successful.

3.2.3. Phenotypic Expression

As for all cell types, there are two methods of control of the phenotypic expression, morphological control and functional control, the latter being the most efficient. HTG cells possess differentiated characteristics that have to be monitored to ascertain whether the cells are completely differentiated. These characteristics include the presence of cytokeratin within the cells, a polarized secretion of the secretory products, which the glands secrete in vivo, and the responsiveness to secretagogs. Figure 4 shows the kinetics of appearance of differentiation characteristics during 30 d of culture. In the exponential growth phase, cells secrete the bronchial inhibitor at a very low level, which is a specific secretory marker of HTG cells. They are also not responsive to secretagogs, nor are they polarized. There are three periods in the stationary phase: This first is 8 d where cells acquire their differentiated characteristics, followed by a plateau of 10–15 d in which cells are fully differentiated as demonstrated by their ability to be stimulated by many pharmacological and physiological agonists *(7,8,13)*. During this period, cells are also highly polarized *(7)*, which is demonstrated by the apical secretion (>98%) of all the secretory products of the cells. All these characteristics (high polarity, higher constitutive secretion, and responsiveness to agonists) appear in unison, and they were only observed when cells were grown on a collagen substratum and in the presence of epinephrine *(7)*. The third period corresponds to cell degeneration in which cells lose their differentiated features and die.

As a consequence, the optimal time to use HTG cells is after 8 d confluence at the third passage (which corresponds roughly to 6 wk after cell isolation).

3.3. Possible Applications

In fundamental as well as in applied research, the HTG cell culture represents an interest in the study of physiology, pathology, and pharmacology of bronchial secretions. Before any investigations are performed, it is important first to check the differentiated functions of HTG cells. After these precautions, HTG cell cultures can be used efficiently.

Three main types of studies are usually performed on these cells: the study of their secretory products (which may be in their native form), the study of their pharmacology, and the study of pathological cells. The aim of these studies is to demonstrate the implication of these cells in the comprehension of human bronchial secretion.

3.3.1. Physiology and Pharmacology of HTG Cells

Until recently, the physiology and pharmacology of bronchial secretion have not been well understood. HTG cells in culture appear to be one of the most interesting models for the study of regulating bronchial secretion mechanisms. Considerable studies have been made on human organ culture, but with the complexity of the human bronchial tissue, it is difficult to assign the action of a physiological agent to a direct or an indirect effect, and also to distinguish

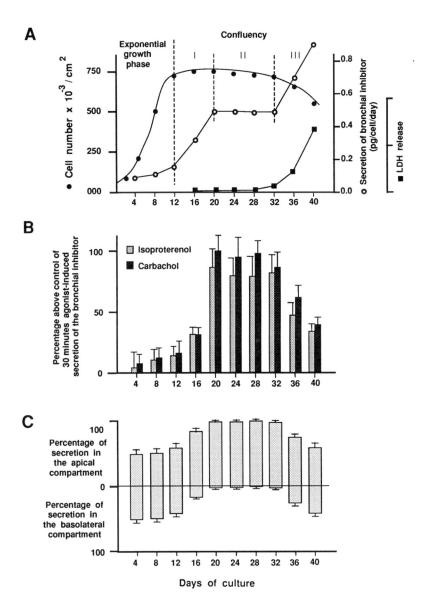

Fig. 4. Phenotypic expression of HTG cells. **(A)** Secretion of the specific secretory serous marker bronchial inhibitor during the growth cycle of HTG cells in culture. **(B)** HTG cells begin to respond to secretagogs (a cholinergic agonist: carbachol and a β-adrenergic agonist: isoproterenol) only after 8 d confluence and stay responsive for 15 d. **(C)** Polarization of HTG cells also appears after 8 d confluence as demonstrated by the almost complete apical secretion of the bronchial inhibitor.

secretion derived from the glands and from other secretory cells in the tissue (i.e., the goblet cells). Despite the great interest in the use of HTG cells in culture to understand the regulation of bronchial secretion, very few studies are available to date. Cultured HTG cells were shown to be responsive to bradykinin and histamine *(14)*, to cholinergic and adrenergic agonists *(7,8)*, and to the purinergic agents ATP and UTP *(13)*. One interesting observation was the evidence of a positive and negative control of secretion by HTG cells. They respond to the physiological neurotransmitters acetylcholine and norepinephrine by an increase or a decrease in secretion, respectively *(15)*. Furthermore, neuropeptide Y was shown to modulate the decrease in secretion induced by norepinephrine *(16)*. HTG cells appear to be particularly responsive to many agents that are present in the airways, the action of which, at cellular and molecular levels, has only recently been documented. These investigations can be relatively easy to perform using the patch-clamp technique, Ussing chamber, or by measuring some secretory products by ELISA measurements (bronchial inhibitor, lysozyme, or lactoferrine, for instance).

3.3.2. The Cell Phenotype of HTG Cells in Culture

HTG cells in culture were shown to secrete lysozyme, lactoferrin, and the bronchial inhibitor, which are proteins specific to the serous gland cells. However, gel-filtration analyses of ^{3}H-fucose and ^{35}S-SO$_4$ labeled secretory products in the medium of the cultures in a Sepharose CL 6B column indicated that the secreted high-mol-wt radioconjugates were partly proteoglycans (chondroitin and heparan sulfates), and also mucins as proved by resistance to all hyaluronidase, chondroitinase, heparitinase, keratanase and also by sensitivity to β-elimination *(17)*. Furthermore, the buoyant density of these hydrolase-resistant radiolabeled macromolecules, as well as the sizes of their glycanic chains are consistent with those expected from mucins, but not from proteoglycans or other glycoproteins. These results suggest that HTG cells are able to produce mucins in vitro.

Tournier et al. *(8)* showed that cultured homogeneous HTG cells are mostly composed of serous cells. Sommerhof and Finkbeiner *(18)* observed, using specific antibodies directed against serous or mucous epitopes, an immunolabeling of all cells with both types of antibodies, suggesting that, at least in culture, HTG cells may carry both the serous and the mucous phenotypes. According to the culture conditions, HTG cells may express one phenotype more than the other. Finkbeiner et al. *(9)*, by varying cell-culture conditions, were able to obtain either HTG cells of the serous or of the mucous phenotype. When HTG cells are cultured onto collagen, they differentiate into serous cells, but when they are cultured onto vitronectin, they differentiate into mucous cells. HTG cell culture therefore permits both cell types present in the glands to be obtained. In addition, it is possible to study the mechanism of transdifferentiation and also the regulation mechanism of secretion of either mucin or protein from the serous cells.

3.3.3. Cystic Fibrosis

Cystic fibrosis derives from mutations in a membrane protein called cystic fibrosis transmembrane conductance regulator (CFTR), which is involved in the cyclic AMP-dependent chloride transport. Engelhardt et al. *(19)*, followed by Jacquot et al. *(20)*, showed that in comparison to the other epithelial cell types of the human bronchus, CFTR is located predominantly in tracheal glands. This may signify that tracheal glands could be one of the principal targets of cystic fibrosis. This disease is an exocrinopathology, the pulmonary syndrome of which is the most serious and which is characterized by mucus hypersecretion and lung infection. It is useful to investigate the implications of CFTR in cultures of HTG cells provided from cystic fibrosis patients. The first data available showed a failure in the chloride transport by HTG cells in culture *(21)*. Becq et al. demonstrated that cultured HTG cells express great quantities of CFTR and the associated cyclic AMP-dependent chloride channel activity *(22)*. A constitutive protein hypersecretion and an insensitivity to secretagogs and physiological neurotransmitters by HTG cells derived from cystic fibrosis patients *(15)* were observed. These alterations in secretory mechanisms can be related to the macroscopic syndrome, but until now could not be explained by the known function of the molecule CFTR.

3.4. Conclusion

HTG cells tend to lose their differentiating functions in culture, but differentiated epithelial cells can be attained in very specific conditions. These conditions were determined by analyzing the growth-supplement requirements, the substrate requirement, and the culture methods. The first objective was to establish conditions that allowed HTG cells to grow and especially differentiate in culture. It was observed that a collagen substrate, an elevated concentration of glucose, and supplementation with epinephrine are important to HTG cell growth and differentiation.

As a result, a 10-cm^2 surgical resected sample, for example, may produce a culture comprising more than 100 24-well plates. This relatively large quantity of cells is sufficient to realize many fundamental or applied studies regarding the action of physiologically or pharmacologically active substances on: receptors, membrane ionic channels, and mechanisms of effector secretion coupling.

The study of human secretion, notably the complex tracheo-bronchial mucosa, remains relatively unexplored, particularly at the cellular level. Primary cultures of HTG cells seem an obvious interesting experimental model.

Cell cultures prove their usefulness in certain investigations and especially in the study of pathologies. Current culture procedures render the HTG cell culture one of the best models available for the study of the human bronchial secretion.

4. Notes

1. A noticeable trend in the technical factors observed in these studies is the crucial importance of the nature of collagenase used in HTG cell isolation procedures. The commercial products, called collagenase, are enriched pooled fractions containing several collagenases, neutral proteases, and clostripain. The composition of "collagenase" is greatly dependent on manufacturing processes, and attention must be given to the lot number. The findings regarding the respective tested lots have been confirmed by other teams involved in HTG cell culture (F. Dupuit, INSERM U314, Reims, France, personal communication, 1993). These results demonstrate the need for an efficient and constant form of collagenase, since an absence of viable cells or any visible cell growth after isolation could be attributed to a toxic "collagenase" lot. In this case, it is necessary to change to another lot number of type IA collagenase.
2. When the ability of different sera to promote HTG cell growth and differentiation was tested, a problem occurred with heterogeneity and the weak efficiency of the widely used fetal calf sera probably owing to some cytotoxic side effects. The serum substitute Ultroser G (Biosepra, Villeneuve-la-Garenne, France) also marketed with the name LPSR-I (Sigma) yielded optimal and reproducible cell growth, and also differentiation at confluence as judged by polarity, cell secretion, and responsiveness to secretagogs. The manufacturer advises the use of LPSR-I at the concentration of 2%. By studying the effects of increasing concentrations of LPSR-I, it was determined that 1% was the most efficient.

   The DMEM/F12 medium mixture contains 3.12 g/L glucose and 0.11 g/L sodium pyruvate. A rapid change in the pH of the culture medium was observed, which was attributed to an active metabolism. From the measurements of the glucose, the pyruvate, and the amino acids that are consumed by HTG cells, it was concluded that the final concentration of these components had to be increased. It is worth noting that cells use many components that are involved in the formation of acyl CoA, indicating the importance of the metabolic activity of these cells.

   Epithelial cells in culture can terminally differentiate, turning into squamous and stratified or secretory and columnar cells. There are factors favoring the differentiation into squamous cells, such as TGFβ1 and an elevated calcium concentration, but there are also other factors favoring differentiation into columnar cells. Epinephrine is countereffective to TGFβ1 in inducing squamous differentiation (23). Epinephrine (at a 3-μM final concentration) was able to promote cell growth and glandular cell differentiation features, such as a high polarity and an optimal responsiveness to secretagogs at confluence (7). Add epinephrine at each medium change (every other day) from aliquots of stock solution of epinephrine stored at –80°C in a 1-mM HCl solution.

3. A principal factor in the successful gathering of viable tracheal gland cells is the time delay at room temperature before the collection of explants. Samples resected during surgery are of a higher quality than those obtained after autopsy owing to the delay that is inevitable between sampling and death. The author rarely succeeded in obtaining satisfactory cultures from autoptic sampling. It is important to put specimens as rapidly as possible into a sterile transport medium at 4°C. When samples are obtained in these conditions, they can be stored at 4°C for several days after surgery without significant reduction in either the viability of cells or the number of isolated cells.

4. The initial seeding of 25,000 cells/cm^2 was determined as the most convenient for growth. A major problem of surgically resected samples is that they are frequently small in size, and the isolation of HTG cells from the obtained tissue is often difficult. However, HTG cells can grow even if their initial quantities or densities after isolation are low, but the presence of other cell types is necessary to generate HTG cell cultures from very little bronchial tissue. In the case of insufficient tissue, instead of destroying the surface epithelium by the ethanol treatment, the surface epithelium is removed with the submucosa, and all the dissected tissue has to be digested. By increasing the amount of viable tissue digested, i.e., the total number of isolated cells, the capacity of HTG cells was augmented to develop despite their much lower percentage. The following partial trypsinizations (*see* Note 7) will be more carefully undertaken. In addition, instead of adding fresh culture medium, a mixture of 50% fresh medium and 50% of an older medium that has already been in contact with cells from a previous cell culture can be added (because of the presence of specific autocrine growth factors). Using these methods, the author succeeded in performing HTG cell cultures from bronchial tissue with as little surface space as 1/2 cm^2.

5. The age of the donor is a significant factor that must be considered when collecting specimens. Results in terms of number and viability of isolated cells were identical whatever the age of the donor. However, the number of population doublings was dramatically dependent on the age—the younger donors giving very satisfactory and useful cultures.

6. Whether the samples are autoptic or surgical resections (most of which have been removed from artificially ventilated patients), there is always a risk of contamination of the material by microorganisms, and care must be taken equally during experiments and in the sterilization of the material. Both the ethanol treatment of the tissue surface and the use of antibiotics in all media and solutions considerably reduce the risk of contamination. However, during the first 5 d of cultures, in addition to the habitually used penicillin G (100 U/mL) and streptomycin (100 µg/mL), add the broad-spectrum antibiotic gentamycin (100 µg/mL), the antifungal agent amphotericine B (5 µg/mL), the antimycoplasma agent neomycine (50 µg/mL), and 5% of unheated rabbit serum, which is also used to eliminate potential mycoplasma contamination. Despite these precautions, for each culture, a classic detection is necessary to verify the absence of mycoplasms. Other rare, but opportunistic contaminations are the multiresistant *Staphylococ-*

cus aureus and *Staphylococcus epidermidis*. Treatment using vancomycin (50 µg/mL) is sufficient to eradicate this contamination, but it can alter HTG cell growth and physiology if treatment is prolonged for more than 3 or 4 d.

7. A significant technical stumbling block in obtaining homogeneous cell culture is the risk of contamination by cells from different origins. After isolation, careful microscopic observation allows different types of bronchial tissue cells to be distinguished. Four types of different cells can be observed:
   a. Surface epithelial cells are big, flattened, and mostly regrouped in clusters. These cells are easily recognizable.
   b. Fibroblasts, the most abundant of contaminating cells, organize themselves in regular parallel bundles. Fibroblasts grow rapidly and spread over all free surfaces.
   c. Endothelial cells, which originate from vessels present in bronchial tissue, are polyedric, joined together, and proliferate at the same rate as the HTG cells. They are difficult to distinguish from epithelial cells and are pinpointed by the absence of labeling by anticytokeratin antibodies and by the presence of the Von Willebrand antigen.
   d. Smooth muscle cells are easily obtained during tissue digestion owing to both the presence of elastase in the digestion medium and the use of collagen type I, which stimulates smooth muscle cell growth.

   The process used to obtain a homogeneous monolayer composed uniquely of HTG cells depends on the partial trypsinization methodology. This technique is based on the fact that HTG cells are fixed firmly to the collagen substratum. A 3-min incubation with trypsin/EDTA detaches the fibroblasts, the myocytes, and most of the endothelial cells. These contaminating cells can then be washed out. Epithelial cells separate after 7 or more minutes of incubation with trypsin/EDTA. This short-time trypsinization has to be performed before each passage and one or two times more during the exponential growth phase of the first two subcultures. This operation not only allows the selection of epithelial cells in the culture, but also an improved cell growth, since trypsinization appears to have a stimulatory effect (probably by an insulin-like effect) on the HTG cells. It was ascertained that the cells obtained under these conditions are HTG cells and not epithelial cells from any remaining surface epithelium, since the cell-culture medium used for the HTG cells is not adapted for the surface epithelial cells and prohibits their growth. In the laboratory, a homogeneous monolayer of HTG cells is obtained at the third passage with four or five previous partial trysinizations. After isolation, the first partial trypsinization must be performed at about 75% of confluence of the cell population of the flasks.

Ackowledgments

The author is on a fellowship from Synthélabo and the "Fondation pour la Recherche Médicale" awards and his works on this subject were supported by grants from the Association Française de Lutte contre la Mucoviscidose. The author thanks Annie Mascall for her help with the English text and Catherine Figarella for advice and fruitful discussions.

References

1. Culp, D. J., Penney, D. P., and Marin, M. G. (1983) A technique for the isolation of submucosal gland cells from cat trachea. *Am. J. Physiol.* **55,** 1035–1041.
2. Finkbeiner, W. E., Nadel, J. A., and Basbaum, C. B. (1986) Establishment and characterization of a cell line derived from bovine tracheal glands. *In vitro* **22,** 561–567.
3. McBride, R. K., Stone, K. K., and Marin, M. G. (1992) Arachidonic acid increases cholinergic secretory responsiveness of ferret tracheal glands. *Am. J. Physiol.* **262,** L694–L698.
4. Read, L. (1960) Measurement of the bronchial mucous gland layer: a diagnostic yardstick in chronic bronchitis. *Thorax* **15,** 132–141.
5. Meyrick, B., Sturgess, J. M., and Read, L. (1969) A reconstruction of the duct system and secretory tubules of the human bronchial submucosal gland. *Thorax* **69,** 729–736.
6. Basbaum, C. B., Jany, B., and Finkbeiner, W. E. (1990) The serous cell. *Annu. Rev. Physiol.* **52,** 97–113.
7. Merten, M. D., Tournier, J. M., Meckler, Y., and Figarella, C. (1993) Epinephrine promotes growth and differentiation of human tracheal gland cells in culture. *Am. J. Respir. Cell Mol. Biol.* **9,** 172–178.
8. Tournier, J. M., Merten, M., Meckler, Y., Hinnrasky, J., Fuchey, C., and Puchelle, E. (1990) Culture and characterization of human tracheal gland cells. *Am. Rev. Respir. Dis.* **141,** 1280–1288.
9. Finkbeiner, W. E., Shen, B. Q., Mrsny, R. J., and Widdicombe, J. H. (1993) Induction of mucous phenotype in cultures of glands from human airways leads to loss of CFTR and chloride secretion. *Pediatr. Pulm.* **9,** 187.
10. Bell, E., Ivarssen, B., and Merril, C. (1979) Production of a tissue like structure by contraction of collagen lattices by human fibroblasts of different proliferative potential in vitro. *Proc. Natl. Acad. Sci. USA* **76,** 1274–1278.
11. Benali, R., Tournier, J. M., Chevillard, M., Zahm, J. M., Klosseck, J. M., Hinnrasky, J., Gaillard, D., Maquart, F. X., and Puchelle, E. (1993) Tubule formation by human surface respiratory epithelial cells cultured in a three-dimensional collagen lattice. *Am. J. Physiol.* **264,** L183–L192.
12. Jacquot, J., Spilmont, C., Burlet, H., Fuchey, C., Buisson, A. C., Tournier, J. M., Gaillard, D., and Puchelle, E. (1994) Glandular like morphogenesis and secretory activity of human tracheal gland cells in a three-dimensional collagen gel matrix. *J. Cell. Physiol.* **161,** 407–418.
13. Merten, M. D., Breittmayer, J. P., Figarella, C., and Frelin, C. (1993) ATP and UTP increase secretion of the bronchial inhibitor by human tracheal gland cells in culture. *Am. J. Physiol.* **265,** L479–L484.
14. Yamaya, M., Finkbeiner, W. E., and Widdicombe, J. H. (1991) Ion transport by cultures of human tracheobronchial glands. *Am. J. Physiol.* **261,** L485–L490.
15. Merten, M. D. and Figarella, C. (1993) Constitutive hypersecretion and insensitivity to neurotransmitters by cystic fibrosis tracheal gland cells. *Am. J. Physiol.* **264,** L93–L99.

16. Merten, M. D. and Figarella, C. (1994) Neuropeptide Y and norepinephrine cooperatively inhibit tracheal gland cell secretion. *Am. J. Physiol.* **266**, L513–L518.
17. Merten, M. D., Tournier, J. M., Meckler, Y., and Figarella, C. (1992) Secretory proteins and glycoconjugates synthesized by human tracheal gland cells in culture. *Am. J. Respir. Cell Mol. Biol.* **7**, 598–605.
18. Sommerhof, C. P. and Finkbeiner, W. E. (1990) Human tracheobronchial submucosal gland cells in culture. *Am. J. Respir. Cell Mol. Biol.* **2**, 41–50.
19. Engelhardt, J. F., Yankaskas, J. R., Ernst, S. A., Yang, Y. P., Marino, C. R., Boucher, R. C., Cohn, J. A., and Wilson, J. M. (1992) Submucosal glands are the predominant site of CFTR expression in the human bronchus. *Nature Genet.* **2**, 240–246.
20. Jacquot, J., Puchelle, E., Hinnrasky, J., Fuchey, C., Bettinger, C., Spilmont, C., Bonnet, N., Dieterle, A., Dreyer, D., Pavirani, A., and Dalemans, W. (1993) Localization of the cystic fibrosis transmembrane conductance regulator in airway secretory glands. *Eur. Respir. J.* **6**, 169–176.
21. Yamaya, M., Finkbeiner, W. E., and Widdicombe, J. H. (1991) Altered ion transport by human tracheal glands in cystic fibrosis. *Am. J. Physiol.* **261**, L491–L494.
22. Becq, F., Merten, M. D., Voelkel, M. A., Gola, M., and Figarella, C. (1993) Characterization of cAMP dependent CFTR-chloride channels in human tracheal gland cells. *FEBS Lett.* **321(1)**, 73–78.
23. Masui, T., Wakefield, L. M., Lechner, J. F., Laveck, M. A., Sporn, M. B., and Harris, C. (1986) Type β transforming growth factor is the primary differentiation-inducing serum factor for normal human bronchial epithelial cells. *Proc. Natl Acad. Sci. USA* **83**, 1438–1442.

19

Human Chondrocyte Cultures as Models of Cartilage-Specific Gene Regulation

Mary B. Goldring

1. Introduction

Chondrocytes comprise the single cellular component of adult hyaline cartilage and are considered to be terminally differentiated cells that maintain the cartilage matrix when turnover is low. The major components of the extracellular matrix synthesized by these specialized cells include highly crosslinked fibrils of the final synthesized and secreted triple helical type II collagen molecule that interact with other cartilage-specific collagens IX and XI, the large aggregating proteoglycan aggrecan, small proteoglycans, such as biglycan and decorin, and other specific and nonspecific matrix proteins that are expressed at defined stages during development and growth *(1,2)*. This matrix confers tensile strength and flexibility to articular surfaces and serves specialized functions in only a few other tissues. Cultured chondrocytes have served as useful models for studying chondrocyte differentiation and the effects of cytokines and growth factors that control the maintenance or suppression of differentiated cartilage phenotype *(3)*.

Freshly isolated human articular or costal chondrocytes express cartilage-specific type II collagen and continue to do so for several days to weeks in primary monolayer culture *(4,5)*. During prolonged culture and serial subculture, these cells begin to express type I and type III collagens. This "switch" can be accelerated by plating the cells at low densities or by treating them with cytokines, such as interleukin-1 (IL-1) *(5,6)*. Early attempts to culture chondrocytes from various animal and human sources were frustrated by the tendency of these cells to "dedifferentiate" to a fibroblast-like phenotype in monolayer culture and their inability to proliferate in suspension culture where cartilage-specific phenotype was maintained *(7–9)*. This loss of phenotype in monolayer culture was found to be reversible if they were placed in suspension cultures in spinner flasks *(10)* or in dishes coated with a nonadherent substrates *(11,12)*, if they were kept at high density in micromass cultures *(13,14)*, or if they were embedded in a solid support matrices, such as collagen gels *(15)*, agarose *(16–18)*, or alginate *(19,20)*. The basement membrane-type matrix commercially known as Matrigel™ has also been shown to support maintenance of chondrocyte phenotype, probably because of presence of growth and differentiation factors, such as IGF-I, that copurify with the matrix proteins *(21)*. Serum-free defined media of varying compositions, but usually including insulin, have also been used, frequently in combination with the other culture systems mentioned *(22)*.

Primary cultures of chondrocytes isolated from young animals that maintain the cartilage-specific phenotype at least throughout primary culture are easily obtained and have been used widely to assess differentiated chondrocyte functions. The use of chondrocytes of human origin has been more problematical, because the source of the cartilage cannot be controlled, sufficient numbers of cells are not readily obtained from random operative procedures, and the phenotypic stability of adult human chondrocytes is lost more quickly on expansion in serial monolayer cultures than that of cells of juvenile human *(4)* or young or embryonic animal origin *(23,24)*. Viral oncogenes have been used to generate immortalized chondrocytes from nonhuman sources that demonstrate high proliferative capacities and retain at least some differentiated chondrocyte properties *(25–29)*. Also, chondrocyte lines have arisen spontaneously from fetal rat calvaria *(30,31)*. Human chondrosarcoma cell lines express some aspects of the chondrocyte phenotype, but are tumorigenic *(32,33)*. The lack of a reproducible source of chondrocytes of human origin has hampered progress in studies of cartilage function relevant to human disease. Recently, the successful immortalization of human chondrocytes was reported using SV40-containing vectors *(34)*. These cells have a high proliferative capacity that can be dissociated from their ability to express chondrocyte-specific phenotype by using permissive culture conditions. These cultures were established and characterized based on conditions and criteria previously defined for culturing normal human chondrocytes *(4,5,35)*, as described in this chapter.

2. Materials
2.1. Isolation and Culture of Human Chondrocytes

1. Cell-culture reagents: Dulbecco's Modified Eagle's medium (DMEM) containing 10% fetal calf serum (FCS); Dulbecco's phosphate-buffered Ca^{2+}- and Mg^{2+}-free saline (PBS); trypsin–EDTA solution. These solutions have shelf lives as recommended by the supplier. If DMEM is prepared from powder, high-quality distilled and deionized water (i.e., using a Milli-Q apparatus), dedicated steril-

ized bottles, and 0.22-μm filters should be used. FCS and trypsin–EDTA are stored at –20°C, but should not be refrozen after thawing for use.
2. Hyaluronidase, 1 mg/mL in PBS. Prepare freshly and filter through a sterile 0.22-μm filter.
3. Trypsin, 0.25% in Hank's balanced salt solution (HBSS) without Ca^{2+} and Mg^{2+}.
4. Collagenase (bacterial, clostridiopeptidase A), 3 mg/mL in DMEM with 10% FCS for articular cartilage or serum-free for costal cartilage. Prepare freshly in ice-cold DMEM, and filter immediately through a sterile 0.22-μm filter.

2.2. Suspension Cultures

1. Agarose-coated dishes: Weigh out 1 g of high-melting-point agarose in an autoclavable bottle, and add 100 mL of dH_2O. Autoclave with cap tightened loosely, allow to cool to ~55°C, and quickly pipet into culture dishes (1 mL/3.5-cm, 3 mL/6-cm, or 9 mL/10-cm tissue-culture or bacteriological dish). Allow the gel to set at 4°C for 30 min, and wash the surface 2–3 times with PBS. Use plates immediately, or wrap tightly with plastic or foil wrap to prevent evaporation, and store at 4°C.
2. Solid suspension culture in agarose gel: Autoclave 2% (w/v) low-gelling-temperature agarose in dH_2O, cool to 37°C, and dilute with an equal volume of 2X DMEM containing 20% FCS either without cells or with a chondrocyte suspension.
3. Suspension culture in alginate beads: Low-viscosity alginate, 1% (w/v) in $0.15 M$ NaCl. Stir to dissolve alginate in NaCl solution and filter sterilize. Prepare sterile-filtered $0.15 M$ NaCl and 102 mM $CaCl_2$.

2.3. Analysis of Matrix Protein Synthesis

1. Staining for metachromatic matrix: 0.05% Alcian Blue 8GX and 2.5% glutaraldehyde in $0.4 M$ $MgCl_2$ and 25 mM sodium acetate, pH 5.6.
2. Biosynthetic labeling of collagens: 10X solution of ascorbic acid (ASC) and β-aminoproprionitrile fumarate (β-APN), 5 mg of each dissolved in 10 mL of serum-free culture medium. Sterile-filter and dilute at 1/10 (v/v) to give the volume of 1X ASC/β-APN solution required for the incubation. Prepare freshly; 25 μCi/mL L-[5-^{3}H]proline (1 mCi/mL; SA > 20 Ci/mmol), 50 μg/mL ASC, and 50 μg/mL β-APN in serum-free culture medium. Add 25 μL of [^{3}H]proline/mL of 1X ASC/β-APN solution using a sterile pipet tip.
3. Collagen typing: 2 mg/mL pepsin in $1 M$ acetic acid. Dissolve 2 mg pepsin/mL dH_2O, then add 58 μL of glacial acetic acid/each mL of solution, and cool on ice. Prepare freshly the volume required for the experiment. Reagents for Laemmli SDS-PAGE are prepared as stock solutions.
4. Proteoglycan synthesis: $8.0 M$ guanidine-HCl, buffered with $0.01 M$ sodium acetate, and containing $0.02 M$ disodium EDTA, $0.20 M$ 6-aminocaproic acid, with 5.0 mM benzamidine HCl, 10 mM N-ethylmaleimide, and 0.5 mM PMSF added immediately before use.
5. Immunocytochemistry: 2% paraformaldehyde in $0.1 M$ cacodylate buffer, pH 7.4. Dissolve 10 g of paraformaldehyde in 150 mL dH_2O in Erlenmeyer flask on hot

plate in fume hood (do not exceed 65°C). Add ~2 mL of 1N NaOH while stirring, and stir until solution is clear. Let solution cool for 15 min. Add 250 mL of 0.2M cacodylate buffer, pH 7.4, and adjust pH if necessary.

2.4. Analysis of mRNA

1. Guanidine lysis buffer: 5M guanidine monothiocyanate, 10 mM EDTA, 50 mM Tris-HCl, pH 7.5. Add stock solutions of 1 mL of 0.5M EDTA, pH 8.0, and 2.5 mL of 1.0M Tris-HCl, pH 7.5, to 5M guanidine thiocyanate to volume of 50 mL. Freshly add ~72 µL of β-mercaptoethanol (β-ME)/10 mL of buffer.
2. 4M LiCl: Dissolve 84.8 g of lithium chloride in ~400–450 mL of DEP-H_2O, allow to cool to room temperature before bringing up to final volume of 500 mL, and sterile-filter using disposable sterile vacuum flask with cap. Store at –20°C.
3. Solubilization buffer: 0.2% SDS, 1.0 mM EDTA, 10 mM Tris-HCl, pH 7.5. Add stock solutions of 0.5 mL of 20% SDS, 0.1 mL of 0.5M EDTA, and 0.5 mL of 1M Tris-HCl, pH 7.5, to final volume of 50 mL in DEP-H_2O. Store at –20°C.
4. Phenol reagent (phenol:chloroform:isoamyl alcohol [25:24:1]). Prepare freshly using phenol saturated with Tris-HCl/EDTA, pH 8.0, or obtain commercially.
5. 75% Ethanol: Dilute bottled absolute ethanol with DEP-treated water in sterile tube. Store at –20°C.

2.5. Analysis of Gene Regulatory Sequences by Transient Transfection Assays

1. 2X HBSS: 50 mM HEPES, 1.5 mM Na_2HPO_4, 10 mM KCl, 280 mM NaCl, 12 mM glucose, pH 7.05. Note: Use an accurately calibrated pH meter, since the pH is critical for formation of a fine $CaPO_4$/DNA precipitate. Filter-sterilize and store at 4°C.
2. 2M $CaCl_2$: Filter sterilize and store at room temperature.
3. 15% Glycerol in HBSS: Combine 30 mL of 50% (w/v) glycerol in dH_2O, 50 mL of dH_2O, and 20 mL of dH_2O. Filter-sterilize and store at 4°C.
4. Prepare $CaPO_4$/DNA precipitate by combining in order:

H_2O	to make final volume of 1 mL
plasmid DNA	10–25 µg
2M $CaCl_2$	62 µL
2X HBSS	500 µL

    Gently add DNA and $CaCl_2$ to H_2O without mixing. Use 1-mL pipet attached to an automatic pipeter with tip placed in bottom of tube to add slowly the 2X HBSS buffer and gently release ~5 bubbles to mix. Allow precipitate to form for 15–30 min at room temperature.

3. Methods

3.1. Isolation and Culture of Human Chondrocytes

1. Human adult articular cartilage is obtained from knee joints or hips after surgery for joint replacement or at autopsy, and dissected free from underlying bone and

Table 1
Culture Vessel Area vs Chondrocyte Number Required for Plating Density of ~2.6 × 10⁴ cells/cm²

Diameter	Area, cm²	No. of cells plated
16-mm well	2	50,000
2.2-cm well	3.8	100,000
3.5-cm plate	10	250,000
6-cm plate	28	750,000
10-cm plate	79	2×10^6

any adherent connective tissue. Juvenile costal cartilage is obtained from ribs removed during pectus excavatum repair and dissected free from perichondrium.

2. Place slices of cartilage in 10-cm dishes containing PBS and use ~10 mL vol of proteinases for digestion of each gram of tissue. Wash slices several times in PBS, and incubate at 37°C in hyaluronidase for 10 min followed by 0.25% trypsin for 30–45 min with 2–3 washes in PBS after each enzyme treatment.
3. Add collagenase solution, chop the cartilage in small pieces using a scalpel and blade, and incubate at 37°C overnight (18–24 h) for articular cartilage and up to 48 h for costal cartilage until the cartilage matrix is completely digested and the cells are free in suspension (*see* Note 1). Break up any clumps of cells by repeated aspiration of the suspension through a 10-mL pipet or a 12-cc syringe without needle.
4. Transfer cell suspension to a sterile 50-mL conical polypropylene tube, wash the plate with PBS to recover remaining cells, and combine in tube. Centrifuge cells at ~1000*g* in benchtop centrifuge for 10 min at room temperature and wash cell pellet three times with PBS, resuspending cells each time and centrifuging.
5. Resuspend the final pellet in culture medium containing 10% FCS, count with a Coulter counter or hemocytometer, and bring up to volume with culture medium to give 1×10^6 cells/mL. For monolayer culture, plate the cells at ~1.3 to 2.6×10^4 cells/cm² (Table 1) in dishes already containing some culture medium, and agitate without swirling to distribute the cells evenly. Incubate at 37°C in an atmosphere of 5% CO_2 in air with medium changes every 3–4 d, as described *(4,5)* (*see* Note 2). Primary cultures of adult articular and juvenile costal chondrocytes incubated in the absence and presence of IL-1 are shown in Fig. 1A–D.
6. Preparation of subcultured cells (*see* Note 3): Remove culture medium by aspiration with a sterile Pasteur pipet attached to a vacuum flask, and wash with PBS. Add trypsin–EDTA (1 mL/10-cm dish), and incubate at room temperature for 10 min with periodic gentle shaking of dish and observation through microscope to assure that cells have come off the plate. If significant numbers of cells remain attached, continue the incubation for a longer time (≤20 min) or higher temperature (37°C) and/or scrape the cell layer with a sterile plastic scraper, Teflon™ policeman, or syringe plunger. Repeatedly aspirate and expel the cell suspen-

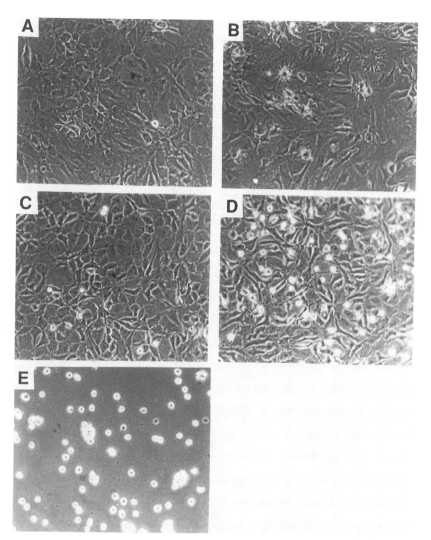

Fig. 1. Phase-contrast micrographs of human adult articular and juvenile costal chondrocytes in primary culture. **(A,B)** Articular chondrocytes at d 13 of culture and **(C,D)** costal chondrocytes at d 14 were photographed after incubation for 24 h in culture medium alone (A and C) or with 5 pM IL-1β (B and D). **(E)** The same costal chondrocyte preparation as shown in C and D was left in primary culture for 2.5 mo, trypsinized, and incubated for 1 wk in suspension culture. Note that most of the cells in clumps have begun to break up and form a single-cell suspension.

sion into the plate using a 5- or 10-mL pipet containing culture medium, and then transfer to a sterile conical 15- or 50-mL polypropylene tube. Perform cell counts, or determine the split ratio required, distribute equal volumes of the cell suspension in dishes or wells that already contain culture medium, and agitate the culture plate immediately after each addition to assure uniform plating density on the culture surface (*see* Note 4).

3.2. Suspension Cultures

1. Fluid suspension culture above agarose (*see* Note 5): Trypsinize monolayer cultures, spin down cells, wash with PBS, centrifuge, and resuspend in culture medium at 1×10^6 cells/mL. Plate chondrocyte suspension in culture medium containing 10% FCS in dishes that have been coated with 1% agarose and culture for 2–4 wk. Change the medium weekly by carefully removing the medium above settled cells while tilting the dish, centrifuging the remaining suspended cells, and replacing them in the dish after resuspension in fresh culture medium. The cells first form large clumps that begin to break up after 7–10 d and eventually form single-cell suspensions (Fig. 1E).
2. Recovery and analysis of agarose suspension cultures: To recover cells for direct experimental analysis, for redistribution in agarose-coated wells, or for culture in monolayer, transfer the cell suspension to 15- or 50-mL conical tubes, wash agarose surface at least twice with culture medium to recover remaining cells, and spin down and resuspend cells in an appropriate volume of culture medium for replating or in wash or extraction buffer for subsequent experimental analysis.
3. Solid suspension culture in agarose: Precoat plastic tissue-culture dishes with cell-free 1% agarose in culture medium (0.5 mL/3.5-cm, 1.5 mL/ 6-cm, or 4.5 mL/ 10-cm dish), and allow to gel at room temperature. Add the same volume of 1% agarose medium containing chondrocytes at a density of $1-4 \times 10^6$ cells/mL of gel, incubate at 37°C for 20–30 min to allow the cells to settle, and leave at room temperature until the agarose is gelled. Add culture medium containing 10% FCS, and incubate at 37°C with medium changes every 3–4 d.
4. Analysis of solid agarose cultures: After incubations with test reagents and/or radioisotopes in minimal volumes of appropriate culture medium, the whole cultures may be stored frozen, or medium and gel can be treated separately. For subsequent analysis, add appropriate guanidine extraction buffer directly to the gel (for proteoglycan or RNA extraction), or whole cultures may be adjusted to $0.5M$ acetic acid, treated with pepsin, and neutralized, as described in Section 3.3., step 3 for analysis of collagens. To remove agarose and debris, the samples are centrifuged at high speed, e.g., in Eppendorf centrifuge at top speed, 4°C.
5. Suspension culture in alginate (*see* Note 6): Resuspend pelleted cells in sterile $0.15M$ NaCl containing low-viscosity alginate gel (1%) at a density of $1-4 \times 10^6$ cells/mL, and then slowly express through a 22-gage needle in a dropwise fashion into a 102-mM $CaCl_2$ solution. Leave the beads to polymerize further in the $CaCl_2$ solution for 10 min at room temperature. Decant $CaCl_2$ solution, and wash three times in $0.15M$ NaCl and once in culture medium, using 10 volumes of each wash solution/volume of packed beads, decanting each wash after the beads have settled. Pipet the beads into culture dishes or wells (70–80 beads/mL of medium), and incubate at 37°C. For medium changes, carefully pipet the culture medium from the top of the settled beads.

6. Analysis of alginate cultures: Decant the growth medium, and solubilize the alginate gel by adding 3 vol of a solution of 55 mM Na citrate in 0.15M NaCl for 10 min at 37°C. Add appropriate guanidine extraction buffer or centrifuge at 500g for 10 min to recover the chondrocytes with pericellular matrix or at 2000g for 5 min followed by trypsin–EDTA treatment to recover dispersed cell suspensions.

3.3. Analysis of Matrix Protein Synthesis

1. Staining for metachromatic matrix: Remove culture medium, and wash with PBS. Add Alcian Blue/glutaraldehyde solution at room temperature for several hours, remove excess stain by washing with 3% acetic acid, and store cultures in 70% ethanol for subsequent examination by light photomicrography.
2. Biosynthetic labeling of collagens (*see* Note 7): Remove serum-containing culture medium, wash with serum-free medium, and add [^{3}H]proline at 25 µCi/mL for a further 24 h in serum-free culture medium supplemented with 50 µg/mL ascorbate and 50 µg/mL β-APN (or without β-APN to retain collagen in the pericellular matrix). Remove culture medium, and store at –20°C. Wash cell layer with PBS, and solubilize by adding equal volumes of serum-free culture medium and 1M ammonium hydroxide (an aliquot may be analyzed for DNA by the diphenylamine method).
3. Collagen typing: To analyze pepsin-resistant collagens, add pepsin/HAc solution to equal volume of either labeled culture medium or solubilized cell solution for 16 h at 4°C, lyophilize, redissolve in 2X SDS sample buffer, and neutralize with 1-µL additions of 2M NaOH to titrate the color of the bromophenol blue in the sample buffer from yellow-green to blue (but not to violet). To analyze procollagens and fibronectin, add 2X SDS sample buffer containing 0.2% β-ME to an equal volume of the culture medium. Heat samples to boiling for 10 min, and load on SDS gels (5% acrylamide running gels or 7–15% gradient gels) that include a radiolabeled rat tail tendon collagen standard in one lane. Perform delayed reduction with 0.1% β-ME on pepsinized samples to distinguish α1(III) from α1(I or II) collagens. Absence of the α2(I) collagen band indicates the virtual absence of type I collagen synthesis. In cultures containing a mixture of type I and type II collagen, definitive identification of these collagens requires Western blotting using specific antibodies and standard techniques.
4. Proteoglycan synthesis (*see* Note 7): Add [^{35}S]sodium sulfate at 20 µCi/mL in culture medium containing 25 µg/mL ASC, and incubate at 37°C for a further 4–24 h. Extract labeled medium and cell layer with equal volume of 8M guanidine hydrochloride solution containing 20 mM EDTA and proteinase inhibitors at 4°C for 48 h. To quantitate ^{35}S-labeled PGs, pass extracts over Sephadex G-25M in PD 10 columns, elute under dissociative conditions, and measure by scintillation counting. Further characterization of proteoglycans may be performed by agarose/polyacrylamide composite gel electrophoresis, Western blotting, or gel-filtration chromatography using standard published techniques.
5. Immunocytochemistry (*see* Note 7): Plate cells in plastic Lab-Tech 4-chamber slides (Nunc, Inc. Naperville, IL) at 6×10^4 cells/chamber in culture medium

containing 10% FCS, and grow for 3–5 d to subconfluence. Change to culture medium containing 25 µg/mL ASC, and test reagents. At the end of the incubation period, carefully wash the chambers three times with PBS, and fix the cells with 2% paraformaldehyde in 0.1M cacodylate buffer, pH 7.4, for 2 h at 4°C. After two rinses with 0.1M cacodylate buffer, add monoclonal antibodies (MAbs) specific for human type II collagen, aggrecan core protein, and so forth to different chambers at concentrations recommended by the supplier. Incubate separate chamber slides with chondroitinase ABC for 30 min at 37°C prior to addition of MAbs against chondroitin sulfates in order to expose epitopes. Visualize the staining by incubation with a gold-conjugated secondary antibody (Auroprobe LM, Amersham, Arlington Heights, IL) followed by silver enhancement (e.g., IntenSE Kit, Amersham).

3.4. Analysis of mRNA

Total RNA for Northern blots has been extracted successfully from human chondrocytes by several methods *(35–37)*. We currently use a procedure modified from that of Cathala et al. *(38)* (*see* Notes 8 and 9).

1. 10-cm Dishes: Remove medium, and wash cell layer with PBS. Trypsinize cells in 0.5 mL of trypsin–EDTA (add 1.5 mL and then remove 1 mL immediately) at room temperature for ≤10 min. Suspend cells, and bring up suspension with 2 mL of DMEM/10% FCS. Transfer to sterile polypropylene 10 × 75 mm tubes (15 mL). Wash plates with 2 mL of DMEM/FCS, and combine in tube with cell suspension. Place tube on ice. Remove aliquot for cell count, if required. Spin down cells at 1200–1500 rpm, and wash with cold PBS. Resuspend and combine pellets from 1–4 dishes (1–5 × 10^6 cells) for each extraction.
2. To final pellet add 500 µL of guanidine lysis buffer. Vortex. Homogenize cell suspension by aspiration 5–10 times through tuberculin syringe with needle. Add 3 mL of 4M LiCl. Vortex. Store at 4°C overnight (can be left for a few days).
3. Spin at 9000 rpm in high-speed centrifuge for 90 min at 4°C. Carefully remove supernatant with heat-treated Pasteur pipet (do immediately, standing at centrifuge). Respin if pellet slips.
4. To pellet add 500 µL of solubilization buffer. Let stand at room temperature for 45 min, vortexing every 10 min until pellet is dissolved, or freeze on dry ice (samples can be left at –80°C at this stage), and vortex while thawing to break up pellet.
5. Transfer sample to sterile 1.2-mL tubes with caps. Add 500 µL of phenol reagent (phenol:chloroform:isoamyl alcohol [25:24:1]). Mix by vortexing and/or shaking. Leave on ice 5 min. Centrifuge at 12,000 rpm (in Eppendorf centrifuge) for 15 min at 4°C. Remove upper aqueous phase to fresh sterile 1.2-mL tube on ice, taking care not to take any of the interphase. (Discard lower organic phase with interphase containing DNA and proteins.)
6. To precipitate the RNA, add 1/10 vol of 4M LiCl, vortex, then add 2.5 vol of ethanol and vortex. Leave overnight at –20°C. Centrifuge for 30 min at 12,000 rpm at 4°C. Discard supernatant.

7. Wash pellet with 800 µL of 75% ethanol, and shake tube or vortex to break up pellet. Centrifuge at 12,000 rpm for 15 min at 4°C. (Note: Samples can be left in ethanol at –20°C, if necessary.)
8. Speed vac final pellet, but not to dryness. Dissolve pellets in 50 µL of diethylpyrocarbonate (DEP)-treated H_2O or 10 mM HEPES. This takes time on ice. Samples may be left overnight at 4°C and/or heated briefly at 65°C to speed up dissolution.
9. Read ODs at 240, 260, and 280. Use OD 260 to calculate RNA concentration. The final preparations should give yields of approx 10 µg of RNA/1 × 10^6 cells with the appropriate A_{260}:A_{280} ratio of approx 2.0. Store at –20°C in nonself-defrosting freezer or at –80°C.
10. Separate RNAs (5–10 µg of total RNA/lane) on 0.8% agarose gels in the presence of 2% formaldehyde. Transfer to nitrocellulose or nylon-supported nitrocellulose membranes in 10X SSC overnight.
11. Label DNA plasmid probes with ^{32}P-dCTP and ^{32}P-dGTP by nick translation or cDNA inserts with ^{32}P-dCTP by random primer labeling. Prehybridize blots in 50% formamide, 5X SSC, 5X Denhardt's solution, and 0.3% SDS at hybridization temperature. Add 100 ng/mL DNA probe, and hybridize for at least 16 h at 54°C for collagen probes and 42°C for noncollagen probes. Staining with ethidium bromide and hybridization with a glyceraldehyde-3-phosphate dehydrogenase (GAPDH) cDNA probe are employed to monitor uniform loading of RNA on Northern blots.

A typical Northern blot of human chondrocyte mRNAs is shown in Fig. 2.

3.5. Analysis of Gene Regulatory Sequences by Transient Transfection Assays

1. Plate chondrocytes at 1 × 10^6 cells/10-cm dish in culture medium containing 10% FCS, and allow to settle down for 16–20 h. Change the medium 2–4 h prior to the transfection.
2. Add 10–25 µg of plasmid DNA in a $CaPO_4$ coprecipitate in a dropwise fashion throughout the dish, and agitate gently. Incubate at 37°C for 4 h, and perform glycerol shock for 2 min. Wash cell layers with serum-free medium, and then allow the cells to recover overnight in culture medium containing 10% FCS. Change to medium containing required serum concentration or serum substitute and test reagents, incubate for 36–48 h, and harvest for assay of reporter gene activity (*see* Note 10).

4. Notes

1. Chondrocytes are quite resilient, and tolerate the prolonged incubation times required for complete dissociation of the matrix and the absence of serum in the costal cartilage digestion. If the digestion is not complete by the end of the allotted time, then more collagenase solution may be added, or the suspension may be recovered and the fragments left behind for further digestion. These conditions result in suspensions that are essentially single-cell, and therefore, it is not necessary to resort to filtration through a nylon mesh, as has been done by others

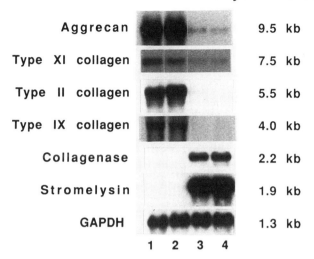

Fig. 2. Expression of mRNAs encoding cartilage-specific matrix proteins and IL-1-induced metalloproteinases by human costal chondrocytes. Costal chondrocytes at d 11 of primary culture were incubated without (lanes 1 and 2) or with 5 pM IL-1β (lanes 3 and 4) for 24 h prior to harvest of the cells for RNA extraction. Total RNAs (5 µg/lane) were electrophoresed on a 0.8% agarose gel in the presence of 2% formaldehyde, blotted on nylon-supported nitrocellulose membranes, and hybridized with the [^{32}P]-labeled cDNA probes encoding large aggregating proteoglycan core protein (aggrecan), α2(XI) procollagen (type XI collagen), α1(II) procollagen (type II collagen), α1(IX) procollagen (type IX collagen), collagenase, stromelysin, and glyceraldehyde-3-phosphate dehydrogenase (GAPDH). The sizes of the mRNA transcripts in kilobases are indicated on the right. Descriptions and sources of the probes are found in ref. *34*.

when shorter digestion times are used *(17)*. These considerations are important for decreasing the loss of chondrocytes during their isolation from valuable human cartilage specimens.

2. After initial plating of the primary cultures, the chondrocytes require 2–3 d before they have settled down and spread out completely. Culture for ~4–7 d is required before reasonable amounts of total RNA may be extracted. Juvenile costal chondrocytes continue to express chondrocyte phenotype (e.g., type II collagen mRNA) for several weeks and will form multilayer cultures. Adult articular chondrocytes are strongly contact-inhibited, and they may lose phenotype within 2–4 wk of monolayer culture. After they are subcultured, both types of chondrocytes stop expressing chondrocyte matrix proteins, but this loss of phenotype is reversible and the cells may be redifferentiated in suspension culture within or on top of a nonadherent matrix.

3. Since chondrocytes are strongly adherent to tissue-culture plastic, possibly because of the strongly charged glycosaminoglycans in their matrix, a trypsin–EDTA solution rather than trypsin alone should be used for full recovery of chondrocytes from tissue–culture plastic during passaging. It is preferable not to use any antibiotics in order that any contamination that arises becomes apparent immediately. If necessary, standard concentrations of penicillin-streptomycin, gentamycin, and so on, that are suggested for fibroblast cultures are acceptable for use in chondrocyte cultures.
4. Primary chondrocyte cultures should be used for experimental analyses immediately before or just after confluence is reached to permit optimal matrix synthesis and cellular responsiveness. If the cells are not used or subcultured, they may be left at confluence for several weeks with weekly medium changes as long as the volume of the culture medium is maintained. If long-term culture results in the deposition of excessive matrix that is not easily digested with trypsin–EDTA, then a single-cell suspension may be obtained by using a dilute solution of collagenase (0.25%) and trypsin (0.25%) in PBS.
5. Although the growth and maintenance of chondrocytes in primary culture or after subculture require the use of 10% FCS, the loss of phenotype that occurs under these conditions may be delayed if the cells are plated at 4–10 times higher density. Since high cell yields are not usually attainable from human cartilage sources, the reversibility of the loss of phenotype may be exploited by expanding the chondrocyte populations in monolayer cultures, redifferentiating the cells in fluid suspension culture, and replating them in monolayer immediately before performing the experimental procedure.
6. Biosynthetic labeling and immunocytochemistry procedures are readily performed on chondrocytes in a solid suspension system, such as alginate, agarose, or collagen gels. Alginate culture may be the method of choice, since the chondrocytes are easily recovered by depolymerization of the alginate with a calcium chelator. For long-term alginate cultures, high-viscosity alginate may provide more stable beads. Concentrations of serum as low as 0.5%, serum substitutes, or combinations of growth and differentiation factors or hormones have been used successfully, depending on the experimental protocol, to permit chondrocyte phenotypic expression.
7. Optimization of extracellular matrix synthesis and deposition: ASC is not routinely added during growth and maintenance culture of chondrocytes, since it is known to inhibit the transcription of cartilage-specific matrix genes in long term incubations *(37)*. It is a requirement, however, for synthesis and secretion of collagen and should be added at 25–50 µg/mL during 1- to 3-d incubations as required for biosynthetic labeling of collagens or proteoglycans, or extracellular deposition of matrix proteins that will be extracted for immunoblotting or analyzed by immunocytochemistry.
8. General recommendations for RNA extractions: Use sterile plastic tubes and pipets and/or heat-treated glassware (not to be touched by human hands). Wear gloves at all times (even when touching outside of tubes). Never use parafilm! All procedures are

done on ice (4°C) unless indicated otherwise. All solutions (except phenol solution) and dH$_2$O are DEP-treated and autoclaved and/or sterile-filtered.

9. Optimization of RNA extraction technique: Use of trypsinization of cell cultures prior to extraction and of a lithium chloride precipitation step during extraction will reduce or eliminate contamination with proteoglycans and other glycoproteins. The amount of total RNA loaded per well on agarose gels should not exceed 10 µg, and sufficiently long gels should be run to resolve collagen and large proteoglycan mRNAs that migrate more slowly than 28S RNA on Northern blots and tend to smear if overloaded. Northern blots on nitrocellulose or nylon-supported nitrocellulose and hybridizations in the absence of dextran sulfate result in virtually no background in the detection of high kilobase transcripts in chondrocyte RNA preparations. High stringency hybridization conditions are used to prevent crossreactivities among some collagen probes.

10. Optimization of transient expression assays in normal primary and subcultured chondrocytes: Chondrocytes are plated at somewhat higher densities than immortalized cell lines, such as NIH/3T3 cells routinely used for transient transfections. Subcultured chondrocytes are transfected the day after passaging. In contrast, primary cultures must be left several (4–7) days, with at least one interim medium change, until the cells have settled down and begun to undergo cell division. Recovery overnight after the transfection by incubation in culture medium also permits optimal expression of the reporter gene and responsiveness to test reagents. Primary or subcultured costal chondrocytes incubated in an insulin-containing serum substitute during transient expression of type II collagen gene regulatory sequences have been used successfully *(35)*.

References

1. Heinegard, D. and Oldberg, A. (1989) Structure and biology of cartilage and bone matrix noncollagenous macromolecules. *FASEB J.* **3**, 2042–2051.
2. Mayne, R. and Brewton, R. G. (1993) Extracellular matrix of cartilage: collagen, in *Joint Cartilage Degradation: Basic and Clinical Aspects* (Woessner, J. F., Jr. and Howell, D. S., eds.), Marcel Dekker, New York, pp. 81–108.
3. Goldring, M. B. (1993) Degradation of articular cartilage in culture: regulatory factors, in *Joint Cartilage Degradation: Basic and Clinical Aspects* (Woessner, J. F., Jr. and Howell, D. S., eds.), Marcel Dekker, New York, pp. 281–345.
4. Goldring, M. B., Sandell, L. J., Stephenson, M. L., and Krane, S. M. (1986) Immune interferon suppresses levels of procollagen mRNA and type II collagen synthesis in cultured human articular and costal chondrocytes. *J. Biol. Chem.* **261**, 9049–9056.
5. Goldring, M. B., Birkhead, J., Sandell, L. J., Kimura, T., and Krane, S. M. (1988) Interleukin 1 suppresses expression of cartilage-specific types II and IX collagens and increases types I and III collagens in human chondrocytes. *J. Clin. Invest.* **82**, 2026–2037.
6. Goldring, M. B. and Krane, S. M. (1987) Modulation by recombinant interleukin 1 of synthesis of types I and III collagens and associated procollagen mRNA levels in cultured human cells. *J. Biol. Chem.* **262**, 16,724–16,729.

7. Holtzer, J., Abbott, J., Lash, J., and Holtzer, A. (1960) The loss of phenotypic traits by differentiated cells in vitro. I. Dedifferentiation of cartilage cells. *Proc. Natl. Acad. Sci. USA* **46**, 1533–1542.
8. Ham, R. G. and Sattler, G. L. (1968) Clonal growth of differentiated rabbit cartilage cells. *J. Cell. Physiol.* **72**, 109–114.
9. Green, W. T., Jr. (1971) Behavior of articular chondrocytes in cell culture. *Clin. Orthopaed. Related Res.* **75**, 248–260.
10. Norby, D. P., Malemud, C. J., and Sokoloff, L. (1977) Differences in the collagen types synthesized by lapine articular chondrocytes in spinner and monolayer culture. *Arthritis Rheum.* **20**, 709–716.
11. Glowacki, J., Trepman, E., and Folkman, J. (1983) Cell shape and phenotypic expression in chondrocytes. *Proc. Soc. Exp. Biol. Med.* **172**, 93–98.
12. Castagnola, P., Moro, G., Descalzi-Cancedda, F., and Cancedda, R. (1986) Type X collagen synthesis during in vitro development of chick embryo tibial chondrocytes. *J. Cell Biol.* **102**, 2310–2317.
13. Kuettner, K. E., Pauli, B. U., Gall, G., Memoli, V. A., and Schenk, R. K. (1982) Synthesis of cartilage matrix by mammalian chondrocytes in vitro. I. Isolation, culture characteristics, and morphology. *J. Cell Biol.* **93**, 743–750.
14. Bassleer, C., Gysen, P., Foidart, J. M., Bassleer, R., and Franchimont, P. (1986) Human chondrocytes in tridimensional culture. *In Vitro Cell Dev. Biol.* **22**, 113–119.
15. Gibson, G. J., Schor, S. L., and Grant, M. E. (1982) Effects of matrix macromolecules on chondrocyte gene expression: synthesis of a low molecular weight collagen species by cells cultured within collagen gels. *J. Cell Biol.* **93**, 767–774.
16. Benya, P. D. and Shaffer, J. D. (1982) Dedifferentiated chondrocytes reexpress the differentiated collagen phenotype when cultured in agarose gels. *Cell* **30**, 215–224.
17. Aulthouse, A. L., Beck, M., Friffey, E., Sanford, J., Arden, K., Machado, M. A., and Horton, W. A. (1989) Expression of the human chondrocyte phenotype in vitro. *In Vitro Cell Dev. Biol.* **25**, 659–668.
18. Aydelotte, M. B. and Kuettner, K. E. (1988) Differences between sub-populations of cultured bovine articular chondrocytes. I. Morphology and cartilage matrix production. *Connect. Tiss. Res.* **18**, 205–222.
19. Guo, J., Jourdian, G. W., and MacCallum, D. K. (1989) Culture and growth characteristics of chondrocytes encapsulated in alginate beads. *Conn. Tiss. Res.* **19**, 277–297.
20. Hauselmann, H. J., Aydelotte, M. B., Schumacher, B. L., Kuettner, K. E., Gitelis, S. H., and Thonar, E. J.-M. A. (1992) Synthesis and turnover of proteoglycans by human and bovine adult articular chondrocytes cultured in alginate beads. *Matrix* **12**, 116–129.
21. Vukicevic, S., Kleinman, H. K., Luyten, F. P., Roberts, A. B., Roche, N. S., and Reddi, A. H. (1992) Identification of multiple active growth factors in basement membrane Matrigel suggests caution in interpretation of cellular activity related to extracellular matrix components. *Exp. Cell Res.* **202**, 1–8.

22. Adolphe, M., Froger, B., Ronot, X., Corvol, M. T., and Forest, N. (1984) Cell multiplication and type II collagen production by rabbit articular chondrocytes cultivated in a defined medium. *Exp. Cell Res.* **155,** 527–536.
23. Gerstenfeld, L. C., Kelly, C. M., Von Deck, M., and Lian, J. B. (1990) Comparative morphological and biochemical analysis of hypertrophic, non-hypertrophic and 1,25(OH)$_2$D$_3$ treated non-hypertrophic chondrocytes. *Connect. Tiss. Res.* **24,** 29–39.
24. Adams, S. L., Pallante, K. M., Niu, Z., Leboy, P. S., Golden, E. B., and Pacifici, M. (1991) Rapid induction of type X collagen gene expression in cultured chick vertebral chondrocytes. *Exp. Cell Res.* **193,** 190–197.
25. Alema, S., Tato, F., and Boettiger, D. (1985) Myc and src oncogenes have complementary effects on cell proliferation and expression of specific extracellular matrix components in definitive chondroblasts. *Mol. Cell. Biol.* **5,** 538–544.
26. Gionti, E., Pontarelli, G., and Cancedda, R. (1985) Avian myelocytomatosis virus immortalizes differentiated quail chondrocytes. *Proc. Natl. Acad. Sci. USA* **82,** 2756–2760.
27. Horton, W. E., Jr., Cleveland, J., Rapp, U., Nemuth, G., Bolander, M., Doege, K., Yamada, Y., and Hassell, J. R. (1988) An established rat cell line expressing chondrocyte properties. *Exp. Cell Res.* **178,** 457–468.
28. Thenet, S., Benya, P. D., Demignot, S., Feunteun, J., and Adolphe, M. (1992) SV40-immortalization of rabbit articular chondrocytes: alteration of differentiated functions. *J. Cell Physiol.* **150,** 158–167.
29. Mallein-Gerin, F. and Olsen, B. R. (1993) Expression of simian virus 40 large T (tumor) oncogene in chondrocytes induces cell proliferation without loss of the differentiated phenotype. *Proc. Natl. Acad. Sci. USA* **90,** 3289–3293.
30. Grigoriadis, A. E., Heersche, J. N. M., and Aubin, J. E. (1988) Differentiation of muscle, fat, cartilage, and bone from progenitor cells present in a bone-derived clonal cell population: effect of dexamethasone. *J. Cell Biol.* **106,** 2139–2151.
31. Bernier, S. M. and Goltzman, D. (1993) Regulation of expression of the chondrogenic phenotype in a skeletal cell line (CFK2) in vitro. *J. Bone Miner. Res.* **8,** 475–484.
32. Block, J. A., Inerot, S. E., Gitelis, S., and Kimura, J. H. (1991) Synthesis of chondrocytic keratan sulphate-containing proteoglycans by human chondrosarcoma cells in long-term cell culture. *J. Bone Joint Surg. Am.* **73,** 647–658.
33. Takigawa, M., Pan, H. O., Kinoshita, A., Tajima, K., and Takano, Y. (1991) Establishment from a human chondrosarcoma of a new immortal cell line with high tumorigenicity in vivo, which is able to form proteoglycan-rich cartilage-like nodules and to respond to insulin in vitro. *Int. J. Cancer* **48,** 717–725.
34. Goldring, M. B., Birkhead, J. R., Suen, L.-F., Yamin, R., Mizuno, S., Glowacki, J., Arbiser, J. L., and Apperley, J. F. (1994) Interleukin-1β-modulated gene expression in immortalized human chondrocytes. *J. Clin. Invest.* **94,** 2307–2316.
35. Goldring, M. B., Fukuo, K., Birkhead, J. R., Dudek, E., and Sandell, L. J. (1994) Transcriptional suppression by interleukin-1 and interferon-γ of type II collagen gene expression in human chondrocytes. *J. Cell Biochem.* **54,** 85–99.

36. Chomczynski, P. and Sacchi, N. (1987) Single step method of RNA isolation by acid guanidinium thiocyanate-phenol-chloroform extraction. *Anal. Biochem.* **162,** 156–159.
37. Sandell, L. J. and Daniel, J. C. (1988) Effects of ascorbic acid on collagen mRNA levels in short term chondrocyte cultures. *Connect. Tiss. Res.* **17,** 11–22.
38. Cathala, G., Savouret, J.-F., Mendez, B., West, B. L., Karin, M., Martial, J. A., and Baxter, J. D. (1983) A method for isolation of intact, translationally active ribonucleic acid. *DNA* **2,** 329–335.

20

Isolation and Culture of Bone-Forming Cells (Osteoblasts) from Human Bone

James A. Gallagher, Roger Gundle, and Jon N. Beresford

1. Introduction

The most conspicuous function of the osteoblast is the formation of bone. During phases of active bone formation, osteoblasts synthesize bone matrix and prime it for subsequent mineralization. Active osteoblasts are plump, cuboidal cells rich in organelles involved in the synthesis and secretion of matrix proteins. Unlike fibroblasts, they are obviously polarized, secreting matrix onto the underlying bony substratum which consequently grows by apposition. Some osteoblasts are engulfed in matrix during bone formation and are entombed in lacunae. These cells are described as osteocytes and remain in the bone matrix in a state of low metabolic activity. At the completion of a phase of bone formation, those osteoblasts which avoided entombment in lacunae lose their prominent synthetic function and become inactive osteoblasts, otherwise known as bone-lining cells. In mature bone, lining cells cover most of the bone surfaces. Osteocytes and bone-lining cells should not be considered as inactive cells since they play a major role in the regulation of bone modeling and remodeling and in calcium homeostasis (1).

Much of our knowledge of the biology of bone tissue has been derived from morphological studies. The heterogeneity of cell types in bone, the highly cross-linked extracellular matrix and the mineral phase, combine to make bone a difficult tissue to study at the cellular and molecular level (2). Consequently the earliest attempts to isolate specific cell populations utilized enzymic digestion of fetal or neonatal tissue from experimental animals which is poorly mineralized and highly cellular. Although these studies undoubtedly furthered our knowledge of bone cell biology, there are major advantages in attempting to investigate cells isolated from human bone. First, there are differences in cell

From: *Methods in Molecular Medicine: Human Cell Culture Protocols*
Edited by: G. E. Jones Humana Press Inc., Totowa, NJ

physiology between species, and also between adults and neonates within a species. Second, the ability to culture human bone cells opens up the prospect of investigating the pathological mechanisms that underlie bone diseases including age-related bone loss.

The earliest report to describe the isolation of viable cells from adult human bone is that of Bard et al. *(3)*. The isolation procedure they employed involved extensive demineralization of the tissue in a solution of ethylenediaminetetraacetic acid (EDTA) followed by digestion with collagenase. The cells obtained expressed low levels of alkaline phosphatase and incorporated radiolabeled proline into macromolecular material, but the hydroxyproline/proline ratio indicated that collagen formed only a minor proportion of the total protein synthesized. The cultured cells remained viable for periods of up to 2 wk, but failed to proliferate leading the authors to conclude that osteocytes were the predominant cell type present *(4)*.

In a study of patients with Paget's disease, Mills et al. succeeded in culturing cell populations from explants of trabecular bone *(5)*. These populations responded to crude extracts of the parathyroid gland with an increase in radiolabeled thymidine incorporation and a proportion of the cells expressed alkaline phosphatase activity.

Encouraged by these reports, in the early 1980s we *(6–15)* and several other groups of investigators *(16–22)* developed systems to isolate human bone-derived cells (HBDCs). We undertook a systematic investigation to identify the phenotypic characteristics of HBDC populations and to determine the conditions of culture that favored their proliferation and differentiation. Within a relatively short space of time it was shown that the cell populations obtained reproducibly expressed an osteoblast-like phenotype and that they represented a viable alternative to the use of normal or neoplastic cell lines of avian or rodent origin.

HBDCs have been widely used to investigate the biology of the human osteoblast. Their use has facilitated several major developments in our understanding of the hormonal regulation of human bone remodeling, including the first demonstration of an effect of cytokines on bone-forming cells *(10)*, identification of direct effects of oestrogen on osteoblasts *(11,23)*, and recently the identification of purinergic receptors in bone *(24,25)*. We are now moving into a new phase of research in which HBDCs, isolated from patients with specific disorders including Paget's disease *(26)*, McCune Albright syndrome *(27)*, and osteoporosis *(28–30)* are being used to investigate the cellular and molecular pathology of bone disease.

Cultured chondrocytes have been used to repair cartilage defects with some success *(31)*. Recently we have started to identify conditions which promote the osteogenic potential of HBDCs in vitro. Now we have the prospect that human bone cell culture may become an important tool in tissue engineering, allowing the autologous transplantation of osteoblastic populations expanded in vitro and seeded onto suitable carrier matrices.

The purpose of this chapter is threefold:

1. To describe in detail the methodology currently in use in the authors' laboratories for the isolation and culture of HBDCs;
2. To demonstrate that, by employing the methods described, it is possible to obtain a cell population that is phenotypically stable and that retains the potential for osteogenic differentiation in vitro and in vivo; and
3. To promote the wider use and continued development of the HBDC culture system.

2. Materials
2.1. Tissue-Culture Media and Supplements

1. Phosphate-buffered saline (PBS) without calcium and magnesium, pH 7.4 (Gibco, Gaithersburg, MD).
2. Dulbecco's modification of minimum essential medium (DMEM) (Gibco) supplemented to a final concentration of 10% with heat-inactivated fetal calf serum (FCS), 2 mM L-glutamine, 25–50 U/mL penicillin, 25–50 µg/mL streptomycin, and 50 µg/mL freshly prepared L-ascorbic acid.
3. Serum-free DMEM (SFM).
4. FCS (see Note 1).
5. Tissue-culture flasks (75 cm^2) or Petri dishes (100-mm diameter) (see Note 2).

2.2. Preparation of Explants

1. Bone rongeurs and/or bone curet from any surgical instrument supplier.
2. Solid stainless steel scalpels with integral handles (BDH Merck).

2.3. Passaging and Secondary Culture

1. Trypsin–EDTA solution: 0.05% trypsin and 0.02% EDTA in Ca^{2+}- and Mg^{2+}-free PBS, pH 7.4 (Gibco).
2. 0.4% Trypan blue in 0.85% NaCl (Sigma Aldrich).
3. 70 µm "Cell Strainer" (Becton Dickinson).
4. Neubauer Hemocytometer (BDH Merck).
5. Collagenase (Sigma type VII from Clostridium histolyticum).
6. DNAse I (Sigma Aldrich).

2.4. Phenotypic Characterization

1. Calcitriol (Leo Pharmaceuticals).
2. Menadione (vitamin K$_3$) (Sigma Aldrich).
3. Staining Kit 86-R for alkaline phosphatase (Sigma Aldrich).
4. Osteocalcin radioimmunoassay (RIA) (Cis UK Ltd., High Wycombe Bucks, UK) (see Note 3).

2.5. In Vitro Mineralization

1. L-Ascorbic acid 2-phosphate (Wako Pure Chemical Industries Ltd.).
2. Dexamethasone (Sigma Aldrich).
3. Hematoxylin (BDH Merck).
4. DPX (BDH Merck).

3. Methods
3.1. Bone Explant Culture System
3.1.1. Establishment of Primary Explant Cultures

A scheme outlining the culture technique is shown in Fig. 1.

1. Transfer tissue, removed at surgery or biopsy, into a sterile container with PBS or serum-free medium (SFM) for transport to the laboratory with minimal delay, preferably on the same day (*see* Note 4). Ideally, the bone used should be radiologically normal. An excellent source is the upper femur of patients undergoing total hip replacement surgery for osteoarthritis. Cancellous bone is removed from this site prior to the insertion of the femoral prosthesis and would otherwise be discarded. The tissue obtained is remote from the hip joint itself, and thus from the site of pathology, and is free of contaminating soft tissue (*see* Note 5).
2. Remove extraneous soft connective tissue from the outer surfaces of the bone by scraping with a sterile scalpel blade. Rinse the tissue in sterile PBS and transfer to a sterile Petri dish containing a small volume of PBS (5–20 mL, depending on the size of the specimen). If the bone sample is a femoral head, remove cancellous bone directly from the open end using a bone curet or a solid stainless steel blade with integral handle. Disposable scalpel blades may shatter during this process. With some bone samples (e.g., rib), it may be necessary to gain access to the cancellous bone by breaking through the cortex with the aid of sterile surgical bone rongeurs.
3. Transfer the cancellous bone fragments to a clean Petri dish containing 2–3 mL of PBS and dice into pieces 3–5 mm in diameter. This can be achieved in two stages using a scalpel blade first, and then fine scissors. Decant the PBS and transfer the bone chips to a 50-mL polypropylene tube containing 15–20 mL of PBS. Vortex the tube vigorously three times for 10 s and then leave to stand for 30 s to allow the bone fragments to settle. Carefully decant off the supernatant containing hematopoietic tissue and dislodged cells, add an additional 15–20 mL of PBS, and vortex the bone fragments as before. Repeat this process a minimum of three times, or until no remaining hematopoietic marrow is visible and the bone fragments have assumed a white, ivory-like appearance.
4. Culture the washed bone fragments as explants at a density of 0.2–0.6 g of tissue/ 100-mm diameter Petri dish or 75-cm^2 flask (*see* Note 2) in 10 mL of medium at 37°C in an humidified atmosphere of 95% air, 5% CO_2.
5. Leave the cultures undisturbed for 7 d after which time replace the medium with an equal volume of fresh medium taking care not to dislodge the explants. Feed again at 14 d and twice weekly thereafter.

Culture of Osteoblasts

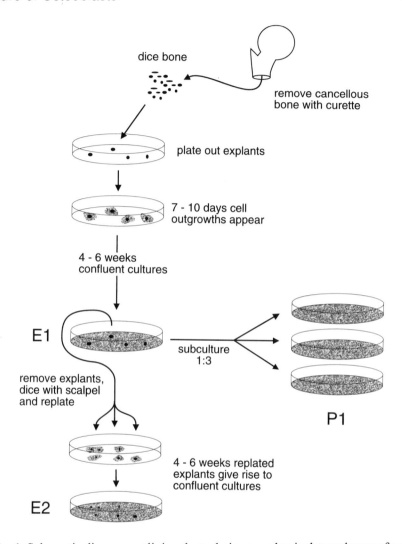

Fig. 1. Schematic diagram outlining the technique used to isolate cultures of osteoblasts from bone.

With the exception of small numbers of isolated cells, which probably become detached from the bone surface during the dissection, the first evidence of cellular proliferation is observed on the surface of the explants, and this normally occurs within 5–7 d of plating. After 7–10 d, cells can be observed migrating from the explants onto the surface of the culture dish (*see* Fig. 2). If care is taken not to dislodge the explants when feeding, and they are left undisturbed between media changes, they rapidly become anchored to the sub-

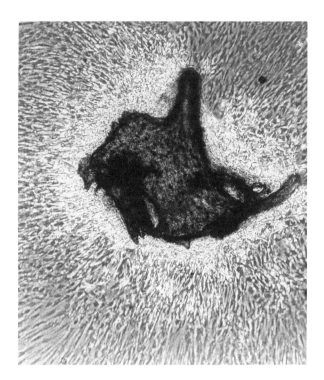

Fig. 2. Photomicrograph of explanted cancellous bone showing migration of osteoblasts.

stratum by the cellular outgrowths. Typical morphology of the cells is shown in Fig. 3, but cell shape varies between donors from fibroblastic to cobblestone-like. Cultures generally attain confluence 4–6 wk postplating, and typically achieve a saturation density of 29,000 ± 9000 cells/cm^2 (mean ± SD, $n = 11$ donors).

3.1.2. Passaging and Secondary Culture

1. Remove and discard the spent medium.
2. Gently wash the cell layers three times with 10 mL of PBS.
3. To each flask add 5 mL of freshly thawed trypsin–EDTA solution at room temperature (20°C). Incubate for 2 min at room temperature with gentle rocking every 30 s to ensure that the entire surface area of the flask and explants is exposed to the trypsin–EDTA solution. Remove and discard all but 2 mL of the trypsin–EDTA solution, and then incubate the cells for an additional 5 min at 37°C.
4. Remove the flasks from the incubator and examine under the microscope. Look for the presence of rounded, highly refractile cell bodies floating in the trypsin–EDTA solution. If none, or only a few, are visible, tap the base of the flask sharply

Culture of Osteoblasts

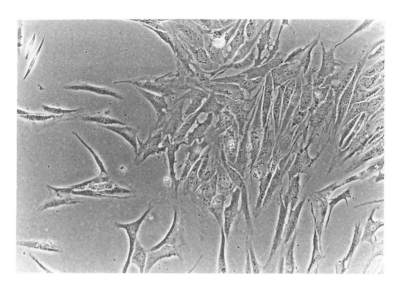

Fig. 3. Photomicrograph showing the typical morphology of HBDCs.

on the bench top in an effort to dislodge the cells. If this is without effect, incubate the cells for a further 5 min at 37°C.

5. When the bulk of the cells has become detached from the culture substratum, transfer the cells to a 50-mL polypropylene tube containing 5 mL of DMEM with 10% FCS to inhibit tryptic activity. Wash the flask two to three times with 10 mL of SFM and pool the washings with the original cell isolate. To recover the cells centrifuge at 250g for 10 min at 15°C.

6. Remove and discard the supernatant, invert the tube, and allow to drain briefly. Holding the top of the tube, sharply flick the base of the tube with the first finger to dislodge and break up the pellet. Add 2 mL of SFM containing 1 µg/mL DNAse I for each dish or flask treated with trypsin–EDTA, and using a narrow bore 2-mL pipet, repeatedly aspirate and expel the medium to generate a cell suspension.

7. Filter the cell suspension through a 70-µm "Cell Strainer" (Becton Dickinson) to remove any bone spicules or remaining cell aggregates. For convenience and ease of handling, the filters have been designed to fit into the neck of a 50-mL polypropylene tube. Wash the filter with 2–3 mL of SFM containing DNAse I, and add the filtrate to the cells.

8. Take 20 µL of the mixed cell suspension and dilute to 80 µL with SFM. Add 20 µL of trypan blue solution, mix, and leave for 1 min before counting viable (round and refractile) and nonviable (blue) cells in a Neubauer Hemocytometer. Using this procedure, typically 1–1.5 × 10^6 cells are harvested per 75-cm^2 flask of which ≥75% are viable.

9. Plate the harvested cells at a cell density suitable for the intended analysis. We routinely subculture at 5 × 10^3–10^4 cells/cm^2 and achieve plating efficiencies measured after 24 h of ≥70% (*see* Note 6).

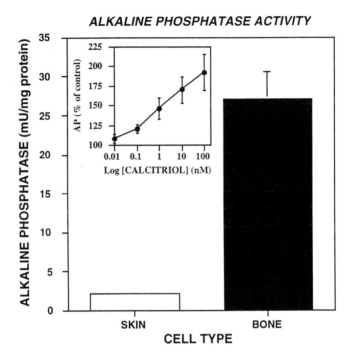

Fig. 4. The expression of alkaline phosphatase by cultures of HBDCs and skin-derived cells at first passage. Confluent cell monolayers were lysed in alkaline buffer (Sigma Aldrich) containing 0.1% Triton X-100, chilled on ice, and sonicated for 10 s three times using an MSE Model 150 Sonicator with a Micro-tip at 20% of full output power. The homogenates were centrifuged to remove insoluble material and an aliquot of the supernatant used for the determination of alkaline phosphatase activity as described previously *(6)*. Open bar, skin fibroblasts ($n = 6$ donors); closed bar, bone-derived cells ($n = 18$ donors). The values shown are the mean ± SD. Inset: Stimulation of HBDC alkaline phosphatase activity by calcitriol. The values shown are the mean ± SD of data from three donors. Over the same dose range there was no effect on the alkaline phosphatase activity of skin fibroblasts derived from the same donors (data not shown).

If dishes have reached confluence but the cells are not required immediately, the cells can be stored by cryopreservation (*see* Note 7).

3.1.3. Phenotypic Characteristics at First Passage

Compared with cultures of skin fibroblasts obtained from the same donor, cultures of HBDCs express high basal levels of the enzyme alkaline phosphatase, a widely accepted marker of early osteogenic differentiation (Fig. 4). Basal activity is initially low, but increases with increasing cell density *(10)* (*see* Note 8). Treatment with calcitriol (1,25[OH]$_2$D$_3$; the active metabolite of vita-

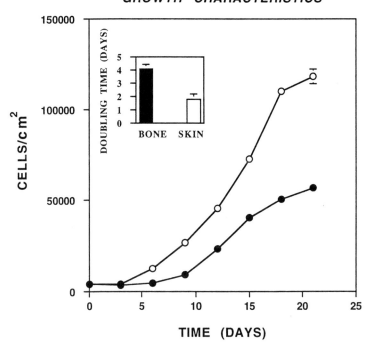

Fig. 5. Representative growth curves from HBDCs and skin-derived cells at first passage. Cells were subcultured into multiwell trays at a density of 5×10^3 cells/cm^2 and cultured for the indicated time periods in medium supplemented with 10% (v/v) heat-inactivated FCS. The media were changed every third day. Open symbols, skin fibroblasts; closed symbols, HBDCs. Tissues were obtained from a 10.5-yr-old male donor undergoing corrective surgery. Inset: Mean ± SD doubling times at first passage for HBDCs and skin-derived cells ($n = 4$ donors) cultured under identical conditions. Closed bars, HBDCs ($n = 6$ donors); open bars, skin fibroblasts ($n = 4$ donors).

min D$_3$) increases HBDC alkaline phosphatase activity, but not that of skin fibroblasts (*see* Fig. 4 inset). The magnitude of this stimulatory effect decreases as cell density and, hence, basal alkaline phosphatase activity increases *(10)*. When plated at similar densities and cultured under identical conditions, compared with skin fibroblasts obtained from the same donor, HBDCs proliferate less rapidly and reach lower saturation densities (Fig. 5).

In common with cells of the osteogenic lineage from all species studied, HBDC respond to PTH with an increase in the levels of intracellular cAMP (Fig. 6) *(13,22)*. As shown, however, a similar response can be observed in cultures of human skin fibroblasts (Fig. 6). This is in agreement with a report that human dermal fibroblasts possess PTH receptors comparable to

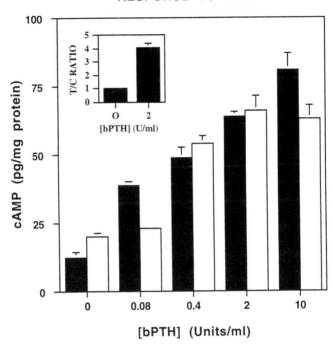

Fig. 6. The effect of bPTH on the production of 3', 5'-cyclic adenosine monophosphate by HBDCs and skin-derived cells at first passage. Confluent cell monolayers were incubated for 10 min at 37°C in medium supplemented with 1% (v/v) charcoal-stripped FCS and 500 μM isobutyl-1-methyl-xanthine in the absence or presence of bPTH (0.08–10 U/mL; National Institute of Biological Standards and Control reagent #77/533, 230 U/100 μg protein) (*see* Note 13). Total cAMP (medium + cell layer) was measured by specific RIA. Cultures were established from tissues obtained from the same donor described in Fig. 1. Open bars, skin fibroblasts; closed bars, HBDCs. Inset: Mean ± SD index of stimulation (treated/control ratio) for cultures of HBDCs treated with 2 U/mL bPTH (n = 6 donors).

those present on cells derived from bone and kidney tissue *(32)*. Over the same dose range, in addition to increasing the level of intracellular cAMP, PTH inhibits basal- and calcitriol-stimulated alkaline phosphatase activity (Table 1).

Over a 24-h period, taking into account the relative abundance of proline in collagen compared to noncollagenous protein, collagen accounts for 19.6 ± 1.0% of the protein secreted into the culture medium (mean ± SEM; n = 16 donors). As with basal alkaline phosphatase activity, the percentage of collagen synthesis increases with increasing cell density *(12)*. Of the total amount of collagen

Table 1
The Effect of PTH on the Basal- and Calcitriol-Stimulated Alkaline Phosphatase Activity of HBDC

bPTH, U/mL	Alkaline phosphatase, mU/mg protein		
	Experiment A[a]	Experiment B[b]	
		Basal	+ 50 nM Calcitriol
0.0	25.2 ± 1.4	63.9 ± 3.0	98.7 ± 2.0[c]
0.04	19.3 ± 0.9[c]	—	—
0.2	15.1 ± 0.7[c]	48 ± 3.4[c]	76 ± 1.6[d]
1.0	13.8 ± 1.3[c]	43 ± 2.4[c]	—
5.0	—	40 ± 0.6[c]	48 ± 3.0[d]

[a]Experiment A. Confluent cells (E1P1) were incubated for 72 h in medium supplemented with 5% (v/v) FCS in the absence or presence of bovine PTH (bPTH) (see Note 13) at the indicated concentrations. Cells were obtained from the femur of a 30-yr-old male undergoing surgery following an industrial accident.

[b]Experiment B. Confluent cells (E1P1) were incubated for 48 h in medium supplemented with 10% (v/v) FCS in absence or presence of bPTH, calcitriol, or both hormones in combination at the indicated concentrations. Cells were obtained from the tibia of a 6-yr-old female undergoing corrective surgery for non-union of a previous fracture. Alkaline phosphatase activity in the solubilized cell layers was measured as described previously (8).

[c]Significantly different ($p < 0.01$) when compared with the control.
[d]Significantly different ($p < 0.01$) when compared with calcitriol alone.

synthesized, <20% (range 12–19%) is retained in the cell layer under these conditions of culture.

Analysis of the radiolabeled collagens synthesized by HBDCs reveals the presence of predominantly type I collagen (ratio $\alpha_1[I]:\alpha_2[I]$ 2.1 ± 0.11; mean ± SD for 3 donors) with small amounts of type V and, in the laboratories of some investigators, type III (Fig. 7) (6,12,31,33). In contrast, skin cell cultures obtained from the same donors synthesize, in addition to type I collagen, significant amounts of type III and synthesize an excess of $\alpha_1(I)$ chains over $\alpha_2(I)$ chains (Fig. 7). This is indicative of the presence of some other collagen type, probably type I trimer.

In the presence of calcitriol in cultures of HBDCs, but not skin fibroblasts obtained from the same donors, there is a dose-dependent increase in the production of osteocalcin (Fig. 8) (8,11,19,21). This is a protein of M_r 5800 containing multiple residues of the vitamin K-dependent amino acid γ-carboxyglutamic acid, and in humans its synthesis is restricted to mature cells of the osteoblast lineage (see Note 9). As such, it is an excellent late stage marker for cells of this series despite the fact that its precise function in bone has yet to be established. The aforementioned represents only a brief descrip-

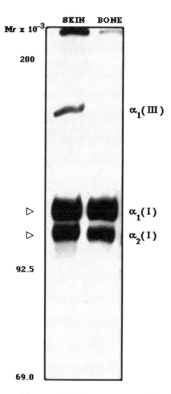

Fig. 7. Analysis of the collagen chain types synthesized by HBDCs and skin-derived cells obtained from the same donor at first passage. Equal amounts of pepsin-resistant, radiolabeled collagens were separated by SDS-PAGE using 7.5% slab gels. After 1 h the current was turned off and the samples reduced *in situ* by the addition to the sample wells of 100 mM dithiothreitol in gel sample buffer. This was done in order to resolve the α_1 (I) chains of collagen type I from the α_1 (III) chains of collagen type III. Electrophoresis was resumed after 30 min and continued until the dye front was ~2 cm from the gel-buffer interface. Following autoradiography, the collagen chain types were identified on the basis of size and, in the case of collagen type I, by their comigration with the α chains of purified, unlabeled human collagen type I (open arrowheads). Tissues were obtained from a 17.5-yr-old female donor undergoing corrective surgery for spina bifida.

tion of the phenotypic characteristics of HBDCs at first passage, and for a more detailed consideration the reader is referred to the original papers.

3.1.4. Phenotypic Stability in Culture

As a matter of routine we perform our studies on cells at first passage. Other investigators have studied the effects of repeated subculture on the phenotypic

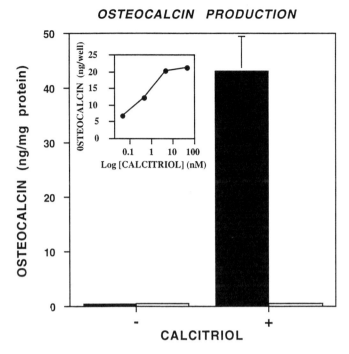

Fig. 8. The effect of calcitriol on the expression of osteocalcin by cultures of HBDCs and skin-derived cells at first passage. Cells at confluence were transferred to medium supplemented with 5% charcoal-stripped FCS (to remove endogenous calcitriol and osteocalcin), 50 µg/mL L-ascorbic acid, and 10 nM menadione (vitamin K$_3$), and then incubated for 48 h in the absence or presence of 10 nM calcitriol. Osteocalcin secreted into the culture medium was measured by RIA. Open bar, skin fibroblasts; closed bar, HBDCs. Under all conditions in cultures of skin fibroblasts, and in the absence of added calcitriol in bone cell cultures, the levels of osteocalcin were below the limits of detection for the assay (<0.5 ng/mL). Tissues were obtained from a 17-yr-male with an osteosarcoma. Normal bone tissue from the femur was used for culture Inset: Typical dose response curve for the production of osteocalcin by HBDCs in response to treatment with calcitriol. Tissue was obtained from the vertebral bone of a 2 yr old undergoing corrective surgery for spina bifida.

stability of HBDCs and found that they rapidly lose their osteoblast-like characteristics *(19,22,34,35)*. In practical terms this presents real difficulties, since it is often desirable to obtain large numbers of HBDCs from a single donor.

As an alternative to repeated subculture, we have investigated the potential of replating the trabecular explants at the end of primary culture into a new flask (*see* Fig 1). Using this technique, it is possible to obtain additional cell populations that continue to express osteoblast-like characteristics, including

Table 2
Phenotypic Stability of the Cell Populations Obtained by Replating Explants of Human Trabecular Bone[a]

Population	Time in culture	Alkaline phosphatase, mU/mg protein		Osteocalcin, ng/well/48h
		Basal	+D	+D
E1P1	21	52 ± 3	79 ± 3[b]	19 ± 2
E2P1	41	—	—	10 ± 4
E3P1	86	64 ± 3	98 ± 3[b]	13 ± 1
E4P1	147	39 ± 2	57 ± 3[b]	16 ± 1

[a]Confluent cell monolayers (P1) obtained following the successive replating of the trabecular explants (E1–E4) were incubated for 48 h in medium supplemented with 10% (v/v) FCS, 50 µg/mL L-ascorbate, and 10 nM menadione in the absence or presence of 10 nM calcitriol. Alkaline phosphatase in the solubilized cell layers was measured as described in the caption to Fig. 4. Osteocalcin secreted into the culture medium was measured by RIA (limit of detection 0.5 ng/mL). Cells were obtained from the tibia of the donor described in the legend to Table 1, experiment B.

[b]Significantly different ($p < 0.01$) when compared with the control.

the ability to mineralize their extracellular matrix (Table 2), and maintain their cytokine expression profile (26). Presumably, these cultures are seeded by cells that are situated close to the bone surfaces, and that retain the capacity for extensive proliferation and differentiation. The continued survival of these cells may be related to the gradual release over time in culture of the myriad of cytokines and growth factors that are known to be present in the extracellular bone matrix, many of which are known to be produced by mature cells of the osteoblast lineage (26,30,36,37).

3.2. Modifications to the Basic Culture System: The Importance of Ascorbate

Cells of the osteoblast lineage secrete an extracellular matrix that is rich in type I collagen, and factors that regulate the synthesis, secretion, and extracellular processing of procollagen have been shown to influence profoundly the proliferation and differentiation of osteoblast precursors in a variety of animal bone-derived cell culture systems (38). L-Ascorbic acid (vitamin C) functions as a cofactor in the hydroxylation of lysine and proline residues in collagen, and is essential for its normal synthesis and secretion. In addition, it increases procollagen mRNA gene transcription and mRNA stability (39).

The addition on alternate days of freshly prepared L-ascorbic acid (50 µg/mL = 250 µM) to HBDCs in secondary culture (E1P1) increases proliferation and produces a sustained increase in the steady-state levels of α_1 (I)-procollagen mRNA (Table 3), and a dramatic increase in the synthesis and secretion of type

Table 3
The Effect of Ascorbate on the Steady-State Levels of Osteoblast-Related mRNAs in Cultures of HBDC at First Passage[a]

	mRNA/GAPDH			
	−Ascorbate		+Ascorbate	
mRNA	−D	+D	−D	+D
Pro α_1 (I) (4.9 and 6.3 Kb)	1.0	1.0	3	3.3
AP (B/L/K; 2.4 Kb)	1.0	1.2	1.3	2.0
BSP (1.8 Kb)	1.0	1.3	3	3.4
OC (0.6 Kb)	—	1.0	—	2.0
OC (ng/mL/24 h)	0.4 ± 0.2	3 ± 0.2	2.0 ± 0.6	15 ± 0.9

[a]Cells were cultured in medium supplemented with 10% (v/v) FCS + 50 μg/mL L-ascorbic acid with changes of media every 48 h. After 8 d the cells were transferred to medium supplemented with 2% FCS and 10 nM menadione, ± ascorbate as before, and incubated for a further 24 h in the absence (−D) or presence (+D) of 10 nM calcitriol. Total RNA and poly-A$^+$ mRNA was isolated and analyzed as described previously *(39)*. Autoradiographs of Northern blots hybridized sequentially with the indicated cDNA probes were scanned using a laser densitometer. To correct for differences in loading, the absorbance values were corrected using the steady-state levels of GAPDH mRNA as an internal standard and then expressed as a treated/control ratio. Osteocalcin mRNA was not detected in the −D cultures, and so the control value for osteocalcin was assigned to −Ascorbate, +D culture. Osteocalcin secreted into the culture medium was measured using a commercially available RIA (Cis U.K. Ltd.).

Pro α_1 (I), pro α_1 (I) collagen mRNA; AP, mRNA for the bone/liver/kidney (B/L/K) isozyme of alkaline phosphatase; BSP, bone sialoprotein mRNA; OC, osteocalcin mRNA. The numbers in parentheses indicate the size(s) of the mRNA transcript(s) determined from plots of R_f vs log size for RNAs of known size.

I collagen (sixfold in the cell layer and fivefold in the culture medium *[40]*). Treatment with ascorbate also increases total noncollagenous protein synthesis (approximately twofold) and the proportion of newly synthesized protein that is retained in the cell layer (67 vs 42%) *(40)*. The increase in noncollagenous protein synthesis is accompanied by an increase in the steady-state levels of the mRNA for bone sialoprotein and osteocalcin, which are characteristically expressed by differentiated cells of the osteoblast lineage (Table 3) *(40)*. Using a specific RIA, it can be shown that the increase in the expression of osteocalcin mRNA following treatment with calcitriol is associated with an increase in protein secretion, and that, in the presence of ascorbate, there is detectable basal production of protein in the absence of added calcitriol (Table 3).

The changes described are consistent with ascorbate acting to promote both the proliferation and differentiation of HBDCs. However, the maintenance of adequate levels of ascorbate in vitro is difficult because the naturally occurring vitamin is labile in solution at neutral pH and 37°C ($t_{1/2}$ in culture variously

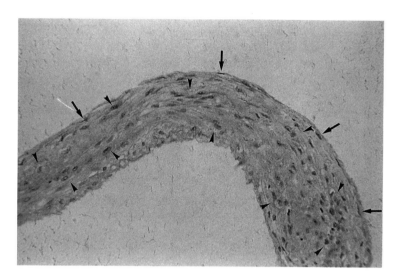

Fig. 9. Section through the cell layer of an HBDC culture (E1P1) grown in the continuous presence of the long-acting ascorbate analog L-ascorbate 2-phosphate. Note the presence of cells with ovoid nuclei (arrowheads) embedded within the dense extracellular matrix and elongated cells with flattened nuclei (arrows) covering its surface. Hematoxylin and eosin. Original magnification ×160.

reported as being between 1 and 6 h *[41]*). This is a particular problem when culturing cells of human origin, since they are incapable of synthesizing ascorbic acid *de novo (42)*.

The problems associated with the use of ascorbate can be overcome by the use of its long-acting analog, L-ascorbic acid 2-phosphate, which has a half-life of 7 d under the conditions that prevail in culture *(43,44)*. This is a particular advantage during the early stages of HBDC culture when the medium is changed weekly. In secondary culture the effects of culture in the presence of L-ascorbic acid 2-phosphate are similar to those of L-ascorbate, but of far greater magnitude. Dose-response studies revealed that these effects are maximal at 100 µM *(40)*, which is close to the levels found in human blood (34–68 µM) *(45)*. The effects of continuous culture in the presence of ascorbic acid 2-phosphate are even more dramatic and include an approximately ninefold increase in cell number (threefold in primary and threefold in secondary) and the elaboration of a dense, collagen-rich extracellular matrix within which many of the cells become enmeshed (Fig. 9) *(46)*.

An observation of particular importance is that the addition of ascorbate in secondary culture, even for periods of up to 28 d, cannot compensate fully for its omission in primary culture *(46–48)*. This suggests that maintaining adequate levels of ascorbate, and hence matrix synthesis, during the early stages of explant culture is of critical importance for the survival of cells that retain the ability to proliferate extensively, and give rise to precursors capable of undergoing osteogenic differentiation.

3.3. Passaging Cells Cultured in the Continuous Presence of Ascorbate

Because of their synthesis and secretion of an extensive collagen-rich extracellular matrix, HBDCs cultured in the continuous presence of ascorbate cannot be subcultured using trypsin–EDTA alone. They can, however, be subcultured if first treated with purified collagenase. The basic procedure is as follows:

1. Rinse the cell layers twice with SFM (10 mL/80-cm^2 flask).
2. Incubate the cells for 2 h at 37°C in 10 mL of SFM containing 25 U/mL purified collagenase (Sigma type VII) and 2 mM additional calcium (1:500 dilution of a filter-sterilized stock solution of 1M CaCl$_2$). Gently agitate the flask for 10–15 s every 30 min.
3. Terminate the collagenase digestion by discarding the medium (check that there is no evidence of cell detachment at this stage). Gently rinse the cell layer twice with 10 mL of Ca^{2+}- and Mg^{2+}-free PBS. To each flask add 5 mL of freshly thawed trypsin–EDTA solution, pH 7.4, at room temperature (20°C). From this point on, cells are treated as described for cultures grown in the absence of the long-acting ascorbate analog.

Typically this procedure yields ~3.5–4 × 10^6 cells/80-cm^2 flask after 28 d in primary culture. Cell viability is generally ≥90%.

3.4. In Vitro Mineralization

The *sine qua non* of the mature osteoblast is the ability to form bone. Despite the overwhelming evidence that cultures of HBDC contain cells of the osteoblast lineage, initial attempts to demonstrate the presence of osteogenic (i.e., bone forming) cells proved unsuccessful *(21)*. Subsequently, several authors reported that culture of HBDCs in the presence of ascorbate and millimolar concentrations of the organic phosphate ester β-glycerophosphate (β-GP) led to the formation of mineralized structures resembling the nodules that form in cultures of fetal or embryonic animal bone-derived cells (reviewed in ref. *34*). The nodules formed by animal-derived bone cells have been extensively characterized and shown by a variety of morphological, biochemical, and immunochemical criteria to resemble embryonic/woven bone formed in vivo. β-GP, however, is not a physiologically available organic phosphate ester and although there is no doubt that it promotes mineralization in vitro, it has yet to be established unequivocally that this process bears any relationship to the process of osteoblast-mediated mineralization that occurs in vivo *(34,48)*.

An alternative to the use of β-GP to provide levels of inorganic phosphate sufficient for supporting the process of cell-mediated mineralization in vitro, and the preferred method when studying HBDCs, is supplementation of the culture medium with 5 mM Pi, as described originally by Bellows et al. *(49)*.

In many bone-derived cell culture systems, osteogenic differentiation is glucocorticoid dependent. The degree of glucocorticoid dependence varies according to the cells state of maturation, which reflects, at least in part, their origin (bone surfaces vs bone marrow) and the species under investigation *(34,40,48,50–53)*.

Early studies using HBDCs revealed that long-term culture in the presence of glucocorticoids, while enhancing some indices of osteoblast maturation, decreased collagen synthesis, and inhibited cell proliferation *(7,8,53)*. However, when HBDCs are cultured in the presence of the long-acting ascorbate analog, these inhibitory effects of glucocorticoids are negated and the cells secrete a dense extracellular matrix that, in the presence of added phosphate, undergoes extensive mineralization *(48,54,55)*.

The protocol for inducing matrix mineralization in cultures of HBDCs is as follows:

Fragments of human trabecular bone are prepared as described and cultured in medium additionally supplemented with 100 μM L-ascorbic acid 2-phosphate and either 200 nM hydrocortisone or 10 nM dexamethasone, which approximates to a physiological dose of glucocorticoid *(56)*. For studies of in vitro mineralization, it is preferable to obtain trabecular bone from sites containing hemopoietic marrow. In practice this is usually from the upper femur or iliac crest.

When the cells have attained confluence and synthesized a dense extracellular matrix, typically after 28–35 d, subculture the cells using the sequential collagenase/trypsin–EDTA protocol and plate the cells in 25-cm^2 flasks at a density of 10^4 viable cells/cm^2. Change the medium twice weekly.

After 14 d supplement the medium with 5 mM inorganic phosphate. This is achieved by adding 1% (v/v) of a 500 mM phosphate solution, pH 7.4, at 37°C, prepared by mixing 500-mM solutions of Na$_2$HPO$_4$ and NaH$_2$PO$_4$ in a 4:1 (v/v) ratio *(49)*. The filter-sterilized stock solution can be stored at 4°C.

After 48–72 h, the cell layers are washed two to three times with 10 mL of SFM prior to fixation with 95% ethanol at 4°C. This can be done *in situ*, for viewing *en face*, or if sections are to be cut following detachment of the cell layer from the surface of the flask using a cell scraper. Great care is needed if the cell layer is to be harvested intact, particularly when mineralized.

For the demonstration of alkaline phosphatase activity, Sigma Staining Kit 86-R is used. 2.5 mL of staining solution is used per flask (or sufficient to cover the section). Place specimens in an humidified chamber and incubate for 1 h at 20°C in the dark. Wash under running tap water and counterstain the

Culture of Osteoblasts

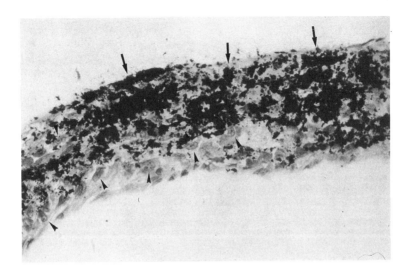

Fig. 10. Mineralization in a culture of HBDCs (E1P1) grown in the continuous presence of the long-acting ascorbate analog L-ascorbate 2-phosphate and 10 nM dexamethasone. Section through the cell layer of a culture supplemented with an additional 5 mM Pi 72 h prior to harvest. Note the presence of viable cells with prominent nucleoli (arrowheads) embedded within a dense extracellular matrix and the abundant mineral deposits (arrows). Von Kossa's stain and toluidine blue. Original magnification ×250.

nuclei for 15 s with hematoxylin prepared according to the method of Bancroft and Stevens *(57)*. Mineral deposits are then stained using a modification of von Kossa's technique. Prior to examination, sections are mounted in DPX and cell layers in flasks covered with glycerol.

HBDCs cultured in the continuous presence of glucocorticoid and the long-acting ascorbate analog produce a dense extracellular matrix that mineralizes extensively following the addition of Pi (Figs. 10–12) This is the case for the original cell population (E1P1) and that obtained following replating of the trabecular explants (E2P1), which further attests to the phenotypic stability of the cultured cells *(48)*.

Cells cultured in the continuous presence of ascorbate and treated with glucocorticoid at first passage show only a localized and patchy pattern of mineralization, despite possessing similar amounts of extracellular matrix and alkaline phosphatase activity *(48)*. Cells cultured without ascorbate, irrespective of the presence or absence of glucocorticoid, secrete little extracellular matrix, and do not mineralize.

The ability of the cells to mineralize their extracellular matrix is dependent on ascorbate having been present continuously in primary culture. The addi-

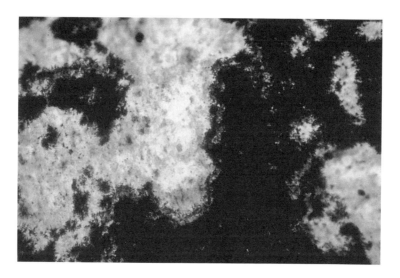

Fig. 11. Mineralization in a culture of HBDCs (E1P1) grown in the continuous presence of the long-acting ascorbate analog L-ascorbate 2-phosphate and 10 nM dexamethasone. Surface view showing extensive mineralization (black) and areas devoid of mineral but covered by alkaline phosphatase positive cells (gray). Culture conditions were as described in the caption to Fig. 10. Original magnification ×160.

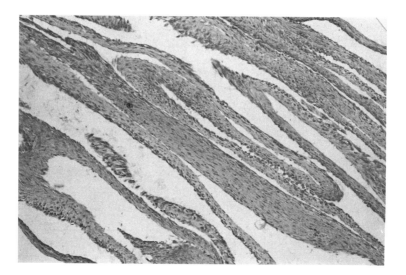

Fig. 12. Absence of mineralization in an HBDC culture (E1P1) grown in the continuous presence of the long-acting ascorbate analog L-ascorbate 2-phosphate but without dexamethasone. Section through the folded cell layer of a culture supplemented with an additional 5 mM Pi 72 h prior to harvest. Note the abundant extracellular matrix, but complete absence of mineral deposits, despite the presence of additional Pi. Von Kossa's stain and toluidine blue. Original magnification ×50.

tion of ascorbate in secondary culture, even for extended periods, cannot compensate for its omission in primary culture. This finding provides further evidence to support the hypothesis that maintenance of adequate levels of ascorbate during the early stages of explant culture is of critical importance for the survival of cells that retain the ability to proliferate extensively and give rise to precursors capable of undergoing osteogenic differentiation.

Although the mineralized tissue formed by HBDCs in vitro has yet to be characterized at the ultrastructural level, it is the authors' contention that its presence is indicative of the true osteogenic nature of the cultured cell population. In support of this contention, it has recently been shown that HBDCs cultured continuously in the presence of ascorbate and glucocorticoid retain the ability to form bone when implanted in vivo within diffusion chambers in athymic mice *(55,58,59)*.

3.5. Other HBDC Culture Systems

Since the description of the original method, many investigators have turned to the use of HBDCs. Most have continued to use the explant technique with only minor modifications. Some, however, have developed techniques for the isolation and culture of HBDCs that differ significantly from the method originally described (*see* Notes 10–12).

4. Notes

1. Batches of serum vary in their ability to support the proliferation and differentiation (in terms of basal alkaline phosphatase activity) of HBDCs. It is advisable to screen batches and reserve a large quantity of serum once a suitable batch has been identified. HBDCs will grow in autologous and heterologous human serum, but as yet no comprehensive studies have been performed to identify the effects on growth and differentiation.
2. The authors have obtained consistent results with plasticware from Becton Dickinson. Smaller flasks or dishes can be used if the amount of bone available is <0.2 g.
3. Several other assays for osteocalcin are commercially available.
4. It is our experience that storage for periods of up to 24 h at 4°C in PBS or SFM does not noticeably alter the ability of the tissue to give rise to populations of osteoblast-like cells.
5. We have cultured bone cells from many anatomical sites including tibia, femur, rib, vertebra, patella, iliac crest, and digits.
6. In our experience the minimum plating density for successful subculture is 3500 cells/cm^2. Below this the cells exhibit extended doubling times and often fail to grow to confluence.
7. If required, these HBDCs can be stored frozen for extended periods in liquid nitrogen or in ultra-low temperature (–135°C) cell freezer banks. For this purpose the cells are removed from the flask using trypsin–EDTA, or collagenase followed by trypsin–EDTA if cultured in the continuous presence of ascorbate,

and following centrifugation, resuspended (1–2 × 10^6 cells/mL) in a solution of (v/v) 90% serum and 10% dimethyl sulfoxide. For best results the cells should be frozen gradually using one of the many devices available that allow the rate of cooling to be controlled precisely. The authors have found the following freezing protocol to give excellent results: 5°C/min to 4°C, 1°C/min to –30°C, and 2°C/min to –60°C. The cells are then transferred directly to liquid nitrogen. Prior to use, frozen cells are rapidly thawed in a 37°C water bath and then diluted into ≥20 volumes of preheated medium containing 10% FCS and the usual supplements. After 12–18 h, the medium is replaced with the normal volume of fresh medium. The efficiency of plating obtained after 24 h using this method is typically ≥70%.
8. The available evidence indicates that cultures of HBDCs contain cells of the osteogenic lineage at all stages of differentiation and maturation. This conclusion is consistent with the expression of both early (alkaline phosphatase) and late (osteocalcin, bone sialoprotein) stage markers of osteoblast differentiation. In addition, in ascorbate-treated cultures there is a small subpopulation (≤5%) of cells that express the epitope recognized by the monoclonal antibody (MAb) STRO-1 (Walsh and Beresford, unpublished observations), which is a cell-surface marker for clonogenic, multipotential marrow stromal precursors capable of giving rise to cells of the osteogenic lineage in vitro *(60,61)*. The presence of other cell types, including endothelial cells and those derived from the hematopoietic stem cell, has been investigated using a large panel of MAbs and flow cytometry and/or immunocytochemistry. The results of these studies reveal that at first passage there are no detectable endothelial, lymphoid, or erythroid cells present. A consistent finding, however, is the presence of small numbers of cells (≤5%) expressing antigens present on cells of the monocyte/macrophage series (Skøjdt and Russell, unpublished observations).
9. It is our experience that in the absence of added calcitriol, particularly in SFM or FCS that has been depleted of endogenous calcitriol by charcoal treatment, the amount of osteocalcin produced by HBDCs is below the limits of detection in most assays *(8,11,34)*. The same applies to the detection of steady-state levels of osteocalcin mRNA (Gundle, Brown, and Beresford, unpublished observations). An exception to this general rule is when HBDCs are cultured for extended periods in the presence of L-ascorbate or its stable analog, L-ascorbate-2-phosphate.
10. The most commonly used alternative to the basic explant culture technique is that described by Gehron Robey and Termine *(20)*. In this method the trabecular bone is minced finely prior to digestion with *Clostridial* collagenase. The treated explants are then cultured in a medium containing reduced levels of calcium (0.2 vs 1.8 m*M*), which has been reported to favor the proliferation and differentiation of cells of the osteoblast lineage *(62)*. The initiating cell types in these cultures have yet to be identified unequivocally. It has been suggested, however, that they may include osteocytes and/or pericytes associated with the microvasculature *(63)*. The principal advantage claimed for this method is that it yields a uniform starting material free of contaminating marrow tissue. This may be of particular importance when attempting to study age-related changes that are

Culture of Osteoblasts

intrinsic to cells of the osteoblast lineage, rather than reflecting age-related changes in the structure and composition of the bone marrow *(64,65)*. The following protocol outlining this technique has been provided by Dr. Pamela Gehron Robey.

Samples of human bone are obtained from local hospitals under internal review board (IRB)-approved protocols that ensure patient confidentiality (only the age and sex of the patient are provided). Trabecular bone from the femoral neck, patella, ribs, iliac crest, or vertebrae, and obtained from donors of any age, is suitable for use with this procedure. Samples are placed in nutrient medium containing between 10 and 20% fetal bovine serum, and stored at 4°C until use (up to 48 h). Fragments of trabecular bone are removed by cureting with a #2 orthopedic curet and placed into a sterile 3-mL Pearce™ Reacti-vial (glass vial with a conical bottom) containing SFM (50% DMEM-50% Ham's F12K) supplemented with 2 mM freshly prepared L-glutamine and penicillin/streptomycin (50 U/mL and 50 µg/mL, respectively). Using straight-edged scissors, the fragments are minced until they are the consistency of sand. They are then washed by allowing the fragments to settle to the base of the vial, aspirating the medium with a suction pipet and then forcibly introducing fresh SFM into the Reacti-vial to resuspend the fragments. This process is repeated until the supernatant is free of blood and lipid and the fragments are white in color. Mincing is repeated to ensure that all fragments are ≤1.0 × 1.0 mm in size and free of blood. Finally, the fragments are washed again to remove the last contaminating traces of blood and lipid. The minced and washed fragments are resuspended (1 mL packed volume) in 10 mL of SFM containing 250 U/mL of *Clostridial* collagenase (Sigma Type IV; cat. no. C5138) in a 50-mL polypropylene tube with a conical base (Becton Dickinson "Blue Max," cat. no. 2098). The collagenase used is preselected to be low in clostripain activity. The fragments are incubated with constant rotation at 37°C for 2 h, and then vortexed vigorously to release soft connective tissue cells into the supernatant. The fragments are allowed to settle at 1g and the supernatant removed with a suction pipet. This process is repeated until the supernatant becomes clear and the fragments rapidly sediment to the base of the tube. Prior to culture, the collagenase-treated fragments are examined using an inverted microscope. If soft connective tissue is still present (visible as interwoven strands of fibrous material on the surface of the mineralized matrix), the collagenase digestion is continued for an additional 30 min using fresh collagenase. After washing, the treated fragments are transferred to 15-cm diameter Falcon dishes (Becton Dickinson, cat. no. 3025; one heaped #5 bone curet scoop per dish) containing 30 mL of growth medium (1:1 [v/v] DMEM and Ham's F12K) supplemented with 10% (v/v) FBS, 2 mM L-glutamine, 0.25 µg/mL L-ascorbic acid, 50 U/mL penicillin, and 50 µg/mL streptomycin sulfate. Cultures are fed three times weekly by gently tipping the plate forward and aspirating the spent medium, taking care not dislodge the fragments, and adding 30 mL of fresh medium. Cells begin to migrate from the collagenase-treated fragments within 2–3 wk, and are confluent in 6–8 wk (~5 × 10^6 cells/dish). For studies of matrix mineralization in primary culture, cells at confluence are transferred to complete medium (prepared as described but using Ham's F12K medium that is replete in calcium) additionally supplemented with 5 mM β-glycerophosphate with daily addition of fresh ascorbate. Alternatively the cells may be passaged using trypsin–EDTA, plated at a density of 40,000 cells/cm^2, and cultured in growth medium as before. When the cells are confluent (~10^5 cells/cm^2; 1–2 wk after plating depending on the age of the donor), they are transferred to complete medium additionally supplemented with 5 mM β-glycerophosphate to promote matrix mineralization.

Like the HBDCs, these cells have been characterized extensively and shown to possess a number of osteoblast-like characteristics, including the ability to mineralize their extracellular matrix *(20,64–67)*. A limited series of comparisons have been made between the cell populations obtained using explants of trabecular bone obtained from the same donor, but prepared and cultured using the different protocols *(33,35,67;* Beresford, Gehron Robey, and Termine, unpublished observations. Both protocols synthesized predominantly type I collagen, produced osteocalcin in response to treatment with calcitriol, and synthesized the bone proteoglycans decorin and biglycan, albeit with dermatan sulphate side chains *(68)*. At this level of analysis, therefore, the two cell populations appear indistinguishable.

11. A method of HBDC culture that has been used in the characterization of skeletal growth factors was developed by Wergedal and Baylink *(17)*. In this method, trabecular bone is minced into small fragments as described, and then treated with collagenase. In contrast to the method of Gehron Robey and Termine *(20)*, however, it is the collagenase released cells that are harvested and cultured. The following protocol has been adapted from one provided by Dr. David Baylink.

Collect the bone in cold DMEM supplemented with penicillin, streptomycin, and Nystatin. Ideally, samples should be processed on the same day, but remain viable if stored overnight in DMEM at 4°C. For surgical specimens it is only necessary to use antibiotics and antimycotics during the initial preparation of the tissue. When dealing with autopsy specimens, however, it is advisable to maintain their presence throughout primary and secondary culture. Remove contaminating soft connective tissue and dice the bone into 2 × 2-mm fragments. This is best achieved in two stages with the final dicing being performed using iris scissors and a Pearce Reacti-vial as described. Wash the diced fragments extensively to remove contaminating marrow cells. Incubate the fragments in 5–15 mL of SFM containing 2 mg/mL crude *Clostridial* collagenase for 2 h at 37°C with constant agitation. Allow the fragments to settle and draw off the supernatant containing the released cells. Wash the fragments three times with vigorous agitation, and add the washings to the original supernatant. Pellet the released cells by centrifugation at 100g. If additional cells are required, the remaining fragments can be cultured for 1–2 wk in DMEM containing 10% FCS and then treated with collagenase as described. Culture the cells in DMEM supplemented with 10% FCS (10^5–10^6 cells/cm^2). The efficiency of plating is typically ≤1%. The culture medium is replaced after 48–72 h and three times weekly thereafter. Cells are subcultured at 1:4 split ratio following treatment with trypsin–EDTA.

Cells harvested using this technique normally stop dividing after 6–12 passages. The cell populations harvested using this technique express high levels of alkaline phosphatase activity, secrete osteocalcin in response to treatment with calcitriol, and respond to a wide variety of osteotropic hormones and growth factors *(69–71)*. They are also capable of metabolizing 25(OH)D$_2$ to 1,25(OH)$_2$D$_3$ and other more polar, dihydroxylated metabolites *(72)*; a property they share with cells isolated by the original explant procedure (Beresford, Gallagher, Lawson-Matthews, and Russell, unpublished observations).

12. The third method in common usage was originally developed for the isolation of cells from the endosteal surfaces of mouse caudal vertebrae. Since its application to the isolation and culture of HBDCs, it has been extensively used to investigate the pathophysiology of bone formation *(73–76)*. The distinguishing feature of this method is that the washed explants are first cultured on a nylon mesh. This has the advantage, it is claimed, of separating endosteal bone cells from the nonadherent bone marrow fraction *(22)*. The method is described briefly in the following, but for a more detailed consideration of the technique and its claimed advantages, the reader is referred to refs. *22* and *77*.

    Following the removal of soft connective tissue, the trabecular bone is dissected away from the cortical and periosteal tissue and minced finely into fragments ~1 mm^3. The minced fragments are washed extensively to remove adherent marrow tissue, and then placed onto a nylon mesh of 80-μm pore size in a nontissue-culture grade plastic Petri dish containing DMEM supplemented with 10% FCS, 292 mg/L L-glutamine, 100 U/mL penicillin, and 100 μg/mL streptomycin sulfate. The fragments are then cultured at 37°C in an humidified atmosphere of 95% air, 5% CO$_2$ with three times weekly changes of medium. After 4 d the nylon mesh supporting the explant and migrating cells is removed and placed face side down in a new tissue-culture grade Petri dish and cultured for an additional 4–7 d. Adherent cells are subcultured using 0.05% trypsin in Ca^{2+}- and Mg^{2+}-free PBS and plated at a density of 10^4 cells/cm^2 in 35-mm diameter Petri dishes or in 25-cm^2 flasks (Becton Dickinson, cat. nos. 3001 and 3013E, respectively). The cells may be subcultured up to six times using a 1:4 split ratio. The cell populations obtained using this method express alkaline phosphatase; secrete osteocalcin; produce type I collagen; and respond to a variety of osteotropic hormones, growth factors, and cytokines *(73,74)*. It is currently not clear whether these cell populations also express osteogenic potential either in vitro or in vivo.

13. Synthetic PTH peptides, including human PTH(1–34) which exhibits full biological activity, are commercially available from Peninsula laboratories.

Acknowledgments

The authors are grateful to Pamela Gehron Robey, David Baylink, and Pierre Marie for providing protocols and manuscripts describing their techniques to isolate osteoblasts from human bone. We also thank the many surgeons who have suplied tissues, and thus have contributed over the years to the development and characterization of the HBDC culture system.

References

1. Gallagher, J. A. (1991) Human bone remodeling, in *Encyclopaedia of Human Biology* vol. 1 (Dulbecco, R., ed.), Academic, San Diego, CA, pp. 881–923.
2. Gallagher, J. A., Beresford, J. N., Caswell, A., and Russell, R. G. G. (1987) Subcellular investigation of skeletal tissue, in *Subcellular Pathology of Systemic Disease* (Peters, T. J., ed.), Chapman and Hall, London, pp. 377–397.
3. Bard, D. R., Dickens, M. J., Smith, A. U., and Zarek, J. M. (1972) Isolation of living cells from mature mammalian bone. *Nature* **236,** 314,315.
4. Bard, D. R., Dickens, M. J., Edwards, J., and Smith, A. U. (1974) Ultra-structure, in vitro cultivation and metabolism of cells isolated from arthritic human bone. *J. Bone Joint Surg.* **56B,** 352–360.
5. Mills, B. G., Singer, F. R., Weiner, L. P., and Hoist, P. A. (1979) Long term culture of cells from bone affected with Paget's disease. *Calcif. Tiss. Int.* **29,** 79–87.
6. Beresford, J. N., Gallagher, J. A., Gowen, M., McGuire, M. K. B., Poser, J. W., and Russell, R. G. G. (1983) Human bone cells in culture. A novel system for the investigation of bone cell metabolism. *Clin. Sci.* **64,** 38,39.
7. Gallagher, J. A., Beresford, J. N., McGuire, M. K. B., Ebsworth, N. M., Meats, J. E., Gowen, M., Elford, P., Wright P., Poser, J., Coulton, L. A., Sharrard, M., Imbimbo, B., Kanis, J. A., and Russell, R. G. G. (1983) Effects of glucocorticoids and anabolic steroids on cells derived from human skeletal and articular tissues in vitro, in *Glucocorticoid Effects and their Biological Consequences* (Imbimbo, B. and Avioli, L. V., eds.), Plenum, New York, pp. 279–292.
8. Beresford, J. N., Gallagher, J. A., Poser, J. W., and Russell, R. G. G. (1984) Production of osteocalcin by human bone cells in vitro. Effects of 1,25(OH)$_2$D$_3$, parathyroid hormone and glucocorticoids. *Metab. Bone Dis. Rel. Res.* **5,** 229–234.
9. MacDonald, B. R., Gallagher, J. A., Ahnfelt-Ronne, I., Beresford, J. N., Gowen, M., and Russell, R. G. G. (1984) Effects of bovine parathyroid hormone and 1,25(OH)$_2$D$_3$ on the production of prostaglandins by cells derived from human bone. *FEBS Lett.* **169,** 49–52.
10. Beresford, J. N., Gallagher, J. A., Gowen, M., Couch, M., Poser, J., Wood, D. D., and Russell, R. G. G. (1984) The effects of monocyte-conditioned medium and interleukin 1 on the synthesis of collagenous and non-collagenous proteins by mouse bone and human bone cells *in vitro*. *Biochim. Biophysica. Acta Gen. Subj.* **801,** 58–65.
11. Skjodt, H., Gallagher, J. A., Beresford, J. N., Couch, M., Poser, J. W., and Russell, R. G. G. (1985) Vitamin D metabolites regulate osteocalcin synthesis and proliferation of human bone cells *in vitro*. *J. Endocrinol.* **105,** 391–396.
12. Beresford, J. N., Gallagher, J. A., and Russell, R. G. G. (1986) 1,25-dihydroxyvitamin D$_3$ and human bone derived cells *in vitro*: effects on alkaline phosphatase, type I collagen and proliferation. *Endocrinology* **119,** 1776–1785.
13. MacDonald, B. R., Gallagher, J. A., and Russell, R. G. G. (1986) Parathyroid hormone stimulates the proliferation of cells derived from human bone. *Endocrinology* **118,** 2445–2449.

14. Vaishnav, R., Gallagher, J. A., Beresford, J. N., Poser, J. W., and Russell, R. G. G. (1984) Direct effects of stanozolol and oestrogen on human bone cells in culture, in *Osteoporosis. Proceedings of Copenhagen International Symposium* (Christiansen, C., ed.), Glostrup Hospital, Copenhagen, Denmark, pp. 485–488.
15. Treble, N. J., Dorgan, J. C., and Gallagher, J. A. (1990) Maintenance of cell viability in stored bone. *Spine* **15**, 830–832.
16. Maurizi, M., Binaglia, L., Donti, E., Ottaviani, F., Paludetti, G., and Venti Donti, G. (1983) Morphological and functional characterisation of human temporal-bone cultures. *Cell Tiss. Res.* **229**, 505–513.
17. Wergedal, J. E. and Baylink, D. J. (1984) Characterization of cells isolated and cultured from human trabecular bone. *Proc. Soc. Exp. Biol. Med.* **176**, 60–69.
18. Crisp, A. J., McGuire-Goldring, M. B., and Goldring, S. R. (1984) A system for culture of human trabecular bone and hormone response profiles of derived cells. *Br. J. Exp. Path.* **65**, 645–654.
19. Auf'mkolk, B., Hauschka, P. V., and Schwartz, R. (1985) Characterisation of human bone cells in culture. *Calcif. Tiss. Int.* **37**, 228–235.
20. Gehron Robey, P. and Termine, J. D. (1985) Human bone cells *in vitro*. *Calcif. Tiss. Int.* **37**, 453–460.
21. Ashton, B. A., Abdullah, F., Cave, J., Williamson, M., Sykes, B. C., Couch, M., and Poser, J. W. (1985) Characterization of cells with high alkaline phosphatase activity derived from human bone and marrow: preliminary assessment of their osteogenicity. *Bone* **6**, 313–319.
22. Marie, P. J., Lomri, A., Sabbagh, A., and Basle, M. (1989) Culture and behaviour of osteoblastic cells isolated from normal trabecular bone surfaces. *In Vitro Cell Dev. Biol.* **25**, 373–380.
23. Eriksen, E. F., Colvard, D. S., Berg, N. J., Graham, M. L., Mann, K. G., Spelsberg, T. C., and Riggs, B. L. (1988) Evidence of estrogen receptors in normal human osteoblast-like cells. *Science* **241**, 84–86.
24. Schoefl, C., Cuthbertson, K. S. R., Walsh, C. A., Cobbold, P., Mayne, C. N., Von zur Muhlen, A., Hesch, R. D., and Gallagher, J. A. (1992) Evidence for P2-purinoceptors on osteoblast-like cells. *J. Bone Min. Res.* **7**, 485–449.
25. Bowler, W. B., Gallagher, J. A., and Bilbe, G. (1995) Identification and cloning of human P_{2U} purinoceptor present in osteoclastoma, bone and osteoblasts. *J. Bone Miner. Res.* **10**, 1137–1145.
26. Birch, M. A., Ginty, A. F., Walsh, C. A., Fraser, W. D., Gallagher, J. A., and Bilbe, G. (1993) PCR detection of cytokines in normal human and pagetic osteoblast-like cells. *J. Bone Miner. Res.* **8**, 1155–1162.
27. Birch, M. A., Taylor, W., Fraser, W. D., Ralston S. H., Hart, C. A., and Gallagher, J. A. (1994) Absence of paramyxovirus RNA in cultures of pagetic bone cells and in pagetic bone. *J. Bone Miner. Res.* **9**, 11–16.
28. Walsh, C. A., Birch, M. A., Fraser, W. D., Lawton, R., Dorgan, J., Walsh, S., Beresford, J. N., Sansom, D., and Gallagher, J. A. (1995) Expression and secretion of parathyroid hormone-related protein by human osteoblasts in vitro: effects of glucocorticoids. *J. Bone Miner. Res.* **10**, 17–25.

29. Walsh, C. A., Birch, M. A., Fraser, W. D., Robinson, J., Lawton, R., Dorgan, J., Klenerman L., and Gallagher, J. A. (1994) Primary cultures of human osteoblasts produce parathyroid hormone-related protein. *Bone Miner.* **27**, 43–51.
30. Chaudhary, L. R., Spelsberg, T. C., and Riggs, B. L. (1992) Production of various cytokines by normal human osteoblast-like cells in response to interleukin-1β and tumour necrosis factor-α: lack of regulation by 17β-estradiol. *Endocrinology* **130**, 2528–2534.
31. Brittberg, M., Lindahl, A., Nilson, A., Ohlsson, C., Isaksson, O., and Peterson, L. (1994) Treatment of deep cartilage defects in the knee with autologous chondrocyte transplantation. *New Engl. J. Med.* **331**, 889–895.
32. Goldring, S. R. (1979) PTH inhibitors: comparison of biological activity in bone and skin derived tissue. *J. Clin. Endocrinol. Metab.* **49**, 655–659.
33. Beresford, J. N., Taylor, G. T., and Triffitt, J. T. (1990) Interferons and bone. A comparison of the effects of interferon-α and interferon-γ in cultures of human bone-derived cells and an osteosarcoma cell line. *Eur. J. Biochem.* **193**, 589–597.
34. Beresford, J. N., Graves, S. E., and Smoothy, C. A. (1993) Formation of mineralized nodules by bone derived cells in vitro: a model of bone formation? *Am. J. Med. Genet.* **45**, 163–178.
35. Chavassieux, P. M., Chenu, C., Valentin-Opran, A., Merle, B., Delmas, P. D., Hartmann, D. J., Saez, S., and Meunier, P. J. (1990) Influence of experimental conditions on osteoblast activity in human primary bone cell cultures. *J. Bone Miner. Res.* **5**, 337–343.
36. Mohan, S. and Baylink, D. J. (1991) Bone growth factors. *Clin. Orthop.* **263**, 30–48.
37. Canalis, E., Pash, J., and Varghese, S. (1993) Skeletal growth factors. *Crit. Rev. Eukaryotic Gene Express.* **3**, 155–166.
38. Franceschi, R. T. and Iyer, B. S. (1992) Relationship between collagen synthesis and expression of the osteoblast phenotype in MC3T3-E1 cells. *J. Bone Miner. Res.* **7**, 235–246.
39. Prockop, D. J. and Kivivrikko, K. I. (1984) Heritable diseases of collagen. *New Engl. J. Med.* **311**, 376–386.
40. Graves, S. E., Francis, M. J. O., Gundle, R., and Beresford, J. N. (1994) Ascorbate increases collagen synthesis and promotes differentiation in human bone derived cell cultures. *Bone* **15**, 133. (abstract)
41. Winkler, B. S. (1987) In vitro oxidation of ascorbic acid and its preservation by GSH. *Biochim. Biophys. Acta* **925**, 258–264.
42. Friedrich, W. (1988) Vitamin C, in *Vitamins* (Friedrich, W., ed.), Walter de Gruyter, Berlin, pp. 931–1021.
43. Nomura, H., Ishiguro, T., and Morimoto, S. (1969) Studies on L-ascorbic acid derivatives. III. Bis(L-ascorbic acid-3,3')phosphate and L-ascorbic acid 2-phosphate. *Chem. Pharm. Bull.* **17**, 387–393.
44. Hata, R.-I. and Senoo, H. (1989) L-ascorbic acid 2-phosphate stimulates collagen accumulation, cell proliferation and formation of a three-dimensional tissue-like substance by skin fibroblasts. *J. Cell Physiol.* **138**, 8–16.
45. *Oxford Textbook of Medicine,* 2nd ed. (1987) Oxford University Press, Oxford, UK.

46. Graves, S. E., Francis, M. J. O., Gundle, R. G., and Beresford, J. N. (1994) Primary culture of human trabecular bone: effects of L-ascorbate-2-phosphate. *Bone* **15,** 132. (abstract)
47. Graves, S. E. (1991) *Studies on Human Bone-Derived Cells In Vitro.* D Phil Thesis, Oxford University, Oxford, UK.
48. Gundle, R. (1995) *Microscopical and Biochemical Studies of Mineralised Matrix Production by Human Bone-Derived Cells.* D Phil Thesis, Oxford University, Oxford, UK.
49. Bellows, C. G., Heersche, J. N. M., and Aubin, J. E. (1992) Inorganic phosphate added exogenously or released from β-glycerophosphate initiates mineralization of osteoid nodules *in vitro. Bone Miner.* **17,** 15–29.
50. Beresford, J. N., Joyner, C. J., Devlin, C., and Triffitt, J. T. (1994) Osteogenic differentiation in human marrow stromal cell cultures: effects of dexamethasone and 1,25-dihydroxyvitamin D_3. *Arch. Oral Biol.* **39,** 941–947.
51. Falla, N., Van Vlasselaer, P., Bierkens, J., Borremans, B., Schoeters, G., and Van Gorp, U. (1993) Characterization of a 5-fluorouracil-enriched osteoprogenitor population of the murine bone marrow. *Blood* **82,** 3580–3591.
52. Benayahu, D., Kletter, Y., Zipori, D., and Weintroub, S. (1989) Bone marrow-derived stromal cell line expressing osteoblastic phenotype *in vitro* and osteogenic capacity *in vivo. J. Cell Physiol.* **140,** 1–7.
53. Wong, M. M., Rao, L. G., Ly, H., Hamilton, L., Tong, J., Sturtridge, W., McBroom, R., Aubin, J. E., and Murray, T. M. (1990) Long-term effects of physiologic concentrations of dexamethasone on human bone-derived cells. *J. Bone Miner. Res.* **5,** 803–813.
54. Gundle, R., Graves, S. E., Francis, M. J. O., and Beresford, J. N. (1994) Osteogenic and adipogenic differentiation in human bone derived cell cultures. *Bone* **15,** 114. (abstract)
55. Gundle, R., Bradley, J., Joyner, C. J., Francis, M. J. O., Triffitt, J. T., and Beresford, J. N. (1993) Bone formation *in vitro* and *in vivo* by cultured adult human bone derived cells. *J. Bone Miner. Res.* **8(Suppl. 1),** 1031.
56. *Davidson's Principles and Practice of Medicine,* 13th ed. (1981) Churchill Livingstone, London.
57. Bancroft, J. D. and Stevens, A. C. (1977) *Theory and Practice of Histological Techniques.* Churchill Livingstone, London.
58. Gundle, R. and Beresford, J. N. (1995) The isolation and culture of cells from explants of human trabecular bone. *Calcif. Tiss. Int.* **56(Suppl.),** 8–10.
59. Gundle, R., Joyner, C. J., and Triffitt, J. T. (1995) Human bone tissue formation in diffusion chamber culture *in vivo* by bone derived cells and marrow stromal cells. *Bone* **16,** 597–601.
60. Simmons, P. J. and Torok Storb, B. (1991) Identification of stromal cell precursors in human bone marrow by a novel monoclonal antibody, STRO-1. *Blood* **78,** 55–62.
61. Gronthos, S., Graves, S. E., Ohta, S., and Simmons, P. J. (1994) The stro-1(+) fraction of adult human bone-marrow contains the osteogenic precursors. *Blood* **84,** 4164–4173.
62. Binderman, I. and Somjen, D. (1992) Serum factors and calcium modulate the growth of osteoblast-like cells in culture, in *Current Advances in Skeletogenesis: Development, Biomineralisation, Mediators and Metabolic Bone Disease* (Silbermann, M. and Slavkin, H. C., eds.), Elsevier, Amsterdam, pp. 338–342.

63. Gehron Robey, P. (1995) Collagenase-treated trabecular bone fragments—a reproducible source of cells in the osteoblastic lineage. *Calcif. Tiss. Int.* **56(Suppl. 1),** 11,12.
64. Fedarko, N. S., Vetter, U. K., and Gehron Robey, P. (1995) Age-related-changes in bone-matrix structure in-vitro. *Calcif. Tiss. Int.* **56(Suppl. 1),** 41–43.
65. Fedarko, N. S., Vetter, U., Weinstein, S., and Robey, P. G. (1992) Age-related changes in the in hyaluronan, proteoglycan, collagen and osteonectin synthesis by human bone cells. *J. Cell Physiol.* **151,** 215–227.
66. Fedarko, N. S., Moerike, M., Brenner, R., Gehron Robey, P., and Vetter, U. (1992) Extracellular matrix formation by osteoblasts from patients with osteogenesis imperfecta. *J. Bone Miner. Res.* **7,** 921–930.
67. Fedarko, N. S., Bianco, P., Vetter, U., and Gehron Robey, P. (1990) Human bone cell enzyme expression and cellular heterogeneity: correlation of alkaline phosphatase enzyme activity with cell cycle. *J. Cell Physiol.* **144,** 115–121.
68. Beresford, J. N., Fedarko, N. S., Fisher, L. W., Midura, R. J., Yanagashita, M., Termine, J. D., and Gehron Robey, P. (1987) Analysis of the proteoglycans synthesised by human bone cells *in vitro*. *J. Biol. Chem.* **262,** 17,164–17,172.
69. Wergedal, J. E., Matsuyama, T., and Strong, D. D. (1992) Differentiation of normal human bone cells by transforming growth factor-beta and $1,25(OH)_2D_3$. *Metabolism* **41,** 42–48.
70. Matsuyama, T., Lau, K. H. W., and Wergedal, J. E. (1990) Monolayer cultures of normal human bone cells contain multiple subpopulations of alkaline phosphatase positive cells. *Calcif. Tiss. Int.* **47,** 276–283.
71. Wergedal, J. E., Mohan, S., Lundy, M., and Baylink, D. J. (1990) Skeletal growth factor and other growth factors known to be present in bone matrix stimulate proliferation and protein synthesis in human bone cells. *J. Bone Miner. Res.* **5,** 179–186.
72. Howard, G. A., Turner, R. T., Sherrard, D. J., and Baylink, D. J. (1981) Human bone cells in culture metabolize $25(OH)D_3$ to $1,25(OH)_2D_3$ and $24,25(OH)_2D_3$. *J. Biol. Chem.* **256,** 7738–7740.
73. Lomri, A. and Marie, P. J. (1990) Bone cell responsiveness to transforming growth factor β, parathyroid hormone and prostaglandin E_2 in normal and postmenopausal osteoporotic women. *J. Bone Miner. Res.* **5,** 1149–1155.
74. Marie, P. J., Sabbagh, A., De Vernejoul, M. C., and Lomri, A. (1988) Osteocalcin and deoxyribonucleic acid synthesis *in vitro* and histomorphometric indices of bone formation in postmenopausal osteoporosis. *J. Clin. Invest.* **69,** 272–279.
75. Marie, P. J., De Vernejoul, M. C., and Lomri, A. (1992) Stimulation of bone formation in osteoporosis patients treated with fluoride associated with increased DNA synthesis by osteoblastic cells *in vitro*. *J. Bone Miner. Res.* **7,** 103–113.
76. Marie, P. J., Hott, M., Launay, J. M., Graulet, A. M., and Gueris, J. (1993) In vitro production of cytokines by bone surface-derived osteoblastic cells in normal and osteoporotic postmenopausal women: relationship with cell proliferation. *J. Clin. Endocrinol. Metab.* **77,** 824–830.
77. Marie, P. J. (1995) Human endosteal osteoblastic cells—relationship with bone-formation. *Calcif. Tiss. Int.* **56(Suppl.),** 13–16.

21

The Isolation of Osteoclasts from Human Giant Cell Tumors and Long-Term Marrow Cultures

Catherine A. Walsh, John A. Carron, and James A. Gallagher

1. Introduction

The osteoclast is a large multinucleate cell formed from the fusion of mononuclear precursor cells of hemopoietic origin. Unique markers of the osteoclast have been difficult to identify. Widely used techniques, such as histochemical location of tartrate-resistant acid phosphatase, and response to calciotropic hormones do not distinguish effectively between osteoclasts and non-osteoclastic cells. The only unequivocal evidence of osteoclast activity is the ability of the cells to excavate authentic resorption lacunae in vivo and in vitro *(1)*.

The study of bone resorption has traditionally involved the measurement of radiolabeled calcium or proline release from intact fetal bone *(2,3)*. Although this system has been used extensively, it has several disadvantages. The release of radiolabel from intact bone may not be an accurate measure of osteoclast activity, particularly in intact fetal bone where the role of the osteoblast and other cells cannot be overlooked.

The recent development of a system in which isolated osteoclasts excavate resorption lacunae in vitro has been a major breakthrough in bone cell biology *(1,4)*. This system has been used extensively to investigate the activity of avian and rodent osteoclasts. The major problem with the study of human osteoclasts in this system arises from difficulties encountered routinely obtaining a large number of osteoclasts for in vitro studies.

One of the sources from which human osteoclasts can be isolated is the giant cell tumor of bone, a benign tumor that produces lytic lesions locally. The tumor is characterized by the large number of multinucleated cells associated with the stroma. These cells have been found to have the capacity to resorb bone and respond to calciotropic hormones in vitro. The rarity of giant cell

From: *Methods in Molecular Medicine: Human Cell Culture Protocols*
Edited by: G. E. Jones Humana Press Inc., Totowa, NJ

tumors means that samples become available very infrequently and, as a result, many researchers have turned to the culture of human bone marrow in attempts to generate osteoclasts from their precursors in vitro *(5)*.

This chapter describes the isolation of osteoclasts from two sources, giant cell tumors and long-term marrow cultures. The subsequent assay of osteoclast activity in vitro is also discussed.

2. Materials
2.1. Tissue-Culture Medium

1. α-Modification of minimum essential medium (α-MEM) (ICN, High Wycombe, UK) supplemented to a final concentration of 10% with fetal calf serum (FCS) (Gibco, Gaithersburg, MD) for the maintenance of osteoclasts isolated from giant cell tumors.
2. α-MEM supplemented with 20% horse serum (Gibco) for long-term culture of bone marrow.
3. Phosphate-buffered saline (PBS) (Gibco) prepared without calcium and magnesium for washing cells and cell layers.
4. 0.05% Trypsin/0.5 mM EDTA solution (Gibco) for detachment of cells from the tissue culture plastic.

2.2. Preparation of Wafers for Tissue Culture

1. Prepare devitalized wafers of bovine bone from femurs obtained from an abattoir or other source. Remove all traces of connective tissue and bone marrow from the shaft of the bone using a large scalpel or knife, and soak the bone overnight in a dilute solution of detergent, e.g., 1% (v/v) solution of Triton X-100 in water.
2. Cut thin wafers of bone (200 μm) using a water-cooled diamond wafering blade. Discard wafers with a high level of remodeling i.e., those wafers with a high number of Haversian systems (Fig. 1). Wafers can be checked for the presence of remodeling using a dissecting microscope.
3. Sonicate wafers of bone in double-distilled water for three 15-min washes. Air-dry the wafers, and then sterilize under a UV light in a flow cabinet. Irradiate for 1 h. Turn the wafers and irradiate for 1 h further.

2.3. Fixation of Bone Wafers

Fixative solution: Use Analar-grade reagents and distilled water for the solutions. The solutions need not be sterile, but should be stored in clean containers.

1. 0.2M Sodium cacodylate solution: Dissolve 8.56 g of sodium cacodylate in 100 mL of water.
2. 8% (v/v) Glutaraldehyde solution in water.

Make up the fixative solution by mixing solutions 1 and 2 in equal volumes. The fixative solution should optimally be made up immediately before use, but if necessary may be stored at 4°C. Discard the solution if discoloration occurs.

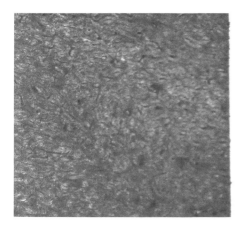 

Fig. 1. Light micrograph of two devitalized bovine bone wafers. The wafer on the left shows extensive Haversian systems. The presence of the Haversian whorls is characteristic of remodeled bone. These areas of bone are not suitable for use in the resorption assay. In our experience, cellular adhesion is reduced in this area, and the quantitation of resorption is substantially more difficult. In contrast, the wafer on the right shows the plywood-like appearance of plexiform lamellae. Such areas are characteristic of unremodeled bone and are optimal for use in the resorption assay. Magnification 25×.

2.4. Staining of Fixed Wafers

Use Analar-grade reagents and distilled water for solutions.

1. 0.5M Sodium tetraborate solution: Dissolve 19.07 g of disodium tetraborate in 100 mL of water.
2. 1% Toluidine blue solution: Dissolve 1 g of toluidine blue in 100 mL of 0.5M sodium tetraborate solution (prepared as in step 1).
3. 70% Ethanol solution: Mix 70 mL of absolute ethanol with 30 mL of water.

2.5. Quantitation of Bone Resorption

The area and volume of resorption lacunae are determined using an Olympus BH2 microscope fitted with a source of incident light, metallurgical objectives, and a drawing arm attachment.

3. Methods

3.1. Isolation of Osteoclasts from Giant Cell Tumors

If a sample of giant cell tumor becomes available, it is relatively easy to isolate osteoclasts from the stroma. The size and cellular composition of the tumor sample will vary, but in our experience, a 1-cm^3 piece of tissue will, in

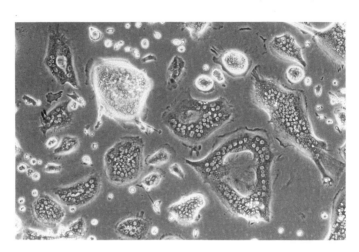

Fig. 2. Phase-contrast light micrograph of osteoclasts isolated from a human giant cell tumor. Magnification 130×.

general, yield sufficient osteoclasts to seed onto forty 5-cm² wafers of bone for resorption studies. Osteoclasts are isolated from the tissue as follows:

1. The wafers to be used should be prewetted by soaking in α-MEM in a six-well plate for 2 h at room temperature or overnight at 4°C (*see* Note 11).
2. Transfer the tissue to a Petri dish, and add 10 mL of α-MEM/10% FCS.
3. Using a sterile scalpel, cut the tissue into 1-mm³ pieces. Chop up any small fragments of bone present in the tissue if possible.
4. Mince the small pieces of tumor finely using a scalpel, and triturate 20–30 times using a sterile blunt-ended pipet or 1-mL syringe. This procedure will release the osteoclasts into the tissue culture medium.
5. Leave the Petri dish to stand for 2 min. This will allow the tissue fragments to settle to the bottom of the dish. The cell suspension may then be drawn off.
6. Seed the cell suspension drop by drop onto wafers of bone. Settle the cells for 30 min at 37°C, and then remove any non-adherent cells by thoroughly washing in PBS. Transfer the wafers to fresh α-MEM/10% FCS, and incubate at 37°C for 24 hr.

The morphology of osteoclasts seeded onto tissue culture plastic is shown in Fig. 2. The cells will remain viable for 2–3 d in culture.

3.2. Long-Term Culture of Human Bone Marrow

3.2.1. Establishing Cell Cultures

Human bone marrow is maintained in long-term culture using a modification of the technique first reported by Testa et al. with feline marrow *(6)*. Human bone marrow can be obtained from both flat and long bones. Tissues

Isolation of Osteoclasts

should be obtained as soon after surgery as possible, since after 6 h, the viability of the cells falls off markedly.

1. Crack open the bone with strong instruments (bone rongeurs are very useful), and expose the marrow cavity.
2. Break the bone into small (2–5 mm^3) pieces, and place the fragments into a sterile Universal container with 10 mL of α-MEM.
3. Shake the container vigorously to separate the marrow from the bone fragments. After 2–3 min of shaking, the medium will become cloudy.
4. Pour off the cell suspension into a fresh container, and discard the bone fragments.
5. Centrifuge the suspension for 5 min at 100g to pellet the cells. Pour off the supernatant, and resuspend the cell pellet in 1 mL of α-MEM with 20% horse serum.
6. Count the number of cells present using a hemocytometer.
7. Adjust the cell number to a final concentration of 10^6 cells/mL using medium supplemented with horse serum.
8. Seed the cells into a 90-mm Petri dish, and place in a humidified incubator at 37°C air/7% CO_2. Optimally, the cultures should be left undisturbed for 10–14 d before examining microscopically for the presence of multinucleate cells. The appearance of cultures at 8, 12, and 14 d is shown in Fig. 3.

3.2.2. Feeding Cultures of Human Bone Marrow

After 14 d in culture, carefully withdraw half of the medium from the Petri dish using a sterile pipet. Replace with fresh α-MEM/20% horse serum (*see* Notes 12 and 13).

3.2.3. Transferring Cells Formed in Long-Term Marrow Culture

After 14–21 d, human bone marrow cultures will contain populations of stromal cells, small mononuclear cells, and large multinucleate cells (MNCs). The appearance of these different cell types is shown in Fig. 3. In order to investigate the resorptive capacity of the cells or to carry out immunohistochemical or enzymic studies on the cells, it is often necessary to remove them for the original culture vessels. This is carried out as follows:

1. Remove the medium from the Petri dish, and wash the cell layer two or three times with PBS (serum present in the medium will inhibit the action of trypsin, therefore all traces must be removed).
2. Add 3 mL of a warm solution of trypsin–EDTA to the culture dish, and return to the incubator. Incubate for 45 min at 37°C. Using a phase-contrast microscope, check that the large MNCs have become detached from the dish. If they have not detached, return the dish to the incubator for a further 15 min.
3. When the cells have detached from the dish, resuspend in α-MEM/10% FCS, and seed into tissue culture vessels containing sterile wafers of bovine bone or sterile glass coverslips. If the cells are to be used for immunohistochemical analysis it is particularly useful to seed the cells into chamber slides (Gibco).

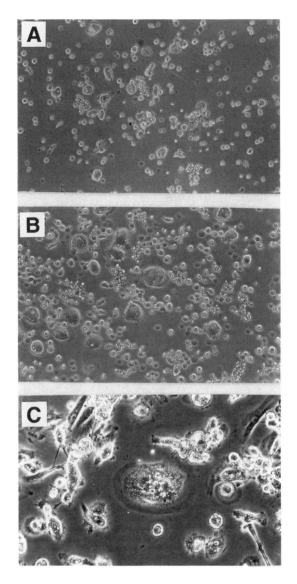

Fig. 3. Long-term culture of human bone marrow. Cells were cultured in α-MEM supplemented with 20% horse serum. **(A)** After 8 d in culture, the formation of mononuclear cells is evident. Magnification 100×. **(B)** After 12 d in culture, the mononuclear cells have increased in number, and in some areas, fusion of these cells can be seen. Magnification 100×. **(C)** After 14 d in culture, some large MNCs are visible. Fusion of mononuclear precursors is also seen. Also present in the culture are stromal cells formed also from precursors in the marrow. Magnification 200×.

4. Return the cells to the incubator, and leave to settle and recover for 48 h.
5. Fix and stain attached cells as appropriate.

3.3. Fixation and Staining

Cells seeded onto bone wafers: At the end of the experiment, discard the incubation medium, and wash the wafers in PBS. Transfer the wafers to clean wells containing enough fixative solution (*see* Section 2.3.) to cover the wafers. Fix for 30 min, wash with distilled water, and then stain for 3 min with a 1% solution of toluidine blue buffered with sodium tetraborate. Remove excess stain by washing in a 70% solution of ethanol for 1 min. Rinse in tap water, and air-dry before examining the wafers under the microscope. Wafers can then be stored indefinitely at room temperature.

3.4. Quantitation of Resorption

The activity of osteoclasts in vitro can be assessed in a number of ways. Some researchers use scanning electron microscopy (SEM) to quantitate osteoclast activity. Stereophotogrammetry of scanning electron micrographs enables the area, volume, and maximum and average depth of resorption lacunae to be determined. The images produced by this technique are of excellent quality, but the procedure requires expertise, expensive equipment, and specifically designed software. In addition, the preparation and scanning of wafers are time-consuming, making the assessment of a large number of wafers impractical. Figure 4 shows scanning electron micrographs of resorption lacunae excavated by cells formed in long-term bone marrow culture.

The use of reflected light microscopy (RLM) avoids many of the problems associated with SEM. RLM provides accurate information on the area, volume, and depth of resorption lacunae using an adapted standard microscope *(7)*. The system used for this procedure is shown diagrammatically in Fig. 5.

3.4.1. Area Measurements

1. Using reflected light and the 10x objective lens, resorption lacunae can be reliably identified even when the wafer is covered by a dense layer of cells. Figure 6A shows a resorption lacuna excavated by osteoclasts isolated from a human giant cell tumor visualized by SEM. For comparison, a light micrograph of the same resorption lacuna is shown in Fig. 6B.
2. Using the microscope and drawing arm, overlay a grid of point onto the image of the bone wafer (Fig. 5). The magnification of the drawing tube should be adjusted to ensure that the grid covers an area of 1 mm^2 on the surface of the bone wafer.
3. Move the stage of the microscope in both X and Y starting at a point selected randomly in the extreme corner of the wafer.
4. Record the number of points of the grid falling onto resorption lacunae on the wafer.

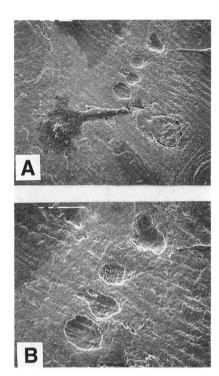

Fig. 4. Scanning electron micrograph of resorption lacuna excavated by MNCs formed from long-term culture of human bone marrow. (A) A low-power micrograph to show the appearance of resorption lacunae. A marrow stromal cell is also visible, but the MNC responsible for the excavation has moved away from the area. The small size of the lacunae would suggest that the excavations were produced by a small cell rather than a large MNC. Magnification 500×. (B) At higher power, the morphology of the resorption lacunae is more clearly visible. Magnification 1000×.

Fig. 5. *(opposite page)* A diagrammatic representation of the system used to determine the area, depth, and volume of resorption lacunae. An Olympus BH2 microscope is fitted with a source of reflected light and metallurgical objectives. A grid of points is then overlaid onto the image of the wafer using the drawing arm attachment shown. The magnification of the drawing tube is adjusted so that the grid of 100 points overlays an area of 1 mm².

Fig. 6. *(opposite page)* (A) Scanning electron micrograph of resorption lacuna excavated by human osteoclasts isolated from a giant cell tumor. Fragments of human giant cell tumor were chopped and minced to release a population of large MNCs which can resorb bone in vivo and in vitro. The resorption lacuna and the resorbing cell are visible. (B) Light micrograph of the resorption lacuna shown in Fig. 6A. The morphology of the resorption lacuna and the resorbing cell are again evident using this more simplified technique. Magnification 250×.

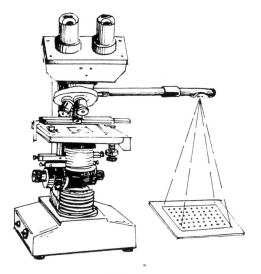

Fig. 5.

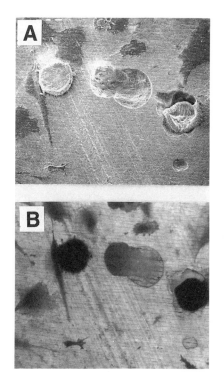

Fig. 6.

5. On a 5-mm² wafer, count the number of points falling onto resorption lacunae in a total of 16 areas.
6. Calculate the mean % surface area of the wafer resorbed by adding the counts from each field and dividing by the number of areas quantified.

3.4.2. Volume Measurements

In our experience, the quantitation of mean area of resorption provides a good indicator of osteoclast activity. If, however, the amount of resorption is minimal, as may be the case with MNCs derived from human bone marrow, a more sensitive technique may be required to quantify resorption. To this end, the measurement of the volume of resorption lacunae may be necessary.

1. Bring the surface of the wafer into focus using the 50× objective lens (numerical aperture 0.75). It is important that the surface of the wafer is brought into precise focus, since this is the reference from which the depth and finally volume of the resorption lacunae will be assessed. The position of the fine focus knob of the microscope at the surface of the wafer should be noted.
2. Using the Olympus BH2 microscope, move the fine focus knob one division at a time representing a change in depth of 2 µm. For other instruments, the manufacturer should be consulted. At each depth, record the area of the lacuna by counting the number of points of the grid falling within the part of the lacuna that is in focus or has yet to come into focus.
3. Repeat this procedure for 10 resorption lacunae chosen at random.
4. Calculate the volume of resorption lacunae by application of the Cavalieri estimator of volume *(8)*. This states that the volume of a sectioned object is "the sum of the products of the section areas × the section separation."

4. Notes
4.1. Optimum pH for Bone Resorption

1. It has been reported that the activity of rodent and avian osteoclasts is increased at low pH in vitro (pH 6.4–6.8). However, in our experience, avian osteoclast activity was optimal at pHs around 7.2 *(9)*.

4.2. Recently Reported Methods

2. Recently, a new model for the study of bone resorption in vitro has been reported *(10)*. The authors report that cells harvested from human giant cell tumors, maintained in culture and passaged several times express many of the characteristics of osteoclasts, including the excavation of resorption lacunae in vitro. This system may provide a model for studying resorption in vitro.
3. In another recent study the use of a modified long-term marrow culture system has been reported *(11)*. Marrow mononuclear cells were isolated and left to form a confluent stroma. The stroma was then seeded onto sterile bone wafers and settled. A second population of mononuclear cells was then seeded onto the

stroma. The authors report a marked improvement in the area of bone resorbed in this system when compared to the MNCs formed by the more usual long-term marrow culture.

4.3. Identification of Osteoclasts by Immunocytochemistry

The identity of the MNCs formed in long-term cultures of human bone marrow has been somewhat contentious with some researchers favoring the view that the cells are not osteoclasts, but are instead macrophage polykaryons. A variety of phenotypic markers have been used to describe cells as "osteoclast-like." Some of these are described in Notes 4–9.

4. The vitronectin receptor (integrin $\alpha v\beta 3$): Davies et al. *(12)* originally described the "Osteoclast Functional Antigen" with the antibody 23C6. This was later shown to be identical to the vitronectin receptor *(13)*. This molecule appears to be universally expressed by osteoclasts and is functionally involved in adhesion and possibly resorption, but it is by no means restricted to osteoclasts and is expressed by many cells of the monocyte/macrophage lineage. Many investigators consider the expression of the vitronectin receptor by MNCs in culture to be indicative of possible osteoclastic differentiation. Certainly, the failure to express this molecule would suggest the absence of osteoclasts.
5. Calcitonin receptors: The ability of MNCs to bind radiolabeled calcitonin or to contract in response to the molecule has also been taken as a sign of osteoclastic phenotype *(14)*.
6. Tartrate-resistant acid phosphatase (TRAP): Enzyme cytochemical studies have indicated that osteoclasts *in situ* and giant cell tumors express TRAP in abundance *(15,16)*. Again, however, numerous other cell types may also express this enzyme.
7. Type II carbonic anhydrase *(17)*: This enzyme is also present in abundance in osteoclasts. It is usually detected by specific antibodies or *in situ* hybridization for mRNA.
8. The enzymes nonspecific esterase and peroxidase are usually present in monocytes and macrophages, but are reported to be absent in osteoclasts *(18,19)*. The absence of these enzymes in cytochemical tests may, therefore, be suggestive of osteoclastic differentiation when supported by other evidence.
9. In our laboratory, we have developed a monoclonal antibody (MAb) that specifically stains the cytoplasm of osteoclasts (OCA1). The antibody does not stain macrophage polykaryons. The staining of osteoclasts obtained from a human giant cell tumor is shown in Fig. 7. The large MNCs of the tumor are densely stained, whereas the stromal cells of the tumor remain unstained.

These phenotypic markers can still only be regarded as suggestive of possible osteoclastic differentiation. MNCs from long-term bone marrow cultures have been varyingly reported to express some or all of the above, whereas macrophage polykaryons from inflammatory lesions in general do not. However, the only definitive marker for osteoclasts remains the ability to resorb bone.

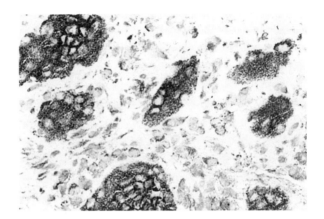

Fig. 7. Light micrograph showing immunoperoxidase staining of a human giant cell tumor with an MAb specific for osteoclasts (OCA1). Frozen sections of the giant cell tumor were prepared, and incubated with the antibody. The antibody binds specifically to the osteoclasts in the sections, whereas other cell types present remain unstained. Magnification 370×.

4.4. Wafers of Devitalized Bovine Bone

10. Wafers can be purchased from Ultrabone (University of Liverpool, UK) Wafers of dentine may also be used in resorption assays. Dentine is a more homogenous tissue than bone, but is difficult to obtain. Dentine wafers can also be purchased from Ultrabone.
11. Bone or dentine wafers must be prewetted in α-MEM immediately prior to use in resorption assays. Six-well plates are most convenient for this. No more than 12 wafers should be placed in 1–2 mL medium in each well to ensure the wafers are kept separate. Care should be taken that the wafers are immersed in the medium and do not float on the top. They can be pushed down with sterile forceps if necessary.

4.5. Long-Term Bone Marrow Culture

12. If the medium becomes acidic before the 10th d in culture, it is necessary to feed before this time. Particular care must be taken when feeding before the 10th d. To ensure that the cell layer is not disturbed, medium must be drawn off from just below the surface.
13. With increasing age, the proportion of adipose tissue in the bone marrow increases. As a result, if marrow is obtained from the bone of an elderly subject, a layer of fat may settle on the surface of the culture. This layer may prevent the exchange of gases at the surface of the culture and should be carefully removed using a fine-ended pipet. Add an equivalent volume of fresh α-MEM to return the final volume of the culture to 10 mL.

4.6. MNCs for Enzymatic and Immunocytochemical Studies

14. After detachment from tissue-culture plastic, MNCs formed from long-term culture of human bone marrow may be seeded onto glass coverslips or multichamber slides (Gibco) and settled in α-MEM/10% FCS. The cells may then be used for enzymatic or immunocytochemical studies following fixation in ice-cold acetone.

References

1. Boyde, A., Ali, N. N., and Jones, S. J. (1984) Resorption of dentine by isolated osteoclasts in vitro. *Br. Dent. J.* **156,** 216–220.
2. Raisz, L. G. and Niemann, I. (1969) Effect of phosphate, calcium and magnesium on bone resorption and hormonal responses in tissue culture. *Endocrinology* **85,** 446–452.
3. Reynolds, J. J. and Dingle, J. T. (1970) A sensitive in vitro method for studying the induction and inhibition of bone resorption. *Calcif. Tiss. Res.* **4,** 339–343.
4. Chambers, T. J., Revell, P. A., Fuller, K., and Athanasou, N. (1984) Resorption of bone by isolated rabbit osteoclasts. *J. Cell Sci.* **66,** 383–399.
5. MacDonald, B. R., Takajashi, N., McManus, L. M., Holahan, J., Mundy, G. R., and Roodman, G. D. (1987) Formation of multinucleated cells that respond to osteotropic hormones in long term human bone marrow cultures. *Endocrinology* **120,** 2326–2333.
6. Testa, N. G., Allen, T. D., Lajtha, L. G., Onions, D., and Jarret, O. (1981) Generation of osteoclasts in vitro. *J. Cell Sci.* **47,** 127–137.
7. Walsh, C. A., Beresford, J. N., Birch, M. A., Boothroyd, B., and Gallagher, J. A. (1991) Application of reflected light microscopy to identify and quantitate resorption by isolated osteoclasts. *J. Bone Miner. Res.* **6,** 661–671.
8. Sterio D. C. (1983) The unbiased estimation of number and sizes of arbitrary particles using the dissector. *J. Microscopy* **134,** 127–136.
9. Walsh, C. A., Dawson, W. E., Birch, M. A., and Gallagher, J. A. (1990) The effects of extracellular pH on bone resorption by avian osteoclasts in vitro. *J. Bone Miner. Res.* **5,** 1243–1247.
10. Flanagan, A. M., Stow, M. D., Kendall, N., and Brace, W. (1994) The role of 1,25 dihydroxycholecalciferol and prostaglandin E2 in the regulation of human osteoclastic bone resorption in vitro. *Int. J. Exp. Pathol.* in press.
11. Grano, M., Colucci, S., De Bellis, M., Zigrino, P., Argentino, L., Zambonin, G., Serra, M., Scotlandi, K., Teti, A., and Zambonin-Zallone, A. (1994) New model for bone resorption study in vitro: human osteoclast-like cells from giant cell tumors of bone. *J. Bone Miner. Res.* **9,** 1013–1020.
12. Davies, J., Warwick, J., and Horton, M. (1988) The osteoclast functional antigen: biochemical and immunological evidence that it is a member of the cell adhesion receptor family recognising Arg-Gly-Asp containing peptides. *Bone* **9,** 264.
13. Davies, J., Warwick, J., Totty, N., Philp, R., Helfrich, M., and Horton, M. (1989) The osteoclast functional antigen, implicated in the regulation of bone resorption, is biochemically related to the vitronectin receptor. *J. Cell Biol.* **109,** 1817–1829.

14. Nicholson, G. C., Horton, M. A., Sexton, P. M., D'Santos, C. S., Moseley, J. M., Kemp, B. E., Pringle, J. A., and Martin, T. J. (1987) Calcitonin receptors of human osteoclastoma. *Hormone Metab. Res.* **19**, 585–589.
15. Minkin, C. (1982) Bone acid phosphatase: tartrate-resistant acid phosphatase as a marker of osteoclast function. *Calcif. Tiss. Int.* **34**, 285–290.
16. Scheven, B. A., Kawilarang-De-Haas, E. W., Wassenaar, A. M., and Nijweide, P. J. (1986) Differentiation kinetics of osteoclasts in the periosteum of embryonic bones in vivo and in vitro. *Anat. Rec.* **214**, 418–423.
17. Hall, G. E. and Kenny, A. D. (1985) Role of carbonic anhydrase in bone resorption induced by 1,25 dihydroxyvitamin D3 in vitro. *Calcif. Tiss. Int.* **37**, 134–142.
18. Hermanns, W. (1987) Identification of osteoclasts and their differentiation from mononuclear phagocytes by enzyme histochemistry. *Histochemistry* **86**, 225–227.
19. Tanaka, T. and Tanaka, M. (1988) Cytological and functional studies of pre-osteoclasts and osteoclasts in the alveolar bones from neonatal rats using peroxidase as a tracer. *Calcif. Tiss. Int.* **47**, 267–272.

22

In Vitro Cellular Systems for Studying OC Function and Differentiation

Primary OC Cultures and the FLG 29.1 Model

Donatella Aldinucci, Julian M. W. Quinn, Massimo Degan, Senka Juzbasic, Angela De Iuliis, Salvatore Improta, Antonio Pinto, and Valter Gattei

1. Introduction

Several pieces of evidence have shown that osteoclasts (OCs) are derived from progenitors originating from hemopoietic stem cells *(1–3)*. More specifically, early OC precursors seem to be closely related to the colony-forming unit for granulocytes and macrophages (CFU-GM) *(3–5)*. However, compared with other bone or marrow cells, OCs are found in extremely low numbers in normal adult bone. In addition, active OCs are strongly adherent to the bone surface. For these reasons, it is impossible to obtain pure or highly enriched cultures of intact OCs, although it is possible to obtain large numbers of OCs if good source tissue is available. OCs are found in large numbers only in bone undergoing extensive physiological remodeling (e.g., fetal bone and growing bone metaphyses) or pathological osteolysis (e.g., fracture callus). Since human tissue is often difficult to obtain, most OC research has employed animal models, notably rabbit, rat, and chick.

Another approach has used human leukemic cell lines, such as HL60 and U937 cells, induced by phorbol esters or 1,25 dihydroxyvitamin D_3 (1,25[OH]$_2D_3$) to differentiate toward monocyte macrophages, which in turn might show some resorbing activity in vitro *(6,7)*. However, such in vitro-derived macrophage-like cells considerably differ from OCs both functionally and phenotypically. The availability of a clonal population of OC precursors would offer an important tool for studying mechanisms that regulate replication and differentiation of OCs. The use of pure populations of progenitors or terminally differentiated

From: *Methods in Molecular Medicine: Human Cell Culture Protocols*
Edited by: G. E. Jones Humana Press Inc., Totowa, NJ

OCs, would further allow the analysis of the interactions among homogeneous, but distinct, classes of bone cells. The establishment of a human continuous cell line (FLG 29.1) of bone marrow-derived OC progenitors *(8)* has recently provided a unique model of OC differentiation to investigate cytokine circuitries mediating the interactions of OCs with specialized or accessory bone cell populations. FLG 29.1 cells display several phenotypic features of OCs, including tartrate-resistant acid phosphatase and a specific immunophenotypic profile, closely resembling that of human fetal OCs *(8)*.

This protocol summarizes the most commonly adopted methods for studies in the field of OC cell biology. Methods for extracting viable OCs from bone or giant cell tumors of bone (GCT-bone) suitable for OC cultures, as well as immunohistochemistry or enzyme histochemical staining are described. OC culture techniques are also discussed with particular regard to preparation of OCs on cortical bone slices. Finally, the FLG 29.1 model is discussed by presenting two in vitro differentiation assays. First, FLG 29.1 pre-OC cells can be induced by phorbol esters or vitamin D_3 derivatives to differentiate further toward mature elements expressing functionally active calcitonin receptors and capable of degrading ^{45}Ca-labeled devitalized bone particles. In a second assay, FLG 29.1 cells in coculture with the osteoblast (OB) cell line Saos-2 or with bone marrow fibroblasts (BMF) are studied with a reverse transcriptase-polymerase chain reaction (RT-PCR) approach to investigate the cytokine/cytokine receptor network governing OC proliferation or activation, as well as the crosstalk among OCs and other specialized cells of the bone microenvironment.

1.1. Extraction of Viable OCs and Preparation of OC Cultures by Curettage of Bone or Tumor

Outgrowth cultures from chick fetal bones have been used for cytochemical studies *(9,10)*, but these result in very low yields, the OCs constituting only a tiny minority of the cells obtained. To improve yield and purity of OCs, Mears *(11)* developed a method of scraping and collagenase digestion of young rat long bones. Nelson and Bauer *(12)* used a system of mechanical disaggregation, followed by unit gravity sedimentation and removal of nonadherent cells by vigorous rinsing. Hefley and Stern *(13)* concluded that these methods irreversibly damage OC membranes and selected for OCs of lower adherence to bone. However, this method has been successfully used by Chambers et al. *(14)* (with rabbit and rat) and Zambonin-Zallone et al. *(15)* (with chick) in morphological and functional studies, and is now widely employed. Other methods that have been developed include the subcutaneous implantation of bone particles *(16)*, the imprinting of rat femurs combined with microdissection to produce pure OC populations *(17)*, and the sequential collagenase digestion of flat bones of rats or other animals *(18–21)*.

In studies of human OCs, imprints of fetal, child, and adult bone, or of adult fracture callus have been used for cytochemical studies *(22)*. Such tissues can yield large numbers of OCs, which can be fixed immediately and used for histochemical and immunohistochemical analysis, as described in Sections 3. and 4. Human OCs can be extracted from these tissues by mechanical or enzymatic disaggregation for use in tissue-culture systems *(23)*; membranous bone surgical waste has also been successfully employed *(24)*. However, OCs extracted from these sources are of relatively low viability, with only a minority surviving in culture conditions, and when assayed, are capable of bone resorption, but only at very low rates.

Many studies on human OCs employ a surrogate, the OC-like giant cell derived from the GCT-bone *(25,26)*. This is a histologically distinctive osteolytic primary tumor. It is characterized by a well-vascularized tissue, containing spindle-shaped and ovoid mononuclear cells, among which are scattered numerous multinucleated OC-like giant cells (GCs). The mononuclear component includes both reactive mononuclear phagocyte system (MPS) cells and proliferating cells that are fibroblast-like or OB-like *(26–28)*. Ultrastructural studies have demonstrated that the GCs display an abundance of mitochondria and rough endoplasmic reticulum, an organelle-free peripheral cytoplasm, a complex cell membrane structure with interdigitating villi, and many vacuoles, all of which are features typical of OCs *(27,28)*. The presence of an osteoclastic ruffled border has been claimed, but disputed by others *(27,29)*. The GCs of this tumor can resorb bone in vitro, express calcitonin receptors, and respond morphologically to calcitonin *(30)*. Histochemical and immunohistochemical studies also show that, like with human OCs from fetal and adult bone, they express tartrate-resistant acid phosphatase, integrins $\alpha_v\beta_3$ (CD51/CD61, vitronectin receptor), and $\alpha_2\beta_1$ (CD49b/CD29), along with CD68 antigens, while lacking expression of other macrophage-associated antigens, such as β_2 integrins and CD14 *(26,31–34)*. GCT-bone are thus commonly used as a source of human OCs; although other types of GCs from certain types of human soft tissue lesions have been shown to resorb bone in vitro, such cells can express phenotypic characteristics more typical of macrophage polykaryons *(33,35)*. Methods for extracting viable OCs are described in Section 3.1. Methods for phenotypic analysis of OCs by immuno- and enzyme histochemistry are described in Sections 3.2. and 3.3.

1.2. Preparation of Cortical Bone Slices for In Vitro Bone Resorption Assay by OCs

Both organ culture and disaggregated cell-culture techniques have been used to investigate bone resorption. In organ culture systems, the effect of hormones on the whole system can be followed, including the effects on

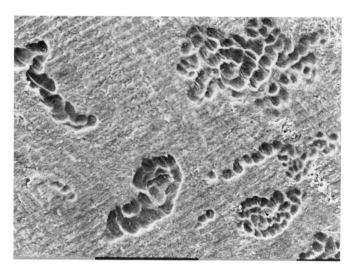

Fig. 1. SEM showing bone resorption pits formed by OCs formed in vitro (in murine macrophage/UMR106 cocultures) on the surface of cortical bone slices. The pits typically overlap to form contiguous areas of resorption with distinctive smooth edges, with exposed collagen fibrils at the base of the pits (large black bar = 100 μm).

OC formation. However, to study the direct effects of hormones and other molecules on OC bone resorption, it is necessary to separate them as much as possible from the other cells of bone. Bone resorption by OCs can be studied in vitro by means of the cortical bone slice assay, in which cells are cultured on the surface of thin wafers of cortical bone (or sperm whale dentine), and after an incubation period of 12 h or more, the cells are removed and the surface of the bone examined for evidence of resorption pits *(36,37)*. The bone resorption pits can be visualized by reflected light microscopy after staining the bone slice with touldine blue. However, a better method for detecting bone resorption pits is by use of scanning electron microscopy (SEM). SEM allows much better discrimination between resorption pits and other features, such as osteocyte lacunae, blood vessels, and imperfectly removed cellular debris (Fig. 1). The bone resorption assay is frequently used in OC differentiation assays, in which hemopoietic cells (such as spleen cells or blood monocytes) are cocultured long term with bone marrow stromal cells or OBs *(38–41)* in the presence of 1,25 $(OH)_2D_3$. In such assays, bone resorption does not commence before 4 d of incubation, and incubation times of 2 or 3 wk are often used. When employing such coculture methodologies, it is advisable to remove the stromal cells from

Osteoclast Function and Differentiation

the bone surface with trypsin solutions before stripping the surface with NH_4OH (*see* Section 3.5.). Methods for preparation of cortical bone slices are reported in Section 3.4., whereas those for in vitro bone resorption are described in Section 3.5.

1.3. The Human Pre-OC Cell Line FLG 29.1 as a Model for OC Differentiation and Function

The FLG 29.1 cell line was established in 1985 in the laboratory of P. A. Bernabei (U.O. Haematology, University of Florence, Italy) from a 38-yr-old female suffering from acute monoblastic leukemia, M5a type according to the FAB classification. The clinical course of the patient was characterized, after a 15-mo complete remission, by a fatal pancytopenic episode associated with a peripheral neuropathic syndrome and generalized increase in bone density, mimicking an osteopetrotic syndrome. Bone marrow mononuclear cells were isolated on Ficoll-Hypaque gradient (Pharmacia, Uppsala, Sweden) and resuspended in RPMI-1640 culture medium (Flow Laboratories Inc., Irvine, UK) supplemented with 10% heat-inactivated FCS (Flow Laboratories). Cells (10^5 cells/mL) were grown in Falcon polystyrene tissue-culture flasks at 37°C in an atmosphere of 5% CO_2/95% air. An adherent cell monolayer developed after 20 d of culture, and on top of the adherent cells, a homogeneous population of floating cells proliferated. This latter cell population was collected, propagated in suspension culture, and cloned by limiting dilution. The cell line was determined to be free of mycoplasma contamination and has been maintained in continuous suspension culture for almost 10 yr. Kinetic characteristics, cytochemical, biochemical, and ultrastructural features along with karyotype analysis and molecular rearrangements of FLG 29.1 cells are described elsewhere *(8)*.

Differentiation of FLG 29.1 cells (*see* Section 3.6.) is usually evaluated by means of the following markers: morphology and cytochemical staining (studies performed on cytospin smears; *see* Sections 3.2., 3.3., and 3.6.), cell-surface and intracytoplasmic specific markers (flow cytometry; *see* Sections 3.7. and 3.8.), and capability to resorb bone particles (liquid culture; *see* Sections 2.8 and 2.9.) and to express functionally active calcitonin receptors (Scatchard plot of binding data, RT-PCR and enzyme immunoassay).

1.4. Detection of Functionally Active Calcitonin Receptors in FLG 29.1 Cells

Calcitonin (CT) responsiveness is considered a reliable and early marker for OC differentiation in vitro, being absent in cells of the mononuclear phagocyte lineage *(42)*. Indeed, CT receptors were shown to identify mononuclear cells committed to the osteoclastic lineage and are expressed before the acquisition of bone resorption capacity by OCs *(42)*. Studies concerning the expression by FLG 29.1 OC-like cells of functionally active CT receptors are essentially carried out:

1. By documenting the presence of CT receptor on the cell surface of FLG 29.1 cells by CT-binding studies (*see* Section 3.10.).
2. By documenting the presence of specific CT receptor mRNA by RT-PCR (*see* Section 3.11.).
3. By measuring the intracellular content of cAMP following FLG 29.1 stimulation by salmon CT (*see* Section 3.12.).

1.5. FLG 29.1 Cells in Coculture with OB Cells (Saos-2) and BMF

A number of studies have suggested that specialized cell populations of the bone microenvironment, such as OBs, as well as accessory stromal cells may directly regulate proliferation, differentiation, and functional recruitment of OC progenitors *(3,43,44)*. These cell types have been shown to mediate some effects of systemic hormones and vitamins on bone resorption *(3,43,45,46)*, being also able to support OC-like cell formation from hemopoietic progenitors *(3,39,47)*. It is therefore apparent that the bone microenvironment plays an essential role in regulating bone resorption processes by acting both on immature OC precursors and on their terminally differentiated progeny *(3,43)*. Even though a number of growth factors, including M-CSF, GM-CSF, interleukin (IL)-1, IL-6, tumor necrosis factor (TNF) *(3,43,48)*, and more recently IL-4 *(49)*, IL-10 *(50)*, and IL-11 *(51)*, have been implicated in osteoclastogenesis, the cytokine network governing OC proliferation or activation, and the growth factor receptor repertoire of these latter cells are still poorly understood *(3,43)*.

To address this point, FLG 29.1 OC-like cells in coculture with the human OB cell line Saos-2 or with BMF are studied with an RT-PCR approach to investigate the cytokine/cytokine receptor network governing OC proliferation or activation, and the crosstalk among OCs and other specialized cells within the bone microenvironment. The Saos-2 cell line represents a useful model for studying OB properties, since it displays several OB-specific features, including alkaline phosphatase, response to parathyroid hormone, and expression of a number of bone-related protein, such as osteonectin and osteocalcin *(52)*. BMF cultures are established from bone marrow mononuclear cells obtained by sternal or iliac crest aspiration. Methods for establishing Saos-2 and BMF monolayers are reported in Sections 3.13. and 3.14. RT-PCR approach is described in Section 3.15.

2. Materials

2.1. Extraction of Viable OCs and Preparation of OC Cultures

1. Tissue-culture medium, such as RPMI (Gibco, Paisley, Scotland) or α minimal essential medium (αMEM, Gibco), supplemented or not with 10% fetal calf serum (FCS, Gibco).
2. Smooth-ended glass pipets, sterile scalpel blade, and small forceps.

Osteoclast Function and Differentiation

3. Culture substrates should be prepared in advance by being placed in 6- or 16-mm plastic tissue-culture wells containing tissue-culture medium containing FCS. Typical substrates include sterile, cleaned 6- or 13-mm glass coverslips or cortical bone slices prepared as described in Section 1.2.
4. CO_2 (5%) gassed tissue-culture incubator.

2.2. Phenotypic Analysis of OCs by Immunohistochemistry

1. Trypsin solution (TS): 0.1% trypsin (Sigma, St. Louis, MO) solution in a 0.1% solution of $CaCl_2$, pH 7.8.
2. Phosphate-buffered saline (PBS) and PBS supplemented with 20% porcine serum (PBS/PS).
3. Methanol.
4. 30% Solution of H_2O_2.
5. Peroxidase-conjugated rabbit antimouse polyclonal antibody (Dakopatts, Glostrup, Denmark).
6. Peroxidase-conjugated swine antirabbit polyclonal antibody (Dakopatts).
7. Diaminobenzidine hydrochloride (DAB; BDH, Poole, UK).
8. Gill's No. 3 hematoxylin (Sigma).

2.3. Phenotypic Analysis of OCs by Enzyme Histochemistry (see Note 5)

1. Citrate-acetone-formaldehyde fixation solution: This is freshly prepared from 65 mL acetone, 8 mL of 37% formaldehyde, and 25 mL citrate buffer; citrate buffer consists of 18 mM citric acid, 9 mM sodium citrate, and 12 mM sodium chloride, pH adjusted to 3.6.
2. Acid phosphatase substrate: This is prepared by adding 0.5 mL of 12.5 mg/mL Naphthol AS-BI phosphate solution, 2 mL 2.5M, pH 5.2 acetate buffer solution, and 1 mL Fast Garnet solution to 45 mL distilled water warmed to 37°C. Fast garnet solution is prepared by adding 0.5 mL of Fast Garnet GBC base (7.0 mg/mL in 0.4M HCl) to 0.5 mL of 0.1M sodium nitrite, and allowed to stand at room temperature for 2 min.
3. Tartrate-resistant acid phosphatase (TRAP) substrate: Acid phosphatase substrate is prepared, and 1 mL of 0.335M tartrate buffer (pH 4.9) is added.
4. Nonspecific esterase substrate: This is prepared by adding 1 mL 0.1M sodium nitrite to 1 mL Fast Blue BB solution (15 mg/mL Fast Blue BB in 0.4M HCl). This is mixed by inversion, allowed to stand for 2 min, and this solution added to 40 mL distilled water (at 37°C), followed by 5 mL 1M Trizma-maleate (pH 6.3) and 1 mL 12.5 mg/mL α-naphthyl acetate (in methanol) solution.

2.4. Preparation of Cortical Bone Slices for In Vitro Bone Resorption Assay by OCs

1. Shafts of unfixed bovine or human postmortem femur are obtained, cleaned of any soft tissue, and cut into small (approx 2-cm long) pieces with a bandsaw. This can be stored long term at –20°C.
2. A low-speed diamond wafer saw (Buehler, Lake Bluff, IL).
3. Ultrasonicator.

2.5. Cortical Bone Resorption Assay

1. Tissue-culture medium, such as RPMI or αMEM, supplemented with 10% FCS.
2. Tissue-culture cluster plates with 16-mm diameter wells (Costar, UK).
3. $0.25M$ NH_4OH solution.
4. Gold sputter coater machine suitable for SEM.
5. SEM aluminum pin stubs.
6. SEM.
7. CO_2 (5%) gassed tissue-culture incubator.

2.6. Culture Conditions for Optimal In Vitro Differentiation of FLG 29.1 Cells

1. RPMI-1640 culture medium (Flow Laboratories) supplemented with 10% heat-inactivated (56°C for 30 min) FCS (Flow Laboratories).
2. 12-*O*-tetradecanoylphorbol-13-acetate (TPA; Sigma) solubilized in acetone at 1.6 mM and stored at –20°C.
3. 1,25(OH)$_2$D$_3$ (Hoffman-La Roche) dissolved in absolute ethanol at 1.0 mM and stored at –20°C.
4. May-Grünwald's solution (Farmitalia Carlo Erba, Milan, Italy).
5. Giemsa's solution (Farmitalia Carlo Erba).

2.7. Phenotypic Characterization of FLG 29.1 Cells by Immunofluorescence Assay for Surface Antigens

1. Binding buffer (BB): 2 mg/mL purified human immunoglobulin (Jackson Immunoresearch Lab. Inc., West Grove, PA), 0.2% bovine serum albumin (Sigma), 0.2% sodium azide in Hank's Balanced Salt Solution (HBSS; Gibco).
2. Washing buffer (WB): HBSS/0.2% sodium azide.
3. Most monoclonal antibodies (MAbs) used are derived from the panels of the IVth and Vth editions of the International Conference and Workshop on Leukocyte Antigens *(53,54)*. Other MAbs (Table 1, pp. 286–287) can be purchased from Ortho (Raritan, NJ), Coulter (Coulter Immunology, Hileah, FL), Becton Dickinson, and Dako (Dakopatts).

2.8. Detection of Intracytoplasmic Antigens in FLG 29.1 Cells

1. Fixing/permeabilizing buffer (FPB): PBS buffer containing 2% paraformaldehyde, 0.1% Tween-20, and 0.2 μg/mL EDTA.
2. WB: 0.5% Tween-20, 2% FCS in PBS.

2.9. Resorption of Devitalized Bone Particles by FLG 29.1 Cells

1. CBA T6T6 mice (4–8 wk old).
2. $^{45}CaCl_2$ (SA 5–50 mCi/mg Ca, Amersham International, Amersham, UK).
3. $1M$ NaOH.
4. $5M$ HCl.
5. HBSS.
6. 10% Trichloroacetic acid (TCA).
7. Materials for cell culture (*see* Section 2.6.).

2.10. CT-Binding Studies

1. ^{125}I-labeled salmon CT (SA 2,000 Ci/mmol, Amersham).
2. Unlabeled salmon CT (Sigma).
3. BB: 100 mM HEPES, 1 mM KCl, 120 mM NaCl, and 1% BSA, pH 7.6.
4. WB: HBSS containing 1% BSA.

2.11. Detection of CT Receptor mRNA by RT-PCR

1. RT-PCR reagents and buffers are described in Section 2.14.
2. Primers specific for CT receptor selected according to published sequence data *(55)* are as follows: sense 5'-CTT CTT CTA AAT CAC CCA ACC-3' (sequence nucleotide 287–308); antisense 5'-AGT AAA CAC AGC CAC GAC A-3' (sequence nucleotide 1058–1039).

2.12. Intracellular cAMP Accumulation by Salmon CT

1. Intracellular cAMP content in FLG 29.1 cells is measured by a standard enzyme immunoassay (EIA) system (Biotrak cAMP EIA system, Amersham) following the manufacturer's procedures. Most reagents are supplied by the manufacturer, including assay (0.05M sodium acetate buffer, pH 5.8, and 0.02% BSA) and washing (0.01M phosphate buffer, pH 7.5, 0.05% Tween-20) buffers, antibodies, unlabeled- and peroxidase-conjugated cAMP, and peroxidase substrate (tetramethylbenzidine/hydrogen peroxide).
2. Salmon CT (Sigma).
3. HBSS (Gibco).
4. Absolute ethanol.
5. 1.0M Sulfuric acid.

2.13. Preparation of Soas-2 and BMF Monolayers and Coculture Conditions

1. McCoy's medium (Gibco) supplemented with 10% FCS (Flow Laboratories).
2. Ficoll-Hypaque (Pharmacia).
3. Iscove's Modified Dulbecco's Medium (IMDM, Gibco)/10% FCS.
4. Basic fibroblast growth factor (bFGF; Genzyme, Cambridge, MA).

2.14. Changes in Gene Expression During Coculture of FLG 29.1 Cells with BMF or Saos-2 Cells: Detection by RT-PCR (see Note 20)

1. AMV RT (Promega, Madison WI).
2. Hexadeoxyribonucleotide random primers (Promega).
3. 20 mM Deoxynucleotides-triphosphate (dNTPs, containing 100 mM of dATP, dCTP, dGTP, dTTP; Promega).
4. Dithiothreitol (DTT) 0.1M (Promega).
5. Recombinant ribonuclease inhibitor (RNasin, Promega).
6. 25 mM MgCl$_2$ (Promega).

Table 1
Phenotypic Profile of FLG 29.1 Cells Before and After Exposure to TPA and Its Relationship with Normal Fetal Bone OCs[a]

CD	Recognized molecule cell distribution[b]	FLG 29.1	FLG 29.1 + TPA	OCs[c]	Macropoly[c]	Induced FLG 29.1
4	Th, Macro	5[d]	12	–	–	–
13	My, M, Macro	15	45	+	+	+
14	M, Macro	4	10	–	+	–
15	X-hapten, M, macro	10	27	±	±	±
31	GpIIa, My, M, plt	5	73	–	+	+
32	FcRII, My, M, plt	92	99	+	+	+
33	My, M, Macro	25	21	–	+	±
34	Precursor cells	36	20	NT	NT	–
35	PMN, M, DRC, plt, RBC	15	20	–	+	–
36	M, Macro, RBC	7	12	–	+	–
68	Macro	8[e]	25[e]	+	+	+
68	Macro	24[e]	66[e]	+	+	+
41a	GpIIb, plt	3	18	–	–	–
42b	GpIb, plt	10	30	+	+	+
44	H-CAM	28	99	+	+	+
45	T200, LCA	31	67	+	+	+
11a	LFA-1α	3	20	–	+	–
11b	Mac1α	3	36	±	+	+
18	LFA-1β	5	59	NT	+	+
54	ICAM-1	99	99	+	+	+

51	VNR-α	25	89	+	+	+
61	GpIIIa, VNR-β	53	75	+	+	+
51/61[f]	VNR	14	80	+	−	+
71	Trf-R	99	65	−	+	+
9	Act, PMN, B, M, Macro, plt	27	9	+	+	+
—	HLA-Class I	2	48	NT	NT	+
—	HLA-Class II	4	10	NT	+	−

[a]Abbreviations: Act, activated cells; B, B-lymphocytes; DRC, dendritic reticulum cells; M, monocytes; Macro, macrophages; Macropoly, macrophage polykaria; My, myeloid cells; NT, not tested; plt, platelets; PMN, neutrophils; RBC, red blood cells; Th, T Helper lymphocytes.
[b]According to the V International Workshop and Conference on Human Leukocyte Differentiation Antigens (43).
[c]Data obtained from ref. 8.
[d]Percent stained cells.
[e]99% of positive cell in intracytoplasmic staining.
[f]13C2 MAb recognizing the CD51/CD61 complex of VNR.

7. 10X PCR reaction buffer (100 m*M* Tris-HCl, pH 9.0, 500 m*M* KCl, 1% Triton X-100; Promega).
8. *Taq* polymerase (Promega).
9. Gene-specific primers (Table 2, pp. 290–291).
10. RNase-free water.

3. Methods

3.1. Extraction of Viable OCs and Preparation of OC Cultures (see Note 1)

1. The animal is killed, its hind legs dissected, and the long bones removed and placed in sterile serum-free medium, as in Section 2.1.
2. All remaining periosteum and attached soft tissue is dissected or scraped off, and articular cartilage is removed.
3. The bones are placed in a small volume (typically 1 mL) of tissue-culture medium containing 10% FCS, and then the metaphyseal regions of the bones are rapidly and finely curetted by a sharp scalpel.
4. The resulting cell suspension and the bone particles released are then agitated by a smooth-ended glass pipet for 30 s.
5. The cell suspension is then added to the culture substrate (such as glass, culture plastic, or cortical bone slice prepared as described in Section 3.4.) in tissue-culture wells.
6. These cultures are incubated at 37°C in a CO_2 gassed incubator for 30–60 min to allow the OCs to adhere to substrate.
7. After this time, the culture substrates are washed vigorously in tissue-culture medium (RPMI or αMEM). At this point, they were either fixed (10 min in cold [–20°C] acetone) or placed into longer term culture (*see* Section 3.5.).
8. As an alternative, bone imprints can be prepared by cutting the bone longitudinally to expose the trabecular bone and marrow cavities, and pressing it against a glass slide. Imprints are fixed for 10 min in cold (–20°C) acetone.

3.2. Phenotypic Analysis of OCs by Immunohistochemistry (see Note 2)

Frozen sections of undecalcified bone can be made of metaphyseal bone with a toughened tungsten blade, but even when the morphology is good, a common problem occurs in the removal of some or all of the bone itself because of the repeated washing steps of immunohistochemistry (IHC) techniques. For this reason, we have often employed imprints of bone (*see* Section 3.1.), which have an additional advantage in containing whole OCs that are more evenly stained by antibodies for membrane-associated antigens.

1. Acetone-fixed frozen sections and imprints stored at –20°C are brought to room temperature (*see* Note 3).
2. Where recommended for a particular antibody (*see* Note 4), paraffin sections are then digested in TS at 37°C for 15 or 30 min. Trypsin is removed from the

tissue by a thorough rinse in running tap water, followed by rinses in distilled water and PBS.
3. Fifty microliters of the antibody, diluted to the recommended concentration in PBS containing 2% bovine serum albumin (Boehringer-Mannheim, Mannheim, Germany) are added to each section or imprint. The slides are incubated in a moist environment for 30–60 min at room temperature, or overnight at 4°C.
4. The slides are rinsed in PBS for 5 min and placed in a glass trough containing methanol/3% H_2O_2 (BDH), at room temperature for 30 min, while being gently agitated.
5. The slides are removed and rinsed in PBS for 10 min, with agitation. Fifty microliters of peroxidase-conjugated rabbit antimouse polyclonal antibody, diluted 1:50 in PBS/PS, are added to each section/imprint and incubated on the slides in a moist environment for 30–60 min.
6. The slides are rinsed in PBS for 5 min. Fifty microliters of peroxidase-conjugated swine antirabbit polyclonal antibody (Dakopatts), diluted 1:50 in PBS/PS, are added to each section/imprint and incubated on the slides in a moist environment for 30–60 min.
7. The slides are rinsed in PBS for 5 min. A solution of 500 µg/mL of DAB containing 0.02% H_2O_2 is added to the section/imprint and incubated on the slides for 4–5 min.
8. The sections/imprints are thoroughly rinsed in running tap water and lightly counterstained by hematoxylin.
9. The sections and imprints are finally dehydrated through graded alcohols and brought to xylene. The slides are then mounted a medium, such as DPX medium, and examined by light microscopy.

3.3. Phenotypic Analysis of OCs by Enzyme Histochemistry

1. Sections or cytological preparations of bone are first fixed at room temperature in freshly prepared citrate-acetone-formaldehyde solution.
2. The slides are rinsed thoroughly in distilled water and then immersed in one of the appropriate substrates for 30 min.
3. The slides are removed and rinsed repeatedly in distilled water to remove all traces of substrate.
4. If desirable, slides stained for acid phosphatase or TRAP can be counterstained by immersion in hematoxylin solution and rinsed in running water for 10 min.
5. The slides can then be mounted in an aqueous mounting medium, such as Kaisers' jelly, or dried and stored unmounted.

3.4. Preparation of Cortical Bone Slices for In Vitro Bone Resorption Assay by OCs

1. The pieces of femoral bone shaft are brought to room temperature.
2. As much bone marrow and trabecular bone as possible is removed.
3. The pieces of bone are clamped onto the arm of the low-speed diamond wafer saw and cut into thin wafers of bone approx 0.2-mm thick. During the cutting

Table 2
Oligodeoxynucleotides Used in RT-PCR for the Detection of Human Cytokine/Cytokine Receptors[a]

Primer	Sequence, 5'–3'	cDNA position	Appendix refs.	Amplified fragment, bp
M-CSF S	ATGACAGACAGGTGGAACTGCCAGTGTAGAGG	518–550	(1)	437
M-CSF AS	TCACACAACTTCAGTAGGTTCAGGTGATGGGC	922–954		
M-CSFR S	TCCAACTACATTGTCAAGGGCAATGCCGCCT	2719–2751	(2)	360
M-CSFR AS	CAGATTGGTATAGTCCCGCTCTCTCCTGTCCT	3046–3078		
G-CSF S	TTGGACACACTGCAGTGACGTCGCCGACTTT	1467–1500	(3)	470
G-CSF AS	ATTGCAGAGCCAGGGCTGGGGAGCAGTCATAGT	2067–2100		
G-CSFR S	AAGAGCCCCTTACCCACTACACCATCTT	1835–1864	(4)	340
G-CSFR AS	TGCTGTGAGCTGGGTCTGGGACACTT	2148–2174		
GM-CSF S	ATGTGGCTGCAGAGCCTGCTGC	622–684	(5)	424
GM-CSF AS	CTGGCTCCCAGCAGTCAAAGGG	2652–2685		
GM-CSFR S	CTGCATGAAGGAGTCACATT	411–431	(6)	422
GM-CSFR AS	CAACGTACGGTGACATTGCT	813–833		
IGF-1 S	ACATCTCCCATCTCTCTGGATTTCCTTTTGC	85–116	(7)	514
IGF-1 AS	CCCTCTACTTGCGTTCTTCAAATGTACTTCC	567–598		
IGF-1R S	GAATGGAGTGCTGTATGCCTCTGTGAACC	2949–2978	(8)	540
IGF-1R AS	GTGAAATCTTCGGCTACCATGCAATTCCG	3519–3548		
IGF-2 S	AGTCGATGCTGGTGCTTCTCACCTTCTTGGC	270–301	(9)	538
IGF-2 AS	TGCGGCAGTTTTGCTCACTTCCGATTGCTGG	746–777		
IGF-2R S	GCTTAGCGGACAAGCATTTCAACTACACC	5396–5425	(10)	768
IGF-2R AS	GCAGGTCGTAGGTTTTGTGTTTCTGGACG	6144–6173		
IL-1α S	CAAGGAGAGCATGGTGGTAGTAGCAACCAACG	259–291	(11)	491
IL-1α AS	TAGTGCCGTGAGTTTCCCAGAAGAAGAGGAGG	717–749		
IL-1R S	ACACATGGTATAGATGCAGC	1031–1051	(12)	300
IL-1R AS	TTCCAAGACCTCAGGCAAGA	1310–1330		

Name	Sequence	Position	Ref	Size
TNFα S	GAGTGACAAGCCTGTAGCCCATGTTGTAGCA	337–362	(13)	444
TNFα AS	GCAATGATCCAAAGTAGACCTGCCCAGACT	749–780		
TNFβ S	ATGACACCACCTGAACGTCTCTTC	1276–1300	(14)	610
TNFβ AS	CGAAGGCTCCAAAGAAGACAGTACT	2193–2218		
TNF Rp55 S	ATTTGCTGTACCAAGTGCCACAAAGGAACC	350–380	(15)	587
TNF Rp55 AS	GTCGATTTCCCACAAACAATGGAGTAGAGC	906–936		
TNF Rp75 S	GAATACTATGACCAGACAGCTCAGATGTGC	219–249	(16)	403
TNF Rp75 AS	TATCCGTGGATGAAGTCGTGTTGGAGAACG	591–621		
TGF-β1 S	GCCCTGGACACCAACTATTGCT	1678–1700	(17)	161
TGF-β1 AS	AGGCTCCAAATGTAGGGGCAGG	1816–1838		
TGF-β2 S	GATTTCCATCTACAAGACCACGAGGGACTTGC	668–700	(18)	503
TGF-β2 AS	CAGCATCAGTTACATCGAAGGAGAGCCATTCG	1138–1170		
IL-6 S	ATGAACTCCTTCTCCACAAGCGC	34–57	(19)	628
IL-6 AS	GAAGAGCCCTCAGGCTGGACTG	639–661		
IL-6R S	CATTGCCATTGTTCTGAGGTTC	1143–1165	(20)	251
IL-6R AS	AGTAGTCTGTATTGCTGATGTC	1371–1393		
PDGF-A S	AGAAGTCCAGGTGAGGTTAGAGGAGCAT	898–926	(21)	304
PDGF-A AS	CTGCTTCACCGAGTGCTACAATACTTGCT	1172–1201		
FGF-R1 S	CAGAATTGGAGGCTACAAGGTCCGTTATGC	690–720	(22)	705
FGF-R1 AS	GATGCACTGGAGTCAGCAGACACTGTTACC	1360–1394		
SCF S	GTGGCATCTGAAACTAGTGAT	649–670	(23)	276
SCF AS	TGCCCTTGTAAGACTTGGCTG	883–904		

[a]AS, antisense primer; bp, base pair; FGF-R1, fibroblast growth factor 1 receptor; G-CSF, granulocyte colony-stimulating factor; GM-CSF, granulocyte-macrophage colony-stimulating factor; GM-CSFR, granulocyte-macrophage colony-stimulating factor receptor; IGF-1, insulin-like growth factor-1; IGF-1R, insulin-like growth factor-1 receptor; IGF-2, insulin-like growth factor-2; IGF-2R, insulin-like growth factor-2 receptor; IL-1α, interleukin 1α; IL-1α R, interleukin 1α receptor; IL-6, interleukin 6; IL-6R, interleukin 6 receptor; M-CSF, macrophage colony-stimulating factor; M-CSFR, macrophage colony-stimulating factor receptor; PDGF-A, platelet-derived growth factor A; S, sense primer; SCF, stem-cell factor; TGF-β1, transforming growth factor β1; TGF-β2, transforming growth factor β2; TNFα, tumor necrosis factor α; TNFβ, tumor necrosis factor β; TNF Rp55, tumor necrosis factor receptor 55 kDa; TNF Rp75, tumor necrosis factor receptor 75 kDa.

process, the saw should be lubricated with tap water running continuously over both the bone and the diamond saw.
4. The resulting bone slices are trimmed with a scalpel blade and rinsed in distilled water.
5. The bone slices are cleaned by 3–5 min sonications in distilled water, with two rinses in distilled water between each sonication.
6. The bone slices are then immersed in acetone for 10 min.
7. The bone slices are sterilized by rinsing in absolute alcohol, and then dried and stored at room temperature (*see* Note 6).

3.5. Cortical Bone Resorption Assay

1. The OC cultures are prepared on cortical bone slices, as described in Section 2.
2. The cultures are incubated at 37°C for at least 12 h.
3. The bone slices are removed from culture, rinsed briefly in distilled water, and placed in NH_4OH solution overnight to remove the cells (*see* Notes 7 and 8).
4. Bone slices are dehydrated in graded alcohols and allowed to dry at room temperature.
5. Bone slices are mounted on the aluminum electron microscopy stubs. Although specialized adhesives exist for mounting specimens for examination by electron microscopy, we use a cheap double-sided adhesive tape, such as sellotape.
6. Bone slices are then sputtered with gold and examined by electron microscopy.
7. Bone resorption pits can be identified by their size (between 5 and 50 μm in diameter depending on species), shallow concave shape, characteristic smooth outline, and exposed collagen fibrils at their base. Also, the pits may be present in small clusters, which may or may not overlap to form large contiguous areas of excavation (*see* Note 9).

3.6. Cultural Conditions for Optimal In Vitro Differentiation of FLG 29.1 Cells

1. For induction experiments with a single agent, TPA or $1,25(OH)_2D_3$ is freshly diluted in subdued light into culture medium at a final concentration of 0.1 μM (TPA) or 10 nM ($1,25[OH]_2D_3$). FLG 29.1 cells (1×10^5 cells/mL) are induced in Falcon polystyrene tissue-culture flasks at 37°C in an atmosphere of 5% CO_2/95% air for 72 h.
2. For cooperation experiments, $1,25(OH)_2D_3$ is added to FLG 29.1 cell cultures after a 12-h preincubation with 0.1 μM TPA, and then the incubation proceeds for 60 h in the dark at the same conditions (*see* Note 10).
3. Control cultures contain equal amounts of acetone and ethanol.
4. After 72 h, FLG 29.1 cells are collected and counted. For cytospin smear preparation, $50–100 \times 10^3$ cells are spun at 500 rpm for 5 min in a Cytospin 3 Shandon cytocentrifuge and stained with May-Grünwald-Giemsa (MGG), acid phosphatase, TRAP, and α-naphthyl acetate esterase.
5. MGG staining: Cytospin smears are stained with undiluted May-Grünwald's solution (5 minutes at room temperature), then with a 35% May-Grünwald's solu-

tion (5 minutes at room temperature), and finally with a 10% Giemsa's solution (15 minutes at room temperature).
6. Acid phosphatase, TRAP, and α-naphthyl acetate esterase staining (*see* Section 3.3.).
7. Cytospins are air-dried and evaluated microscopically (Fig. 2; *see* Note 11).

3.7. Phenotypic Characterization of FLG 29.1 Cells by Immunofluorescence Assay for Surface Antigens (see Note 12)

The cellular antigenic pattern of both unstimulated and differentiated FLG 29.1 cells is analyzed with a standard immunofluorescence method using a FACScan cytofluorograph (Becton Dickinson and Immunocytometry Systems, San Jose, CA). Viable, antibody-labeled cells are identified according to their forward and right-angle scattering, and electronically gated and assayed for surface fluorescence.

1. FLG 29.1 cells are resuspended in BB (4×10^6 cells/mL) and preincubated for 1 h on ice.
2. Cells are then distributed (2×10^5 cells/50 µL) in test tubes and incubated with saturating concentration of the first MAb.
3. After 15–30 min of incubation on ice, cells are washed twice with WB and analyzed in flow cytometry (direct immunofluorescence assay).
4. For indirect immunofluorescence assay, the cells are incubated for an additional 15–30 min with a fluoresceine isothiocyanate (FITC)-conjugated F(ab')$_2$ fragments of goat antimouse immunoglobulins (H + L) (Technogenetics, Milan, Italy), washed twice in WB, and analyzed in flow cytometry.
5. Nonspecific binding of MAbs is assessed by labeling cells with unconjugated or phycoerythrinated and fluoresceinated isotype-matched control mouse immunoglobulins (Coulter). Antibody staining is usually considered positive if more then 20% of cells exhibit fluorescence intensity greater than that of 95% of cells stained with negative control antibodies (*see* Note 13).

3.8. Detection of Intracytoplasmic Antigens in FLG 29.1 Cells

For the detection of intracytoplasmic antigens (typically CD68), the procedure is as follows:

1. FLG 29.1 cells (1×10^6 cells/mL) are fixed and permeabilized in FPB for 10 min at room temperature.
2. Cells are then washed in WB, centrifuged (400*g*, 5 min), and incubated for 15–30 min on ice with a saturating concentration of the first MAb.
3. After two additional washing in WB, cells are analyzed in flow cytometry (direct immunofluorescence), or incubated with second fluorochrome-labeled F(ab')$_2$ fragments of goat antimouse immunoglobulins (indirect immunofluorescence), washed twice and analyzed in flow cytometry.

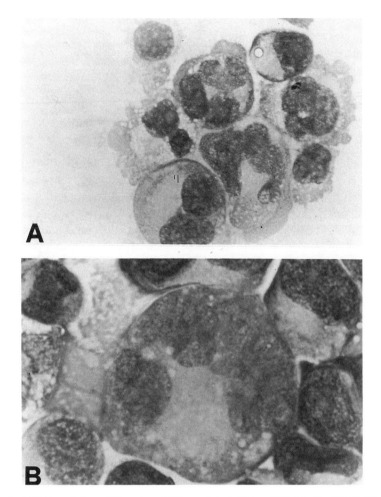

Fig. 2. MGG staining of in vitro differentiated (0.1 μM TPA for 72 h) FLG 29.1 OC-like cells. Original magnifications: 600x **(A)**, 1000x **(B)**.

3.9. Resorption of Devitalized Bone Particles by FLG 29.1 Cells

Resorption of devitalized bone by FLG 29.1 is quantified using particles of ^{45}Ca-labeled bone.

1. CBA T6T6 mice (4–8 wk old) receive ip injections of ^{45}CaCl$_2$ (160 μCi/injection dissolved in 100 μL of normal saline solution) twice weekly for 2 mo.
2. Mice are then sacrificed by cervical dislocation, and long bones are removed, cleaned of adherent tissues, fragmented to expose bone marrow, and soaked overnight in a solution of 1M NaOH.

3. Bone fragments are subsequently washed twice in PBS, sonicated, and ground to a coarse powder (45–50 μm diameter) in a mortar.
4. To evaluate the specific activity of ^{45}Ca-labeled bone particles, a 10-mg sample is dissolved in 5M HCl and counted with a liquid scintillation analyzer (Tri-carb 1600 TR, Canberra Packard). A specific activity (SA) of at least 8.0×10^5 cpm/10 mg bone is requested for this assay.
5. Ten-milligram aliquots of ^{45}Ca-labeled bone, sterilized by autoclaving, are then added to FLG 29.1 cells cultured in 16-mm wells (10^6 cells/well) in RPMI-1640 medium containing 10% FCS in the presence or in the absence of 0.1 μM TPA.
6. After 72 h of incubation, the upper one-half of the medium is collected, centrifuged at 15,000g, and assayed for radiocalcium.
7. The pellet along with the remaining half of the incubation medium is solubilized in 10% TCA for 24 h and assayed for radiumcalcium, as in step 6.
8. Net ^{45}Ca release is calculated from the isotope released into the medium expressed as percent of total after correction for the spontaneous loss of isotope from the bone particles owing to simple exchange with culture media (*see* Note 14).
9. To account for bone resorption owing to release of lytic enzymes from cultured cells, similar experiments are also carried out in a cell-free system, by incubating groups of bone particles with conditioned media obtained from FLG 29.1 cells exposed to different stimuli (*see* Note 15).
10. This procedure is run in parallel by assessing negative control cell lines (*see* Note 14).

3.10. CT-Binding Studies

1. FLG 29.1 cells differentiated with 0.1 μM TPA for 72 h are transferred to test tubes (10^6 cells/tube) and incubated for 2 h with increasing concentrations (0.1–1.0 nM) of ^{125}I-labeled salmon CT with or without unlabeled salmon CT (0.1 mM) at 22°C in BB.
2. After incubation, samples are washed in WB and centrifuged (400g for 15 min).
3. Cell pellets are counted in a γ-counter, and Scatchard analysis of binding data is performed using the computer program "Ligand" (*see* Note 16).

3.11. Detection of CT Receptor mRNA by RT-PCR

1. Total RNA (1 μg) from both untreated and TPA-treated FLG 29.1 cells is extracted and reverse-transcribed as described in Section 3.15.
2. Five microliters of cDNA are amplified in a 50-μL vol of final reaction mix (*see* Section 3.15.) using a Perkin Elmer (Norwalk, CT) thermal cycler (mod. 9600), with 25 pmol of primer pairs specific for the CT receptor. The conditions for PCR reaction are: 4 min at 94°C followed by 35 cycles of 45 s at 94°C, 45 s at 53°C, 1 min at 72°C, and a final extension of 10 min at 72°C.
3. Fifteen microliters of amplified cDNA are run in a 1.5% agarose gel, stained with ethidium bromide, and observed under UV light to detect a 771-bp amplified band specific for CT receptor cDNA (*see* Note 17).

3.12. Intracellular cAMP Accumulation by Salmon CT

1. Quadruplicate samples of untreated or TPA-treated FLG 29.1 cells are incubated with increasing concentration (1 pM to 0.1 µM) of salmon CT for 30 min.
2. Cells are collected, washed with cold HBSS, and covered with absolute ethanol overnight at –20°C.
3. After removing of ethanol under vacuum, samples (2 × 10^5 cells/sample) are reconstituted with assay buffer.
4. Measurement of cAMP content is performed with a standard EIA system by sequentially pipeting into wells precoated with donkey antirabbit antibodies, 100–200 µL aliquots of the following reagents: assay buffer, known concentrations of cAMP (into the "standard" wells; concentrations generally ranging from 12.5–3200 fmol) or of unknown samples (into the "sample" wells), anti-cAMP antisera.
5. Wells are then incubated at 4°C for 2 h, supplemented with 50 µL of peroxidase-conjugated cAMP, and incubated again for 1 h at 4°C.
6. After washing (four times with 400 µL washing buffer), colorimetric reaction is finally developed by adding 150 µL/well of substrate followed by a final 1 h of incubation at room temperature.
7. After stopping the reaction with 100 µL of 1.0M sulfuric acid, wells are read in a plate ELISA reader at 450 nm within 30 min (*see* Note 19).

3.13. Preparation of Saos-2 and BMF Monolayers

1. Saos-2 cells (obtained from the American Tissue Culture Collection, Rockville, MD), growing as adherent monolayer, are cultured in McCoy's/10% FCS.
2. Heparinized bone marrow aspirates (2–3 mL) obtained after informed consent from solid tumor patients undergoing staging procedures are usually utilized as a source of BMF.
3. To establish BMF-enriched cultures, mononuclear bone marrow cells are separated by centrifugation (400g for 30 min) through Ficoll-Hypaque (Pharmacia), and seeded onto 24-well plates containing IMDM/10% FCS.
4. After an overnight incubation (37°C in a CO_2 gassed incubator), nonadherent cells are removed, cultures are maintained in the same medium supplemented with 0.1 ng/mL bFGF, replenished with fresh medium every 3 d, and split when confluent.
5. After four to five passages, a virtually pure population of BMF is usually obtained and utilized for further experiments. The absence of contaminating endothelial cells and macrophages may be verified by immunostaining for von Willebrand's factor, nonspecific esterase, and CD14.

3.14. Coculture Conditions of Saos-2 and BMF with FLG 29.1 OC-Like Cells

1. FLG 29.1 OC-like cells (5×10^4 cells/mL) are seeded in 24-well plates coated with a subconfluent monolayer of Saos-2 cells or BMF in IMDM supplemented with 20% FCS.
2. After 72 h, FLG 29.1 cells are carefully collected by repeated washing of wells with culture medium, and RNA from each cell fraction (e.g., aspirated cells [FLG 29.1] and the adherent layer [Saos-2 or BMF]) is separately extracted and analyzed in RT-PCR for cytokine/cytokine receptor expression.

3.15. Changes in Gene Expression During Coculture of FLG 29.1 Cells with BMF or Saos-2 Cells: Detection by RT-PCR (see Note 20)

Total RNA is isolated from BMF, FLG 29.1, and Saos-2 cell lines utilizing an acid guanidium thiocyanate-phenol-chloroform single-step method *(56)*. The amount and integrity of RNAs extracted are checked by electrophoresis in a standard formaldehyde-agarose gel.

1. One microgram of total RNA is denaturated (5 min at 95°C) in a 10-µL vol with 0.5 µg of hexadeoxyribonucleotide random primers and rapidly chilled on ice.
2. 20 U RNasin, 2 µL 10X PCR reaction buffer, 5 µL 25 m*M* MgCl$_2$, 1 µL dNTPs, and 1 µL DTT are sequentially added to the samples and incubated for 10 min at 23°C.
3. Samples are then reverse transcribed by AMV RT (8–10 U) for 1 h at 42°C and denaturated for 5 min at 95°C.
4. Five microliters of cDNA bulk are amplified in a 50-µL vol of final reaction mix using a Perkin Elmer thermal cycler (mod. 9600), with 25 pmol of specific primers pairs (Table 2), 1 µL 25 m*M* MgCl$_2$, and 2.5 U *Taq* polymerase. The conditions for PCR reaction are 4 min at 94°C followed by 35 cycles of 45 s at 94°C, 45 s at 60°C (55°C for GM-CSFR and 57°C for SCF), 90 s at 72°C, and a final extension of 7 min at 72°C (*see* Note 21).
5. Ten microliters of amplified cDNA are run in a 1.5–2.0% agarose gel, and the amplified bands are visualized by UV transilluminator (*see* Notes 22 and 23).

3.16. Conclusion

Studies on OC development and functional regulation have been hampered by the lack of suitable in vitro cellular systems. On the other hand, the improving knowledge of the critical role exerted by OC populations in a number of human pathologic conditions has underscored the need for appropriate study models.

The described developments in the primary culture of osteoprogenitor cells or of their differentiated progeny, along with the improvement of isolation and purification procedures of OCs from adult fetal bone, have deeply modified the current knowledge of the regulatory processes underlying the functional crosstalk among specialized bone cell subpopulations. In addition, the establishment and characterization of a human cell line (FLG 29.1) of human OC progenitors has further widened the potential of studying OC functions in vitro. The application of such models will certainly improve our knowledge of the cellular basis of several human diseases. Owing to the emerging role of OCs in mediating the pathologic circuitries of abnormal bone resorption and remodeling in osteoporosis as well as in the metastatic bone, it is probable that information gained by exploiting the aforementioned in vitro models will rapidly improve the clinical management of such conditions. The unveiling of the complex network of interactions between some types of tumor cells and residing bone OCs, will also represent an important step for the understanding of several neoplastic diseases, such as multiple myeloma. Finally, the possibility of analyzing the crosstalk of primary tumor cells or tumor cell lines and OC cell lines, such as FLG 29.1, may in turn lead to the development of innovative therapeutic strategies for human malignant diseases.

4. Notes

1. OC-like GCs from human GCT of bone can be prepared in large numbers by curettage of small tumor pieces in a similar way. It is notable that, although metaphyseal bone specimens must always be employed when very fresh, these GCs from GCT-bone can retain their viability even after many hours of storage in normal saline at 4°C.
2. This is according to Gatter et al. *(57)*.
3. We have previously employed sections of paraffin-embedded decalcified bone for IHC studies of OCs *(58)*. These have the advantages of good morphological preservation, and a histopathology department would have access to large numbers of archive specimens.
4. A range of MAbs exist that are able to immunostain paraffin-embedded archive material, although they often require prior enzymatic digestion of the sections *(57)*. Many of these antibodies are also capable of immunostaining tissues decalcified with EDTA, 5% TCA, and sometimes even stronger agents, such as 10% HCl *(57)*. However, the great majority of antibodies are able to immunostain only frozen sections fixed in cold (–20°C) acetone for 10 min.
5. Kits for making these enzyme substrates are commercially available (e.g., from Sigma).
6. Although cortical bone slices prepared as described do not contain bone resorption pits unless they are cultured with OCs, their surfaces are not completely featureless, since the diamond saw cutting exposes osteocyte lacunae and blood vessels. However, these are not difficult to distinguish from bone resorption pits.
7. If preferred, a weak detergent solution (1% Triton X-100) can be used to remove the cells from the cortical bone slice, but this is less effective in removing contaminating stromal cells, such as fibroblasts. Hypochlorite can also be used, and although this is quicker (10 min immersion time), it also renders the surface of

the cortical bone anorganic, removing the exposed collagen fibrils, and thus making the resorption pits harder to distinguish from exposed osteocyte lacunae and blood vessels.
8. If it is preferred to visualize the OCs on the surface of the cortical bone (rather than just the resorption pits), then the bone slices can be fixed in EM-grade glutaraldehyde (4%), dehydrated though graded alcohols, and critical point dried prior to sputtering.
9. The bone resorption can be quantified by several methods. Measuring pit volume is the ideal, but requires specialist equipment and a lot of time. Bone surface area resorbed can be estimated by point counting or image processing methods by SEM or reflected light microscopy.
10. Optimal FLG 29.1 cell differentiation (Fig. 2) is obtained by adding 0.1 μM TPA to the cell suspension for 72 h. No differences are usually observed between FLG 29.1 cells treated with 0.1 μM TPA alone or with 0.1 μM TPA and 10 nM 1,25(OH)$_2$D$_3$. Conversely, only a suboptimal differentiation (i.e., increase of multinucleated and TRAP-positive cells without any evidence of immunophenotypic changes) is obtained on treatment of FLG 29.1 cells with 10 nM 1,25 (OH)$_2$ D$_3$.
11. Unstimulated FLG 29.1 cells stained with MGG appear like undifferentiated blasts, with cell diameter ranging from 15–20 µm and <3% of multinucleated cells. Acid phosphatase and sodium fluoride inhibitable α-naphthyl acetate esterase are present in 100% of the cells with about 10% of TRAP-positive cells. After 72 h of TPA treatment, more than 40% of the cells become bi-, tri-, or multinucleated with size ranging from 50–100 mm (Fig. 2). In addition, acid phosphatase and sodium fluoride inhibitable α-naphthyl acetate esterase appear increased, along with the percent of TRAP-positive cells (to about 40–50%).
12. FLG 29.1 cells display a significant autofluorescence when assayed by flow cytometry. It is therefore important to optimize MAb-binding procedures to reduce maximally nonspecific binding of first- and second-layer antibodies. This can be achieved by the use of a binding solution containing human immunoglobulins (see Section 9.1.).
13. The phenotypic profile of FLG 29.1 cells before and after TPA exposure, along with its relationship with normal fetal bone OCs is reported in Table 1.
14. By utilizing such assay, TPA-induced FLG 29.1 cells display a bone-resorbing activity at least three- to fourfold higher then unstimulated FLG 29.1 cells or TPA-stimulated myeloid cells introduced in the assay as negative controls. We routinely use myeloid leukemic cell lines (e.g., HL60 or U937), which although differentiating into monocyte-macrophages on exposure to TPA, do not show bone resorbing activity (8).
15. Experiments with cell supernatants demonstrate a role of both cell-bone contact and soluble factor(s) released into the culture medium in mediating FLG 29.1-induced resorption of devitalized bone particles.
16. Differentiated FLG 29.1 cells exhibit both high- (kDa about 1.0 nM) and low- (kDa about 6.6 µM) affinity receptors, whereas no binding sites can be documented on undifferentiated FLG 29.1 cells.

17. A strong 771-bp amplified band specific for CT receptor mRNA is detected in TPA-treated FLG 29.1 cells.
18. The assay is based on the competition between unlabeled cAMP and a fixed quantity of peroxidase-labeled cAMP for a limited number of binding sites on a anti-cAMP-specific antibody. The amount of peroxidase-labeled ligand bound by the antibody will be inversely proportional to the concentration of added unlabeled ligand.
19. Although no significant increase in cAMP content is observed in unstimulated FLG 29.1 cells incubated with up to 0.1 μM of salmon CT, a dose-dependent increase of cAMP content is observed in TPA-treated cells, reaching a maximum at 1 nM salmon CT.
20. Such a strategy is based on the RT-PCR analysis of genes differentially expressed by each cell type following cell contact-dependent interaction in the coculture system. Genes known to be selectively expressed by a single cell type and not modulated following coculture are used as a tool to check for the purity of each cell population. Since FLG 29.1 cells strongly express IGF-1, but not FGF-R or exon 6$^+$SCF mRNAs (for references, see Table 2 and the Appendix), which are conversely expressed by Saos-2 and BMF, respectively, such primer pairs are utilized as tools for ruling out a possible crosscontamination of cultures. Sequences, cDNA positions, and amplified fragment lengths of such primers, along with other primers utilized in our model, are listed in Table 2. With the exclusion of primer sets specific for GM-CSF-R and SCF, all primers can be obtained from Clontech Laboratories (Palo Alto, CA). References of cDNA sequences utilized for primers selection are listed in the Appendix.
21. For each set of primers, internal reaction standards for PCR control are performed, which include primers without cDNA (negative control) and primers with target DNA provided by the manufacturer (positive control). In addition, each cDNA bulk is analyzed with a β-actin primer pair (Clontech Laboratories) as a check for first-strand synthesis.
22. Expected sizes of amplified cDNA bands specific for each of the analyzed genes are listed in Table 2.
23. By using such an approach, the expression of a number of cytokines or cytokine receptor is differentially modulated in FLG 29.1 or Saos-2 cells following their direct cell–cell interaction in our coculture system. In particular, IL-6-specific mRNA is upregulated both in FLG 29.1 pre-OCs and Saos-2 cells, whereas TNF-α, TNF-Rp55, PDGF-A, and IL-1α mRNAs are strongly induced in FLG 29.1 cells cocultured with Saos-2 cells for 72 h.

Acknowledgments

This work is supported by Associazione Italiana per la Ricerca sul Cancro (AIRC), Milan, Italy; Consiglio Nazionale per le Ricerche (CNR) Progetto Finalizzato-Applicazioni Cliniche della Ricerca Oncologica, grant no. 92.02347.PF39, Rome, Italy; and Ministero della Sanita', Ricerca Finalizzata IRCCS, Rome, Italy.

References

1. Scheven, B. A. A., Visser, J. W. M., and Nijweide, P. J. (1986) In vitro osteoclast generation from different bone marrow fractions, including a high enriched hematopoietic stem cell population. *Nature* **321,** 79–81.
2. Kurihara, N., Suda, T., Miura, Y., Nakauchi, H., Kodama, H., Hiura, K., Hakeda, Y., and Kumegawa, M. (1989) Generation of osteoclasts from isolated hematopoietic progenitor cells. *Blood* **74,** 1295–1302.
3. Manolagas, S. C. and Jilka, R. L. (1992) Cytokines, hematopoiesis, osteoclastogenesis, and estrogens. *Calcif. Tissue Int.* **50,** 199–202.
4. Kurihara, N., Civin, C., and Roodman, G. D. (1991) Osteotropic factor responsiveness of highly purified populations of early and late precursors for human multinucleated cells expressing the osteoclast phenotype. *J. Bone Miner. Res.* **6,** 257–261.
5. Hattersley, G., Kerby, J. A., and Chambers, T. J. (1991) Identification of osteoclast precursors in multilineage hemopoietic colonies. *Endocrinology* **128,** 259–262.
6. Yoneda, T., Alsina, M. M., Garcia, J. L., and Mundy, G. R. (1991) Differentiation of HL-60 cells into cells with the osteoclast phenotype. *Endocrinology* **129,** 683–689.
7. Hewison, M., Barker, S., Brennan, A., Katz, D. R., and O'Riordan, J. L. H. (1989) Modulation of myelomonocytic U937 cells by vitamin D metabolites. *Bone and Mineral* **5,** 323–333.
8. Gattei, V., Bernabei, P. A., Pinto, A., Bezzini, R., Ringressi. A., Formigli, L., Tanini, A., Attadia, V., and Brandi, M. L. (1992) Phorbol ester induced osteoclast-like differentiation of a novel human leukemic cell line (FLG 29. 1). *J. Cell Biol.* **116,** 437–447.
9. Warner, S. P. (1964) Hydrolytic enzymes in osteoclasts cultured in vitro. *J. Roy. Microscop. (Soc. Series 111)* **83,** 397–403.
10. Osdoby, P., Martini, M. C., and Kaplan, A. I. (1982) Isolated osteoclasts and their presumed progenitor cell, the monocyte, in culture. *J. Exp. Zool.* **224,** 331–344.
11. Mears, D. C. (1971) Effects of parathyroid hormone and thyrocalcitonin on the membrane potential of osteoclasts. *Endocrinology* **88,** 1021–1028.
12. Nelson, R. L. and Bauer, G. E. (1977) Isolation of osteoclasts by velocity sedimentation at unit gravity. *Calcif. Tissue Res.* **22,** 303–313.
13. Hefley, T. J. and Stern, P. H. (1982) Isolation of osteoclasts from fetal rat long bones. *Calcif. Tissue Int.* **34,** 480–487.
14. Chambers, T. J., Athanasou, N. A., and Fuller, K. (1984) Effect of parathyroid hormone and calcitonin on the cytoplasmic spreading of isolated osteoclasts. *J. Endocrinol.* **102,** 303–313.
15. Zambonin-Zallone, A., Teti, A., and Primavera, M. V. (1984) Monocytes from circulating blood fuse in vitro with purified osteoclasts in primary culture. *J. Cell Sci.* **66,** 335–342.
16. Glowacki, J., Jasty, M., and Goldring, S. (1986) Comparison of multinucleated cells elicited in rats by particulate bone, polyethylene or polymethylmethacrylate. *J. Bone Miner. Res.* **1,** 327–331.
17. Walker, D. G. (1972) Congenital osteopetrosis in mice cured by parabiotic union with normal siblings. *Endocrinology* **91,** 916–920.

18. Wong, G. L. and Cohn, D. V. (1974) Separation of parathyroid hormone and calcitonin-sensitive cells from non-responsive bone cells. *Nature* **252**, 713–715.
19. Wong, G. L. and Cohn, D. V. (1975) Target cells in bone for parathormone and calcitonin are different: enrichment for each cell type by sequential digestion of mouse calvaria and selective adhesion to polymeric surfaces. *Proc. Natl. Acad. Sci. USA* **72**, 3167–3171.
20. Luben, R. A., Wong, G. L., and Cohn, D. V. (1976) Biochemical characterisation with parathormone and calcitonin and isolated bone cells. Provisional identification of osteoclasts and osteoblasts. *Endocrinology* **99**, 526–534.
21. Peck, W. A., Burks, J. K., Wilkins, J., Rodan, S. B., and Rodan, G. A. (1977) Evidence for preferential effects of parathyroid hormone, calcitonin and adenosine on bone and periosteum. *Endocrinology* **100**, 1357–1364.
22. Athanasou, N. A., Heryet, A., Quinn, J., Gatter, K. C., Mason, D. Y., and McGee, J. O.'D. (1986) Osteoclasts contain macrophage and megakaryocyte antigens. *J. Pathol.* **150**, 234–246.
23. Murrills, R. J., Shane, E., Lindsay, R., and Dempster, D. W. (1989) Bone resorption by isolated human osteoclasts in vitro: effects of calcitonin. *J. Bone Miner. Res.* **4**, 251–268.
24. Pensler, J. M., Radosevich, J. A., Higbee, R., and Langman, C. B. (1990) Osteoclasts isolated from membranous bone in children exhibit nuclear estrogen and progesterone receptors. *J. Bone Miner. Res.* **5**, 797–802.
25. Chambers, T. J., Fuller, K., and McSheehy, P. M. J. (1985) The effect of calcium regulating hormones on bone resorption by isolated human osteoclastoma cells. *J. Pathol.* **145**, 297–305.
26. Athanasou, N. A., Bliss, E., Gatter, K., Heryet, A., Woods, C. G., and McGee, J. O.'D. (1985) An immunohistochemical study of giant cell tumour of bone: evidence for an osteoclast origin of the giant cells. *J. Pathol.* **147**, 153–158.
27. Gothlin, G. and Ericsson, J. L. E. (1976) The osteoclast. *Clin. Orthop.* **120**, 201–231.
28. Kashara, K., Yamamuro, T., and Kashara, A. (1979) Giant-cell tumour of bone: cytological studies. *Br. J. Cancer* **40**, 201–209.
29. Aparisi, T., Arbough, B., and Ericsson, J. (1979) Giant cell tumour of bone—variations in the pattern of appearance of different cell type. *Virchows Arch A. Pathol. Anat. Histol.* **382**, 159–178.
30. Athanasou, N. A., Pringle, J. A. S., Revell, P. A., and Chambers, T. J. (1983) Resorption of bone by human osteoclastoma cells. *J. Pathol.* **141**, 508–511.
31. Horton, M. A. and Chambers, T. J. (1985) Human osteoclast specific antigens are expressed by osteoclasts in a wide range of non-human species. *Br. J. Exp. Pathol.* **67**, 97–104.
32. Athanasou, N. A., Quinn, J., and McGee, J. O.'D. (1988) Immunocytochemical analysis of the human osteoclast: phenotypic relationship to other marrow derived cells. *Bone Miner.* **3**, 317–333.
33. Quinn, J., Joyner, C., Triffitt, J. T., and Athanasou, N. A. (1992) Polymethyl methacrylate-induced inflammatory macrophages resorb bone *J. Bone Joint Surg. (B)* **74**, 652–658.

34. Quinn, J., Puddle, B., West, L., McGee, J. O.'D., and Athanasou, N. A. (1995) Myeloid antigens on osteoclasts and macrophage polykaryons, in *Proceedings of the 5th International Workshop on Human Leukocyte Antigens* (Schlossmann, S., Boumsell, L., Gilks, W., Harlan, J., Kishimoto, T., Morimoto, C., Ritz, J., Shaw, S., Silverstein, R., Springer, T., Tedder, T, and Todd, R., eds.), Oxford University Press, Oxford, UK, pp. 1012,1013.
35. Quinn, J. M. W. and Athanasou, N. A. (1992) Tumour infiltrating macrophages are capable of bone resorption. *J. Cell Sci.* **101,** 681–686.
36. Chambers, T. J., Revell, P. A., Fuller, K., and Athanasou, N. A. (1984) Resorption of bone by isolated rabbit osteoclasts. *J. Cell Sci.* **66,** 383–399.
37. Jones, S. J., Ali, N. N., and Boyde, A. (1986) Survival and resorptive activity of chick osteoclasts in culture. *Anat. Embryol. (Berl.)* **174,** 265–275.
38. Takahashi, N., Yamana, H., Yoshiki, S., Roodman, G. D., Mundy, G., and Jones, S. J. (1988) Osteoclast-like cell formation and its regulation by osteotropic hormones in mouse marrow cultures. *Endocrinology* **122,** 1373–1382.
39. Udagawa, N., Takahashi, N., Akatsu, T., Sasaki, T., Yamaguchi, A., Kodama, H., Martin, T. J., and Suda, T. (1989) The bone marrow-derived stromal cell line MC3T3-G2/PA6 and ST2 support osteoclast-like cell differentiation in cocultures with mouse spleen cells. *Endocrinology* **125,** 1805–1813.
40. Yamashita, T., Asano, K., Takahashi, N., Akatsu, T., Udagawa, N., Sasaki, T., et al. (1990) Cloning of an osteoblastic cell line involved in the formation of osteoclast-like cells. *J. Cell Physiol.* **145,** 587–595.
41. Udagawa, N., Takahashi, N., Akatsu, T., Tanaka, H., Sasaki, T., Nishihara, T., Koga, T., Martin, J., and Suda, T. (1990) Origin of osteoclasts: mature monocytes and macrophages are capable of differentiation into osteoclasts under a suitable microenvironment prepared by bone marrow-derived stromal cells. *Proc. Natl. Acad. Sci. USA* **87,** 7260–7264.
42. Hattersley, G. and Chambers, T. J. (1989) Calcitonin receptors as a marker for osteoclasic differentiation: correlation between generation of bone-resorptive cells and cells that express calcitonin receptors in mouse bone marrow cultures. *Endocrinology* **125,** 1606–1612.
43. Mundy, G. R. (1992) Cytokines and local factors which affect osteoclast function. *Int. J. Cell Cloning* **10,** 215–222.
44. Takahashi, N., Akatsu, T., Udagawa, N., Sasaki, T., Yamaguchi, A., Moseley, J. M., Martin, T. J., and Suda, T. (1988) Osteoblastic cells are involved in osteoclast formation. *Endocrinology* **123,** 2600–2602.
45. McSheehy, P. M. and Chambers, T. J. (1986) Osteoblastic cells mediate osteoclastic responsiveness to parathyroid hormone. *Endocrinology* **118,** 824–828.
46. McSheehy, P. M. and Chambers, T. J. (1987) 1,25-dihydroxyvitamin D3 stimulates rat osteoblastic cells to release a soluble factor that increases osteoclastic bone resorption. *J. Clin. Invest.* **80,** 425–429.
47. Hattersley, G. and Chambers, T. J. (1989) Generation of osteoclasts from hemopoietic cells and a multipotential cell line in vitro. *J. Cell Physiol.* **140,** 478–485.

48. Jilka, R. L., Hangoc, G., Girasole, G., Passeri, G., Williams, D. C., Abrams, J. S., Boyce, B., Broxmeyer, H., and Manolagas, S. C. (1992) Increased osteoclast development after estrogen loss: mediation by interleukin-6. *Science* **257,** 88–91.
49. Lewis, D. B., Liggitt, D. H., Effmann, E. L., Motley, S. T., Teitelbaum, S. L., Jepsen, K. J., Goldstein, S. A., Bonadio, J., Carpenter, J., and Perlmutter, R. M. (1993) Osteoporosis induced in mice by overproduction of interleukin 4. *Proc. Natl. Acad. Sci. USA* **90,** 11,618–11,622.
50. Van Vlasselaer, P., Borremans, B., Van Den Huevel, R., Van Gorp, U., and de Waal Malefyt, R. (1993) Interleukin-10 inhibits the osteogenic activity of mouse bone marrow. *Blood* **82,** 2361–2370.
51. Girasole, G., Passeri, G., Jilka, R. L., and Manolagas, S. C. (1994) Interleukin-11: a new cytokine critical for osteoclast development. *J. Clin. Invest.* **93,** 1516–1524.
52. Rodan, S. B., Imai, Y., Thiede, M. A., Wesolowski, G., Thompson, D., Bar-Shavit, Z., Shull, S., Mann, K., and Rodan, G. A. (1987) Characterization of a human osteosarcoma cell line (Saos-2) with osteoblastic properties. *Cancer Res.* **47,** 4961–4966.
53. Knapp, W., Dorken, B., Gilks, W. R., Rieber, E. P., Schmidt, R. E., Stein, H., and von dem Borne, A. E. G. K. (eds.) (1989) *Leukocyte Typing IV. White Cell Differentiation Antigens.* Oxford University Press, Oxford, UK.
54. Schlossmann, S., Boumsell, L., Gilks, W., Harlan, J., Kishimoto, T., Morimoto, C., Ritz, J., Shaw, S., Silverstein, R., Springer, T., Tedder, T, and Todd, R. (eds.) (1995) *Proceedings of the 5th International Workshop on Human Leukocyte Antigens.* Oxford University Press, Oxford, UK.
55. Lin, H. Y., Harris, T. L., Flannery, M. S., Aruffo, A., Kaji, E. H., Gorn, A., Kolakowski, L. F., Lodish, H. F., and Goldring, S. R. (1991) Expression cloning of an adenylate cyclase-coupled calcitonin receptor. *Science* **254,** 1022–1026.
56. Chomczynski, P. and Sacchi, N. (1987) Single step method of RNA isolation by acid guanidinium thiocyanate-phenol-chloroform extraction. *Anal. Biochem.* **162,** 156–159.
57. Gatter, K. C., Falini, B., and Mason, D. Y. (1984) The use of monoclonal antibodies in histological diagnosis, in *Recent Advances in Histopathology, No 12.* (Anthony, P. P. and MacSween, R. N. M., eds.), Churchill Livingstone, Edinburgh, pp. 35–67.
58. Athanasou, N. A., Puddle, B., Quinn, J., and Woods, C. G. (1991) Use of monoclonal antibodies to recognise osteoclasts in routinely processed bone biopsies. *J. Clin. Pathol.* **44,** 664–666.

Appendix—References of cDNA Sequences Utilized for Primer Selection

1. Kawasaki, E. S., Ladner, M. B., Wang, A. M., Van Arsdell, J., Warren, M. K., Coyne, M. Y., Schweickart, V. L., Lee, M.-T., Wilson, K. J., Boosman, A., Stanley, R., Ralph, P., and Mark, D. F. (1985) Molecular cloning of a comple-

mentary DNA encoding human macrophage-specific colony-stimulating factor (CSF-1). *Science* **230,** 291–296.
2. Coussens, L., Van Beveren, C., Smith, D., Chen, E., Mitchell, R. L., Isacke, C. M., Verma, I. M., and Ullrich, A. (1986) Structural alteration of viral homologue of receptor proto-oncogene *fms* at carboxyl terminus. *Nature* **320,** 277–280.
3. Nagata, S., Tsuchiya, M., Asano, S., Yamamoto, O., Hirata, Y., Kubota, Y., Oheda, M., Nomura, H., and Yamazaki, T. (1986) The chromosomal gene structure and two mRNAS for human granulocyte colony-stimulating factor. *EMBO J.* **5,** 575–581.
4. Fukunaga, R., Seto, Y., Mizushima, S., and Nagata, S. (1990) Three different mRNAs encoding human granulocyte colony-stimulating factor receptor. *Proc. Natl. Acad. Sci. USA* **87,** 8702–8706.
5. Miyatake, S., Otsuka, T., Yokota, T., Lee, F., and Arai, K. (1985) Structure of the chromosomal gene for granulocyte-macrophage colony stimulating factor: comparison of the mouse and human genes. *EMBO J.* **4,** 2561–2568.
6. Gearing, D. P., King, J. A., Gough, N. M., and Nicola, N. A. (1989) Expression cloning of a receptor for human granulocyte-macrophage colony-stimulating factor. *EMBO J.* **8,** 3667–3676.
7. Steenbergh, P. H., Koonen-Reemst, A. M., Cleutjens, C. B., and Suusenbach, J. S. (1991) Complete nucleotide sequence of the high molecular weight human *IGF-I* mRNA. *Biochem. Biophys. Res. Commun.* **175,** 507–514.
8. Ullrich, A., Gray, A., Tam, A. W., Yang-Feng, T., Tsubokawa, M., Collins, C., Henzel, W., Le Bon, T., Kathuria, S., Chen, E., Jacobs, S., Francke, U., Ramachandran, J., and Fujita-Yamaguchi, Y. (1986) Insulin-like growth factor I receptor primary structure: comparison with insulin receptor suggests structural determinants that define functional specificity. *EMBO J.* **5,** 2503–2512.
9. Rall, L. B., Scott, J., and Bell, G. I. (1987) Human insulin-like growth factor I and II messenger RNA: isolation of complementary DNA and analysis of expression. *Methods Enzymol.* **146,** 239–248.
10. Morgan, D. O., Edman, J. C., Standring, D. N., Fried, V. A., Smith, M. C., Roth, R. A., and Rutter, W. J. (1987) Insulin-like growth factor II receptor as a multifunctional binding protein. *Nature* **329,** 301–307.
11. Furutani, Y., Notake, M., Yamayochi, M., Yamagishi, J., Nomura, H., Ohue, M., Furuta, R., Fukui, T., Yamada, M., and Nakamura, S. (1985) Cloning and characterization of the cDNAs for human and rabbit interleukin-1 precursor. *Nucleic Acids Res.* **13,** 5869–5882.
12. Sims, J. E., Acres, B., Grubin, C. E., McMahan, C. J., Wignall, J. M., March, C. J., and Dower, S. K. (1989) Cloning the interleukin 1 receptor from human T cells. *Proc. Natl. Acad. Sci. USA* **86,** 8946–8950.
13. Pennica, D., Nedwin, G. E., Hayflick, J. S., Seeburg, P. H., Derynck, R., Palladino, M. A., Kohr, W. J., Aggarwal, B. B., and Goeddel, D. V. (1984) Human tumour necrosis factor: precursor structure, expression and homology to lymphotoxin. *Nature* **312,** 724–728.
14. Nedwin, G. E., Naylor, S. L., Sakaguchi, A. Y., Smith, D., Jarrett-Nedwin, J., Pennica, D., Goeddel, D. V., and Gray, P. W. (1985) Human lymphotoxin and

tumor necrosis factor genes: structure, homology and chromosomal localization. *Nucleic Acid Res.* **13,** 6361–6373.
15. Schall, T. J., Lewis M., Koller, K. J., Lee, A., Rice, G. C., Wong, G. H. W., Gatanaga, T., Granger, G. A., Lentz, R., Raab, H., Kohr, W. J., and Goeddel, D. V. (1990) Molecular cloning and expression of a receptor for human tumor necrosis factor. *Cell* **61,** 361–370.
16. Smith, C. A., Davis, T., Anderson, D., Solam, L., Beckmann, M. P., Jerzy, R., Dower, S. K., Cosman, D., and Goodwin, R. G. (1990) A receptor for tumor necrosis factor defines an unusual family of cellular and viral proteins. *Science* **248,** 1019–1023.
17. Derunck, R., Jarrett, J. A., Chen, E. Y., Eaton, D. H., Bell, J. R., Assoian, R. K., Roberts, A. B., Sporn, M. B., and Goeddel, D. V. (1985) Human transforming growth factor-β complementary DNA sequence and expression in normal and transformed cells. *Nature* **316,** 701–705.
18. Madisen, L., Webb, N. R., Rose, T. M., Marquardt, H., Ikeda, T., Twardzik, D., Seyedin, S., and Purchio, A. F. (1988) Transforming growth factor-β 2: cDNA cloning and sequence analysis. *DNA* **7,** 1–8.
19. Hirnao, T., Yasukawa, K., Harada, H., Taga, T., Watanabe, Y., Matsuda, T., Kashiwamura, S., Nakajima, K., Koyama, K., Iwamatsu, A., Tsunasawa, S., Sakiyama, F., Matsui, H., Takahara, Y., Taniguchi, T., and Kishimoto, T. (1986) Complementary DNA for a novel human interleukin (BSF-2) that induces B lymphocytes to produce immunoglobulin. *Nature* **324,** 73–76.
20. Yamasaki, K., Taga, T., Hirata, Y., Yawata, H., Kawanishi, Y., Seed, B., Taniguchi, T., Hirano, T., and Kishimoto, T. (1988) Cloning and expression of the human interleukin-6 (BSF-2/IFNb 2) receptor. *Science* **241,** 825–828.
21. Hoppe, J., Schumacher, L., Eichner, W., and Weich, H. A. (1987) The long 3'-untranslated regions of the *PDGF-A* and *-B* mRNAs are only distantly related. *FEBS Lett.* **223,** 243–246.
22. Dionne, C. A., Crumley, G., Bellot, F., Kaplow, J. M., Searfoss, G., Ruta, M., Burgess, W. H., Jaye, M., and Schlessinger, J. (1990) Cloning and expression of two distinct high-affinity receptors cross-reacting with acidic and basic fibroblast growth factors. *EMBO J.* **9,** 2685–2692.
23. Martin, F. H., Suggs, S. V., Langley, K. E., Lu, H. S., Ting, J., Okino, K. H., Morris, C. F., McNiece, I. K., Jacobsen, F. W., Mendiaz, E. A., et al. (1990) Primary structure and functional expression of the rat and human stem cell factor DNAs. *Cell* **63,** 203–211.

23

Isolation, Purification, and Growth of Human Skeletal Muscle Cells

Grace K. Pavlath

1. Introduction

Skeletal muscle cells can be used in vitro for the study of myogenesis, as well as in vivo as gene-delivery vehicles for the therapy of muscle and nonmuscle diseases *(1–9)*. These skeletal muscle cells are derived from muscle satellite cells, which lie between the basal lamina and the sarcolemma of differentiated muscle fibers *(10)*. Normally quiescent after the period of muscle development and growth during fetal life and the early postnatal period, these cells are induced to proliferate on muscle damage and fuse with existing muscle fibers. Satellite cells isolated and grown in vitro are called myoblasts. Myoblasts proliferate in mitogen-rich media, but on reaching high cell density followed by exposure to mitogen-poor media, are induced to differentiate and become postmitotic. Muscle differentiation is characterized by the fusion of myoblasts to form multinucleated myotubes, which express differentiation-specific proteins. In this chapter, methods are given for the isolation of myoblasts from human muscle tissue using two different techniques: flow cytometry *(11)* and cell cloning *(12–14)*. These methods are applicable to muscle tissue from both fetal and postnatal donors, as well as from normal and diseased individuals.

2. Materials

All solutions are prepared using double-distilled water. All solutions and materials are sterile. Whenever possible, reagents that have been tissue-culture-tested by the manufacturer are recommended. Specific vendors are indicated for certain items. Alternate sources for these particular items have not been tested.

2.1. Cells/Tissue

1. MRC-5 (ATCC Accession Number CRL 171): These cells are used in preparation of the conditioned media required at the early stages of cell growth in vitro (*see* Note 1). The preparation of conditioned media is described in Section 3.1.
2. Muscle tissue:
   a. Biopsy material: Muscle biopsies should be 0.5–1.0 cm^3 (approx 0.5–1 g).
   b. Autopsy material: Collect muscle sample as soon as possible within the first 12–24 h after death.

2.2. Transport, Tissue Dissociation, Growth, and Freezing

1. Transport media: Ham's F10 + 50 µg/mL gentamicin (store at 4°C). Store the muscle sample in a plastic 50-mL centrifuge tube in transport media at 4°C. Under these conditions, the cells are fine up to 24 h. Fill the tube completely with transport media to prevent desiccation of the muscle sample.
2. Tissue dissociation: Ham's F10; mixture of 0.05% trypsin, 0.53 m*M* EDTA; calf serum.
3. Growth media:
   a. MRC-5 maintenance media (MM; store at 4°C, use within 2 wk). Fetal bovine serum (FBS) is directly added to DMEM (formulation containing 1000 mg/L glucose) to give a final concentration of 10%.
   b. Human muscle growth media (HuGM; store at 4°C, use within 2 wk) (*see* Note 2): The following are added to Ham's F10 to give the indicated final concentrations: 10% FBS; 5% bovine calf serum, defined, supplemented with iron (CS) (#A-2151, Hyclone, Logan, UT); 0.5% chick embryo extract (CEE) (*see* Note 3) (#15115-017, Gibco, Grand Island, NY); 100 U/mL penicillin (store frozen at –20°C); 100 µg/mL streptomycin (store frozen at –20°C).
   c. Human muscle fusion medium (FM): The following are added to DMEM (formulation containing 1000 mg/L glucose) to give the indicated final concentration: 2% horse serum; 100 U/mL penicillin (store frozen at –20°C); 100 µg/mL streptomycin (store frozen at –20°C).
4. Freezing media: 90% calf serum, 10% dimethyl sulfoxide (DMSO).

2.3. Flow Cytometry

1. 5.1H11 monoclonal antibody (MAb) against muscle-specific isoform of neural cell adhesion molecule (NCAM) *(15)* (*see* Note 4). Use hybridoma supernatant neat or purified antibody at 1 µg/10^6 cells.
2. Biotinylated antimouse IgG (#BA 2000, Vector Labs, Burlingame, CA). Use sterile water to reconstitute the contents of the vial. Use at 7 µg/mL in PBS + 0.5% BSA for immunostaining.
3. Streptavidin conjugated to either FITC (#43-4311, Zymed, South San Francisco, CA) or Texas Red (#43-4317, Zymed, South San Francisco, CA). These reagents are light sensitive. Use sterile water to reconstitute the contents of the vial. Use at 0.5 µg/mL in PBS + 0.5% BSA for immunostaining.

4. Dulbecco's phosphate-buffered saline without calcium and magnesium (PBS-CMF) containing 0.5% BSA (sterile-filtered through 0.22-μm filter; store at 4°C and use cold in the immunostaining procedure).
5. Propidium iodide (1 mg/mL stock solution prepared in sterile water; aliquot and store at –20°C; light sensitive). Use at 1 μg/mL.
6. Gentamicin (use at 50 μg/mL).
7. Sterile 1.5-mL microcentrifuge tubes.
8. Sterile polystyrene tubes (12 × 75 mm) (*see* Note 5)

3. Methods
3.1. Preparation of Conditioned Media (see Note 6)

MRC-5 cells are grown in MRC-5 MM in a humidified 37°C incubator containing 10% CO_2.

1. Remove a cryogenic vial from liquid nitrogen storage, and rapidly thaw it in a 37°C water bath.
2. Transfer the contents of the vial to a 100-mm dish containing 10 mL of MM.
3. On the following day, aspirate the old media and refeed the cells with 10 mL of fresh MM.
4. Culture the cells in 100-mm dishes in 10 mL MM/dish. Feed the cultures with fresh media every 2–3 d, subculture the cells using trypsin when cells reach approx 80% confluence, and dilute them 1:5.
    To subculture MRC-5 cells:
    a. Aspirate the old media and gently rinse the cells twice with PBS-CMF.
    b. Add 1 mL of trypsin (0.05% trypsin, 0.53 mM EDTA)/100-mm dish. Incubate for a few minutes at room temperature until the cells round up or lift off the dish.
    c. Add 9 mL of MM to the dish, and gently pipet the cells to disperse.
    d. Seed the cells into 100-mm dishes in 10 mL of MM.
5. When the cells in the desired number of 100-mm dishes (usually 5–10) reach 70–80% confluence, refeed the cultures with 10 mL of HuGM.
6. Leave HuGM on the MRC-5 cells overnight.
7. Collect the conditioned HuGM (CM) and store at 4°C. Refeed the cultures with 10 mL of fresh HuGM, and incubate overnight again.
8. Collect the CM again. Pool all the CM, filter through 0.45-μm filter units (*see* Note 7), and freeze at –20°C in 5-mL aliquots.
9. When needed, thaw aliquots of CM at 37°C, and mix 1:1 with fresh HuGM prior to use.

3.2. Tissue Preparation/Dissociation/Initial Plating

Dissociates of human muscle tissue are grown initially in CM (*see* Section 3.1.), but after adaptation to culture conditions, they are grown in HuGM. All culturing is done in a humidified 37°C incubator containing 5% CO_2.

1. Add approx 10 mL of F10 into each of four 100-mm dishes.
2. Wash the muscle sample four times by gently agitating with forceps in each of the four dishes containing F10.
3. In another 100-mm dish containing F10, remove any obvious connective or fatty tissue (*see* Note 8).
4. Transfer the tissue sample to 5 mL of trypsin in a 100-mm dish.
5. Mince the muscle tissue, while in trypsin, into pieces no larger than 1 mm^3 with a pair of sterile razor blades (*see* Note 9).
6. Transfer the minced muscle suspension to a sterile trypsinization flask (#355803, Wheaton, Millville, NJ) containing a stir bar, and add trypsin/EDTA to a final volume of 20–25 mL.
7. At 37°C, very gently stir the minced muscle suspension for 20 min.
8. Let the tissue pieces settle for a few minutes and decant (*see* Note 10) the supernatant to a 50-mL plastic centrifuge tube on ice.
9. To the contents of the 50-mL tube add calf serum to a final concentration of 10% in order to neutralize the trypsin.
10. Repeat the trypsinization steps with 20–25 mL of fresh trypsin each time until no pink tissue pieces remain, up to a maximum of three times.
11. Centrifuge the supernatants at 800–900*g* for 5 min.
   If proceeding with **myoblast isolation using flow cytometry,** continue as follows:
12. Pool and resuspend the cell pellet(s) in 4 mL of 1:1 HuGM/CM. Transfer the cell suspension (consisting of myofiber debris, red blood cells, myoblasts, and fibroblasts) to a 60-mm dish. Proceed to Section 3.3.
   If proceeding with **myoblast isolation using cloning procedures,** continue as follows:
13. Pool and resuspend the cell pellet(s) in 10 mL of 1:1 HuGM /CM. Assuming the following viable cell yields *(11–14)*:
   a. 5×10^3 cells/0.1 g tissue from normal donors;
   b. 200 cells/0.1 g tissue from Duchenne muscular dystrophy patients; and
   c. 5×10^5 cells/ 0.1 g tissue from fetal sources.
   Plate 2 rows each of a 96 well plate at 1, 0.5, 0.25, and 0.125 cells/well. Pellet and freeze the remaining cells at 0.1 g tissue/mL freezing medium in 0.2-mL aliquots. Skip to Section 3.6.

3.3. Growth and Expansion of Primary Dissociates Containing Mixed-Cell Populations (Bulk Cultures)

1. Feed with 1:1 HuGM/CM at d 1 or 2 if the cells are 30–40% confluent. Otherwise, feed with 1:1 HuGM/CM at d 4 or 5 (*see* Note 11) or when cells reach 40% confluence (whichever is sooner).
2. Feed the cells every 2–3 d with HuGM. If the cells are <40% confluent, feed the cultures with 1:1 HuGM/CM.
3. When cells are about 70–80% confluent, subculture as follows. To subculture human muscle cells (*see* Note 12):

a. Aspirate the old media, and gently rinse the cells twice with PBS-CMF.
b. Add 0.5 mL of trypsin (0.05% trypsin, 0.53 mM EDTA)/60-mm dish. Incubate for a few minutes at room temperature until the cells round up or lift off the dish.
c. Add 3–4 mL of HuGM to the dish, and gently pipet the cells to disperse.
d. Seed the cells into 100-mm dishes at 5–10 × 10^5 cells/dish in 10 mL of HuGM. Dishes should be approx 20% confluent once the cells attach to the dish.

4. Feed cells every 2–3 d with HuGM, subculturing when the cells reach 70–80% confluence (see Note 13).
5. Continue feeding and subculturing until there are at least four 100-mm dishes at 70–80% confluence. At this point, one-half of the cells should be frozen (see Note 14), and the other half used for flow cytometric purification of the myoblasts.

To freeze cells:
a. Rinse and trypsinize the cells as described in step 3, but use 1 mL of trypsin/EDTA/100-mm dish. Collect into a 15-mL tube.
b. Determine the number of cells using a hemocytometer.
c. Centrifuge the cells at 800–900g for 2 min.
d. Gently aspirate the media, and resuspend in freezing media to give approx 2 × 10^6 cells/mL.
e. Aliquot the cell suspension at 0.5 mL/2 mL cryogenic freezing vial.
f. Place the freezing vials in a foam-filled box (see Note 15), and transfer to a –70°C freezer for up to 1 wk. Transfer to liquid nitrogen cryogenic unit for long-term storage.

3.4. Flow Cytometry (Bulk Cultures)

1. Trypsinize the cells and transfer to a 15-mL centrifuge tube (see Note 16); add PBS-CMF containing 0.5% BSA (see Note 17). Centrifuge at 800–900g for 2 min at room temperature.
2. Mix the pellet by tapping the tube with your finger several times, then resuspend in 1 mL of PBS-CMF + 0.5% BSA. Transfer the cells to a 1.5-mL microcentrifuge tube, and centrifuge for 3 s in a microcentrifuge. Gently aspirate the supernatant, but avoid bringing the tip of the Pasteur pipet close to the cell pellet.
3. Loosen the pellet by tapping the tube as in step 2. Then add either 5.1H11 hybridoma supernatant neat or 1 μg of 5.1H11 antibody/10^6 cells in PBS-CMF + 0.5% BSA. Mix by gently pipeting up and down.
4. Incubate for 20 min on ice.
5. Wash two to three times with 1 mL of ice-cold PBS-CMF + 0.5% BSA, and centrifuge for 3 s in a microfuge at room temperature, loosening the pellet each time.
6. Add 1 mL of biotinylated antimouse IgG, and gently mix the cells by pipeting up and down.
7. Incubate for 20 min on ice.
8. Wash two to three times as in step 5.

Working with the light in the tissue-culture hood **off** from here to the end:

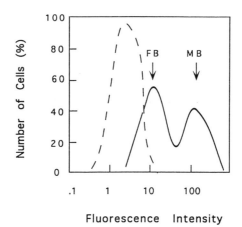

Fig. 1. Schematic of a typical FACS analysis of a mixed primary culture of myoblasts (MB) and fibroblasts (FB) labeled with an αN-CAM antibody (5.1H11). Dotted line: control staining in which the 5.1H11 antibody was omitted from the staining procedure and the cell mixture was exposed to biotinylated antimouse IgG antibody and Texas Red-streptavidin. Solid line: the cell mixture was labeled with 5.1H11, biotinylated antimouse IgG antibody, and Texas Red-streptavidin.

9. Loosen the pellet. Add 1 mL of streptavidin conjugated to desired fluorophore (see Note 18) in PBS-CMF + 0.5% BSA. Add propidium iodide stock to 1 µg/mL (see Note 19).
10. Incubate for 10–15 min on ice.
11. Wash two times as in step 5.
12. Resuspend in 1 mL of ice-cold PBS + 0.5% BSA, and filter through a 70-µm disposable nylon filter to remove cell clumps (see Note 20). Wash the filter with several milliliters of PBS + 0.5% BSA, and centrifuge the tube for 2 min at 800–900g at room temperature.
13. Resuspend the cell pellet in approx 0.2 mL of ice-cold PBS + 0.5% BSA containing 0.5 µg/mL propidium iodide. Cells should be at a density of 10^7/mL for flow cytometry. Transfer to a 5-mL polystyrene tube.
14. Keep the cells on ice and in the dark prior to flow cytometry.
15. Collect the purified myoblasts (see Fig. 1) into a 5-mL tube containing HuGM. Keep the cells on ice after collection (see Note 21).
16. Centrifuge the myoblasts at 800–900g for 2 min at room temperature, resuspend in fresh HuGM + 50 µg/mL gentamicin (see Note 22), seed into 100-mm dishes at 5–10×10^5/dish.

3.5. Culturing of Purified Human Myoblasts After Flow Cytometry

1. Feed myoblast cultures every 2–3 d with HuGM, subculturing when the cells reach 70–80% confluence (see Note 23). Remove gentamicin from the HuGM media at the first refeeding after the flow cytometry step.

2. Continue feeding and subculturing until the desired number of myoblasts is obtained (*see* Note 24).
3. After several weeks in culture, test the following for each muscle sample:
   a. Purity of culture: Restain a small number of cells (approx 5×10^5) with the NCAM antibody, and determine the percentage of myoblasts after several weeks in culture (*see* Note 25). The purity of the culture should be maintained at >98% *(11)*.
   b. Differentiative capacity:
      i. Collagen-coat some 35-mm dishes (*see* Note 26).
      ii. Grow myoblasts to near confluence in HuGM.
      iii. Aspirate the old media and refeed with FM (*see* Note 27).
      iv. Refeed cultures every day with fresh FM. Multinucleated myotubes will form typically within 2–3 d (*see* Note 28).
      v. Determine the fusion index: Count the number of nuclei inside and outside of myotubes in a number of different fields. The fusion index is defined as:

$$(\text{number of nuclei inside myotubes/total number of nuclei}) \times 100 \qquad (1)$$

3.6. Growth of Clonal Cultures

1. Feed the 96-well plates with 1:1 HuGM/CM at 4–5 d intervals. Clones should be ready to transfer to 24-well plates by d 11. Myoblasts and fibroblasts can be distinguished by morphologic criteria (*see* Fig. 1 in ref. *13*).
2. Keep feeding with 1:1 HuGM/CM every 2–3 d, every 3 d if the cultures are sparse, and every 2 d when denser. Subculture the cells when they reach 70–80% confluence, and transfer to a 60-mm dish containing HuGM.
3. At the next subculture, plate some cells in one well of a collagen-coated (*see* Note 26), 24-well dish containing HuGM to test the fusion index of the clone (*see* Section 3.5. and Note 29) and the remainder in an uncoated 100-mm dish.
4. Feed the cells every 2–3 d with HuGM, freezing (*see* Section 3.3.) when the cells reach 70–80% confluence.

4. Notes

1. Initially one can use conditioned media prepared from MRC-5 cultures, but later, for simplicity, one may want to collect the spent HuGM directly from cultures of either the primary human muscle dissociates or the purified human myoblasts.
2. Serum-free media may be substituted if desired *(16)*.
3. Reconstitute the lyophilized chick embryo extract with 10 mL of sterile water. Transfer the slurry to a 15-mL conical plastic centrifuge tube. Spin for 10 min at 800–900g at room temperature. Transfer the supernatant to a new tube and store at –20°C. Discard the brown pellet in the original tube. As an alternative to the use of chick embryo extract, 1 ng/mL of basic fibroblast growth factor (bFGF) can be used. bFGF is added directly to the media in the culture dish at each feeding.

4. The MAb 5.1H11 may be substituted with an MAb against Leu19 (Becton-Dickinson, Mountain View, CA), which crossreacts with the NCAM isoform expressed by human myoblasts and regenerating fibers *(17–22)*. This Leu19 antibody has been used for immunofluorescence analyses of cultured human myoblasts, but not for flow cytometry. In theory, the anti-Leu19 antibody should also work in the flow cytometric procedure outlined here.
5. Check with the operator of the FACS machine to be used to determine which manufacturer's brand of tube fits the particular instrument. Slight differences among manufacturers can render the 12 × 75-mm polystyrene tube unable to maintain an adequate seal with the FACS machine.
6. It is not known whether toxic compounds from the media are removed or whether stimulatory compounds are added to the media during the conditioning process.
7. The filtration rate through the 0.45-μm filter is very slow. Only about 25–50 mL will be collected before the filter clogs up and prevents further filtration.
8. Removing as much of the connective tissue as possible is advantageous, because the yield of contaminating fibroblasts is reduced in the primary dissociates.
9. One should be able to pipet easily the minced muscle tissue suspension with a 25-mL pipet.
10. If one holds a magnet to the bottom of the trypsinization flask while decanting, the metal stir bar will not fall out.
11. There is a lag phase of several days before the cells start dividing in vitro.
12. Bulk cultures should never be plated at lower than 20% confluence.
13. Cells should be evenly plated in the dishes. Otherwise, the cells will become denser in certain areas and the myoblasts may start to fuse with each other to form differentiated myotubes, which are postmitotic. One should maintain the cultures with spaces between the cells, never approaching close to confluence. At high cell densities, the cells will start to become elongated, indicating that differentiation has been triggered. During the isolation and expansion process in culture, one wants small, mononucleated cells with a good mitotic index. If this is not observed, then something is wrong with the culture conditions.
14. If there are problems with the immunostaining or the FACS machine or the purified myoblasts become microbially contaminated after collection, then the remainder of the cells can be thawed, expanded, and used for flow cytometric purification again.
15. The vials must be frozen while wrapped in insulating material. Without insulation, the viability of the cells decreases dramatically.
16. Start with as many cells as possible. Approximately $5–6 \times 10^6$ is a good starting point, but scale up or down as needed. The number of myoblasts obtained after FACS depends on how much of the total starting cell population was expressing NCAM (i.e., myoblasts). Fetal muscle tissue has a much higher proportion of fibroblasts than myoblasts *(11)*.
17. Do not substitute serum for the BSA anywhere in this protocol. The use of serum during the staining procedure leads to clumping of cells.

18. Texas Red is excited at 590 nm by a tunable dye laser. Not all FACS machines are outfitted with this type of laser. The more commonly found machines are equipped with an argon-ion laser, which excites FITC at 488 nm. *See* ref. *11* for further details about the FACS machine itself.
19. Propidium iodide stains only dead cells. Dead cells will nonspecifically react with the antibodies in this procedure, resulting in cells with high fluorescence levels. Unless the dead cells (propidium iodide positive) are gated out electronically on the FACS machine from the analysis, these dead cells will mistakenly be identified as 5.1H11-positive myoblasts. Less than 5% of the starting cell population should be stained with propidium iodide.
20. Cell clumps clog up the nozzle of the FACS machine.
21. The viability of the cells decreases if they are not kept on ice.
22. Occasionally, one encounters problems with microbial contamination after the FACS purification step. Even though HuGM contains penicillin and streptomycin, the addition of gentamicin to the cultures immediately after plating in culture helps to prevent the loss of cultures to bacterial contamination.
23. The doubling time of human myoblasts is approx 24 h.
24. Freeze a number of vials at low passage number after sorting. Myoblasts from normal postnatal donors are capable of approx 40 doublings, whereas fetal myoblasts can undergo 60 doublings. Myoblasts from individuals with neuromuscular disease have a limited proliferative capacity in culture *(14)*. As myoblasts begin to senesce in culture, they become bigger and contain many stress fibers, the doubling time decreases, and the fusion index decreases.
25. If the fibroblasts were not sufficiently eliminated during the cell-sorting procedure owing either to inadequate setting of sort parameters or to cell clumping, the contaminating fibroblasts will overgrow rapidly and the percentage of myoblasts will decrease dramatically.
26. Calf skin collagen (Sigma #C8919) is prepared according to the manufacturer's suggestions. The solution is added to tissue-culture dishes to cover the bottom and placed overnight at 37°C. The next morning, remove the collagen solution (store at 4°C, can be reused many times), and put the dish back at 37°C to dry for a few hours. When dry, the dishes can be taped shut and stored in a drawer at room temperature indefinitely. Rinse the dishes once with PBS-CMF before adding any cells.
27. It takes on average of 2–3 d for myotubes to form. Some muscle samples take a shorter time, and some longer. Some muscle samples form long, thin myotubes, whereas others tend to form shorter, fatter ones.
28. As an alternative, serum-free media may be substituted *(23)*.
29. The fusion rate and the morphology of the myotubes derived from different myoblast clones can vary dramatically within the same muscle sample. If the clone does not fuse, it is most likely a fibroblast clone, but it could be a poor fusing myoblast clone. To prove definitively the muscle origin of clones, one can immunostain clones in small dishes with a muscle-specific antibody (5.1H11, Leu19, desmin).

Acknowledgment

I am grateful for the encouragement and support of Helen Blau in whose laboratory these methods were developed and optimized.

References

1. Barr, E. and Leiden, J. M. (1991) Systemic delivery of recombinant proteins by genetically modified myoblasts. *Science* **254**, 1507–1509.
2. Dhawan, J., Pan, L. C., Pavlath, G. K., Travis, M. A., Lanctot, A. M., and Blau, H. M. (1991) Systemic delivery of human growth hormone by injection of genetically engineered myoblasts. *Science* **254**, 1509–1512.
3. Dai, Y., Roman, M., Naviaux, R. K., and Verma, I. M. (1992) Gene therapy via primary myoblasts: long-term expression of factor IX protein following transplantation in vivo. *Proc. Natl. Acad. Sci USA* **89**, 10,892–10,895.
4. Gussoni, E., Pavlath, G. K., Lanctot, A. M., Sharma, K. R., Miller, R. G., Steinman, L., and Blau, H. M. (1992) Normal dystrophin transcripts detected in Duchenne muscular dystrophy patients after myoblast transplantation. *Nature* **356**, 435–438.
5. Jiao, S., Gurevich, V., and Wolff, J. A. (1993) Long-term correction of rat model of Parkinson's disease by gene therapy. *Nature* **362**, 450–453.
6. Karpati, G., Ajdukovic, D., Arnold, D., Gledhill, R. B., Guttmann, R., Holland, P., Koch, P. A., Shoubridge, E., Spence, D., Vanasse, M., Watters, G. V., Abrahamowicz, M., Duff, C., and Worton, R. (1993) Myoblast transfer in Duchenne muscular dystrophy. *Ann. Neurol.* **34**, 8–17.
7. Law, P. K., Goodwin, T. G., Fang, Q., Duggirala, V., Larkin, C., Florendo, J. A., Kirby, D. S., Deering, M. B., Li, H. J., Chen, M., Cornett, J., Li, L. M., Shirzad, A., Quinley, T., Yoo, T. J., and Holcome, R. (1992) Feasibility, safety, and efficacy of myoblast transfer therapy on Duchenne muscular dystrophy boys. *Cell Transplant.* **1**, 235–244.
8. Tremblay, J. P., Malouin, F., Roy, R., Huard, J., Bouchard, J. P., Satoh, A., and Richards, C. L. (1993) Results of a triple blind clinical study of myoblast transplantations without immunosuppressive treatment in young boys with Duchenne muscular dystrophy. *Cell Transplant.* **2**, 99–112.
9. Yao, S.-N., Smith, K. J., and Kurachi, K. (1994) Primary myoblast-mediated gene transfer: persistent expression of human factor IX in mice. *Gene Therapy* **1**, 99–107.
10. Mauro, A. (1961) Satellite cells of skeletal muscle fibers. *J. Biophys. Biochem. Cytol.* **9**, 493–495.
11. Webster, C., Pavlath, G. K., Parks, D. R., Walsh, F. S., and Blau, H. M. (1988) Isolation of human myoblasts with the fluorescence-activated cell sorter. *Exp. Cell. Res.* **174**, 252–265.
12. Blau, H. M. and Webster C. (1981) Isolation and characterization of human muscle cells. *Proc. Natl. Acad. Sci. USA* **78**, 5623–5627.
13. Blau, H. M., Webster, C., and Pavlath, G. K. (1983) Defective myoblasts identified in Duchenne muscular dystrophy. *Proc. Natl. Acad. Sci. USA* **80**, 4856–4860.

14. Webster, C. and Blau, H. M. (1990) Accelerated age-related decline in replicative life-span of Duchenne muscular dystrophy myoblasts: implications for cell and gene therapy. *Somat. Cell Mol. Genet.* **16,** 557–565.
15. Walsh, F. W. and Ritter, M. A. (1981) Surface antigen differentiation during myogenesis in culture. *Nature* **289,** 60–64.
16. Ham, R. G., St. Clair, J. A., Webster, C., and Blau, H. M. (1988) Improved media for normal human muscle satellite cells: serum-free clonal growth and enhanced growth with low serum. *In Vitro Cell Dev. Biol.* **24,** 833–844.
17. Lanier, L. L., Testi, R., Binal, J., and Phillips, J. H. (1989) Identity of Leu-19 (CD56) Leukocyte differentiation antigen and neural cell adhesion molecule. *J. Exp. Med.* **169,** 2233–2238.
18. Schubert, W., Zimmerman, K., Cramer, M., and Starzinski-Powitz, A. (1989) Lymphocyte antigen Leu-19 as a marker of regeneration in human skeletal muscle. *Proc. Natl. Acad. Sci. USA* **86,** 307–311.
19. Illa, I., Leon-Monzon, M., and Dalakas, M. C. (1992) Regenerating and denervated human muscle fibers and satellite cells express neural cell adhesion molecule recognized by monoclonal antibodies to natural killer cells. *Ann. Neurol.* **31,** 46–52.
20. Michaelis, D., Goebels, N., and Hohlfeld, R. (1993) Constitutive and cytokine-induced expression of human leukocyte antigens and cell adhesion molecules by human myotubes. *Am. J. Pathol.* **143,** 1142–1149.
21. Hohlfeld, R. and Engel, A. G. (1990) Induction of HLA-DR expression on human myoblasts with interferon-gamma. *Am. J. Pathol.* **136,** 503–508.
22. Hohlfeld, R. and Engel, A. G. (1990) Lysis of myotubes by alloreactive cytotoxic T cells and natural killer cells. Relevance to myoblast transplantation. *J. Clin. Invest.* **86,** 370–374.
23. St. Clair, J. A., Meyer-Demarest, S. D., and Ham, R. G. (1992) Improved media with EGF and BSA for differentiated human skeletal muscle cells. *Muscle and Nerve* **15,** 774–779.

24

Cultures of Proliferating Vascular Smooth Muscle Cells from Adult Human Aorta

Heide L. Kirschenlohr, James C. Metcalfe, and David J. Grainger

1. Introduction

Abnormal proliferation of human vascular smooth muscle cells (hVSMCs) is a central event in the development of atherosclerosis *(1–3)*. As a result, there is considerable interest in the establishment of hVSMC cultures as a model of this disease process. However, it has been noted in the past *(4,5)* that hVSMCs, especially when cultured by the enzyme-dispersal technique (hVSMC$_{ED}$), grow poorly in culture compared to VSMCs from other species (e.g., rat). This has limited their use for cell-culture studies. We have recently reported that the reduced proliferative capacity of hVSMC$_{ED}$ from adult aorta can be attributed to the endogenous production of active TGF-β *(6,7)*. We *(6–8)* and others *(9–11)* have shown that TGF-β is a potent inhibitor of smooth muscle cell proliferation. Moreover, recent studies in animal models of atherosclerosis *(12)* have suggested that TGF-β plays a pivotal role in regulation of vessel wall architecture *(13)*. We have also shown that hVSMCs derived by the alternative method of explanting (hVSMC$_{EX}$) have a greater proliferative capacity than the hVSMC$_{ED}$ *(14)*. In accordance with our hypothesis, the cells grown from explanted tissue did not produce TGF-β *(14)*.

This chapter describes the two alternative methods for preparation of hVSMC cultures. This account is followed by a brief comparison of the properties of the two cell-culture types, including growth characteristics, morphology, and expression of protein markers. Finally, we discuss the relative merits of both methods for various applications.

2. Materials

2.1. Enzyme Solutions and Media for Establishing Cultures

1. Serum-free Hank's balanced salt solution (SF-HBSS): This is an inorganic salt solution buffered for use in atmospheric CO_2. Use a bicarbonate-free powdered medium (ICN/Flow, Costa Mesa, CA; store at 4°C). Dissolve powdered medium as indicated by the supplier. To 1 L of medium add 2 mL of stock solution (prepared as in item 7) of antibiotics (100 µg/mL streptomycin, 100 U/mL penicillin final concentration) and 0.35 g sodium bicarbonate. Once dissolved, sterilize the medium by passing through a 0.22-µm sterile Millipore filter and store at 4°C. SF-HBSS is used for dissecting the tissue.
2. Serum-free Medium 199 (SF-M199): Use endotoxin-tested, sterile Medium 199 (modified) containing HBSS (1XH199) and sodium bicarbonate (0.35 g/L), but without L-glutamine (liquid from ICN/Flow; store at 4°C). The medium is supplemented with antibiotics, L-glutamine (2 mM final concentration), and additional sodium bicarbonate (17 mM final concentration). For each 500-mL bottle, mix stock solutions (prepared as in item 7) of antibiotics (1 mL), L-glutamine (5.5 mL of 200 mM solution), and sodium bicarbonate (7.6 mL of 7.5% [w/v] solution), and filter directly into the bottle through a 0.22-µm sterile Millipore filter. SF-M199 is used during enzymatic digestion of the tissue.
3. Collagenase: Use collagenase type I (Sigma [St. Louis, MO] C-0130 with >125 collagen digestion units/mg solid; store at –20°C). Add 300 mg collagenase to 100 mL SF-M199 (3 mg/mL). Stir on ice for 45 min until dissolved. Remove residual particulate material by filtering the enzyme solution through a 0.45-µm Millipore filter, and then sterilize by passing through a 0.22-µm Millipore filter. Divide the enzyme solution into 5- or 10-mL aliquots, and freeze immediately at –20°C until use (*see* Note 1).
4. Elastase: Use elastase from porcine pancreas (Sigma E-0258, Type IV; store at –20°C). Immediately before use, dissolve elastase (1 mg/mL) in SF-M199 (*see* procedure for volume required in Section 3.2.1.). Adjust the pH to ~6.8 with 1M HCl, and then filter through a 0.22-µm Millipore filter (*see* Note 1).
5. Sodium bicarbonate: Use cell-culture-tested sodium bicarbonate (powder from Sigma; store at room temperature). Make a 7.5% (w/v) stock solution in MilliQ water (for SF-M199), or use powder (for SF-HBSS).
6. L-Glutamine: Cell-culture-tested L-glutamine solution (Gibco-BRL [Gaithersburg, MD]; 200 mM solution) is used for SF-M199. Store frozen at –20°C.
7. Antibiotics: Use penicillin G sodium salt (Gibco-BRL, nonsterile powder, 1650 U/mg solid; store at 4°C) and streptomycin sulfate (Sigma; cell-culture-tested powder; store at 4°C). Make a stock solution (in MilliQ water) containing penicillin (30 mg/mL) and streptomycin (50 mg/mL). Split stock solution in convenient aliquots (e.g., 2 mL of stock solution needed for 1 L of SF-M199 or SF-HBSS). Store aliquots frozen at –20°C (*see* Note 2).

2.2. Growth Media and Supplements for Maintaining Cultures

1. Serum-free Dulbecco's modified Eagle's medium (SF-DMEM): Use DMEM with L-glutamine (584 mg/L) and high glucose (4500 mg/L) without sodium

bicarbonate (ICN/Flow; powdered medium; store at 4°C). Dissolve medium as indicated by supplier. Adjust the pH of the medium to 6.9–7.0 with 1M HCl if necessary. To 1 L of final medium add 2 mL stock solution of antibiotics (described in Section 2.1., item 7) and 3.7 g sodium bicarbonate, and stir to dissolve. Sterilize the medium by passing through a 0.22-μm Millipore filter. Filter the medium directly into sterile bottles (usually 500 mL). Store medium at 4°C for up to 1 mo. After 1 mo, the medium may still be used, but should be supplemented with fresh L-glutamine (see Note 3). This medium is used to arrest cell growth.
2. DMEM + 10% fetal calf serum (FCS): Prepare SF-DMEM as described in item 1, but prior to filter sterilization, add FCS to give a final concentration of 10% (v/v). Store medium at 4°C for up to 1 mo. This medium is used for passaging cells. This medium is always prewarmed to 37°C prior to use.
3. DMEM + 20% FCS: Prepare SF-DMEM as in step 1, but supplemented with 20% (v/v) FCS. This medium is used for growing cells, and is always prewarmed to 37°C prior to use.
4. FCS: Use FCS screened for myoplasma and adventitious viruses (e.g., from ICN/Flow or TCS Biologicals [Buckingham, UK]; store frozen at –20°C). Thaw only one bottle (500 mL) at a time, and divide into smaller aliquots (100 mL). Add these aliquots to SF-DMEM (for DMEM + 10% FCS or DMEM + 20% FCS, see items 2 and 3). Freeze remaining aliquots at –20°C (see Note 4).
5. Trypsin/EDTA solution (1X): The enzyme solution is purchased already prepared containing 0.5 g/L trypsin and 0.2 g/L EDTA in Modified Puck's Saline A (Gibco-BRL; 1X liquid; shelf life = 18 mo; store at –20°C). Thaw only one bottle of trypsin/EDTA (500 mL) at a time. Divide trypsin into convenient aliquots (e.g., 50 mL) to avoid loss of activity owing to repeated freeze-thawing, and store aliquots at –20°C until use (see Note 5).

2.3. Equipment

All procedures using human tissue should be carried out in a Class II laminar flow hood using aseptic technique. It is strongly recommended for safety reasons to wear a surgical mask, gloves, and a tissue-culture coat during all procedures involving handling of human tissue. In addition to a fully equipped cell-culture laboratory, the following dissecting tools and materials are required:

1. One pair small curved scissors.
2. Two pairs of sharp watchmakers' forceps.
3. Two hypodermic needles (0.8 × 40 mm).
4. Two scalpel handles and blades (Nos. 10 and 11).
5. One pair of curved forceps (with flat points).
6. Screw-cap universals (various sizes).
7. Plastic 90-mm Petri dish.
8. Conical flasks (various sizes).
9. Plugged wide-mouth pipets.
10. Cell scrapers.

All these items must be sterile. Nondisposable items should be thoroughly cleaned, washed, and dried between uses; wrapped in aluminum foil; and autoclaved. Dissection of the tissue is performed in a 90-mm glass Petri dish half-filled with silicone-based gel into which needles can be pushed to position the tissue. This dissecting Petri dish should be sterilized by autoclaving between use.

3. Methods
3.1. Collection of Tissue Samples

We receive our human tissue samples from a heart transplant program. The specimens are tissue samples taken from the thoracic aorta of heart transplant donors, apparently healthy at the time of death.

1. Provide labeled universal tubes containing DMEM + 10% FCS to store harvested tissue. Depending on the size of the tissue specimens obtained (up to 2 × 4 cm), use either 20- or 50-mL universal tubes with leak-proof caps, about 2/3 full with medium. These tubes are placed in a fridge (4°C) by the team harvesting the donor tissue. The tube is labeled with a brief data sheet including date of transplant, time the tissue is harvested, and the sex and age of the donor.
2. Tissue samples are harvested any time of the day or night. Since the success of the procedure depends on rapid handling of the tissue sample, a pager or mobile phone is essential, so you can be contacted as soon as tissue becomes available.
3. Transfer the tissue sample to the cell-culture laboratory as soon as possible after the tissue is harvested (preferably within 12 h, but under no circumstances later than 24 h). Transport the universal tube containing the tissue sample on ice (*see* Note 6).

3.2. Enzyme-Dispersed VSMCs
3.2.1. Establishing Primary Cultures of Enzyme-Dispersed VSMCs

1. Transfer the tissue sample to a new universal tube containing SF-HBSS. Wash the specimen by gently agitating it using long, curved forceps. Repeat twice using fresh tubes and medium. Transfer to the dissecting Petri dish (*see* Section 2.3.) containing enough HBSS to keep the tissue moist.
2. Open the aorta longitudinally. Fix the tissue at either end with two hypodermic needles onto the dissecting dish with the luminal surface upward (*see* Note 7). Remove the endothelium by scraping the cell layer off with a sterile no. 11 scalpel blade. If a clearly defined intima is present, this is removed with watchmaker's forceps before proceeding. Next harvest the smooth muscle cell-containing media by stripping it from the adventitia. Gently peel off thin transverse muscle strips (approx 1–2 mm in width). Transfer the muscle strips into a plastic 90-mm Petri-dish containing SF-HBSS. Repeat this process until approx 2/3 of the media has been removed from the whole surface of the vessel. Discard the remaining media and adventitia.

3. When all muscle strips have been collected, tilt the Petri dish so all the pieces fall to one side, and remove most of the medium keeping the pieces moist. Cut them into smaller pieces (cubes 1–2 mm on a side) using scissors. The enzymatic digestion will be quicker and the viability of the culture improved if the cubes are as small as possible. Alternatively, two scalpel blades can be used to chop the tissue strips into similar cubes (*see* Note 8).
4. Wash the cubes twice in fresh SF-HBSS, and then aspirate most of the medium.
5. Transfer the cubes into a sterile conical flask (10, 25, or 50 mL, depending on the amount of tissue) of known weight (top covered with sterile aluminum foil) and weigh quickly. This will allow you to estimate the volume of enzyme solutions needed for digestion. It will also allow the cell yield per gram of tissue to be estimated.
6. Add 1–2 mL of collagenase in SF-M199 to the flask, and then wash the tissue by gently shaking the flask for 10–20 s. Aspirate and add fresh collagenase in SF-M199 to give a final ratio of tissue (g) to enzyme solution (mL) of 1:5 (w/v). Transfer the flask to a water bath with a shaking platform at 37°C, and gently rock for 30 min. Meanwhile, weigh the elastase and dissolve in SF-M199 to give a solution of 1 mg/mL. Use the same volume of elastase solution as collagenase solution. Lower the pH of the elastase solution to ~6.8 with $1M$ HCl. Add elastase to the flask containing the tissue and collagenase solution by passing the enzyme solution through a 0.22-µm Millipore disposable filter directly into the flask. Mix the tissue with the enzyme solutions using a plugged wide-mouth pipet. Close the flask with sterile aluminum foil.
7. Return the conical flask to the water bath (37°C). Every 30 min, remove the conical flask from the water bath, and mix cell suspension gently using a plugged wide-mouth pipet in the sterile hood. At each time, remove a sample (10 µL) and look for the appearance of single cells in a hemocytometer. Repeat every 30 min until all the tissue has been digested and there are no large aggregates of cells visible in the hemocytometer. At this time, remove the flask from the water bath. The time taken for enzymatic digestion is variable and can take between 2 and 5 h. Do not leave tissue digesting for longer than 5 h (*see* Note 9).
8. Divide the cell suspension equally between two 10-mL centrifuge tubes and centrifuge for 3 min at $900g$. Aspirate the supernatants carefully without disturbing the pellet. Add 0.5 mL medium (DMEM + 10% FCS; prewarmed to 37°C) to each tube, and resuspend the cell pellets gently using a plugged wide-mouth pipet. Calculate your cell yield as follows: Fill one chamber of a Neubauer hemocytometer. Count the cells in the large central square (outlined with triple lines). Multiply the number of cells by 10^4 to give cells/mL. The number of cells/mL multiplied by the number of milliliters of cell suspension will give the total number of cells harvested. Determine the cell viability (the percentage of cells excluding Trypan Blue; *see* Note 10). Adjust the volume of the cell suspension with DMEM + 10% FCS to give a cell density of 8×10^5 cells/mL. Seed the cells at a density of 1.5×10^4 cells/cm^2 onto plastic tissue-culture-grade Petri dishes or 25 cm^2 tissue-culture flasks. Plating efficiency (defined as the number of cells

adhering to the plastic at 48 h, divided by the number of cells plated out) is low (<20%), and the cell density for plating has to be kept high (see Note 11). Transfer dishes or flasks into a humidified incubator (37°C; 5% CO_2 in air). Check the cultures for infections each day (see Notes 2 and 12). The cells will adhere within 12 h, and are fully attached and spread by 48 h. After 48 h, remove medium containing nonadherent cells and discard. Wash the cells once with DMEM + 10% FCS, and then refeed with DMEM + 20% FCS. Aspirate and feed with fresh DMEM + 20% FCS every 48 h up to 10 d when the primary culture is passaged for the first time.

3.2.2. Maintenance and Subculture of Enzyme-Dispersed VSMCs

Enzyme-dispersed hVSMCs can be subcultured (passaged). We find that in order to keep these cultures growing, it is important to maintain a high cell density. At lower cell densities, the $hVSMC_{ED}$ change phenotype, and the proliferation rate decreases. Furthermore, experiments have shown different effects using inhibitors or growth factors (e.g., for TGF-β, ref. *15*) depending on the plating density. We therefore keep the initial plating densities for all experiments constant.

1. Before subculturing the cells, examine the culture microscopically. The primary cell culture should be confluent within 10 d. Even if the cells are subconfluent, do not leave the primary culture longer than 10 d, because the cells become difficult to detach from the plastic substrate when left longer.
2. To subculture the cells, wash the flask twice with a small amount of trypsin/EDTA. Add just enough trypsin/EDTA to cover the bottom of the flask. Agitate the flask gently for a few seconds for each wash. Work quickly so that the cells do not detach during washing. Add fresh trypsin/EDTA solution (e.g., 1.5 mL for 25-cm^2 flask) to the culture vessel, and incubate it at 37°C for 3–4 min. Agitate the culture vessel vigorously to detach the cells fully. Examine the flask under the microscope to ensure the cells have all detached (see Note 13). The action of the trypsin is stopped by rapidly transferring the released cells into a sterile tube containing an equal volume of DMEM + 10% FCS.
3. Mix the cell solution by gentle agitation, and count an aliquot using a hemocytometer. Adjust the volume of the cell suspension with DMEM + 10% FCS to achieve a plating density of approx 6000 cells/cm^2. Return the culture vessels to the incubator at 37°C.
4. After 24 h, change the medium to DMEM + 20% FCS. Change the medium to fresh DMEM + 20% FCS every 48 h.
5. Subcultured cells should reach confluence after the same interval at each passage, usually within 5–6 d. Even if the cells are still subconfluent, do not leave the culture longer than 7 d without subculturing, because the cells become difficult to detach from the plastic substrate.

We find that $hVSMC_{ED}$ of early passages (passage 1–10) dilute approx 1:1.5 every 5–6 d. Cells of later passages (passage 10–20) require 1–2 additional days to achieve a 1:1.5 dilution. Beyond passage 20, the proliferation rate decreases and phenotypic changes occur. We use hVSMC$_{ED}$ cultures for experiments up to the 20th passage.

3.2.3. Recommended Seeding Densities for Experiments

3.2.3.1. CELLS IN EXPONENTIAL GROWTH

For experiments on cells in exponential growth, seed cells at a density of 1500 cells/cm^2. This lower plating density should ensure that cells do not reach confluence and cease to proliferate for at least 96 h, during which time the experiment can be performed. Initial cell densities lower than 1500 cells/cm^2 are not recommended since the cells may stop proliferating owing to low cell density.

3.2.3.2. CELLS SYNCHRONIZED BY SERUM DEPRIVATION

For experiments on synchronized cells, seed cells at 3000 cells/cm^2. Grow cells for 48 h in DMEM + 10% FCS. Wash the cells thereafter three times with SF-DMEM, add fresh SF-DMEM, and leave the cells for 48 h. During this period, the cells will stop proliferating, and early passage (1–5) cells may show signs of cell death. If this is a problem, quiescence may instead be induced in DMEM + 0.5% FCS. To reactivate the cells, replace the medium with either fresh SF-DMEM or DMEM + 20% FCS, and add growth factors or other agents.

3.3. Explant Culture

3.3.1. Establishing Explant-Derived VSMC Cultures

The explant technique is particularly suitable if only small amounts of tissue are available, since the majority of cells are lost during enzymatic digestion in preparing hVSMC$_{ED}$. The tissue sample is treated exactly as described in steps 1–3 of Section 3.2.1. Then proceed as follows:

1. Wash the chopped muscle cubes twice with SF-HBSS. Aspirate the medium, leaving the tissue moist.
2. Divide the muscle cubes into 25-cm^2 culture flasks with a minimum of 20 cubes/flask (*see* Note 14). Add a few milliliters of DMEM + 10% FCS to the flasks, and wash pieces by gently shaking the flask. Then aspirate the medium. Repeat twice.
3. Distribute the cubes evenly on the bottom of the culture flask using a sterile cell scraper or Pasteur pipet (about 2–4 pieces/cm^2; *see* Note 15).
4. Add a minimum volume of DMEM + 10% FCS to the flask (about 1–2 mL), just enough to keep the tissue moist. This is a critical step, because if too much liquid is added, the cubes will float, but if not enough liquid is present, the tissue will eventually dry out.
5. Transfer flasks into a humidified incubator (37°C/5% CO_2 in air), and leave for 2 h. During this time, the tissue cubes should adhere to the plastic. Once the explant

cubes have attached, carefully add additional DMEM + 10% FCS dropwise to a depth of 1–2 mm (*see* Note 16).
6. Leave the explant cubes for 3 d, inspecting regularly and adding more DMEM + 10% FCS as required to maintain a depth of 1–2 mm. Cells will not start to migrate prior to 4 d in culture, and the appearance of cells may take up to a week.
7. After 3 d, aspirate the medium. Place the flask on its long side and tilt. Any unattached tissue cubes will move to the bottom corner of the flask. Remove any remaining medium and unattached explant cubes with a Pasteur pipet, and add DMEM + 20% FCS to a depth of 1–2 mm.
8. Repeat the medium changes every 3 d until a significant number of cells have migrated from the explant cubes, and then change the medium every 2 d.
9. After 2–3 wk, most of the plastic substrate (50–70%) will be covered with cells, although the cell density will be highest next to the pieces. At this time, the explant cubes should be removed. Approximately 18 h prior to removing the explant tissue, aspirate the medium and add 5 mL of DMEM + 20% FCS. This larger volume of medium will result in most of the cubes detaching from the substrate. Remove the pieces that have detached with a Pasteur pipet attached to a vacuum pump. Add more DMEM + 20% FCS, and gently agitate or rotate the flask to detach more cubes. Then aspirate again. Repeat until all the cubes have been removed. Add fresh DMEM + 20% FCS to the cells and return to the incubator.
10. Leave the cells for at least 24 h and up to 96 h after removing the explant cubes. Change the medium every 2 d. In this period, the cells will recover following removal of the explant cubes, and also migrate and proliferate to form a monolayer. Do not leave the cells longer in total than 1 mo after seeding the explant cubes prior to passaging the cells for the first time.

3.3.2. Maintenance and Subculture of Explant VSMC Cultures

As for the hVSMC$_{ED}$ cultures, the hVSMC$_{EX}$ cultures proliferate more slowly and show phenotypic changes at low cell densities. Therefore, seeding densities for maintaining the cultures and for experiments are kept high.

1. Before subculturing the hVSMC$_{EX}$ cells for the first time, examine all flasks microscopically. Some flasks may be less confluent than others, but do not leave the cells longer than 1 mo prior to subculturing for the first time.
2. Detach the cells in all flasks with trypsin/EDTA as described in Section 3.2.2., steps 1 and 2 for hVSMC$_{ED}$ cultures, and transfer cell suspensions to a sterile tube containing an equal volume of DMEM + 10% FCS to inactivate the trypsin.
3. Count a sample (10 µL) of cell solution using a hemocytometer, and adjust the volume of the cell suspension with DMEM + 10% FCS to give a plating density of approx 12000 cells/cm^2.
4. After 24 h, change the medium to DMEM + 20% FCS, and maintain the culture in DMEM + 20% FCS with medium changes every 48 h.
5. Subcultured hVSMC$_{EX}$ should reach confluence after the same interval at each passage, usually 4–5 d. Even if the cells are still subconfluent, do not leave longer than 6 d between subculturing. Treat as described in steps 2–4.

The $hVSMC_{EX}$ dilute 1:2 every 4–5 d, up to passage 20, but as with $hVSMC_{ED}$ cultures, beyond passage 20, the proliferation rate decreases and phenotypic changes occur.

3.3.3. Recommended Seeding Densities for Experiments

3.3.3.1. CELLS IN EXPONENTIAL GROWTH

For experiments on cells in exponential growth, seed cells at a density of 3000 cells/cm^2.

3.3.3.2. CELLS SYNCHRONIZED BY SERUM DEPRIVATION

For experiments on synchronized cells, seed cells at 3000 cells/cm^2 or 6000 cells/cm^2, depending on length of experiment.

3.4. Cell Freezing and Thawing Procedure

3.4.1. Freezing Cells

The following freezing procedure is suitable for both $hVSMC_{ED}$ and $hVSMC_{EX}$ cultures. It is also suitable for many other adherent cell lines, e.g., rat VSMCs, 3T3 fibroblasts. Cultures can be frozen at any time beyond passage 3. Cells from a 75-cm^2 flask are released with 2 mL trypsin/EDTA 24 h prior to confluence and incubated for 3–4 min at 37°C. Check that cells have detached, and add the cell suspension to 8 mL DMEM + 10% FCS. Centrifuge the cell suspension for 3 min at 900g, aspirate the medium, and resuspend the cell pellet in 1 mL DMEM + 10% FCS containing 10% glycerol (pH 7.3). Transfer 1-mL aliquots to cryogenic tubes (e.g., Nunc [Roskilde, Denmark] Cryo Tubes, 1.8–2.0 mL). Allow the cells to equilibrate with the glycerol-containing medium (30 min at room temperature). Freeze the cells in a commercial cell-freezing box (e.g., Nalgene™, cryocontainer, Sigma) filled with isopropyl alcohol. Leave the cells overnight in the freezing box at –80°C. This ensures that the cells are frozen at a rate of approx 1°C/min. The cryotubes are stored at –80°C, and the cells will remain viable for several years. However, any rise in temperature during storage will affect the viability of the culture on thawing. Important cultures may also be stored in liquid nitrogen to improve viability after very long-term storage (>3 yr).

3.4.2. Thawing of Frozen Cells

Thaw out cryotubes rapidly in a water bath at 37°C, agitating the tubes continuously to ensure rapid, even thawing. Transfer the thawed cell suspension from the vial into one (for $VSMC_{ED}$) or two (for $VSMC_{EX}$) 25-cm^2 flasks each containing 5 mL of DMEM + 10% FCS. The flasks should be warmed and equilibrated with CO_2 in the incubator for approx 30 min, prior to adding the thawed cell suspension. Recovery of the cells should be >50%, resulting in a density of adherent cells >50% confluent after incubation at 37°C overnight. Discard any culture where the viability is significantly lower after freezing. Cells recovered from cryopreservation should be passaged at least once prior to use in experiments.

The hVSMC$_{EX}$ cultures recovered from cryopreservation were not distinguishable from cultures that had not been frozen. In contrast, hVSMC$_{ED}$ cultures recovered from cryopreservation exhibited reduced plating efficiency and proliferated through fewer passages compared to hVSMC$_{ED}$ cultures that had not been frozen.

3.5. Comparison of Human VSMC$_{EX}$ and VSMC$_{ED}$ Cultures

Both culture protocols have been used successfully in our laboratory to obtain cultures of smooth muscle cells with reproducible properties. However, cultures of the two types differed significantly in their properties. The hVSMC$_{ED}$ cultures proliferated more slowly and became senescent rapidly at low cell density. By establishing a number of cultures of both types from the same donor tissue, we excluded any contribution from age, sex, clinical history, or genetic background of the donor. In light of the difficulty in obtaining aortic samples from donors with similar characteristics, it is significant that the parameters that vary between donors do not affect the behavior of the cultures. Various properties of the two culture types are summarized here.

3.5.1. Cell Yield

Substantially more cells are obtained by the explant technique compared to enzyme dispersal. Complete dispersal of the aortic tissue samples to a suspension of single cells may result in low plating efficiency owing to the relatively harsh treatment with enzymes. Furthermore, the number of cells present in human aorta per gram of tissue is substantially lower than for commonly used animal species, such as rat. We find that 10^6 cells/g aortic tissue can be obtained from human explanted tissue compared to 10^7 cells/g obtained from rat aorta.

3.5.2. Morphology

We have observed three major cell populations in our cell cultures that differ in size and morphology: small cells with cobblestone morphology, medium-sized cells (spindle-shaped), and large cells (stellate cells). The majority of cells in the hVSMC$_{EX}$ cultures are medium-sized and spindle-shaped, although patches of the other types are often noticed in the primary explant culture. They grow to confluence with a characteristic "hills and valleys" pattern (Fig. 1A,B). Stationary phase is reached at high saturation densities (2.0–4.0×10^4 cells/cm^2). By contrast, the majority of cells in the hVSMC$_{ED}$ cultures are large cells which have a stellate

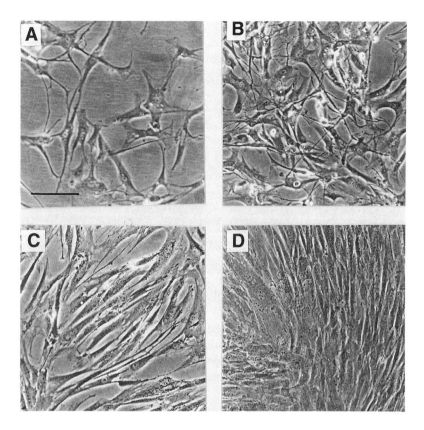

Fig. 1. Phase-contrast photomicrographs of human aortic VSMCs. Micrographs of hVSMC$_{ED}$ in exponential growth **(A)** and at stationary phase **(B)**, and hVSMC$_{EX}$ in exponential growth **(C)** and at stationary phase **(D)**. Bar = 10 μm.

morphology with numerous long cytoplasmic projections. They reach stationary phase at low saturation densities ($0.7–2.0 \times 10^4$ cells/cm^2) without reaching monolayer coverage of the substrate (Fig. 1C,D). The relevance of this heterogeneity of smooth muscle to the interpretation of cell-culture data and to possible VSMC heterogeneity in vivo is only now being investigated *(16–18)*.

3.5.3. Contractile Proteins

The hVSMC$_{ED}$ cultures contain high levels of both smooth muscle α-actin and smooth-muscle-specific myosin heavy chain, whereas cells of the hVSMC$_{EX}$ cultures contain much lower levels of both protein markers *(14)*. Furthermore, even after many passages in culture, the hVSMC$_{ED}$ remain more differentiated than the hVSMC$_{EX}$.

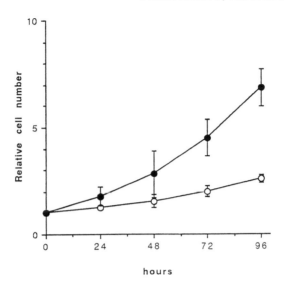

Fig. 2. Proliferation of hVSMC$_{EX}$ and hVSMC$_{ED}$. Cell numbers were determined every 24 h for 96 h by releasing the cells with trypsin/EDTA and counting on hemocytometer. Cell numbers are expressed relative to the cell number present at 24 h after subculturing ($t = 0$ h), which was normalized to 1.0. Values are means of triplicate determinations on three different hVSMC$_{EX}$ (●) and hVSMC$_{ED}$ (○) cultures. Error bars are SEM ($n = 3$). Data were adapted from ref. *14*.

3.5.4. Growth Characteristics

After 6 wk of subculturing both types of cultures, the total yield of cells was fourfold higher per gram wet weight of aorta in the hVSMC$_{EX}$ cultures than the hVSMC$_{ED}$ cultures. Consistent with this observation, the hVSMC$_{ED}$ cultures have a longer population doubling time of 68 ± 2 h ($n = 5$; ref. *14*) compared to hVSMC$_{EX}$ cultures for which doubling time is 35 ± 2 h ($n = 5$, ref. *14*) (Fig. 2).

3.5.5. DNA Synthesis

We investigated whether all of the cells in the two different cultures were proliferating. The fraction of cells that entered DNA synthesis in a 6-d period was measured by bromodeoxyuridine (BrdU) incorporation. The proportion of nuclei staining for BrdU was 80–90% in hVSMC$_{EX}$ cultures, but only 40–50% in the hVSMC$_{ED}$ cultures *(14)*. The population doubling time of the hVSMC$_{ED}$ cultures is therefore increased because less of the cells are cycling. Measured by [^{3}H]thymidine incorporation into DNA, synchronized hVSMC$_{EX}$ cells reach peak rate of DNA synthesis between 16 and 24 h after restimulation with 20% FCS, whereas hVSMC$_{ED}$ reach peak rate of DNA synthesis slightly later, between 24 and 32 h *(14)*.

3.6. Production of Active TGF-β by Human VSMCs

Many of the properties just outlined can be explained by the different production of active TGF-β by the two cultures. Only the hVSMC$_{ED}$ secrete the latent form of TGF-β, which is then activated by plasmin *(19)*. Plasmin is produced from plasminogen (abundant in serum) by the action of tissue plasminogen activator (tPA), which is also produced by hVSMC$_{ED}$ *(6,7)*. We found 15.2 ± 1.6 ng/mL total TGF-β in medium conditioned for 48 h on hVSMC$_{ED}$ of which 64 ± 12% was activated *(6,7)*. In contrast, hVSMC$_{EX}$ produce <1 ng/mL total TGF-β *(14)*.

1. The rate of proliferation of hVSMC$_{ED}$ can be substantially increased by addition of a neutralizing antibody against TGF-β to the cultured cells *(6,7)*. Conversely, the VSMC$_{EX}$ proliferate more slowly, if TGF-β is added to the medium *(14)*.
2. The prevalence of large cells in the hVSMC$_{ED}$ cultures is consistent with previous data showing that TGF-β caused hypertrophy in rat VSMCs *(11)*. Addition of TGF-β to hVSMC$_{EX}$ changes their morphology from spindle-shaped cells to large cells similar to hVSMC$_{ED}$ (Kirschenlohr et al., unpublished observations).
3. The maintenance of differentiation, reflected by the high levels of smooth muscle-specific proteins, in hVSMC$_{ED}$ cultures is consistent with the observation that TGF-β inhibits dedifferentiation of smooth muscle cells from other species *(8,10)*.

3.7. Final Remarks

In summary, we found that the method used to establish hVSMC cultures has a dramatic effect on the proliferation and other properties of the cultures, whereas donor age, sex, clinical history, or even genetic background appeared to contribute relatively little variation *(14)*. The different properties of the two types of cultures can be attributed to the production of TGF-β by hVSMC$_{ED}$ *(6,14)*, although the mechanism underlying this difference has yet to be established. The major differences between the two types of hVSMC cultures must be taken into account when designing experiments involving hVSMCs.

4. Notes

1. A number of different protocols have been described for the enzymatic dispersion of human smooth muscle tissue using elastase and/or collagenase. However, cell yield and viability critically depend on factors, such as specific activity and impurities with nonspecific proteases in the enzyme preparations. It is therefore advisable to test small quantities of different enzyme batches and to reserve batches with optimal properties.
2. Antibiotics are generally added to the media to reduce the risk of bacterial infections. Most laboratories supplement their media with penicillin G and/or streptomycin. We add a mixture of both antibiotics to all our media. The use of other antibiotics should be avoided unless resistant strains of bacteria arise. If this occurs, we use gentamicin (5 mg/mL), but only until the infection is cleared.

3. We recommend purchasing DMEM in powder form. Although powdered medium is stable for several years, reconstituted DMEM is only stable for a short period of time, mainly because glutamine is unstable in solution with a half-life of approx 3 wk at 4°C. Medium should therefore only be made up as required, stored at 4°C, and used within 1 mo. If medium must be stored for longer, it should be supplemented with fresh glutamine.
4. FCS stored at −20°C is stable for at least 2 yr. Batches of FCS vary considerably. Therefore the batch of FCS used should not be changed during a particular series of experiments, and all related subculture work should be done using the same serum batch. New batches of FCS are compared with the previous one before the FCS in use is changed. Consequently, we retain some bottles of each batch of FCS used in case earlier experiments need to be repeated or expanded.
5. Thaw aliquots of trypsin/EDTA only as required. Unused portions of the aliquot may be refrozen and used once or twice more, but further freeze–thawing cycles should be avoided and any remaining solution discarded.
6. For maximum cell viability, the tissue must be processed as soon as possible after excision. If stored at 4°C, tissue samples will remain viable for up to 24 h, although autolysis and degeneration occur progressively with time and a lower cell yield will result. In addition, since tissue samples cannot be harvested and handled under strictly sterile conditions, infections of the subsequent primary cell culture are more likely to occur if the tissue is not processed immediately after harvesting.
7. Examine the luminal surface by eye for any diseased areas (e.g., plaques, fatty streaks). Tissue samples from transplant donors will often be macroscopically free of diseased regions. If diseased tissue is present, excise the area, which may be cultured separately.
8. To use scalpel blades to chop the tissue, transfer muscle strips to a glass Petri dish (to avoid adding plastic to the culture). Pile up the tissue in the middle of the dish, and repeatedly draw the scalpel blades across the tissue in a crosschopping fashion.
9. Adult human aortae contain large amounts of extracellular matrix protein, and hence, the tissue may take hours to disperse completely. However, overdigestion or overheating can lead to cell lysis and viscous DNA release. Culture viability will then be very low.
10. Cell viability should be tested by Trypan Blue exclusion. Make a 0.5% stock solution of Trypan Blue in phosphate-buffered saline. Dilute cells 1:1 with 0.5% Trypan Blue. Cells with intact cell membranes will exclude the stain, whereas nonviable cells appear blue under the hemocytometer. We found that >80% of the cells in our primary cultures are viable.
11. Plating efficiency can be improved slightly by coating cell-culture plastics with collagen, fibronectin, or poly-L-lysine. However, we do not plate our cells on matrix proteins, because it is well recognized that different extracellular matrices may effect the behavior of the cells in a variety of ways.

12. *See* Note 2 for bacterial infections. If yeast infections occur, use nystatin (240 U/mL final concentration) as an antimycotic drug. However, severely infected cultures are discarded into bleach straight away to prevent the infections from propagating.
13. Do not incubate the cells with trypsin/EDTA for longer than 5 min. If most of the cells are still attached, remove the cells that have already detached, and add fresh trypsin/EDTA to the flask. Repeat the incubation at 37°C. Combine all the cells in trypsin/EDTA and then proceed as usual.
14. For tissue samples from the healthy vessel wall, there should be not <20 cubes/flask. However, with tissue samples from diseased vessel wall (such as plaque material), when the amount of tissue available is much smaller, we find that as few as five cubes are sufficient to establish an explant culture.
15. Explant cubes can also be maintained in culture dishes (e.g., 35-mm dish) or in 12-well plates. However, the risk of infection is reduced using culture flasks. Additionally, we found that the explant cultures were established more successfully in flasks, although the reason for this difference is unclear.
16. If the cubes detach from the plastic substrate after a few days, an alternative technique may be used. Prewet the flask with 1–2 mL of DMEM + 10% FCS, and agitate the flask to distribute the liquid over the surface. Aspirate excess medium leaving a thin film on the plastic surface. Distribute the tissue cubes as previously.

Acknowledgments

This research was supported by grants from Glaxo Research and Development Ltd., the British Heart Foundation, and the Wellcome Trust. D. J. Grainger is a Royal Society University Research Fellow. We are grateful to the transplant team at Papworth Hospital for the human aorta specimens. We also thank Christine Witchell for excellent assistance with the cell cultures.

References

1. Ross, R. (1986) The pathogenesis of atherosclerosis: an update. *N. Engl. J. Med.* **314,** 488–500.
2. Ross, R. and Glomset J. A. (1976) The pathogenesis of atherosclerosis I. *N. Engl. J. Med.* **295,** 369–377.
3. Ross, R. (1993) The pathogenesis of atherosclerosis: a perspective for the 1990s. *Nature* **362,** 801–809.
4. Dartsch, P., Voisard, R., Bauriedel, G., Höfling, B., and Betz, E. (1990) Growth characteristics and cytoskeletal organisation of cultured smooth muscle cells from human primary stenosing and restenosing lesions. *Arteriosclerosis* **10,** 62–75.
5. Dartsch, P., Weiss, H.-D., and Betz, E. (1990) Human vascular smooth muscle cells in culture: growth characteristics and protein pattern by use of serum-free media supplements. *Eur. J. Cell Biol.* **51,** 285–294.
6. Kirschenlohr, H. L., Metcalfe, J. C., Weissberg, P. L., and Grainger, D. J. (1993) Adult human aortic smooth muscle cells in culture produce active TGF-β. *Am. J. Physiol.* **265,** C571–C576.

7. Grainger, D. J., Kirschenlohr, H. L., Metcalfe, J. C., Weissberg, P. L., Wade, D. P., and Lawn, R. M. (1993) Proliferation of human smooth muscle cells promoted by lipoprotein(a). *Science* **260**, 1655–1658.
8. Grainger, D. J., Kemp, P. R., Witchell, C. M., Weissberg, P. L., and Metcalfe, J. C. (1994) Transforming growth factor β decreases the rate of proliferation of rat vascular smooth muscle cells by extending the G_2 phase of the cell cycle and delays the rise in cyclic AMP before entry into M phase. *Biochem J.* **299**, 227–235.
9. Assosian, R. K. and Sporn, M. B. (1986) Type β transforming growth factor in human platelets: release during platelet degranulation and action on vascular smooth muscle cells. *J. Cell Biol.* **102**, 1217–1223.
10. Björkerud, S. (1991) Effects of transforming growth factor β1 on human arterial smooth muscle cells in vitro. *Arteriosclerosis Thromb.* **11**, 892–902.
11. Owens, G. K., Geisterfer, A. A., Yang, Y. W., and Kamoriya, A. (1988) Transforming growth factor β-induced growth inhibition and cellular hypertrophy in cultured vascular smooth muscle cells. *J. Cell Biol.* **107**, 771–780.
12. Grainger, D. J., Kemp, P. R., Liu, A. C., Lawn, R. M., and Metcalfe, J. C. (1994) Activation of transforming growth factor β is inhibited in transgenic apolipoprotein(a) mice. *Nature* **370**, 460–462.
13. Grainger, D. J. and Metcalfe, J. C. (1995) A pivotal role for TGF-β in atherogenesis? *Biol. Rev. Camb. Philos. Soc.*, in press.
14. Kirschenlohr, H. L., Metcalfe, J. C., Weissberg, P. L., and Grainger, D. J. (1995) Proliferation of human aortic VSMCs in culture is modulated by active TGF-β. *Cardiovasc. Res.* **29**, 848–855.
15. Majack, R. A. (1987) Beta-type transforming growth factor specifies organizational behaviour in vascular smooth muscle cell cultures. *J. Cell. Biol.* **105**, 465–471.
16. Lemire, J. M., Covin, C. W., White, S., Giachelli, C. M., and Schwartz, S. M. (1994) Characterization of cloned aortic smooth muscle cells from young rats. *Am. J. Pathol.* **144**, 1068–1081.
17. Jonasson, L., Holm, J., Skalli, O., Gabbiani, G., and Hansson, G. K. (1985) Expression of class II transplantation antigen on vascular smooth muscle cells in human atherosclerosis. *J. Clin. Invest.* **76**, 125–131.
18. Grainger, D. J., Witchell, C. M., Weissberg, P. L., and Metcalfe, J. C. (1992) Heterogeneity, dedifferentiation and proliferation of primary rat vascular smooth muscle cells and clones. *J. Mol. Cell. Cardiol.* **24(suppl. I)**, 236.
19. Lyons, R. M., Gentry, L. E., Purchino, A. F., and Moses, H. L. (1990) Mechanism of activation of latent recombinant transforming growth factor β1 by plasmin. *J. Cell Biol.* **110**, 1361–1367.

25

Human Myometrial Smooth Muscle Cells and Cervical Fibroblasts in Culture

A Comparative Study

Françoise Cavaillé, Dominique Cabrol, and Françoise Ferré

1. Introduction

Uterine contractility and cervical tonicity change throughout the menstrual cycle and pregnancy in response to modifications in hormonal environment and tissue receptivity to hormones. The uterine wall consists of a smooth muscle (myometrium) organized into three layers: the inner and outer layers, mainly composed of smooth muscle cells, and the richly vascularized intermediate layer. The musculature is thick in the corpus uteri and vanishes at the level of the corpus/cervix junction. The cervix itself is formed mainly by connective tissue.

During pregnancy, both uterine muscle and cervical connective tissue undergo intense, but reversible structural and biochemical changes. These modifications allow the uterus to ensure two functions: the maintenance of pregnancy followed by the normal delivery of the fetus at a programmed time.

The uterine wall is the target of numerous hormones and growth factors acting simultaneously to permit growth of myometrial cells and modifications of the extracellular matrix. In the myometrium, extensive hypertrophy of the smooth muscle cells, which occurs mainly during the first part of pregnancy, is accompanied by intracellular organite reorganization *(1)*, changes in proteins of the contractile apparatus *(2)*, the appearance of numerous cell junctions in the membranes *(3)*, and modifications in cell receptivity to hormones *(4)*. Softening of the cervix at the end of pregnancy depends on modifications in both collagen and proteoglycan lattices that occur throughout pregnancy *(5)*. It should be noted that similar modifications in the extracellular matrix also occur in the myometrium *(6)*.

Although numerous works analyzing modifications of uterine contractility and morphological and biochemical modifications of myometrium and cervix have been performed on tissue and in vivo, the exact mechanisms that sustain such phenomena remain to be elucidated. This is of critical importance, since a comprehensive analysis of the factors sustaining uterine maturation will enable more efficient treatment of such disorders as premature labor, postdate pregnancy, or labor dystocia, which cause maternal and newborn morbidity.

Several authors have shown that culture of uterine cells constitutes a useful model for in vitro study of hormonal and growth factor effects on the uterus (e.g., 7–11). Two methods for obtaining smooth muscle cells in culture are currently used: enzymatic dispersion of myometrial cells (most of the following studies are carried out on primary culture, some on passaged cells) and cells grown from tissue explants, studied mainly after several passages. There appears to be no difference between results obtained with these two techniques.

In our laboratory, we developed cultures of human myometrial smooth muscle cells and of human cervical fibroblasts in order to study the mechanisms that allow physiological adaptation of the uterine wall to pregnancy.

We report here the simple method we adopted, which is identical for myometrial smooth muscle cells and cervical fibroblasts, indicating the structural features that distinguish the two cell types, as well as the cytoskeletal markers, detected by immunochemistry, that can be used to characterize the cells. Cell response to prostaglandin E_2 (PGE_2), in terms of cAMP production, permits, moreover, the assessment that cells conserve normal physiological properties. We propose a culture system that allows the preservation, almost to the 12th passage, of the expression of smooth muscle markers of differentiation in cultured myometrial cells.

In summary, using the simple explant method described here, we found that the cells that grew were representative of their tissues of origin, i.e., that cells originating from the external layer of the uterus exhibit smooth muscle phenotype, whereas cells originating from the cervix correspond to a special type of fibroblast, namely myofibroblasts (12–14), the two cell types responding differently, as the different parts of the uterus (15,16), to PGE_2. They should therefore be useful for investigating the actions (and interactions) of the multiple factors that modulate uterine and cervical contractile activity and that participate in uterine and cervical biochemical evolution during gestation and parturition.

2. Materials

1. Scalpels and fine curved scissors (sterile).
2. Growth medium (Dulbecco's modified Eagle's medium [DMEM]) containing 20% fetal calf serum (FCS) for the beginning of the culture and 0.1% penicillin/streptomycin.

3. Growth medium containing 10% FCS and 0.1% penicillin/streptomycin.
4. Calcium- and magnesium-free phosphate-buffered saline (PBS).
5. Trypsin–EDTA solution (0.25 and 0.02%, respectively) in PBS.
6. DMEM, without phenol red, supplemented with transferrin (5 µg/mL) and ascorbate (0.2 mM).
7. Acetone/ethanol (1/1 [v/v]), maintained at –20°C.
8. PBS containing 1 mM phenylmethyl sulfonyl fluoride (PMSF) (added extemporanelly from a stock solution, 100 mM in ethanol).
9. Cell lysis buffer (electrophoresis sample buffer): 2.3% sodium dodecyl sulfate (wet w/v), 10% glycerol (v/v), 10 mM EDTA, 2 mM PMSF, 20 µM leupeptin, 20 µg/mL pepstatin, 20 µg/mL aprotinin, 130 µg/mL benzamidine, 0.5% β-mercaptoethanol (v/v), 62.5 mM tris(hydroxymethyl) aminomethane, pH 6.8.

3. Methods

3.1. Collection of Biopsy

1. Tissue samples are removed from uteri after hysterectomy for benign gynecologic indications (mainly fibromas and uterine prolapse); women should be <45 yr old; biopsies can also be obtained from pregnant uteri at the time of Caesarean section (*see* Note 1).
2. Tissue samples are dissected in a locale close to the operating theater and immediately placed in sterile tubes containing culture media supplemented with antibiotics (*see* Note 2). They can be kept at 4°C for several hours if culturing is not immediately possible.

3.2. Primary Explant Culture

1. Tissue is transferred to a Petri dish, rinsed with fresh sterile medium, and cut with fine curved scissors (*see* Note 3). Processing of the explant preparation and cell culturing was the same for myometrial and cervical tissues.
2. Using fine curved forceps, tissue pieces (≤1 mm^2) are put into 6-cm dishes. Using a 1-mL pipet, a drop of medium is carefully placed above each of the tissue pieces (approx 20 pieces/dish). The medium consists of DMEM supplemented with 20% FCS and antibiotics. The dishes are then kept at 37°C in a humidified atmosphere of 5% CO_2–95% air (*see* Note 4).
3. After 3 d, 2 mL of DMEM containing 20% FCS are carefully added to the explants. The first cells growing from the tissue explants can be seen after 7 d. They are spindle-shaped and migrate from the tissue (Fig. 1A). The medium is then changed every 3 d.
4. When a significant number of cells (the cells form a crown of confluent cells around the explants, Fig. 1B) are obtained (approx 15 d), the medium is changed for DMEM containing 10% FCS, and the cells are left for a further 7-d period with one change of medium. At that time, approx 100,000–250,000 cells/dish can be expected (*see* Note 5).

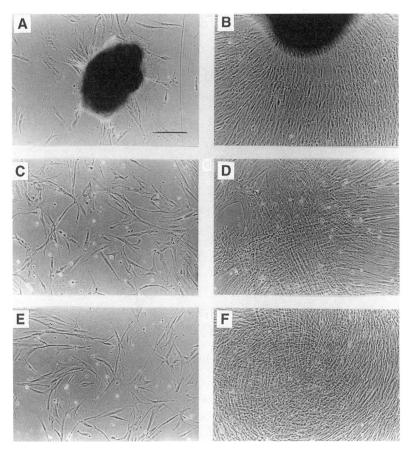

Fig. 1. Phase-contrast micrographs of cells grown from myometrial (**A–D**) and cervical (**E,F**) explants. A and B: myometrial cells 7 and 21 d after the beginning of the culture, respectively; C and D: myometrial cells (4th passage) 2 and 7 d after plating, respectively; E and F: cervical cells (4th passage) 2 and 7 d after plating, respectively. Original magnification ×87.5; bars (A) = 250 µm.

3.3. Subculture and Obtainment of a "Differentiated" Phenotype

1. Experiments can be performed directly on the cells that grew from the explants, but generally more cells are needed; in that case, cell number is amplified by subculture. Cells, rinsed with calcium- and magnesium-free PBS, are harvested with trypsin/EDTA, plated at 2×10^5 cells/6-cm Petri dish, and grown to confluence in DMEM containing 10% FCS (*see* Note 6).
2. Cells are further maintained for a 3- to 5-d period in serum-depleted medium. Medium is phenol red-free DMEM, supplemented with ascorbate and transferrin (*see* Notes 7 and 8).

A) α-Smooth-Muscle Actin

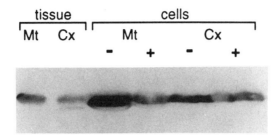

B) Smooth Muscle Myosin

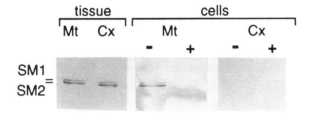

Fig 2. Reactivity on Western blots of protein extracts of myometrial (Mt) and cervical (Cx) tissues and cultured cells with anti α-smooth-muscle actin (**A**) and antismooth muscle myosin (**B**) monoclonal antibodies. SM1 and SM2: the two heavy chains of smooth muscle type. Cells (4th passage) were grown to confluence and left for a further period of 5 d either in serum-depleted medium (–) or in the presence of serum (+).

3.4. Criteria Used to Characterize Cell Phenotype

1. Cells differ in their population doubling times; 7 d after plating, the number of myometrial and cervical cells are 5 and 10 times that of plated cells, respectively.
2. The organization of the cells at confluence is very characteristic: Cervical cells form concentric bundles (Fig. 1F), whereas myometrial cells are grouped into bundles that cross each other (Fig. 1D) (*see* Note 9).
3. More precise characterization of the nature of the cells arising from myometrium (presumed to be smooth muscle cells) and cervix (presumed to be fibroblasts) can be done by using biochemical criteria (*see* Note 10), which also can be used to follow cell phenotype with subculturing *(17)*. Cytoskeletal proteins expressed in the cells can be visualized either by Western blotting (Fig. 2; *see* Note 11) or immunocytochemistry (Fig. 3; *see* Note 12).
4. Response to hormones is also specific to each type of cells, as shown by the measure of cAMP production under prostaglandin stimulation (Fig. 4, *see* Note 13).

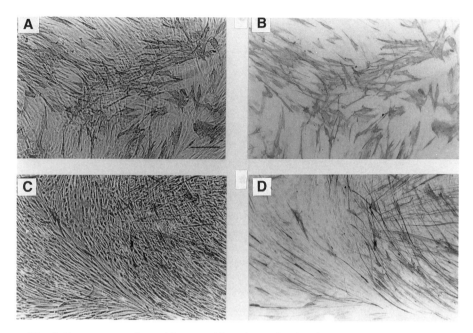

Fig. 3. Immunoreaction with an antidesmin antibody of myometrial (**A,B**) and cervical (**C,D**) cells (4th passage). Cells were grown to confluence and then left for a further period of 5 d in serum-depleted medium. A and B: Phase-contrast micrographs of postconfluent myometrial and cervical cells, respectively. C and D: Same fields as in A and B, respectively; direct illumination enabled visualization of cells in which staining was developed after the peroxidase reaction. Original magnification ×100, bar (A) = 200 μm.

4. Notes

1. Local ethics committee approval is needed for work on human tissues, as well as patients' informed voluntary consent.
2. Small pieces of normal myometrium are to be taken in the external muscle layer to avoid contamination by vascularization and endometrium; tiny strips of cervical stroma are cut at the level of the external uterine os. Epithelial elements are discarded with a scalpel.
3. Scissors are more convenient than scalpels, which wear very rapidly.
4. Tissue pieces do not have to float into the medium to obtain their adhesion to plastic. Each day examination is necessary to ensure that the medium has not evaporated. In that case, small quantities of medium are to be carefully added.
5. As reported in Section 1., two methods can be used to obtain smooth muscle cells in culture. One involves the enzymatic dissociation of the cells from the tissue, whereas the other consists of growing the cells from tissue explants. Human uterine muscle, as well as cervical tissue, are particularly rich in extracellular matrix.

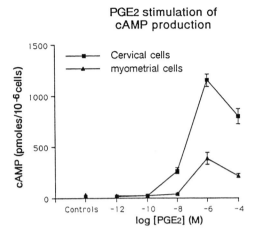

Fig. 4. Stimulation of cAMP accumulation in confluent myometrial and cervical cells. Cell content in cAMP was measured by radioimmunoassay after 10 min of incubation at 37°C in the presence of 0.5 mM IBMX with increasing doses of PGE2, as described in Negishi et al. *(18)*.

Therefore, cell dispersion from the tissue requires the use of extensive digestion times with collagenase and trypsin, which nevertheless result in a low yield of cell recovery. Moreover, it is difficult to obtain large quantities of normal human uterine tissue, particularly from gravid uteri, for obvious ethical reasons.

6. Trypsinized cells can be conserved in liquid nitrogen after freezing in DMSO at –80°C. In our hands, cell viability after a period of 6 mo was about 20%.
7. Estrogens are potential mitogens and affect protein synthesis in the uterine cells. The medium employed at this time was chosen without phenol red, because of the potential estrogenic action of this component.
8. High cell density and cell quiescence are necessary for obtaining a differentiated phenotype. Cell quiescence is obtained by maintenance of cells for a minimum of 2 d in serum-free medium.
9. Cervical cell organization at confluence corresponds to that described for skin fibroblasts *(19)*, whereas confluent myometrial cells show smooth muscle cells typical of a "hill-and-valley" pattern. Cells were passaged (a maximum of 12 passages in this study) without changes in their pattern of growth. Any variations of these cell aspects at confluence indicate that cells either have undergone transformation with subculturing or, alternatively, have been contaminated by other cell types (epithelioid cells were found to be occasional contaminants).
10. Smooth muscle cells could be distinguished from fibroblasts by their content in muscle-specific cytoskeletal proteins and isoforms of contractile proteins *(17,20–22)*. We chose three proteins abundant in human myometrium: desmin (a component of the intermediary filaments), actin (a component of the thin filaments), and

myosin (a component of the thick filaments). Desmin and vimentin are the two intermediary filament major proteins in human myometrium, with desmin strongly predominant *(2)*. Thin filaments of smooth muscles are mostly formed of α- and γ-smooth muscle actin (about 57 and 37% in nonpregnant, and 23 and 57% in pregnant myometrium, respectively *[23]*). In smooth muscles, the myosin hexameric protein contains two specific smooth muscle heavy-chain isoforms, SM1 (204 kDa) and SM2 (200 kDa), whose ratio of expression depends on the tissue origin *(24)*. In human myometrium, SM1 and SM2 represent about 30 and 40%, respectively, of total myosin heavy chains *(13)*.

11. For Western blotting, cells rinsed with cold PBS containing PMSF are detached from the dishes with a rubber policeman, also called a cell scraper. They are pelleted by centrifugation (12,000g) in a microtube, weighted, and dissolved in electrophoresis sample buffer containing proteases inhibitors.

12. For immunocytochemistry, cells grown on plastic are first rinsed with PBS, and then dehydrated, and their membranes permeabilized by impregnation with cold (–20°C) acetone/ethanol for 3 min followed by air-drying at room temperature.

13. Human cervical cells were plated onto 35-mm diameter plastic culture dishes, at a cell density of 10^5 cells/dish and allowed to grow to confluence. The cell layers were washed and preincubated 30 min with HEPES-buffered-saline solution, 125 mM NaCl, 4.7 mM KCl, 2.2 mM CaCl$_2$, 1.2 mM MgCl$_2$, 1.2 mM NaH$_2$PO$_4$, 15 mM NaHCO$_3$, 11 mM glucose, and 15 mM HEPES, pH 7.4. Reactions were started by the addition of test agents (PGE$_2$ or sulprostone from 0–$10^{-4} M$) along with 0.5 mM 3-isobutyl-methylxanthine (IBMX). Experiments were conducted in quadruplicate with each agent.

After incubation for 10 min at 37°C, reactions were terminated by the addition of 10% trichloracetic acid solution (TCA). Content of cAMP in the cells was determined after TCA extraction with ether, by radioimmunoassay using Amersham Corp's (Les Ulis, France) cAMP assay kit.

Acknowledgments

The authors thank Emmanuelle Dallot and Thérèse Fournier for their excellent technical assistance.

References

1. Laguens, R. and Lagrutta, J. (1964) Fine structure of human uterine muscle in pregnancy. *Am. J. Obstet. Gynecol.* **89**, 1040–1048.
2. Cavaillé, F., Janmot, C., Ropert S., and D'Albis A. (1986) Isoforms of myosin and actin in human, monkey and rat myometrium. Comparison of pregnant and non-pregnant uterus proteins. *Eur. J. Biochem.* **160**, 507–513.
3. Tabb, T., Thilander, G., Grover, A., Hertzberg, E., and Garfield, R. (1992) An immunochemical and immunocytologic study of the increase in myometrial gap junctions (and connexin 43) in rats and humans during pregnancy. *Am. J. Obstet. Gynecol.* **167**, 559–567.

4. Wray, S. (1993) Uterine contraction and physiological mechanisms of modulation. *Am. J. Physiol.* **264**, C1–C18.
5. Uldbjerg, N., Ekman, G., Malmström, A., Olsson, K., and Ulmsten, U. (1983) Ripening of the human cervix related to changes in collagen, glycosaminoglycans, and collagenolytic activity. *Am. J. Obstet. Gynecol.* **147**, 662–666.
6. Cabrol, D., Dallot, E., Cédard, L., and Sureau, C. (1985) Pregnancy-related changes in the distribution of glycosaminoglycans in the cervix and the corpus of the human uterus. *Eur. J. Obstet. Gynecol. Reprod. Biol.* **20**, 289–295.
7. Chen, L., Lindner, H. R., and Lancet, M. (1973) Mitogenic action of estradiol-17β on human myometrial and endometrial cells in long-term tissue cultures. *J. Endocrinol.* **59**, 87–97.
8. Casey, M. L., MacDonald, P. C., Mitchell, M. D., and Snyder, J. M. (1984) Maintenance and characterization of human myometrial smooth muscle cells in monolayer culture. *In Vitro* **20**, 396–403.
9. Kawaguchi, K., Fujii, S., Konishi, I., Okamura, H., and Mori, T. (1985) Ultrastructural study of cultured smooth muscle cells from uterine leiomyoma and myometrium under the influence of sex steroids. *Gynecol. Oncol.* **21**, 32–41.
10. Környei J. L., Lei, Z. M., and Rao, C. V. (1993) Human myometrial smooth muscle cells are novel targets of direct regulation by human chorionic gonadotropin. *Biol. Reprod.* **49**, 1149–1157.
11. Heluy, V., Breuiller-Fouché, M., Cavaillé, F., Fournier, T., and Ferré, F. (1995) Characterization of a type A endothelin receptors in cultured human myometrial cells. *Am. J. Physiol.* **268**, E825–E831.
12. Gabbiani, G., Ryan, G. B., and Majno, G. (1971) Presence of modified fibroblasts in granulation tissue and their possible role in wound contraction. *Experientia* **27**, 549.
13. Skalli, O., Schürch, W., Seemayer, T., Lagacé, R., Montandon, D., Pittet, B., and Gabbiani, G. (1989) Myofibroblasts from diverse pathologic settings are heterogeneous in their content of actin isoforms and intermediate filament proteins. *Lab. Invest.* **60**, 275–285.
14. Eyden, B. P., Ponting, J., Davies, H., Bartley, C., and Torgersen, E. (1994) Defining the myofibroblast: normal tissues, with special reference to the stromal cells of Wharton's jelly in human umbilical cord. *J. Submicrosc. Cytol. Pathol.* **26**, 347–355.
15. Wiqvist, N., Linblom, B., Wikland, M., and Wilhelmsson, L. (1983) Prostaglandins and uterine contractility. *Acta Obstet. Gynecol. Scand.* **Suppl. 113**, 23–29.
16. Bryman, I., Sahni, S., Norström, A., and Linblom, B. (1984) Influence of prostaglandins on contractility of the isolated human cervical muscle. *Obstet. Gynecol.* **63**, 280–284.
17. Cavaillé, F., Fournier, T.,. Dallot, E., Dhellemmes, C., and Ferré, F. (1995) Myosin heavy chain isoform expression in human myometrium; presence of an embryonic nonmuscle isoform in leiomyomas and in cultured cells. *Cell Motil. Cytoskel.* **30**, 183–193.
18. Negishi, M., Ito, S., and Hayaishi, O. (1989) Prostaglandin E receptors in bovine adrenal medulla are coupled to adenylated cyclase via Gi and to phosphoinositiole metabolism in a pertussis toxin-insensitive manner. *J. Biol. Chem.* **264**, 3916–3923.

19. Chamley-Campbell, J., Campbell, G. R., and Ross, R. (1979) The smooth muscle cell in culture. *Physiol. Rev.* **59,** 1–61.
20. Osborn, M. and Weber K. (1986) Intermediate filament proteins: a multigene family distinguishing major cell lineages. *Trends Biochem. Sci.* **11,** 469–472.
21. Skalli, O., Vandekerckhove, J., and Gabbiani, G. (1987) Actin-isoform pattern as a marker of normal or pathological smooth-muscle and fibroblastic tissues. *Differentiation* **33,** 232–238.
22. Frid, M. G., Printseva, O. Y., Chiavegato, A., Faggin, E., Scatena, M., Kotelianski, V. E., Pauletto, P., Glukhova, M. A., and Sartore, S. (1993) Myosin heavy-chain isoform composition and distribution in developing and adult human aortic smooth muscle. *J. Vascular Res.* **30,** 279–292.
23. Ewoane, C. and Cavaillé, Г. (1990) ATPase activity of reconstituted actomyosin from pregnant and non-pregnant human uterus; its dependence on the actin isoforms. *Biochem. (Life Sci. Adv.)* **9,** 5–10.
24. Somlyo, A. P. (1993) Myosin isoforms in smooth muscle: how may they affect function and structure? *J. Muscle Res. Cell Motil.* **14,** 557–563.

26

Primary Culture of Human Antral Endocrine Cells

Alison M. J. Buchan

1. Introduction

The mucosal endocrine cells in the antrum are found as individual elements interspersed among the other epithelial cells (i.e., mucin cells) (Fig. 1). In order to establish the factors regulating endocrine cell function, these cells have to be separated not only from the surrounding epithelial cells, but also from circulating and neuronal elements within the stomach.

A major problem in obtaining cultures of gastric endocrine cells is their diffuse distribution in the stomach and the nonsterile nature of the lumen. To overcome these problems, we have used a combination of collagenase digestion of the mucosal layer with centrifugal elutriation to remove small particles, such as bacteria and fungi, and provide an enriched preparation of endocrine cells. The technique represents a modification of the methodology originally developed to isolate endocrine cells from the canine stomach (1).

Unfortunately, none of the techniques so far developed produce a 100% pure culture of an individual endocrine cell type. In the antrum, cultures with a 40% content of gastrin cells can be obtained, the majority of the remaining cells being gastric mucin cells.

The cell cultures obtained have been used for a number of different techniques; release studies examining regulation of hormonal secretion, intracellular ion flux in response to stimulation (e.g., Fura-2 measurement of intracellular calcium levels) (2), immunocytochemical studies at light and electron microscopical levels, and molecular studies of receptor and ion-channel expression (3–8).

2. Materials

2.1. Tissue Collection

A 100-mL screw-topped container with 50 mL of chilled buffer, 1 pair scissors (8–12 in.), 1 pair forceps (6–8 in.).

From: *Methods in Molecular Medicine: Human Cell Culture Protocols*
Edited by: G. E. Jones Humana Press Inc., Totowa, NJ

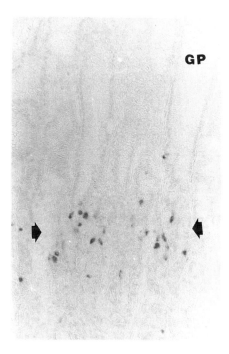

Fig. 1. A section of human antrum immunostained using the peroxidase method to localize the gastrin-immunoreactive endocrine cells (arrows). Note that the cells are grouped in the mid-third of the mucosa with no immunoreactive cells at the top of the gastric pits (GP) (×100).

2.2. Dissection of Stomach

1. Plastic container full of ice.
2. 500-mL Dish of Hank's balanced salt solution (HBSS) containing 0.1% bovine serum albumin (BSA) and 10 mM N-2-hydroxyethyl piperazine-N-2-ethane sulfonic acid (HEPES) buffer on ice (referred to as HBSS/BSA from now on).
3. 50-mL Beaker of HBSS/BSA buffer on ice.
4. Scissors (6–8 in.).
5. Sharp, fine scissors (4–6 in.).
6. Forceps (6–8 in.).
7. Fine forceps (6 in.).
8. 100-mm Glass Petri dish.
9. Latex gloves (mandatory for handling human tissue).
10. Face mask if uncertain of the status of the patient (i.e., Hepatitis, virus infection, and so forth).

2.3. Digestion of Mucosa

1. Shaking water bath at 37°C.
2. 250-mL Nonsterile conical flasks (1/10 g of stripped mucosa).
3. Sigma (St. Louis, MO) type I collagenase (stored as aliquots of 100 mg dry powder at –20°C).
4. 1 L Basal medium Eagle (BME) containing 0.1% BSA, 10 mM HEPES.
5. 1 L HBSS/BSA + 0.01% Dithiothreitol + 0.001% DNase.
6. 5 mL 0.5M EDTA.
7. 5% CO_2 in O_2 gas cylinder.
8. Sharp scissors (4–6 in.).
9. 25-mL Beaker.
10. 500 mL Glass/plastic beaker *(1)*.
11. 250-mL Plastic beaker *(2)*.
12. Squares of 400 µm of Nytex™ mesh cut to fit over the 250-mL beaker.
13. Plastic ring cut from a 250-mL beaker to fit into the rim of a 250-mL beaker.
14. 50-mL nonsterile centrifuge tubes *(6–8)*.
15. 10-mL Gilson pipet *(1)*.
16. 0.4% Trypan Blue.
17. Hemocytometer

2.4. Centrifugal Elutriation

1. Beckman centrifuge with elutriator rotor.
2. 5-mL Standard separation chamber.
3. Pump (Cole Parmer [Niles, IL], Masterflex Model 7520-20 or similar).
4. Laminar flow hood adjacent to centrifuge/rotor assembly.
5. 70% Ethanol.
6. 100-mL Measuring cylinder.
7. 1 L Sterile water.
8. 2 L Sterile HBSS/0.1% BSA.
9. 50-mL Sterile centrifuge tubes (Falcon) (4/elutriation load).

2.5. Tissue Culture

1. 12- or 24-Well Costar plates coated with sterile rat tail collagen are optimal for release experiments.
2. 35-mm Falcon or Costar dishes are optimal for immunocytochemistry.
3. 35-mm Falcon or Costar dishes with collagen-coated sterile coverslips are optimal for imaging of intracellular ion concentration (Fura-2 ratio measurement), and confocal microscopy. Coverslips are sterilized by flaming in 70% ethanol and allowing to cool before coating with collagen.
4. 90-mm Falcon plates are used to collect cells for ultrastructural studies and to isolate mRNA for molecular studies.
5. 300-mL Growth medium.

2.6. Growth Medium

Dulbecco's modified Eagle's medium (DMEM) containing 5.5 mM glucose, 10 mM HEPES, 2 mM glutamine, 8 µg/mL insulin, 20 ng/mL epidermal growth factor (stored at 100X final concentration in sterile DMEM; 0.1-mL aliquots at –70°C), 50 µg/mL gentamycin, 1 µg/mL hydrocortisone, 5% heat-inactivated human serum, 1 mL/100 mL penicillin/streptomycin (Gibco, Burlington, Ontario, Canada), 1 mM sodium pyruvate.

2.7. Eurocollins Buffer

Buffer contains 10.0 mM NaCl, 57.7 mM Na$_2$PO$_4$, 115.0 mM KCl, 19.0 mM glucose, and 10.0 mM NaHCO$_3$. Osmolarity = 330 mosM/kg, pH 7.0.

3. Methods

3.1. Collection of Antral Material

The distal portion of the stomach, 3–4 in. proximal to the pyloric sphincter, is required. The optimal situation would be to obtain material from the retrieval program of the local Transplant Society (Note: Ethical permission forms must be in place for the university/research institute, and the permission of the next-of-kin for use of material for research should be obtained where required).

The critical factor in the tissue collection is to minimize warm ischemic time. In this case because the tissue was collected during an organ harvest for transplantation, the venous supply was replaced by ice-cold Eurocollins buffer prior to clamping the aorta. Thus, there was little, if any, warm ischemia. If the material has to be collected from surgery, care must be taken to ensure that as soon as the antral material is removed from the abdominal cavity, it is immediately immersed in ice-cold buffer.

3.2. Transport to the Laboratory

The transport time should be kept to a minimum. If material has been collected from surgery, once the tissue is out of the operating room, open up the antrum and flush out the lumen with ice-cold buffer. This is vitally important if there is bile present. The bile acts as a detergent and digests the tissue even at 4°C. Once cleaned, it should be transferred to an intracellular buffer at 4°C (such as Eurocollins), which slows enzyme activity and minimizes damage from free radicals. Antral tissue once cleaned and immersed in the ice-cold buffer can be stored at 4°C for up to 10 h prior to digestion (*see* Note 1).

3.3. Initial Separation of Mucosa (see Fig. 2)

1. The antral mucosa is blunt dissected from the underlying submucosa and muscle layers by taking the whole piece of tissue and removing any adherent fat from the

Human Antral Endocrine Cells

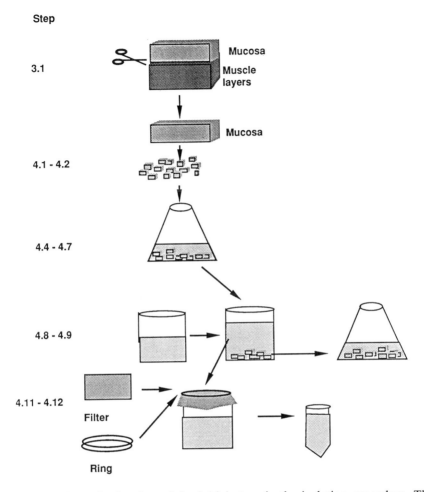

Fig. 2. A schematic drawing of the initial steps in the isolation procedure. The numbers refer to the individual steps in the protocol.

serosal surface. With the mucosal face downward in a Petri dish with approx 5 mL of HBSS/BSA, the muscle layer is gripped using forceps and sharp scissors are used to cut through the submucosal layer. This should be done as close to the mucosal layer as possible without cutting into the mucosa. Once all of the muscle layer is removed, any remaining adherent submucosa appears as a whitish layer over the beige mucosa. As far as possible, this white layer should be removed using sharp fine scissors and fine forceps. It is imperative to remove the submucosa with muscularis mucosae to ensure that no nerve or muscle cells are retained in the preparation. The effectiveness of the dissection can be checked by taking a small sample of the tissue, freezing for cryostat sectioning and staining 20-µm

sections with hematoxylin and eosin. It takes several practice sessions to perfect the removal process.
2. The muscle and submucosal tissue are placed in an autoclave bag (all waste from the isolation procedure contaminated with human cells, including Nytex filters, must be autoclaved prior to disposal; check the regulations governing disposal of human tissue at your institution) and the mucosa turned face upward. At this point, the antral region can be discriminated from the corpus by color. The corpus mucosa is thicker and is a darker beige compared to the thinner antral region. The darker regions are cut away and discarded, leaving the stripped antral mucosa. Again this can be checked by taking a small piece of mucosa, freezing, and cryostat sectioning to ensure no corpus tissue is included.
3. The antral tissue is then weighed; it is usually around 9–11 g.

3.4. Isolation of Single Cells

1. Place the mucosa in a small beaker with no buffer, and mince with sharp scissors.
2. Transfer the minced tissue into a Petri dish, and continue to chop using scalpels (No. 4) until the average size of the pieces is 5 mm^3.
3. If the tissue is <14 g in weight, use a single 250-mL flask for the digestion; if 15–20 g, use two flasks.
4. For the first digest, add 50 mL BME containing 600 U/mL of Sigma type I collagenase to each flask.
5. Gas with 5% CO_2 in O_2 for 1 min prior to capping.
6. Incubate in the 37°C shaking water bath at 200 Hz for 45 min.
7. Add 300 µL 0.5M EDTA for 15 min.
8. Remove flask(s), empty contents into the 500-mL beaker, and double volume with HBSS/BSA/DDT/DNase (*see* Note 2). Allow nondigested material to settle to the bottom of beaker and discard supernatant from first digest (*see* Note 3). Leave a maximum of 10 mL buffer in the beaker.
9. Replace the undigested material in the flask(s), and add 50 mL of BME/BSA containing 300 U/mL Sigma collagenase type I. Gas as before and return to shaking water bath for 45 min.
10. Add 750-µL 0.5M EDTA for 15 min.
11. Remove flask(s), empty contents into the 500-mL beaker, and double the volume with HBSS/BSA/DDT/DNase. Allow nondigested material to settle, and pour supernatant into a second beaker leaving a maximum of 5 mL of buffer with the undigested material in the original beaker. Filter the supernatant through 400-µm Nytex mesh suspended over a plastic 250-mL beaker. The Nytex mesh can be anchored over the beaker using a ring made from part of another plastic beaker. The undigested material should be returned to the flask as in step 9.
12. Separate supernatant from second digest into 50-mL centriguge tubes (nonsterile) and centrifuge at 200g for 10 min. Discard supernatant, and resuspend pellet in 50 mL HBSS/BSA. Cells should always be resuspended using a 10-mL Gilson pipet initially in a small volume (i.e., 5 mL) to aid dispersion of the clumped cells and then brought to final volume.

13. Repeat steps 11 and 12 to ensure collagenase is removed from the cell suspension.
14. Repeat steps 8–13 until no undigested material remains (usually three to four times) (*see* Note 4).
15. Take the 50-mL resuspensions from digests 2–5, centrifuge at 200g for 10 min, resuspend each in 5 mL, and combine. Add HBSS/BSA to take total volume to 75 mL. Filter through 40-µm Nytex mesh to remove cell clumps. To aid filtration through the fine mesh, a hole can be made in the side of the 250-mL plastic beaker at the 225-mL level and a 20-gage needle inserted attached to a 50-mL syringe. Air drawn into the syringe creates suction and speeds up the process. If the digest was not effective at producing a single-cell preparation, the 40-µm mesh will clog rapidly. Replace with a fresh square.
16. Check viability of cells, and count cells using Trypan Blue. A 100-µL sample of the cell suspension is added to 100 µL of 0.4% Trypan Blue, placed in a hemocytometer, and examined under the microscope. Viable cells are unstained. Dead cells have a blue nucleus. At this point, count all cells alive or dead.
17. Dilute the cell suspension to 1.5×10^8 cells/20 mL in HBSS/BSA/DDT/DNase. The number of elutriation loads is calculated by dividing the total number of cells by 1.5×10^8.

3.5. Elutriation

1. Centrifugal elutriation: rotor assembly and calibration of flow rates.
    a. Assemble rotor; ensure separation chamber free of debris.
    b. Sterilize rotor, chamber, and input and output lines with 70% alcohol. Leave in 70% alcohol until 30 min prior to separation.
    c. Replace 70% alcohol with sterile water, and set flow rates. Flow rates are calibrated at 900g for 25, 40, and 55 mL/min. Care was taken to ensure that at low flow rates (25–35 mL/min), pressure in the line was <3 psi. If higher, this indicated that bubbles were trapped either in the separation chamber or in the spindle assembly of the rotor. To remove bubbles, the elutriator rotor was run at 150g with a flow rate at 75 mL/min, the rotor was stopped while the flow rate was maintained, and all bubbles were forced out of the chamber and spindle.
    d. Replace sterile water with sterile HBSS/BSA.
2. Elutriation of antral cells
    a. 1.5×10^8 cells are loaded into the chamber at a flow rate of 25 mL/min and a speed of 900g. Loading volume should be between 20 and 30 mL.
    b. Once cells are in the chamber (checked using the stroboscope), wash for 3 min at 25 mL/min.
    c. Increase the flow rate to 40 mL/min while decreasing the centrifuge speed to 655g, and collect 2 × 50 mL fractions in sterile tubes. This fraction (F1) contains the majority of the G-cells.
    d. Increase the flow rate to 55 mL/min while decreasing the centrifuge speed to 450g, and collect 2 × 50 mL fractions in sterile tubes. This fraction (F2) contains the majority of the somatostatin containing D-cells.

Note: From this point on, care should be taken to ensure sterile techniques are followed.

e. Repeat a–d until all cells have been run through the elutriator (*see* Note 5).
f. Centrifuge all collected fractions at 200g for 10 min.
g. Resuspend in 10 mL growth medium (for composition, *see* Section 2.6.) and complete a cell count using Trypan Blue as before. This time count only live cells (*see* Note 6).

3.6. Tissue Culture

1. The F1 fraction containing the G-cells is resuspended at a final concentration of 1×10^6 cells/mL in growth medium.
2. The cell suspension is plated on 24-well plates coated with rat tail collagen at 1 mL/well for release experiments, 35-mm dishes with or without coverslips at 2 mL/dish for immunocytochemical and intracellular ion measurement (e.g., Fura-2) experiments, and 90-mm dishes at 10 mL/dish. The average yield of cells in the F1 fraction is between 60 and 200×10^6 cells.
3. The F2 fraction containing the D-cells is resuspended at a final concentration of 2×10^6 cells/mL in growth medium.
4. The F2 cell suspension is plated on 12-well plates coated with rat tail collagen at 1 mL/well for release experiments, and 35-mm dishes with or without coverslips at 2 mL/dish for immunocytochemical and intracellular ion measurement (e.g., Fura-2) experiments. The average yield of cells in the F2 fraction is between 40 and 100×10^6 cells (*see* Note 7).

3.7. Characteristics of Resultant Cultures

The F1 and F2 cells adhere to the collagen substrate overnight and are phase bright. Their viability in culture remains >95% for up to 4 d as shown by Trypan Blue exclusion experiments conducted throughout this period. After 4 d, the viability decreases, and the cells detach from the collagen. Cells aggregate into small clusters containing a mixture of endocrine and mucous cells (Fig. 3). In the author's experiments, 48 h was chosen for the majority of the experimental protocols, this time period having been demonstrated to allow reformation of cell-surface receptors and cell polarity.

In studies of receptor and ion-channel expression by the cultured cells, it has proved useful to collect mRNA samples from the cells immediately after elutriation and prior to the culture period. In many cases, the addition of specific growth factors (insulin or epidermal growth factor) at the high concentrations used in these cultures can induce expression of receptors or channels not normally present on the cells in vivo. To control for this possibility, the presence of the receptor or ion channel in the cells prior to culture can be determined using mRNA collected from the postelutriation samples for the polymerase chain reaction.

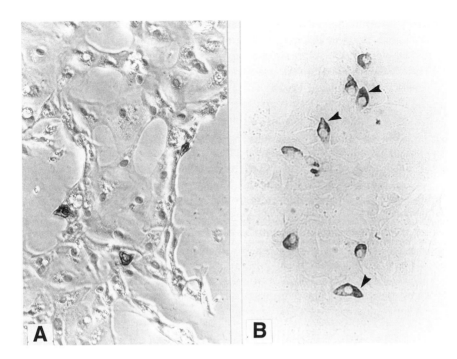

Fig. 3. **(A)** A representative area of the antral cultures showing the clustering of the isolated cells after 2 d in culture (×400). **(B)** A detail of a culture immunostained by the peroxidase method to demonstrate the gastrin containing cells. Note that the cells have repolarized during the 2-d culture period, with the majority of the immunoreactivity located at one pole of the cell (arrows) (×750).

4. Notes

1. One advantage of the studies carried out in the author's laboratory was the availability of tissue from the B.C. Transplant Society in the province of British Columbia. Patients were screened before acquiring tissue to ensure their suitability as organ donors and did not have any known pathophysiological conditions. However, as is often the case with human studies, and unlike studies of laboratory animals, such as rats, there was variability with respect to age, size, and sex of the donors. This information should be noted and reported in any subsequent publications.

    Antral tissue is not as heavily vascularized as the small intestine, and once the tissue is placed in ice-cold buffer, the cells can withstand a delay of up to 6 h before starting the isolation procedure. Longer storage times (>8 h) result in a lower yield of viable cells and should be avoided if possible.

2. The volume of the digestion solution is at least doubled with cold HBSS/BSA/ DDT/DNAse at the end of each 1-h incubation period prior to the centrifugation step to wash the tissue properly. This quickly reduces the activity of the collage-

nase. The DDT and DNase are added to minimize clumping of the single cells. DDT counteracts the aggregation of the mucin released from the gastric mucous cells, and DNase stops DNA released from lysed cells sticking the live cells together.

3. The reason for discarding the cells collected from the first digest is that these represent the surface epithelial cells, which have a reduced viability. You can check the viability by Trypan Blue exclusion. In addition, there are few if any endocrine cells in the upper layers of the human antral mucosa. Therefore, this cell population can be discarded without affecting the total yield of endocrine cells.

4. The isolation protocol described produced the highest yield of viable cells. The choice of collagenase can be difficult. Sigma type I collagenase gave a better yield of antral G cells than either Sigma Type XI or Worthington collagenase type I. However, each batch of collagenase should be tested prior to its use, since the activity and level of different additional enzymes in the preparation vary between batches. It is well worth testing several batches from Sigma and then putting a large amount of a good batch on hold. It is important to remember that if you change batches of collagenase, problems may be encountered either with a loss of viable cells in the initial digest or the isolated cells may not survive in culture.

5. The Beckman elutriator centrifuge is equipped with a stroboscope, which allows the separation chamber to be monitored during each run. To speed up the process, it is usual not to stop the rotor completely between runs, but to wash the cells remaining in a pellet after the second fraction has been collected out of the chamber by decreasing the rotor speed to $85g$ and increasing the flow rate to 100 mL/min. However, this does not always work, and the pellet remaining in the chamber can be seen when the rotor is returned to $900g$ before the next load. If the pellet is retained, **do not attempt to load the next batch of cells**. Stop the rotor with a high flow rate (>100 mL/min), open the centrifuge, and check that the pellet has been washed clear. If stopping the rotor still does not remove the pellet, then the entire rotor assembly will have to be taken out of the centrifuge, the chamber removed, and the pellet dislodged with a fine needle. Once the pellet is removed, the elutriation process can be restarted.

6. This procedure increases the number of viable cells, and removes the majority of the other cell types and the usual bacterial and fungal contaminants introduced through the nonsterile enviroment of the antral lumen.

Receptor damage owing to hyperosmolality *(9)* is avoided since cell separation in these experiments was carried out using elutriation, which permits the use of isotonic solutions *(10)*. Centrifugal elutriation utilizes centrifugal force and flow, which act in opposing directions, to separate cells on the basis of their volume. Different fractions of cells can be removed by altering the flow rate of fluid passing through the elutriation chamber by altering the pump rate or by changing the speed of the centrifuge. Appropriate flow rates, centrifugation speeds, and washing times can be determined empirically, and can be easily altered to choose different populations of cells for study.

Peptide containing cells can be used as indicators for the enrichment by taking samples before and after elutriation centrifugation and carrying out radioimmu-

noassay. For example, gastrin content of 10 mg of undigested antral mucosa can be compared to the content of the cell suspension prior to elutriation and the two cell fractions collected during elutriation (usually determined for 1×10^6 cells). The elutriation procedure can then be altered to maximize enrichment of G-cells.
7. An initial plating density of 1×10^6 cells/mL/plate was chosen for the F1 fraction. This gives optimal attachment of the cells to the collagen substrate, and the content of gastrin is well within the range required for detection by the available radioimmunoassays. This plating density has to be doubled for the F2 fraction if it is intended to complete release experiments examining the regulation of somatostatin release. The number of somatostatin containing cells and their content are lower than that of gastrin. Therefore, the plating density has to be increased to bring the peptide content up to the detection level of the available radioimmunoassays. This is vital to detect basal somatostatin levels to provide the control level for comparision with subsequent stimulation or inhibition of the D-cells.

References

1. Soll, A., Amirian, D., Park, J., Elashoff, J., and Yamada, T. (1985) Cholecystokinin potently releases somatostatin from canine fundic mucosal cells in short-term culture. *Am. J. Physiol.* **248**, G569–G573.
2. Buchan, A. M. J. and Meloche, R. M. (1994) Signal transduction events involved in bombesin-stimulated gastrin release from human G cells in culture. *Can. J. Physiol. Pharmacol.* **72**, 1060–1065.
3. Buchan, A. M. J., Curtis, S. B., and Meloche, R. M. (1990) Release of somatostatin-immunoreactivity from human antral D cells in culture. *Gastroenterology* **99**, 690–696.
4. Buchan, A. M. J., Meloche, M., and Coy, D. H. (1990) Inhibition of bombesin-stimulated gastrin release from isolated G cells by bombesin analogs. *Pharmacology* **41**, 237–245.
5. Buchan, A. M. J. (1991) Effect of sympathomimetics on gastrin secretion from antral cells in culture. *J. Clin. Invest.* **87**, 1382–1386.
6. Buchan, A. M. J., MacLeod, M. D., Meloche, R. M., and Kwok, Y. N. (1992) Muscarinic regulation of somatostatin release from primary cultures of human antral epithelial cells. *Pharmacology* **44**, 33–40.
7. Buchan, A. M. J., Meloche, R. M., Kwok, Y. N., and Kofod, H. (1993) Effect of CCK and secretin on somatostatin release from cultured antral cells. *Gastroenterology* **104**, 1414–1419.
8. Campos, R. V., Buchan, A. M. J., Meloche, M., Pederson, R. A., Kwok, Y. N., and Coy, D. H. (1990) Gastrin secretion from isolated human G cells in primary culture. *Gastroenterology* **99**, 36–44.
9. Guarnieri, M., Krell, L. S., McKhann, G. M., Pasternak, G. W., and Yamamura, H. I. (1975) The effects of cell isolation techniques on neuronal membrane receptors. *Brain Res.* **93**, 337–342.
10. Meinstrich, M. L. (1983) Experimental factors involved in separation by centrifugal elutriation, in *Cell Separation Methods and Selected Applications*, vol. 2 (Pretlow, T. G. and Pretlow, T. P., eds.), Academic, New York, pp. 33–61.

27

Primary Cell Culture of Human Enteric Neurons

Submucosal Plexus

Eric A. Accili and Alison M. J. Buchan

1. Introduction

The enteric nervous system consists of two major plexi, the submucous (Meissner's) lying between the circular muscle and mucosal layers, and the myenteric (Auerbach's) between the circular and longitudinal muscle. In recent years, it has become apparent that the chemical coding of the neurons in these two plexi differs significantly within a single species *(1)* and between species *(2)*. The functional significance of the chemical coding of the enteric neurons is difficult to assess in vivo because of the absence of knowledge about the electrophysiological properties of the neurons and the regulation of neuropeptide/neurotransmitter release.

Cell culture of neurons from a single plexus allows these cells to be maintained in a controlled environment, and provides access for electrophysiological, physiological, and pharmacological studies. Such cultures have been used to determine the regulation of neuropeptide release from cultured neurons, the molecular form of peptides produced in the neurons, and the precise chemical coding of individual neurons within the ganglia *(3–5)*.

A major problem in obtaining cultures of enteric neurons is the nonsterile nature of the intestinal lumen. Human cell cultures are generally plagued by contamination with fungal and bacterial growths. To overcome this problem, we have used a combination of collagenase digestion of the matrix of the submucosal plexus with elutriation to remove small particles, such as bacteria and fungi, although during the initial stages (48 h) of the cultures, antifungal and antibacterial agents have to be added to the medium.

2. Materials
2.1. Tissue Collection from Operating Room
1. A sterile 2-L screw-topped container with 1 L of chilled buffer.
2. 1 pair scissors (8–12 in.).
3. 1 pair forceps (6–8 in.).

2.2. Separation of Plexus
1. Plastic container full of ice.
2. 500-mL Dish of Hank's balanced salt solution (HBSS) containing 0.1% bovine serum albumin (BSA) and 20 mM N-2-hydroxyethyl piperazine-N-2-ethane sulfonic acid (HEPES) buffer on ice (referred to as HBSS/BSA from now on).
3. 50-mL Beaker of HBSS/BSA buffer on ice.
4. Scissors (6–8 in.) (1/person).
5. Forceps (6–8 in.) (1/person).
6. 2 Glass slides (1 mm, 25 × 75 mm) back to back (2/person).
7. 100-mm Glass Petri dish (1 per person).
8. Latex gloves (mandatory for handling human tissue).
9. Face masks if uncertain of the status of the patient (i.e., Hepatitis, virus infection, and so forth).

2.3. Isolation of Neurons
1. Shaking water bath at 37°C.
2. 250-mL Nonsterile conical flasks (1/10 g of stripped plexus).
3. Sigma (St. Louis, MO) type XI collagenase (stored as aliquots of 50 mg).
4. 1 L Basal medium Eagle (BME) containing 0.1% BSA, 20 mM HEPES.
5. 200 mL BME + 4.4 mM calcium chloride.
6. 1 L HBSS/BSA.
7. 5% CO_2 in O_2 gas cylinder.
8. Very sharp scissors (4–6 in.).
9. 25-mL Beaker.
10. 500-mL Plastic beaker *(2)*.
11. Squares of 400-μm Nytex™ mesh cut to fit over the 500-mL beaker.
12. Plastic ring cut from a 500-mL beaker to fit into the rim of the 500-mL beaker.
13. 50 mL nonsterile centrifuge tubes *(6–8)*.
14. 10-mL Gilson pipet *(1)*.
15. 4% Trypan Blue.
16. Hemocytometer.

2.4. Elutriation of Resultant Cell Suspension
1. Beckman centrifuge with elutriator rotor.
2. 5-mL Standard separation chamber.
3. Pump (Cole Parmer [Niles, IL] Masterflex Model 7520-20 or similar).
4. Laminar flow hood adjacent to centrifuge/rotor assembly.

5. 70% Ethanol.
6. 100-mL Measuring cylinder.
7. 1 L Sterile water.
8. 2 L Sterile HBSS/BSA.
9. 50-mL Sterile centrifuge tubes (Falcon) *(4–12)*.

2.5. Culture Preparation of Isolated Neurons

1. 12-Well Costar plates coated with sterile rat tail collagen are optimal for release experiments.
2. 35-mm Falcon or Costar dishes are optimal for immunocytochemistry.
3. 35-mm Falcon or Costar dishes with collagen-coated sterile coverslips are optimal for electrophysiology and imaging of intracellular ion concentration (Fura-2 ratio measurement). Coverslips are sterilized by flaming in 70% ethanol and allowing to cool before coating with collagen.
4. 100-mL Growth medium.

2.6. Growth Medium

Dulbecco's modified Eagle's medium (DMEM) containing 5.5 mM glucose, 10 mM HEPES, 2 mM glutamine, 200 mM cytosine-D-arabinofuranoside, 8 g/mL insulin, 20 ng/mL nerve growth factor (NGF) CS-7S (stored at 100X final concentration in sterile DMEM; 0.1-mL aliquots at $-70°C$), 100 g/mL gentamycin, 1 g/mL hydrocortisone, 4 g/mL fungizone, 5% fetal calf serum.

2.7. Eurocollins Buffer

Buffer contains 10.0 mM NaCl, 57.7 mM Na$_2$HPO$_4$, 115.0 mM KCl, 19.0 mM glucose, and 10.0 mM NaHCO$_3$. Osmolarity = 330 mosM/kg, pH 7.0.

3. Methods

3.1. Collection of Submucosal Plexus

At least 12–15 in. of upper small intestine are required. The optimal situation would be to obtain material from the retrieval program of the local Transplant Society. (Note: Ethical permission forms must be in place for the university/research institute and the permission of the next-of-kin for use of material for research should be obtained where required.) The critical factor in the tissue collection is to minimize warm ischemic time. In our case, because the tissue was collected during an organ harvest for transplantation, the venous supply was replaced by ice-cold Eurocollins buffer prior to clamping the aorta. Thus, there was little if any warm ischemia. If the material has to be collected from general surgery, care must be taken to ensure that as soon as the intestine is removed from the abdominal cavity, it is immediately immersed in ice-cold buffer (*see* Note 1).

3.2. Transport to the Laboratory

The transport time should be kept to a minimum. If material has been collected from surgery, if possible, once the tissue is out of the operating room, open the intestine and flush out the lumen. This is vitally important if there is bile present. The bile acts as a detergent and digests the tissue even at 4°C. Once the intestine has been cleaned, it should be transferred to an intracellular buffer at 4°C (such as Eurocollins; for composition, *see* Section 2.), which slows enzyme activity and minimizes damage from free radicals. It is then safe to transport to the laboratory.

3.3. Initial Separation of Plexus

1. Place the opened intestine in a large pyrex dish in ice-cool HBSS/BSA buffer.
2. The duodenal bulb and the initial portion (1 in.) of duodenum are dissected and discarded owing to the presence of Brunner's glands, since the submucosa cannot be properly separated in these areas.
3. Dissect off any adherent fat and mesenteric tissue, and wash luminal surface clear of debris.
4. Cut intestine into 3 in. lengths, and place in the 500-mL container of buffer. Areas containing obvious ulcerations or surface erosions should be excised.
5. Place a segment mucosal surface upward in a Petri dish containing approx 3 mL of ice-cold HBSS/BSA. Hold one end of the segment with forceps, and use the glass slides held horizontal to the surface to scrape off the muscularis mucosae and the mucosa.

   The underlying submucosal plexus has a higher reflective index and "shines" once the muscularis mucosae has been stripped off. The segment is then inverted, and the same procedure used to remove the two muscle layers and myenteric plexus; again the stripped submucosal region "shines" once it is clear of muscle (*see* Fig. 1). To confirm that the manual dissection is removing all the unwanted tissue layers, hematoxylin staining of samples of stripped submucosa can be completed.
6. Place the stripped submucosa into the 50-mL beaker of HBSS/BSA.
7. Repeat steps e and f until all segments are stripped. The total amount of submucosa obtained from the 12 in.–15 in. length varies from 15–40g.

3.4. Collagenase Digestion of Submucosal Plexus

1. Remove strips from HBSS/BSA buffer, and place in 25-mL beaker with no buffer.
2. Mince finely using very sharp scissors. This requires a lot of effort owing to the amount of collagen in the plexus. Attempts to use either a tissue chopper or mechanical blender at this stage result in a drastically reduced yield of viable cells.
3. Place 10 g of minced plexus in a 250-mL conical flask containing 50 mL BME/BSA with 600 U/mL collagenase type XI and 4.4 mM calcium chloride (*see* Note 2). (Note: The pH of the incubation and washing media should be maintained between 7.0 and 7.2, low enough to minimize collagen reassembly, yet within physiological limits, *see* Note 3.) Watch the phenol red indicator dye in

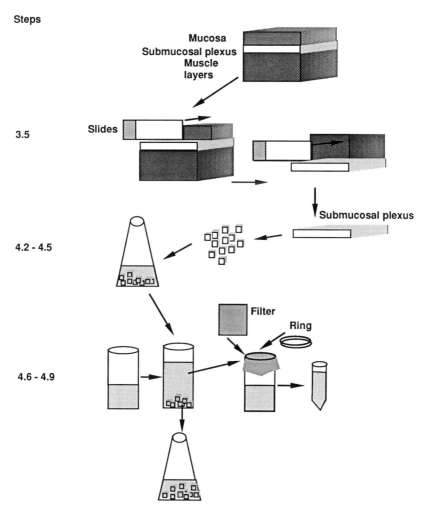

Fig. 1. A diagram of the separation procedure, demonstrating the use of the slides to remove the mucosal and muscle layers from the submucosal region. The numbers refer to individual steps in the protocol.

the BME and HBSS buffers. If it starts to turn deep red/purple, either change the buffer immediately or add 50 µL of 0.1N HCl to bring down the pH.
4. Gas with 5% CO_2 in O_2 for 1 min prior to capping.
5. Incubate in the 37°C shaking water bath at 200 Hz for 30 min.
6. Remove flasks, combine contents into the 500-mL beaker, and double volume with HBSS/BSA (*see* Note 4). Allow nondigested material to settle to bottom of beaker, and discard supernatant. Leave a maximum of 20 mL buffer in the beaker.

7. Divide equally the undigested material between the flasks (if the collagen has started to reaggregate, use scissors to rechop), and add 50 mL of BME/BSA containing 600 U/mL collagenase type XI. Gas as before and return to shaking water bath for 60 min.
8. Remove flasks, combine contents into the 500-mL beaker, and double the volume with HBSS/BSA. Allow nondigested material to settle. Filter supernatant through 400-μm Nytex mesh suspended over a second 500-mL beaker. The nytex mesh can be anchored over the beaker using a ring made from part of another plastic beaker.
9. Separate supernatant into 50-mL centrifuge tubes (nonsterile), and centrifuge at $200g$ for 10 min. Discard supernatant, and resuspend pellet in 50 mL HBSS/BSA. Cells should always be resuspended using a 10-mL Gilson pipet initially in a small volume (e.g., 5 mL) to aid dispersion of the clumped cells and then brought to final volume.
10. Repeat steps 7–9 until no undigested material remains (usually twice).
11. Take the 50-mL resuspensions, centrifuge at $200g$ for 10 min, resuspend each in 5 mL, and combine. Total volume at this point will be between 10 and 20 mL, depending on the number of digestions completed.
12. Check viability of cells, and count cells using Trypan Blue. A 100-μL sample of the cell suspension is added to 100 μL of 4% Trypan Blue, placed in a hemocytometer, and examined under the microscope. Viable cells are unstained; dead cells have a blue nucleus. At this point, count all cells alive or dead.
13. Dilute the cell suspension to 1×10^8 cells/20 mL in HBSS/BSA. At this point, the isolation yields small groups of neurons (2–4 cells), single neurons, partially digested blood vessels, fibroblasts, and satellite cells (*see* Note 5).

3.5. Centrifugal Elutriation

1. Assemble rotor, and ensure separation chamber is free of debris.
2. Sterilize rotor, chamber, and input and output lines with 70% alcohol. Leave in 70% alcohol until 30 min prior to separation.
3. Replace 70% alcohol with sterile water and set flow rates. Flow rates are calibrated at $900g$ for 25, 35, and 100 mL/min. Care was taken to ensure that at low flow rates (25–35 mL/min), pressure in the line was <3 psi. If higher, this indicated that bubbles were trapped either in the separation chamber or in the spindle assembly of the rotor. To remove bubbles, the elutriator rotor was run at $150g$ with a flow rate of 75 mL/min. The rotor was stopped while the flow rate was maintained, and all bubbles were forced out of the chamber and spindle assembly.
4. Replace sterile water with sterile HBSS/BSA.
5. 1×10^8 Cells are loaded into the chamber at a flow rate of 25 mL/min. Loading volume should be between 20 and 30 mL.
6. Wash for 3 min at 25 mL/min once all the cells have entered the chamber. This takes approx 2 min to load cells (the stroboscope allows the separation chamber to be monitored during each run).

7. Increase the flow rate to 35 mL/min, decrease the centrifuge speed to 655$1g$, and collect 2 × 50 mL fractions in sterile tubes. This fraction contains a few small neurons, fungi, and cell fragments, and is not used for the culture experiments. The fraction is collected to monitor the number of smaller neurons. If in a particular preparation a large number of neurons are present in this first fraction, they can be cultured, but the concentration of fungizone used in the initial 48-h period has to be doubled and the cultures closely monitored for contamination.
8. Increase the flow rate to 100 mL/min, decrease the centrifuge speed to $90g$, and collect 2 × 50 mL fractions in sterile tubes. This fraction contains the majority of the neurons.
   Note: From this point on care should be taken to ensure sterile techniques are followed.
9. Repeat steps 5–8 until all cells have been run through the elutriator.
10. Centrifuge all collected fractions at $200g$ for 10 min.
11. Resuspend in 10 mL sterile HBSS/BSA, and check content of fractions using Trypan Blue as before. The first fraction should contain very few neurons, some red blood cells (RBC), small satellite cells, fungi, and debris. The second fraction should contain the majority of the neurons, no RBCs, and few cell fragments. The neurons are distinguished by size and characteristic shape. Neuronal viability should be approx 95% (*see* Note 6).

3.6. Tissue Culture

1. The fraction containing the neurons of interest is centrifuged at $200g$ for 5–10 min.
2. Cells are resuspended in growth medium at a density of $1-2 \times 10^6$ cells/mL.
3. The cell suspension is plated on 12-well plates coated with rat tail collagen at 1 mL/well for release experiments or 35-mm dishes with or without coverslips at 2 mL/dish for immunocytochemical, electrophysiological, and intracellular ion measurement (e.g., Fura-2) experiments. The average yield of neurons is 2.4×10^7 cells (*see* Note 7).

3.7. Characteristics of Submucosal Cultures

The neurons adhere to the collagen substrate overnight and are phase bright. Their viability in culture remained >90% for up to 7 d as shown by Trypan Blue exclusion conducted throughout this period. After 7–10 d, the viability decreased, and the cells became detached from the collagen. Cells aggregated around individual ganglia, and there was abundant neurite outgrowth after 72 h in culture (Fig. 2). The individual clusters of cells were linked by neurite extensions and resembled the submucosal plexus *in situ*. In our experiments, 72 h was chosen for studies of peptide secretion, immunocytochemistry, and determination of the molecular form of the peptides produced by the cultured neurons *(3,4)*. The cultures were checked daily, and the growth medium was changed when necessary.

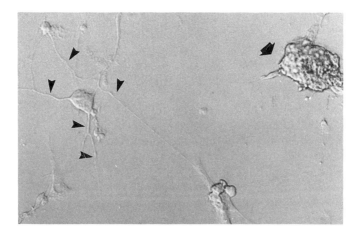

Fig. 2. A detail of a typical culture preparation. Some of the neurons remain within ganglionic clusters (large arrow), whereas others attach to the collagen matrix as single cells. The greatest outgrowth of neurites (small arrows) was seen from the individual neurons (×400).

4. Notes

1. One advantage of the studies carried out in our laboratory was the availability of tissue from the B.C. Transplant Society in the province of British Columbia. Patients were screened before acquiring tissue to ensure their suitability as organ donors and did not have any known pathophysiological conditions. However, as is often the case with human studies, and unlike studies of laboratory animals, such as rats, there was variability with respect to age, size, and sex of the donors. These factors are related to differences in the proportion of collagen to neurons in the submucosal plexus *(6)*. It is important to have this information, preferably before carrying out the dissociation, so that the process may be monitored and necessary adjustments to the protocol can be made.

    The surgeon involved must understand precisely what and how much tissue should be taken. Thus, the appropriate piece of tissue has Brunner's glands (which act as an upper benchmark) and as much tissue as is required after this point.

2. The choice of collagenase can be difficult. Type IX collagenase has smaller amounts of contaminating enzymes (clostrapain, pepsin, and so forth) and a higher activity of collagenase as compared to Type I. It is usually used for isolation of pancreatic islets. We found Sigma type XI collagenase to give a better yield of neurons than either Sigma Type I or Worthington collagenase type I. However, each batch of collagenase should be tested prior to its use, since the activity and level of different additional enzymes in the preparation vary between batches. It is well worth testing several batches from Sigma and then putting a large amount of a good batch on hold. It is important to remember that if you

change batches of collagenase, problems may be encountered either with a loss of viable neurons in the initial digest or the isolated neurons may not survive in culture. Extracellular calcium levels are usually between 1.8 and 2 mM. In the isolation experiments, the initial digest was completed with 4.4 mM Ca^{2+}. There were two reasons for this. First, the efficiency of the collagenase enzyme was increased, and second, there was less variation in the pH of the medium during the digestion.

3. Gelation and reaggregation of collagen occur owing to the large amount of collagen present in the human submucosa. Dimeric collagen has been shown to undergo a reversible gelation as temperature and pH increase (with significant changes in turbidity at pH = 7.4 and T = 28°C), the process also being dependent on the concentration of free collagen (7). Therefore, the pH of the washing and incubation medium is kept below 7.4, and the cells are washed and centrifuged to reduce the concentration of free collagen. The presence of phenol red in the incubation medium is useful, since changes in pH can be noted and corrected quickly (the digest usually becomes basic with time). Likewise, incubations can be stopped earlier if the digest become particularly turbid, which often occurs during the first period of incubation.

   The minced tissue often reaggregates to various degrees forming large masses despite these precautionary measures. Therefore, the tissue is minced again with scissors at various times throughout the digest, for example, just prior to the second and third periods of incubation in the enzyme solution.

4. The volume of the digestion solution is at least doubled with cold HBSS/BSA at the end of each 1-h incubation period, prior to the centrifugation step to wash the tissue properly. This allows a better separation of cells from fat and debris, reduces the concentration of collagen, and quickly reduces the activity of the collagenase. If the digest is particularly turbid, this volume can be further increased to ensure proper washing of the cells.

5. The isolation protocol described produced the highest yield of viable cells, and was designed to isolate single neurons or small clusters. If the intention is to isolate intact ganglia and clusters of neurons, solutions containing other enzymes (such as elastase or protease) or increased concentrations of calcium should be avoided. Using less collagenase (300 U/mL) and increasing the time of digestion always resulted in more undigested tissue and a smaller yield of viable cells. Increasing the collagenase (1000 U/mL) and decreasing the time also resulted in a lower yield of viable cells.

6. This procedure increases the number of viable neurons and removes the majority of the other cell types and the usual bacterial and fungal contaminants introduced through the nonsterile environment of the duodenal lumen.

   Receptor damage resulting from hyperosmolality (8) is avoided, since cell separation in these experiments was carried out using elutriation, which permits the use of isotonic solutions (9). Centrifugal elutriation utilizes centrifugal force and flow, which act in opposing directions, to separate cells on the basis of their volume. Different fractions of cells can be removed by altering the flow rate of fluid passing through the elutriation chamber by altering the pump rate or by changing the speed of the centrifuge. Appropriate flow rates, centrifugation speeds, and washing times can be determined empirically, and can be easily altered to choose different populations of cells for study.

Peptide containing neurons can be used as indicators for the enrichment by taking samples before and after elutriation centrifugation and carrying out radio-immunoassay. For example, somatostatin content of 10 mg of undigested submucosal plexus can be compared to the content of the cell suspension prior to elutriation and the two cell fractions collected during elutriation (usually determined for 1×10^6 cells). The elutriation procedure can then be altered to maximize enrichment of somatostatin containing neurons.

Human tissue, particularly the submucosa, because of the large proportion of collagen, can reaggregate during the elutriation process, blocking the flow of buffer through the chamber and drastically reducing the efficiency of the separation process. The Beckman elutriator centrifuge is equipped with a stroboscope, which allows the separation chamber to be monitored during each run. If a large solid pellet of material builds up in the chamber (most common during the initial stage of the separation when there is a high centrifuge speed with a low flow rate), then the run should be abandoned, the centrifuge stopped with a high flow rate (>100 mL/min) and the material exiting the chamber collected. The material collected can be recentrifuged and resuspended in 15 mL of buffer to reload once the chamber is cleaned. In this way, there is a minimum loss of material.

Unfortunately, to clean the chamber, the rotor has to be disassembled, the chamber removed, and the blockage washed out. If the blockage is not promptly dealt with, a plug of the inlet line can develop, which requires the use of a fine needle to remove. If a plug develops in addition to the presence of a solid pellet in the chamber, the pressure in the in-flow line will increase and serves as an additional warning.

7. Isolation methods utilizing collagenase have been shown to damage receptors specifically on neurons (e.g., muscarinic receptors) *(8)*. Once the cells have been isolated, receptors regenerate in tissue culture, as has been shown for the nicotinic acetylcholine receptor *(10,11)*. Further support comes from experiments using canine submucosal neurons, which have been isolated and maintained in tissue culture in a similar fashion and were able to respond to receptor-dependent secretagogues, such as SP *(4,12)*.

An initial plating density of $1-2 \times 10^6$ cells/mL/plate was chosen and found to be optimal for the survival of the neurons for 72 h. At densities of $5-8 \times 10^5$ cells/mL, the majority of cells would not adhere, whereas at densities of $3-5 \times 10^5$ cells/mL, the cells would detach from the collagen substrate in the 1–3-d period.

The mitotic inhibitor cytosine arabinoside was included in the growth medium and effectively prevented the overgrowth of fibroblasts. A sheet of fibroblasts to support the attachment of neurons was not required with the use of plates coated with rat tail collagen. Rat tail collagen can easily be prepared in the laboratory.

References

1. Costa, M., Furness, J. B., and Llewellyn-Smith, I. (1987) Histochemistry of the enteric nervous system, in *Physiology of the Gastrointestinal Tract*, 2nd ed. (Johnson, L. R., ed.), Raven, New York, pp. 1–40.
2. Pataky, D. M., Curtis, S. B., and Buchan A. M. J. (1990) The co-localization of neuropeptides in the enteric nervous system of normal Wistar and non-diabetic BB rats. *Neuroscience* **36**, 247–254.
3. Accili, E. A., McIntosh, C. H. S., and Buchan, A. M. J. (1992) The release of somatostatin-14 from human submucosal ganglia in tissue culture. *Can. J. Physiol. Pharmacol.* **71**, 619–624.
4. Buchan, A. M. J., Doyle, A. D., and Accili, E. A. (1990) Canine jejunal submucosa cultures: characterization and release of neural somatostatin. *Can. J. Physiol. Pharmacol.* **68**, 705–710.
5. Nishi, R. and Willard, A. L. (1985) Neurones dissociated from rat MYP retain differentiated properties when grown in cell culture. I. Morphological properties and immunocytochemical localization of transmitter candidates. *Neuroscience* **16**, 187–199.
6. Gabella, G. (1990) On the plasticity of form and structure of enteric ganglia. *J. Auton. Nerv. Syst.* **30**, S59–S66.
7. Yurchenko, P. D. and Furthmayr, H. (1984) Self-assembly of basement membrane collagen. *Biochemistry* **23**, 1839–1850.
8. Guarnieri, M., Krell, L. S., McKhann, G. M., Pasternak, G. W., and Yamamura, H. I. (1975) The effects of cell isolation techniques on neuronal membrane receptors. *Brain Res.* **93**, 337–342.
9. Meinstrich, M. L. (1983) Experimental factors involved in separation by centrifugal elutriation, in *Cell Separation Methods and Selected Applications*, vol. 2 (Pretlow, T. G. and Pretlow, T. P., eds.), Academic, New York, pp. 33–61.
10. Hartzell, F. C. and Fambrough, D. H. (1973) Acetylcholine receptor production into membranes of developing muscle fibres. *Dev. Biol.* **30**, 153–165.
11. Willard, A. L. and Nishi, R. (1985) Neurones dissociated from rat MYP retain differentiated properties when grown in cell culture. II. Electrophysiological properties and responses to neurotransmitter candidates. *Neuroscience* **16**, 201–211.
12. Barber, D. L., Buchan, A. M. J., Leeman, S. E., and Soll, A. H. (1989) Canine enteric submucosal cultures: transmitter release from neurotensin immunoreactive neurons. *Neuroscience* **32**, 245–253.

28

Isolation and Culture of Human Hepatocytes

Martin K. Bayliss and Paul Skett

1. Introduction

The liver performs a wide range of physiologically important functions, including the synthesis and secretion of albumin, fibrinogen, and other plasma proteins; the synthesis of cholesterol and bile acids; and the metabolism of drugs, steroids, and amino acids. The liver has a central role in energy metabolism as the major store of glycogen, as the site of gluconeogenesis, and in the synthesis of fatty acids and triglycerides. The liver is, therefore, a vital organ, but it is difficult to study specific liver functions in vivo owing to interfering influences from other organs, e.g., the kidney, gut, and lungs, which metabolize drugs and the muscle involvement in glucose homeostasis. An isolated liver preparation seems necessary, and the isolated human hepatocyte appears to be a suitable experimental model for the study of liver-specific functions.

The preparation of isolated hepatocytes from a heterogeneous tissue, such as the liver, poses several problems. Initially, the cells must be dissociated from the fibroconnective skeleton. Tissue dispersion requires the disruption of reticulin fibrils along with adhesion proteins, such as fibronectin and laminin, which constitute the framework of the liver lobule *(1)*. Hepatocyte isolation also requires the disruption of cell–cell adhesions, i.e., the junctional complexes and finally the recovery of the hepatocyte preparation for culture.

Hepatocytes represent 60–80% of the number of cells in the liver, but occupy some 80% of the liver volume *(2)*. Enzymatic hepatocyte isolation from animal liver was first introduced in 1967 *(3)*. Prior to this time, mechanical and chemical dissociation of the liver parenchyma had been used. This latter isolation yielded very poor-quality hepatocytes. The chemical and mechanical dissociation methods are discussed at length in a review by Berry and coworkers *(4)*. The development of a technique utilizing enzymes to digest the liver providing a high yield of functionally active rat hepatocytes led to a rapid increase in the utility of the isolated hepatocyte as a model system for the study of liver functions (5). The perfusion technique originally developed for the preparation of hepatocytes from small animal species over the last 10–15 yr has been adapted and applied to human material.

Human hepatocytes have been prepared from a variety of liver samples, including a whole liver (6,7), a portion of the whole liver (8), an end of lobe wedge biopsy sample (9–14), and small biopsy fragments (15,16). One of the major problems encountered when working with human liver is the limited and irregular availability of material for research purposes. The preservation of these livers in a relatively healthy state from donor to perfusion has been aided by the development of the University of Wisconsin (UW) solution for organ preservation. This, together with the recent advances in surgical technique, has resulted in significant increases in successful therapeutic liver transplants. This problem has spurred on developments in maximizing the use of human hepatocytes by developing isolation, culture, coculture, and cyropreservation techniques. The limited availability of human material requires a system that can provide a high yield of functional isolated hepatocytes rapidly utilizing all the available material. The majority of human liver made available for research is in the form of surgical biopsies, which are usually obtained from patients undergoing partial hepatectomy. Another major source is the surgical resection of adult livers prepared for pediatric recipients. This material is usually 5–10 g in size and normally has only one cut surface. Therefore, the wedge biopsy perfusion method (9,10) for preparing hepatocytes appears to be the most suitable method for utilizing the available material. This is the method of choice for a number of workers (9–14). However, cell yields vary depending on the size of the biopsy and the ability of the operator to cannulate the exposed vessels.

The preparation of human hepatocytes using this method (9–14) has produced yields of between 10 and 40 × 10^6 hepatocytes/g wet liver with cell viability in the range of 70–95% as determined by Trypan Blue exclusion.

Another major consideration when using human material to prepare hepatocytes is the variation in the functional activity between cell preparations—a problem of lesser importance when using genetically identical and identically housed laboratory animals. This variability may be the result of the length of time taken in obtaining the sample, sample storage conditions, or the sample history, as well as the genetic and environmental factors affecting the donor.

As mentioned, the majority of human tissue used currently to prepare hepatocytes is from surgical waste that has been stored in UW solution. A recent study (17) compared the attachment efficiency and specific metabolic function of human hepatocytes in culture after the tissue had been transported and stored in the presence or absence of this solution. The results demonstrated that human hepatocytes isolated from surgical biopsies that had not been perfused with UW solution had an increased attachment efficiency and maintained liver-specific and nonspecific functions better than hepatocytes isolated from material stored in UW solution. The UW solution may not, therefore, be ideal as a preserving medium for subsequent hepatocyte isolation. Information regarding the patient history, social history, dietary habits, and genetic polymorphism is important for interpreting results. For example, a liver sample removed from a patient who had taken a barbiturate overdose can have drug-metabolizing enzymes that are significantly induced *(18)*. These enzymes remain elevated in isolated hepatocytes that are cultured for several days. Furthermore, studies by Houssin and coworkers *(19)* have also suggested that variable intracellular ATP and glycogen content in hepatocyte populations may be attributed to the nutritional status of the donor.

Human hepatocytes used in suspension with no attachment to any form of extracellular matrix are viable for only a few hours. Hepatocyte survival can be extended considerably when cells are maintained in monolayer culture conditions in the presence of an extracellular matrix and/or a supplemented media *(20)*. Thus, human hepatocytes in culture have been shown to maintain plasma protein production, glycolysis, and urea synthesis for periods of several days *(21)*. Drug-metabolizing enzymes, on the other hand, particularly the cytochromes P450, are maintained during the initial stages of culture, but decline with time *(21,22)*. However, these enzymes can be elevated in culture to some extent by inducing agents, such as phenobarbital, 3-methylcholanthrene, and rifampicin *(23,24)*.

Therefore, a number of factors, including the extracellular matrix, seeding densities, culture media, media supplements, and cell–cell interactions, need to be considered when culturing human hepatocytes. The isolation and culture of animal and human hepatocytes and the evaluation of many of these factors have been discussed and reviewed by a number of workers *(4,20,21,25–27)*. Particular reference should be made to the studies of Guillouzo and coworkers *(23,25,28)*.

It should be emphasized that the ideal conditions for isolating and culturing human hepatocytes have not been evaluated, and this chapter represents one of the best accepted practices at the time of writing.

2. Materials

1. Human liver samples are obtained from either hemi-hepatectomy or from surgical resections of adult livers prepared for pediatric transplant after approval from an Ethical Review Board has been obtained (*see* Notes 1 and 3).
2. A supply of good-quality water should be used. Double-distilled water followed by filtration through a 0.22-μm filter can be used, or alternatively, culture-quality water can be purchased commercially.

3. Preparation of buffers and solutions:
   a. Earle's balanced salt solution (EBSS; calcium- and magnesium-free), sodium bicarbonate solution (7.5%), L-glutamine (200 mM), and penicillin/streptomycin (lyophilized; 10,000 U/10,000 µg/mL) are available from Life Technologies Limited (Paisley, Renfrewshire, UK).
   b. Ethylene glycol bis-(β-aminoethyl ether) N,N,N',N'-tetra acetic acid (EGTA), 4-(2-hydroxyethyl) piperazine-2-ethanesulfonic acid (HEPES), bovine serum albumin (Fraction V, fatty-acid-free), deoxyribonucleate 5' oligonucleotidohydrolase (DNase I; EC 3.2.21.1) from bovine pancreas, trypsin inhibitor (from soyabean, Type II-S), collagen type I (soluble rat tail or calf skin), δ-aminolevulinic acid, insulin (from bovine pancreas), and transferrin (bovine) are available from Sigma (Poole, Dorset, UK).
   c. Ethylenediaminetetraacetic acid (EDTA) and Triton X-100 detergent are available from BDH Chemicals Ltd. (Poole, Dorset, UK).
   d. Collagenase (EC 3.4.24.3) from *Clostridium histolyticum* of specific activity, 0.17–0.44 Wunsch U/mg lyophilisate is available from Boehringer Mannheim (Lewes, Sussex, UK).
   e. Williams' Medium E (WME) without phenol red or L-glutamine and Trypan Blue (4,4'-bis[8-amino-3,6-disulfo-1-hydroxy-2-naphthalazo] 3,3'-dimethylbiphenyl tetrasodium salt) formulated as a 0.4% (w/v) solution in 0.85% saline are available from Flow Laboratories (Rickmansworth, Hertfordshire, UK).
   f. Renex 690 detergent is obtained from Atlas (ICI, Surrey, UK).
4. Perfusion buffer: EBSS (magnesium- and calcium-free), 117 mM NaCl (6.8 g/L), 5.4 mM KCl (0.4 g/L), 0.9 mM NaH$_2$PO$_4$ · H$_2$O (0.125 g/L), 26.16 mM NaHCO$_3$ (2.2 g/L), 5.56 mM glucose (1 g/L), and 0.05 mM phenol red (0.02 g/L). EBSS can be purchased commercially as a sterile preparation. This preparation can be purchased as a 10X concentrate (100 mL) without sodium bicarbonate. If the concentrate is used, add to culture-quality water (870 mL). Sodium bicarbonate (7.5% solution; 30 mL) is added to give a final concentration of 26 mM. The pH is adjusted to 7.4. Commercially prepared EBSS can be stored for up to 6 mo at room temperature. Otherwise, this solution should be prepared on the day of isolation.
5. Chelating solution: A stock solution of 25 mM EGTA is prepared in 0.1M NaOH and the pH adjusted to 7.4. An aliquot of the stock 25 mM EGTA solution (10 mL) is added to 490 mL EBSS solution (pH 7.4) to give a final concentration of 0.5 mM EGTA (*see* Note 4). The stock EGTA solution can be stored at 4°C for up to 6 mo. The chelating solution in EBSS should be prepared on the day of isolation.
6. Enzyme solution: 0.24 U/mL Collagenase is dissolved in EBSS (pH 7.4), (100 mL). Units of activity according to Wunsch and Heidrich *(29)* are determined by the supplier. Standardization on a fixed number of units allowed for the varying specific activities of different batches of enzyme. To this solution is added 10 mg trypsin inhibitor and 2 mM calcium chloride. The enzyme solution is best prepared on the day of isolation.

7. Hepatocyte dispersal buffer: 10 mM HEPES, 142 mM NaCl, 7 mM KCl. Adjust pH to 7.4. Slowly add bovine serum albumin to give a 1% (w/v) solution. Stir continuously, but slowly. Prepare on the day of isolation.
8. Culture medium: WME (*see* ref. *30* for composition) can be prepared with a good-quality water supply (culture or double distilled) and then sterilized. WME can also be purchased commercially in powdered form or as 1X liquid. Both are sterile preparations. The powdered form contains L-glutamine, the liquid form is without. If the commercially available liquid form is used, L-glutamine is required. It is recommended that culture media be purchased commercially.

    L-Glutamine (200 mM) is added to WME (1X liquid preparation) to give a final concentration of 4 mM. Prepare on day of hepatocyte preparation. WME without L-glutamine as a 1X liquid can be stored for up to 6 mo at 2–8°C. WME with L-glutamine, but without sodium bicarbonate, in powder form can be stored for up to 12 mo at room temperature.

    L-Glutamine can be purchased commercially as a 200 mM stock solution and should be stored frozen at –20°C. L-Glutamine is stable for up to 9 mo at –20°C. Above –10°C, the stability of the product is variable. L-Glutamine can also be purchased as a powder with a shelf life of up to 9 mo. For discussion of alternative media, *see* Note 8.
9. Attachment media: WME containing 4 mM L-glutamine, 100 U/100 µg/mL penicillin/streptomycin solution, 0.25 U/mL insulin, 5 nM ZnSO$_4$, 5 µg/mL bovine transferrin, 1 µM δ-aminolevulinic acid, and 10% fetal calf serum. The penicillin/streptomycin, insulin, ZnSO$_4$, bovine transferrin, and δ-aminolevulinic acid are prepared as 100X stock solutions in WME without L-glutamine and stored in 1-mL aliquots at –20°C. The stock solutions described above are added to WME in a final volume of 100 mL. The attachment media should be prepared as required.
10. Incubation media: The same as the attachment media, excluding fetal calf serum.
11. Culture plates can be purchased commercially precoated with collagen (0.4–1.4 µg type I collagen/cm^2). Alternatively, plastic culture plates or flasks can be coated with soluble collagen. A stock solution of collagen is prepared by dissolving either soluble rat tail or calf skin collagen type I in 0.1 M acetic acid. This solution is diluted with sterile water and 100 µg applied to a 35-mm culture well. This gives a coating of 5–10 µg collagen/cm^2. Plates are air-dried in sterile conditions, preferably in a laminar flow cabinet. The dried plates are washed well with either culture-quality water or media to remove the acetic acid. A variety of attachment factors have been used to culture human hepatocytes and are discussed in Note 7.
12. The P450 buffer is based on that of Warner and coworkers *(31)*: 2 mL Renex 690 and 5 g sodium cholate are dissolved separately in 200 mL double-distilled water. 0.5M Sodium phosphate buffer (pH 7.4; 20 mL) and 5 mM EDTA (20 mL) are prepared separately. The Renex 690, sodium cholate, EDTA, and sodium phosphate buffer are mixed, and the volume adjusted to 800 mL with double-distilled water. The pH is adjusted to 7.4 and 200 mL glycerol added. The final concentrations are 0.29% Renex 690, 0.5% sodium cholate, 0.1 mM EDTA, 10 mM sodium phosphate, and 20% glycerol. The Renex buffer can be stored at 4°C for up to 1 mo.

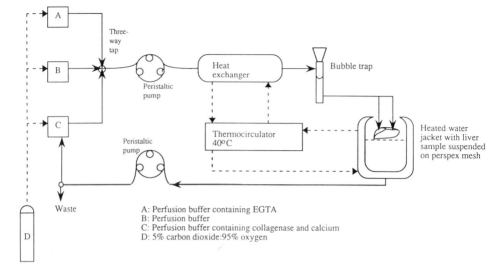

Fig. 1. Apparatus used for the perfusion of human liver biopsy samples. Reproduced with kind permission of D. Cross.

13. Trypan Blue solution is obtained commercially as a 0.4% (w/v) solution in 0.85% saline. The solution is filtered (0.45 µm) prior to use. Trypan Blue solution can be stored at room temperature. For further discussion on the use of Trypan Blue, see Note 5.
14. Perfusion apparatus: The equipment used for the perfusion of liver samples to prepare isolated hepatocytes is shown in Fig. 1. All perfusion equipment should either be autoclaved or washed with 70% ethanol, and rinsed with sterile distilled water prior to use. The perfusion system consists of three reservoirs, each of which contains one of the three solutions used to perfuse the liver biopsy sample. Reservoir A contains perfusate (EBSS, pH 7.4) with chelating agent (EGTA, 0.5 mM), reservoir B contains perfusate only, and reservoir C contains perfusate, collagenase, and calcium ions. The reservoirs are linked via a three-way tap system. The flow of the perfusate is maintained by a Watson-Marlow 502S peristaltic pump (Watson-Marlow Ltd., Falmouth, Cornwall, UK). The temperature of the perfusate is maintained at 37°C by passage through a heat exchanger. The water jacket around the heat exchanger is maintained at 40°C by a recirculating thermoregulator (Harvard Apparatus Ltd, Edenbridge, Kent, UK) to accommodate a 3°C loss in perfusate temperature between the heat exchanger and the polyethylene catheters used to cannulate the liver sample.

A bubble trap is placed between the heat exchanger and the cannulae. Silicon tubing (1.6 mm id and 1.6 mm wd) is used to connect the system. All perfusion buffers are continually gassed with a mixture of oxygen and carbon dioxide (95:5% [v/v]), and the pH of each solution monitored throughout the perfusion

using a pH stick electrode (Jenway Ltd., Dunmow, Essex, UK). For further discussion, *see* Note 2. Similar types of apparatus allowing constant perfusion can be designed as long as the main components (water-jacketed temperature-controlled reservoirs, constant flow, nonpulsatile pump, and bubble trap) are present. Many variants exist.
15. Other equipment requirements include 18-gage catheters (Becton Dickinson, Cowley, Oxford, UK), boulting cloth (64 µm; Henry Simon Ltd., Stockport, Cheshire, UK), a refrigerated benchtop centrifuge (Hereaus Omnifuge 2.0RS), an Improved Neubauer counting chamber (Weber Scientific International Ltd., Lancing, Sussex, UK), a standard laboratory and an ID03 inverted laboratory binocular microscope (400X and 320X magnification, respectively; Carl Zeiss Ltd., Welwyn, Hertfordshire, UK), collagen-coated or noncoated culture plates (Bibby Sterilin, Staffordshire, UK), a thermostatically controlled CO_2 incubator, an autoclave, a class II laminar flow cabinet, and a sterile form of dispensing media, i.e., an automatic Pipettus system (Flow Laboratories, Rickmansworth, Hertfordshire, UK).
16. **Safety note:** A solution of medical disinfectant, i.e., 2% Virkon solution should always be available to disinfect spills. All waste media should be disinfected prior to disposal, and any tissue remaining after the perfusion must be autoclaved before disposal. Local safety rules for the handling of human material need to be carefully observed.

3. Methods

Human hepatocytes are isolated from liver using a three stage perfusion modified from the method of Oldham and coworkers *(32)*, which is derived from the method of Strom and coworkers *(10)*. The major modifications are the substitution of EBSS for HEPES as the perfusion buffer and an additional washing step to remove residual EGTA prior to collagenase perfusion, as adopted later by the same group of workers *(13,33)*.

1. Sterile techniques should be employed during the isolation of human hepatocytes for culture. All solutions are either filtered (0.22 µm) or autoclaved, and all instruments autoclaved when possible.
   Human tissue is usually perfused with UW solution prior to receipt. Perfusion of the liver with UW solution has been shown to preserve liver tissue for up to 20 h prior to transplantation *(34)*. However, *see* discussion of UW solution in Section 1.
   Human liver wedge biopsy (5–10 g) samples should be obtained with only one cut surface where possible. Polyethylene catheters (18 gage) are placed into the lumen of the exposed vessels. It is essential to use several catheters (preferably four) to obtain a good perfusion of the tissue sample (*see* Note 1 and 3).
2. The tissue sample is perfused initially with the chelating solution (reservoir A) at a flow rate of 8 mL/min/cannular for up to 10 min. This should result in rapid blanching of the liver sample. Areas not perfusing will not yield isolated hepatocytes. The perfusate is allowed to drain to waste. The EGTA chelates extracellu-

lar calcium and this step is critical in the dissociation of cellular junctions *(4)* (*see* Note 4).
3. The second step is to perfuse with EBSS (reservoir B) for a further 10 min to remove the chelator from the tissue; EGTA inhibits the action of collagenase in the subsequent perfusion. The perfusate is allowed to drain to waste.
4. The third stage is to perfuse with enzyme solution (reservoir C). The sample is perfused with collagenase in the presence of calcium ions and trypsin inhibitor. This inhibitor is required since commercial collagenase preparations may be contaminated with trypsin, which if active on the cells during perfusion is detrimental to cell receptors. The perfusion of the liver sample with collagenase solution should continue for up to 40 min or until the surface structure of the liver sample obviously changes owing to softening of the tissue. This solution is continuously recirculated (*see* Note 3).

    The perfusion buffers are continuously gassed with oxygen, carbon dioxide (95:5% [v/v]), and the pH of each buffer monitored. Adjustments should be made to keep the pH constant as required. Perfusion of the liver acidifies the buffers, which will be observed by a change in the color of the phenol red indicator (*see* Note 2).
5. The liver sample is removed carefully into ice-cold dispersal buffer containing 4 mg/100 mL DNase I *(35)*. DNase helps prevent cell clumping by acting on nucleic acids released by damaged cells during the isolation procedure. Care should be taken not to rupture the capsule of the perfused sample during transfer to the dispersal buffer.

    Once in the buffer, the capsule surrounding the liver is ruptured with forceps and the hepatocytes gently combed from the tissue mass. The cell suspension is filtered through prewetted boulting cloth, and the filtrate gassed with oxygen and carbon dioxide (95:5% [v/v]). The cell suspension is then centrifuged at 50g and 4°C for 2 min. Round-bottom tubes (20 mL; Nunc, Gibco Ltd., Paisley, Scotland) rather than conical-shaped tubes should be used to minimize cell damage. The supernatant containing cell debris and nonviable cells is removed by aspiration and the cell pellet resuspended in dispersal buffer containing DNase I by gentle swirling of the tube. The centrifugation procedure is repeated. The third and final wash of the hepatocyte preparation is repeated in the absence of DNase I. After centrifugation, the viable hepatocyte preparation is resuspended in pregassed WME (30–100 mL) without phenol red and albumin, but containing L-glutamine and maintained at 4°C.
6. Plasma membrane integrity and hepatocyte number should be determined using Trypan Blue exclusion *(4,36,37)* in the absence of supplemented protein (*see* Note 5). Ensure that no albumin is present at this stage. A 50-µL aliquot of Trypan Blue solution is mixed with 250 µL of human hepatocyte suspension and allowed to stand for 2 min at room temperature. The cell suspension is then analyzed using an improved Neubauer counting chamber. The sample is mixed and applied carefully to both grids using a glass Pasteur pipet. With the coverslip in the correct position, the chamber has a depth of 0.1 mm. The cells are counted in the central grid in the squares surrounded by triple lines. The area of this grid is 1 mm^2, and

the volume is therefore 0.1 mL. The number of cells in 1 mL is the number of cells counted in the grid × 10^4. Any dilution used (e.g., the addition of Trypan Blue) needs to be taken into account, in this case 1:2. Cells will always be found overlapping the outer lines of the grids, and the usual practice is to include the top and left-hand lines in the count, and ignore those cells overlapping the bottom and right-hand lines.

Cellular yields should be approx 18.8 ± 9.6 × 10^6 cells/g tissue *(14)*, which is in keeping with other workers *(9,10,11,21)* for human hepatocytes. Viable cells are recognized by their ability to exclude Trypan Blue from the nucleus *(36,37)*. Therefore, cell viability is determined by counting the number of stained and unstained cells:

% Viability = (number of cells unstained/total cell number) × 100%

All counts are completed in duplicate. The counting chambers should be viewed using a standard laboratory 16-binocular microscope (400× magnification). Typically 80–90% of isolated human hepatocytes exclude Trypan Blue. Suspensions of isolated human hepatocytes can be used for up to 4 h. If longer term experiments are required, cells should be cultured.

7. If viability has been established, only hepatocyte preparations of >80% should be used for culture. The final cell suspension is centrifuged, and the medium aspirated. The isolated hepatocytes are then resuspended in the attachment medium. The final cell suspension should be in the region of 1 × 10^6 cells/mL.

8. For cell culture, all cell manipulations are performed under sterile conditions in an Class II laminar flow cabinet. All liquid handling is performed under sterile conditions using a Pipettus system (Flow Laboratories) or similar equipment.

9. Isolated hepatocytes sediment rapidly on standing. Therefore, to achieve a homogenous cell suspension, the cell mixture should be swirled prior to pipeting. The hepatocyte suspension (2 mL) is dispensed into the wells of the culture plates; either commercially precoated or "in-house" prepared plates should be used. A seeding density of approx 10 × 10^4 cells/cm^2 should be used as recommended by a number of workers *(8,10,13,17,18,23)*. Depending on the size of culture plate used, the initial hepatocyte suspension should be diluted or concentrated to achieve the seeding density. For further discussion, *see* Notes 6–8 and 11.

10. After plating, the culture plates should be placed in a CO_2 incubator at 37°C to allow hepatocytes to attach to the extracellular matrix. The incubator should contain a humidified atmosphere of 5% carbon dioxide in air. Attachment of human hepatocytes can take up to 16 h.

11. Culture plates containing hepatocytes should be viewed under the inverted microscope. Flattening of the hepatocytes indicates cell attachment. Once hepatocyte attachment is complete, the media above the attached monolayer is removed by aspiration using a sterile glass pipet or the Pipettus system. The culture plates should be tilted slightly, and care must be taken not to damage the hepatocyte monolayer. The monolayer is then covered with serum-free incubation media (2 mL), and the plates returned to the CO_2 incubator. If the hepato-

cyte culture is to be kept longer than 24 h, the media should be replaced every 24 h. The morphology of the hepatocyte monolayers and the attachment index should be inspected and determined using an inverted microscope.

12. For metabolism or induction studies, compounds are added to the media after cell attachment (*see* Notes 9 and 10). Metabolism experiments are usually of up to 24 h duration. However, induction experiments can last significantly longer. On completion of the incubation, the incubation medium is removed and rapidly frozen. The monolayer is then washed with 2 mL phosphate-buffered saline (PBS). The hepatocyte monolayer is disrupted by either a detergent solution (Triton 100X, 1% solution in PBS) or mechanical action. The supernatant containing the disrupted cell contents is placed in cyrovials and rapidly frozen.

13. Cytochrome P450 in hepatocyte monolayer cultures is determined by removing the incubation medium and adding 2 mL Renex P450 buffer. The hepatocyte monolayer is disrupted by mechanical action. The resulting supernatant is removed and stored at −70°C until analysis.

4. Notes

The preparation and culture of animal hepatocytes, particularly rat, have been extensively evaluated and refined *(4,20,26)*. This evaluation and refinement have led to the proposal of standard protocols for the isolation and culture of rat hepatocytes *(38)*. In contrast, the limited availability of human material for the preparation and culture of hepatocytes means that the human system has not yet been fully evaluated, and as such, the protocols used for the isolation and culture of human hepatocytes are diverse. The protocol given here is just one of many that have been reported to work for human hepatocytes. The different protocols may give differing results, as happens with rat hepatocytes, but these differences have not been evaluated for human hepatocytes. Some alternative protocols and approaches are reviewed in the following sections.

4.1. Isolation of Human Hepatocytes

A number of factors are important in ensuring a functional hepatocyte preparation with a high yield.

1. It is essential that the perfused tissue has only one cut surface and that the capsule surrounding the tissue is otherwise intact. Careful placing of the cannulae is crucial in achieving the optimum perfusion of the tissue sample.
2. It is important to monitor pH and temperature. The advantage of the bicarbonate buffering system is that oxygenation of the perfusate with a mixture of oxygen and carbon dioxide (95:5%) is combined with a very high buffering capacity. A disadvantage of the system is the formation of microbubbles. It is important not to allow bubbles to enter the tissue, since this will prohibit the perfusion of the tissue. A small drop in temperature of 1–2°C during the perfusion can prolong the perfusion time and increase the cellular damage caused by the perfusion. The temperature at the liver sample should be monitored and maintained at 37°C.

Human Hepatocytes

3. The size of the liver wedge biopsy sample to be perfused is important. Gomez-Lechon and coworkers *(21)* reported that smaller human liver samples up to 5 g gave better yields than samples over 5 g. Furthermore, these workers have indicated that the quality of the collagenase used for the perfusion is important. Collagenase with a high specific activity provides a better yield of viable hepatocytes. This increase in yield was thought to be due to a reduced perfusion time, which in turn leads to a reduction in cell destruction.

    Other workers *(15,16)* have shown that hepatocyte yield and viability can be increased if either hyaluronidase (0.05%) or dispase (500 U/mL) is added to the hepatocyte dissociation medium. However, for the preparation of rat hepatocytes, hyaluronidase is thought to be detrimental *(38)*.

4. The use of a calcium chelator to remove calcium ions, which are essential for desmosome integrity, is recommended. Perfusion with EGTA brings about cleavage of the desmosomes. It has been suggested that perfusion without a chelator reduces hepatocyte yield *(4)*.

5. Trypan Blue exclusion or leakage of certain cytoplasmic enzymes is a relatively quick simple assay of cell viability. Care must be taken with the Trypan Blue assay since it has been reported that a decrease in media pH can lead to increased dye uptake in rat hepatocytes *(36)*, which is reversible. This phenomenon has not been reported for human hepatocytes. Furthermore, it should be noted that staining intensity can vary depending on the batch of Trypan Blue or media components *(36,37)*. Protein binds Trypan Blue strongly, and thus, there should be no protein in the medium when analyzing cell viability by this method. Otherwise a falsely high value for viability will be obtained.

4.2. Culture of Human Hepatocytes

6. Confluent monolayers have been established with seeding densities of $4-17 \times 10^4$ cells/cm^2 *(8,15–17,22,33,39)*. The attachment efficiency (number of cells attached/number of cells seeded) should be determined and be in the region of 70–80% *(21)*. Attachment periods of up to 20 h for human hepatocytes have been reported *(17,33,39)*. Human hepatocytes prepared from the livers of older donors have been reported to require increased attachment time *(39)*.

7. The influence of various matrix components on hepatocytes in culture has been widely studied. In the liver, the hepatocyte is surrounded by a matrix that contains collagen, laminin, fibronectin, and heparan sulfate proteoglycan *(1,4)*. An ideal cell matrix should contain some of these components and should also be able to support hepatocyte morphology over long periods, allow formation of the monolayer and intercellular contacts, protect cells from sheer forces of fluids, and allow free exchange of salts within the media. However, in general, these organic matrices do not delay the occurrence of phenotypic changes. Moreover, the substrata can enhance alteration of liver gene expression.

    Hepatocytes in culture have been shown to have the ability to secrete several types of collagen and noncollagenous glycoproteins *(40,41)*. The use of plastic culture plates coated with an extracellular matrix, such as collagen, fibronectin, laminin, collagen gels, collagen gels embedded on a nylon mesh, or a laminin-rich matrix termed Matrigel®, has been shown to improve the efficiency of hepatocyte attachment, increasing the duration of functional activity and the morphology of the cells in culture. The use of attachment factors in animal hepatocyte culture is extensively reviewed by Berry and coworkers *(4)*, Guillouzo and coworkers *(20)*, Skett *(26)*, and Reid and coworkers *(42,43)*. The evaluation of the effects of extracellular attachment factors on the functionality of human hepatocytes is more limited.

Human hepatocytes in culture have been reported to attach well to extracellular matrices, such as uncoated plastic *(8,12,17)*, soluble rat tail or calf skin collagen type I *(9,10,13,15,33)*, fibronectin *(8,21,44,45)*, collagen gel immobilized hepatocytes *(46)*, and Matrigel *(47)*. It is worth discussing Matrigel in detail, since the hepatocytes do not become attached and flattened into a monolayer as with other extracellular factors, but remain spherical in shape. Matrigel is a complex biomatrix prepared from mouse sarcoma tissue (Engelbreth-Holm-Swarm sarcoma), and contains mainly laminin, type IV collagen, and heparan sulfate proteoglycan.

Schuetz and coworkers *(47)* demonstrated that cells attached to Matrigel exhibit enhanced expression of liver-specific functions. Furthermore, synthesis of gap junction proteins, indicating that intracellular communication has been restored, has been reported.

Matrigel appears to be more advanced than other extracellular matrices used for hepatocyte culture. However, it should be noted that the use of Matrigel may have some limitations. Matrigel is thought to contain some liver growth factors that may interfere with some assays and uses for the cells. Cell recovery is difficult, and the number of workers using this system are at present few.

8. The effects of various hepatocyte culture media and supplements on the maintenance of animal hepatocyte function have been the subject of a number of reviews *(4,20,26)*. Many different media have been used, and little work has been published comparing media effects on animal hepatocytes. In one recent study, however, Skett and Roberts *(48)* showed that WME was the best of the commercially available media for rat hepatocytes, but that Chee's medium is better, although not yet available in quantity through the media suppliers. Even less evaluation of the effects of media and supplements on human hepatocyte culture has taken place. Therefore, the systems employed for human hepatocyte culture are very diverse and tend to be extensions of conditions used for the culture of animal hepatocytes. A diverse variety of media have been used for human hepatocyte culture, including Weymouth's 752, minimal essential media mixed with media 199 (75/25% [v/v]), Ham's F12, Leibovitz L-15, WME, and Isoms. A full evaluation of this aspect is required, since it is seen in rat hepatocytes that the different media can give quite different results.

Once the basic media has been selected, a number of media supplements can be added. These supplements range from natural compounds, including hormones, minerals, and vitamins to nonphysiological factors, such as dimethyl sulfoxide.

a. Serum—composition unknown, but contains proteins and growth factors.
   b. Hormones—insulin, glucagon, glucocorticoids, thyroxine, and sex steroids.
   c. Growth supplements—selenium, zinc, EGF, and HGF.
   d. Nonphysiological factors—dimethyl sulfoxide, ammonium chloride, and metyrapone.

   The role of each of these supplements is, at best, unclear, even for the well-evaluated rat hepatocyte culture *(26)*. It is best, where possible, to keep medium supplements to a minimum. Any additions should be evaluated in your own system using the relevant end point required for the study.

9. Studies of the functionality of human hepatocytes in culture are limited when compared to data on animal hepatocytes *(18,21,40)*. These workers have shown that gluconeogenesis and glycolysis could be stimulated in human hepatocytes by physiological concentrations of glucagon and insulin. The gluconeogenic rate in human hepatocytes was similar to that estimated for the fasted human liver, whereas basal glycolysis is higher in cultured human hepatocytes than in vivo.

   Urea synthesis from ammonia has been used as an indicator of mitochondrial function. Under basal conditions, human hepatocytes produce 2–4 nmol urea/mg cell protein/min. Hepatocytes can be stimulated to produce significantly more urea, up to 11 nmol/min/mg. The maximal rate of ureagenesis decreased by 50% after 1 d in culture, but was then stable for several days thereafter *(21)*.

   Protein production by human hepatocytes has been used as a measure of cell function *(23,49,50)*. Human hepatocytes in culture have been shown to secrete albumin, α-antitrypsin, α-antichymotrypsin, α-acid glycoprotein, serum amyloid A, fibronectin, hepatoglobulin, and $α_2$-macroglobulin. Maximal rates were found during the first 2 d in culture, decreasing thereafter *(8,12,21,51,52)*.

10. One of the primary objectives of human hepatocytes in culture is to provide a model for the study of metabolism and toxicity of foreign compounds prior to administration to humans. These areas of research have attracted the attention of many workers *(9,10,13,24,33,39,44,49,50,53–58)*.

    However, one of the major problems of both animal and human hepatocyte cultures is the degradation of drug-metabolizing enzymes, most notably the cytochrome P450 *(18,22,23,25,47)*. The cytochromes P450 present in cultured human hepatocytes are thought to be more stable than that in rat hepatocytes *(18,59)*. However, cytochrome P450 is vulnerable in human hepatocytes and declines to approx 60% after 24 h of culture and 20% after 5 d of culture *(18,21,22)*. Human hepatocytes cocultured with rat epithelial cells maintain a relatively stable cytochrome P450 content for up to 8 d *(18,23,49)*. The culture conditions (e.g., medium chosen and medium supplements) can affect the levels of cytochrome P450.

    Although the levels of cytochrome P450 decrease in hepatocyte culture, the protein can still be expressed by induction. The cytochromes P450 can be induced by certain compounds in human hepatocyte cultures. Incubation of human hepatocyte cultures in the presence of either 1.5–3.2 mM phenobarbital, 50 μM rifampacin, 25–50 μM 3-methylcholanthrene, 12.5 μM benzanthracene, or 200 mM

ethanol, induces either the mRNA and/or the protein of specific cytochrome P450 isozymes *(22–24,44,60–62)*. Furthermore, the human hepatocyte culture system has been used successfully to examine the potential of new therapeutic agents, such omeprazole and lansoprazole, to induce cytochrome P450 in vitro *(63,64)*. Whether the induction observed in vitro can be extrapolated to humans is a key question. However, the induction observed in the human hepatocyte system with omeprazole was also detected in vivo *(63)*. The human hepatocyte culture has also been used to model a number of drug interactions observed with cyclosporin A in vivo. These interactions result from changes in the metabolism of cyclosporin caused by either induction or inhibition of the enzymes responsible for its metabolism *(65)*.

Cytokines produced during inflammation or infection have been shown to affect the expression of the major human cytochrome P450 enzymes in hepatocyte cultures. The interleukins-1β, -4, -6, tumor necrosis factor-α, and interferon-α and -γ have been shown to downregulate specific isozymes of cytochrome P450 in human hepatocyte cultures *(66,67)*. The observed in vitro effects of cytokines on drug-metabolizing enzymes may be important clinically with respect to chronic hepatitis patients who receive high doses of interferon as well as antiviral compounds.

The main focus of many researchers has been the cytochrome P450 drug-metabolizing enzymes. However, the conjugating enzymes have also been shown to be altered in human hepatocyte culture. UDP glucuronyltransferase increased during culture, whereas sulfotransferase activity decreased. Glutathione content remained relatively stable *(22,45)*. Again, the phase II enzyme content of the cells is affected by medium composition and other culture conditions.

Investigations and evaluation of the drug-metabolizing activities present in human hepatocyte cultures have demonstrated that these cells in culture appear to be a good model for predicting drug metabolism in vivo *(13,25,33,54–57,68,69)*. Human hepatocyte cultures provide an opportunity to study both the routes and rates of metabolism of new drugs and may prove to be a key in vitro model to improve the drug development process *(25,55)*. Several workers have shown that the metabolic routes and rates of a number of drugs in human hepatocyte cultures are similar to those observed in vivo *(13,33,54–56,64,68,69)*. Furthermore, human hepatocyte cultures have been used to help identify species differences in the metabolism of a number of drugs *(13,33,68,69)*. In addition to metabolism studies, human hepatocytes in culture have been evaluated as screens for hepatotoxicity *(44,49,50,57,58)* or to elucidate mechanisms of toxicity *(39,53)*.

4.3. Coculture of Human Hepatocytes with Rat Epithelial Cells

11. Hepatocyte survival and maintenance of specific hepatocyte function have been extended by attaching human hepatocytes to extracellular matrices or by modifying the medium. The attachment of hepatocytes in culture has been discussed previously. However, cell–cell interactions are worthy of more detailed consideration.

    Early attempts at coculturing hepatocytes with epithelial cells were not successful. However, culturing hepatocytes in the presence of untransformed epithelial cells from 10-d-old rat liver has been shown to increase hepatocyte survival and functional capability significantly *(20)*. Originally, these studies were performed with rat hepatocytes, but later were repeated with human hepatocytes.

Long-term culture of more than a few days has been achieved by coculturing human hepatocytes with rat epithelial cells in an attempt to recreate a more suitable environment, thus stabilizing hepatocyte function. Several workers have shown a marked improvement in the maintenance of differentiated function of human hepatocytes in coculture and drug-metabolizing enzymes, particularly the cytochrome P450 *(27,40,49,56,69–71)*. Since specific functions, including drug-metabolizing activity and the characteristic morphology of human hepatocytes, appear to be maintained in coculture for several days, this system may be suited to investigating the hepatotoxicity of xenobiotics *(49,50)*. One problem is the preparation of the epithelial cells. This is a long process that does not always yield functional cells.

4.4. Isolation and Culture of Human fetal Hepatocytes

12. The methods for the preparation of human fetal hepatocytes have been reviewed by different workers *(4,27,28)*. Briefly, minced tissue was digested with either collagenase or a mixture of collagenase with hyaluronidase, trypsin, or dispase. Yields of $4–5 \times 10^6$ cells/g liver with viabilities in the region of 80% have been achieved. Fetal hepatocytes were plated on plastic culture dishes in the presence of serum with an efficiency of 80%. Cultures were maintained in serum-free medium supplemented with hormones. Fetal hepatocyte cultures have been used to study the metabolism of caffeine and theophylline, confirming the routes of metabolism observed in vivo *(54)*. It should be remembered, however, that fetal hepatocytes do not possess all of the liver-specific functions of the adult liver in vivo and, thus, the usefulness of fetal hepatocytes is limited.

4.5. The Cryopreservation of Human Hepatocytes

13. Because of the infrequent supply of human material, methods of storing hepatocytes prior to culture have been investigated. Human hepatocytes have been cryopreserved in mixtures of buffers containing cryoprotectants, such as dimethyl sulfoxide, propylene glycol, acetamide, polyethylene glycol 8000, polyvinyl pyrrolidone, and fetal calf serum. Hepatocytes were then stored at $-80°C$ *(72–75)*. The effects of cryopreservation on hepatocyte functionality in culture are discussed *(72–75)*. It is not considered that cryopreservation techniques are sufficiently well advanced to allow a standard protocol to be presented, but this method holds out hope that the inadequate supply of human hepatocytes can be preserved and, thus, used as efficiently as possible.

Acknowledgments

The authors acknowledge the technical expertise of D. Herriott, A. Woodrooffe, and S. Khan.

References

1. Reid, L. M. (1990) Stem cell biology, hormone/matrix synergies and liver differentiation. *Trends Pharmacogenics* **2**, 121–130.
2. Fry, J. R. and Bridges, J. W. (1979) Use of primary hepatocyte cultures in biochemical toxicology, in *Reviews in Biochemical Toxicology* (Hogson, E., Bend, J. R., and Philpot, R. N., eds.), Elsevier, Amsterdam, pp. 201–248.
3. Howard, R. B., Christensen, A. K., Gibbs, F. A., and Pesch, L. A. (1967) The enzymatic preparation of isolated intact parenchymal cells from rat liver. *J. Cell Biol.* **35**, 675–684.
4. Berry, M. N., Edwards, A. M., and Barritt, G. J. (1991) Isolated hepatocytes preparation; properties and applications, in *Laboratory Techniques in Biochemistry and Molecular Biology*, vol. 21 (Burdon, R. H. and van Knippenberg, P. H., eds.), Elsevier, Amsterdam, pp. 1–460.
5. Berry, M. N. and Friend, D. S. (1969) High yield preparation of isolated rat liver parenchymal cells. *J. Cell Biol.* **43**, 506–520.
6. Bojar, H., Basler, M., Fuchs, F., Dreyfurst, R., and Staib, W. (1976) Preparation of parenchymal and non-parenchymal cells from adult human liver—morphological and biochemical characteristics. *J. Clin. Chem. Clin. Biochem.* **14**, 527–532.
7. Fabre, G., Rahmani, R., Placidi, M., Combalbert, J., Covo, J., Cano, J.-P., Coulange, C., Ducros, M., and Rampal, M. (1988) Characterisation of midazolam metabolism using human hepatic microsomal fractions and hepatocytes in suspension obtained by perfusing whole human livers. *Biochem. Pharmacol.* **37**, 4389–4397.
8. Guguen-Guillouzo, C., Campion, J. P., Brissot, P., Glaise, D., Launois, B., Bourel, M., and Guillouzo, A. (1982) High yield preparation of isolated human adult hepatocytes by enzymic perfusion of the liver. *Cell Biol. Int. Rep.* **6**, 625–628.
9. Reese, J. A. and Byard, J. L. (1981) Isolation and culture of adult hepatocytes from liver biopsies. *In Vitro* **17**, 935–940.
10. Strom, S. C., Jirtle, R. L., Jones, R. S., Novicki, D. L., Rosenberg, M. R., Novotny, A., Irons, G., Mclain, J. R., and Michalopoulos, G. (1982) Isolation, culture and transplantation of human hepatocytes. *J. Natl. Cancer Inst.* **68**, 771–778.
11. Tee, L. B. G., Seddon, T., Boobis, A. R., and Davies, D. S. (1985) Drug metabolising activity of freshly isolated human hepatocytes. *Br. J. Clin. Pharmacol.* **19**, 279–294.
12. Ballet, F., Bouma, M.-E., Wang, S.-R., Amit, N., Marais, J., and Infante, R. (1984) Isolation, culture and characterisation of adult human hepatocytes from surgical liver biopsies. *Hepatology* **4**, 849–854.
13. Seddon, T., Michelle, I., and Chenery, R. J. (1989) Comparative drug metabolism of diazepam in hepatocytes isolated from man, rat, monkey and dog. *Biochem. Pharmacol.* **38**, 1657–1665.
14. Bayliss, M. K. (1991) *Comparative Evaluation of the Drug Metabolising Activity of Rat, Dog and Human Hepatocytes Using Two Model Substrates and Loxtidine.* PhD thesis, University of Hertfordshire, UK.

15. Maekubo, H., Ozaki, S., Mitmaker, B., and Kalant, N. (1982) Preparation of human hepatocytes for primary culture. *In Vitro* **18**, 483–491.
16. Miyazaki, K., Takaki, R., Nakayama, F., Yamauchi, S., Koga, A., and Todo, S. (1981) Isolation and primary culture of adult human hepatocytes. *Cell Tissue Res.* **218**, 13–21.
17. Vons, C., Pegorier, I. P., Ivanov, M. A., Girard, J., Melcion, C., Cordier, A., and Franco, D. (1990) Comparison of cultured human hepatocytes isolated from surgical biopsies or cold-stored organ donor livers. *Toxicol. In Vitro* **4**, 432–434.
18. Guillouzo, A., Beaune, P., Gascoin, M.-N., Begue, J.-M., Campion, J.-P., Guengerich, F. P., and Guguen-Guillouzo, C. (1985) Maintenance of cytochrome P450 in cultured adult human hepatocytes. *Biochem. Pharmacol.* **34**, 2991–2995.
19. Houssin, D., Capron, M., Celier, C., Cresteil, T., Demaugre, F., and Beaune, P. (1983) Evaluation of isolated human hepatocytes. *Life Sci.* **33**, 1805–1809.
20. Guillouzo, A., Morel, F., Ratanasavanh, D., Chesne, C., and Guguen-Guillouzo, C. (1990) Long term culture of functional hepatocytes. *Toxicol. In Vitro* **4**, 415–427.
21. Gomez-Lechon, M. J., Lopez, P., Donato, T., Montoya, A., Larrauri, A., Gimenez, P., Trullenque, R., Fabra, R., and Castell, J. V. (1990) Culture of human hepatocytes from small surgical liver biopsies. Biochemical characterisation and comparison with in vivo. *In Vitro Cell. Dev. Biol.* **26**, 67–74.
22. Grant, H. M., Burke, D. M., Hawksworth, G. M., Duthie, S. J., Engeset, J., and Petrie, J. C. (1987) Human adult hepatocytes in primary monolayer culture. *Biochem. Pharmacol.* **36**, 2311–2316.
23. Morel, F., Beaune, P. H., Ratanasavanh, D., Flinois, J.-P., Yang, C. S., Guengerich, F. P., and Guillouzo, A. (1990) Expression of cytochrome P450 enzymes in cultured human hepatocytes. *Eur. J. Biochem.* **191**, 437–444.
24. Morel, F., Beaune, P., Ratanasavanh, D., Flinois, J.-P., Guengerich, F. P., and Guillouzo, A. (1990) Effects of various inducers on the expression of cytochromes P450 2C8, 9, 10 and 3A in cultured human hepatocytes. *Toxicol. In Vitro* **4**, 458–460.
25. Guillouzo, A., Morel, F., Fardel, O., and Meunier, B. (1993) Use of human hepatocyte cultures for drug metabolism studies. *Toxicology* **82**, 209–219.
26. Skett, P. (1994) Problems in using isolated and cultured hepatocytes for xenobiotic metabolism/metabolism based toxicity testing—solutions? *Toxicol. In Vitro* **8**, 491–504.
27. Rogiers, V. (1993) Cultures of human hepatocytes in *in vitro* pharmaco-toxicology, in *Human Cells in In Vitro Pharmaco-Toxicology* (Rogiers, V., Sonck, W., Shephard, E., and Vercruysse, A., eds.), Vubpress, Brussels, Belgium, pp. 77–115.
28. Guguen-Guillouzo, C. and Guillouzo, A. (1986) Methods for preparation of adult and fetal hepatocytes, in *Isolated and Cultured Hepatocytes* (Guillouzo, A. and Guguen-Guillouzo, C., eds.), John Libbey Eurotext, London, pp. 1–12.
29. Wunsch, E. and Heidrich, H. G. (1963) Zur quantitativen bestimmung der kollagenase. *Hoppe-Seylers Z. Physiol. Chem.* **333**, 149–151.
30. Butler, M. (1992) Culture media, in *Labfax: Cell Culture* (Butler, M. and Dawson, M., eds.), BIOS Scientific and Blackwell Scientific Publications Ltd., UK, pp. 85–106.

31. Warner, M., La Marca, M. V., and Neims, A. H. (1978) Chromatographic and electrophorectic heterogeneity of the cytochromes P450 solubilised from untreated rat liver. *Drug Metab. Dispos.* **6**, 353–362.
32. Oldham, H. G., Norman, S. J., and Chenery, R. J. (1985) Primary cultures of adult rat hepatocytes—a model for the toxicity of histamine H_2-receptor antagonists. *Toxicology* **36**, 215–229.
33. Oldham, H. G., Standring, P., Norman, S. J., Blake, T. J., Beattie, I., Cox, P. J., and Chenery, R. J. (1990) Metabolism of temelastine (SKF93944) in hepatocytes from rat, dog, cynomologous monkey and man. *Drug Metab. Dispos.* **18**, 146–152.
34. Kalayoglu, M., Sollinger, H. W., Stratta, R. J., D'Alessandro, A. M., Hoffman, R. M., Pirsch, J. D., and Belzer, F. O. (1988) Extended preservation of the liver for clinical transplantation. *Lancet* **1**, 617–619.
35. Bellemann, P., Gebhardt, R., and Mecke, D. (1977) An improved method for the isolation of hepatocytes from liver slices. *Anal. Biochem.* **81**, 408–415.
36. Baur, H., Kasperek, S., and Pfaff, E. (1975) Criteria of viability of isolated liver cells. *Hoppe-Seyler's Z. Physiol. Chem.* **356**, 827–838.
37. Jauregui, H. O., Hayner, N. T., Driscoll, J. L., Williams-Holland, R., Lipsky, M. H., and Galletti, P. M. (1981) Trypan blue dye uptake and lactate dehydrogenase in adult rat hepatocytes—freshly isolated cells; cell suspensions, and primary monolayer cultures. *In Vitro* **17**, 1100–1110.
38. Blaauboer, B., Boobis, A. R., Castell, J. V., Coecke, S., Groothuis, J. M. M., Guillouzo, A., Hall, T. J., Hawksworth, G. M., Lorenzon, G., Miltenberger, H. G., Rogiers, V., Skett, P., Villa, P., and Weibel, F. (1994) The practical applicability of hepatocyte cultures in routine testing. *ATLA* **22**, 231–242.
39. Butterworth, B. E., Smith-Oliver, T., Earle, L., Loury, D. J., White, R. D., Doolittle, D. J., Working, P. K., Cattley, R. C., Jirtle, R., Michalopoulos, G., and Strom, S. (1989) Use of primary cultures of human hepatocytes in toxicology studies. *Cancer. Res.* **49**, 1075–1084.
40. Clément, B., Guguen-Guillouzo, C., Campion, J.-P., Glaise, D., Bourel, M., and Guillouzo, A. (1984) Long term co-cultures of adult human hepatocytes with rat liver epithelial cells: modulation of active albumin secretion and accumulation of extracellular material. *Hepatology* **4**, 373–380.
41. Diegelmann, R. F. (1986) Synthesis of extracellular matrix components by cultured hepatocytes, in *Isolated and Cultured Hepatocytes* (Guillouzo, A. and Guguen-Guillouzo, C., eds.), John Libbey Eurotext, London, pp. 209–224.
42. Reid, L. M. and Jefferson, D. M. (1984) Culturing hepatocytes and other differentiated cells. *Hepatology* **4**, 548–559.
43. Reid, L. M., Narita, M., Fujita, M., Murray, Z., Liverpool, C., and Rosenberg, L. (1986) Matrix and hormonal regulation of differentiation in liver cultures, in *Isolated and Cultured Hepatocytes* (Guillouzo, A. and Guguen-Guillouzo, C., eds.), John Libbey Eurotext, London, pp. 225–258.
44. Jover, R., Ponsoda, X., Castell, J. V., and Gomez-Lechon, M. J. (1994) Acute cytotoxicity of ten chemicals in human and rat cultured hepatocytes and in cell lines: correlation between in vitro data and human lethal concentrations. *Toxicol. In Vitro* **8**, 47–54.

45. Iqbal, S., Elcombe, C. R., and Elias, E. (1991) Maintenance of mixed function oxidase and conjugation enzyme activities in hepatocyte cultures prepared from normal and diseased human liver. *J. Hepatol.* **12,** 336–343.
46. Koebe, H. G., Pahernik, S., Eyer, P., and Schildberg, F. W. (1994) Collagen gel immobilisation: a useful cell culture technique for long term metabolic studies on human hepatocytes. *Xenobiotica* **24,** 95–107.
47. Schuetz, E. G., Schuetz, J. D., Strom, S. C., Thompson, M. T., Fisher, R. A., Molowa, D. T., Li, D., and Guzelian, P. (1993) Regulation of human liver cytochromes P450 in family 3A in primary and continuous culture of human hepatocytes. *Hepatology* **18,** 1254–1262.
48. Skett, P. and Roberts, P. (1994) Effect of culture medium on the maintenance of steroid metabolism in cultured adult rat hepatocytes. *In Vitro Toxicol.* **7,** 261–267.
49. Guillouzo, A., Begue, J. M., Campion, J.-P., Gascoin, M.-N., and Guguen-Guillouzo, C. (1985) Human hepatocyte cultures: a model of pharmaco-toxicological studies. *Xenobiotica* **15,** 635–641.
50. Ratanasavanh, D., Baffet, G., Latinier, M. F., Rissel, M., and Guillouzo, A. (1988) Use of hepatocyte co-cultures in the assessment of drug toxicity from chronic exposure. *Xenobiotica* **18,** 765–771.
51. Munck Petersen, C., Christiansen, B. S., Heickendorff, L., and Ingerslev, J. (1988) Synthesis and secretion of α_2-macroglobulin by human hepatocytes in culture. *Eur. J. Clin. Invest.* **18,** 543–548.
52. Christiansen, B. S., Ingerslev, J., Heickendorff, L., and Petersen, C. M. (1988) Human hepatocytes in culture synthesize and secrete fibronectin. *Scand. J. Clin. Lab. Invest.* **48,** 685–690.
53. Monteith, D. K., Novotny, A., Michalopoulos, G., and Strom, S. C. (1987) Metabolism of benzo(a)pyrene in primary cultures of human hepatocytes: dose–response over a four log range. *Carcinogenesis* **8,** 983–988.
54. Berthou, F., Ratanasavanh, D., Alix, D., Carlhant, D., Riche, C., and Guillouzo, A. (1988) Caffeine and theophylline metabolism in new-born and adult human hepatocytes; comparison with adult rat hepatocytes. *Biochem. Pharmacol.* **37,** 3691–3700.
55. Fabre, G., Combalbert, J., Berger, Y., and Cano, J.-P. (1990) Human hepatocytes as a key in vitro model to improve preclinical drug development. *Eur. J. Drug Metab. Pharmacokinet.* **15,** 165–171.
56. Guillouzo, A., Begue, J.-M., Maurer, G., and Koch, P. (1988) Identification of metabolic pathways of pindalol and fluperlapine in adult human hepatocyte cultures. *Xenobiotica* **18,** 131–139.
57. Begue, J.-M., Koch, P., Maurer, G., and Guillouzo, A. (1993) Maintenance of the biotransformation capacity by cultured human hepatocytes after several daily exposures to drugs. *Toxicol. In Vitro* **7,** 493–498.
58. Le Bot, M. A., Begue, J.-M., Kernaleguen, J., Robert, J., Ratanasavanh, D., Airiau, J., Riche, C., and Guillouzo, A. (1988) Different cytotoxicity and metabolism of doxorubicin, daunorubicin, epirubicin, esorubicin and idarubicin in cultured human and rat hepatocytes. *Biochem. Pharmacol.* **37,** 3877–3887.

59. Guguen-Guillouzo, C., Baffet, G., Clement, B., Begue, J.-M., Glaise, D., and Guillouzo, A. (1983) Human adult hepatocytes: isolation and maintenance at high levels of specific function in a co-culture system, in *Isolation, Characterisation and Use of Hepatocytes* (Harris, R. A. and Cornell, N. W., eds.), Elsevier, Amsterdam, pp. 105–110.
60. Guguen-Guillouzo, C. and Guillouzo, A. (1983) Modulation of functional activities in cultured rat hepatocytes. *Mol. Cell. Biochem.* **53/54,** 35–56.
61. Donato, M. T., Gomez-Lechon, M. J., and Castell, J. V. (1990) Effect of xenobiotics and monooxygenase activities in cultured human hepatocytes. *Biochem. Pharmacol.* **39,** 1321–1326.
62. Monteith, D. K., Ding, D., Chen, Y. T., Michalopoulos, G., and Strom, S. C. (1990) Induction of cytochrome P450 RNA and benzo(a)pyrene metabolism in primary human hepatocyte cultures with benzanthracene. *Toxicol. Appl. Pharmacol.* **105,** 460–471.
63. Diaz, D., Fabre, I., Daujat, M., Saint-Aubert, B., Borries, P., Michel, H., and Maurel, P. (1990) Omeprazole is an aryl hydrocarbon like inducer of human hepatic cytochrome P450. *Gastroenterology* **99,** 737–747.
64. Curi-Pedrosa, R., Daujat, M., Pichard, L., Ourlin, J. C., Clair, P., Gervot, L., Lesca, P., Domergue, J., Joyeux, H., Fourtanier, G., and Maurel, P. (1994) Omeprazole and lansoprazole are mixed inducers of CYP1A and CYP3A in human hepatocytes in primary culture. *J. Pharmacol. Exp. Ther.* **269,** 384–392
65. Pichard, L., Fabre, J., Fabre, G., Domergue, J., Saint-Aubert, B., Mourad, G., and Maurel, P. (1990) Cyclosporin A drug interactions: screening for inducers and inhibitors of cytochrome P450 (cyclosporin A oxidase) in primary cultures of human hepatocytes and in liver microsomes. *Drug Metab. Dispos.* **18,** 595–606.
66. Abdel-Razzak, Z., Loyer, P., Fautrel, A., Gautier, J. C., Corcos, L., Turlin, B., Beaune, P., and Guillouzo, A. (1993) Cytokines down-regulate expression of major cytochrome P450 enzymes in adult human hepatocytes in primary culture. *Mol. Pharmacol.* **44,** 770–775.
67. Donato, M. T., Herrero, E., Gomez-Lechon, M. J., and Castell, J. V. (1993) Inhibition of monooxygenase activities in human hepatocytes by interferons. *Toxicol. In Vitro* **7,** 481–485.
68. Le Bigot, J. F., Begue, J.-M., Kiechel, J. R., and Guillouzo, A. (1987) Species differences in metabolism of ketotifen in rat, rabbit and man: demonstration of similar pathways in vivo and in cultured hepatocytes. *Life Sci.* **40,** 883–890.
69. Dauphin, J. F., Graviere, C., Bouzard, D., Rohou, S., Chesne, C., and Guillouzo, A. (1993) Comparative metabolism of tosufloxacin and BMY43748 in hepatocytes from rat, dog, monkey and man. *Toxicol. In Vitro* **7,** 499–503.
70. Begue, J.-M., Le Bigot, J. F., Guguen-Guillouzo, C., Kiechel, J. R., and Guillouzo, A. (1983) Cultured human adult hepatocytes: a new model for drug metabolism studies. *Biochem. Pharmacol.* **32,** 1643–1646.
71. Ratanasavanh, D., Berthou, F., Dreano, Y., Mondine, P., Guillouzo, A., and Riche, C. (1990) Methylcholanthrene but not phenobarbital enhances caffeine and theophylline metabolism in cultured adult human hepatocytes. *Biochem. Pharmacol.* **39,** 85–94.

72. Coundouris, J. A., Grant, M. H., Engeset, J., Petrie, J. C., and Hawskworth, G. W. (1993) Cryopreservation of human adult hepatocytes for use in drug metabolism and toxicity studies. *Xenobiotica* **23,** 1399–1409.
73. Chesne, C., Guyomard, C., Grislain, L., Clerc, C., Fautrel, A., and Guillouzo, A. (1991) Use of cryopreserved animal and human hepatocytes for cytotoxicity studies. *Toxicol. In Vitro* **5,** 479–482.
74. De Sousa, G., Dou, M., Barbe, D., Lacarelle, B., Placidi, M., and Rahmani, R. (1991) Freshly isolated or cryopreserved human hepatocytes in primary culture: influence of drug metabolism on hepatotoxicity. *Toxicol. In Vitro* **5,** 483–886.
75. Moshage, H. J., Rijntjes, P. J. M., Hafkenscheid, J. C. M., Roelofs, H. M. J., and Yap, S. H. (1988) Primary culture of cyropreserved adult human hepatocytes on homologous extracellular matrix and the influence of monocytic products on albumin synthesis. *J. Hepatol.* **7,** 33–44.

29

Culture of Human Pancreatic Islet Cells

Stellan Sandler and Décio L. Eizirik

1. Introduction

The pancreatic islet is a microorgan composed of four types of hormone-producing cells. In the human pancreas, the islets comprise about 1–2% of the pancreatic mass, and the islet number is approx $1-2 \times 10^6$. The majority of cells, which are mainly located in the center of the islet, are the insulin-producing β-cells. In the periphery, α-cells and δ-cells synthesizing glucagon and somatostatin, respectively, are found. Finally, a small number of cells containing pancreatic polypeptide (PP) are scattered within the islets (1). In addition, in the intact pancreatic islet nerve cell fibers, vascular tissue as well as a few residing immune cells can also be observed. Owing to the requirements to study the in vitro function of the β-cells for various aspects of diabetes research, methods have been developed to isolate and culture free pancreatic islets.

Over the last 30 years, a large number of protocols have been described for isolating the pancreatic islets from the remaining exocrine part of the gland. First, the method of microdissection of the islets was described (2). Later, this was substituted by the collagenase isolation technique (3,4), whereupon an enzymatic digestion of the pancreas leads to a separation of the endocrine and exocrine pancreas. Presently, collagenase digestion is still the crucial component of all islet isolation methods for both adult animal and human pancreas. Description of protocols for isolation of human adult islets is beyond the aim of this chapter, but according to the view of the authors, the isolation methods so far published providing the most consistent results in terms of viability and yield of human islets are those described by Ricordi et al. (5) and Warnock et al. (6).

This chapter deals with methods for tissue culture of human islet cells. Both fetal and adult human islet preparations are used in this context, and since the procedures thus differ markedly, we describe and discuss the tissues separately.

From: *Methods in Molecular Medicine: Human Cell Culture Protocols*
Edited by: G. E. Jones Humana Press Inc., Totowa, NJ

Moreover, we give an account for suitable tests to evaluate specific islet function in conjunction with the culture.

2. Materials

2.1. Assessment of Glucose Oxidation Rate and Insulin Release

1. Krebs-Ringer buffer: 114.3 mM NaCl, 4.74 mM KCl, 1.15 mM KH$_2$PO$_4$, 1.18 mM MgSO$_4$, 25.00 mM NaHCO$_3$, 10.00 mM HEPES, 4.26 mM NaOH, 2.54 mM CaCl$_2$, pH 7.4. This buffer is modified from ref. 7, and it is stable at 4°C for about 3 wk. Check pH each time before use, and if necessary, correct it by gassing the buffer with CO$_2$:O$_2$ (5:95).
2. Buffer in step 1 with addition of 2 mg/mL of bovine serum albumin (BSA). The solution is stable at 37°C for about 3 h, and it should not be stored overnight.

2.2. Assessment of Islet (Pro)Insulin Biosynthesis Rate

1. 50 mM Glycine, 6 mM NaOH, 2.5 mg/mL BSA, pH 8.8.
2. 50 mM Glycine, 6 mM NaOH, 0.1% Triton X-100 (1 mL/1000 mL), 2.5 mg/mL BSA, pH 8.8.
3. 1M Acetic acid, 2.5 mg/mL BSA.
4. 50 mM Glycine, 6 mM NaOH, pH 8.8.
5. Allow protein A-Sepharose CL-4B (1.5 g; Pharmacia, Uppsala, Sweden) to swell in a glass filter funnel for 15 min in 50 mM glycine, wash with 200 mL of the same solution, and finally suspend in 30 mL of solution 50 mM glycine containing BSA and Triton X-100.
6. Anti-insulin serum: Suspend an ampule of guinea pig antibovine insulin serum (Code 5506, BioMakor, Rehovot, Israel) in 2.3 mL of redistilled water.
7. Guinea pig serum: Mix 1.15 mL normal guinea pig serum with 1.15 mL of redistilled water.

The items described can be stored at least for a month at 4°C.

2.3. Measurement of Total Protein Biosynthesis Rate

1. 50 mM Glycine, 6 mM NaOH, 2.5 mg/mL BSA, pH 8.8.
2. 0.61 mM Trichloroacetic acid.
3. 0.15M NaOH.

The items described can be stored at least for a month at 4°C.

3. Methods

3.1. Collection and Tissue Culture of Human Fetal Pancreas

3.1.1. Registration of the Fetal Pancreas

Record the following data on each occasion of human fetal pancreas collection. It is advisable to design a special form for this purpose, which secures that no information is omitted.

Human Pancreatic Islet Cells

1. Estimated gestational age of the fetus.
2. The fetal crown–heel length.
3. Clinic and hospital where the abortion was performed.
4. The indication for the abortion, with regard to medical complications in either the fetus or the mother. If there is any indication that the tissue might be bacteriologically or virally contaminated, the fetal pancreas is not dissected.
5. Method used for induction of the abortion and the time interval between induction and the actual delivery. For instance, regarding prostaglandin-induced abortions, there is a direct correlation between deterioration in pancreatic tissue viability and a prolonged time for the fetal delivery, probably owing to intrauterine ischemia and death of the fetus.
6. Time of the delivery and time when the dissection of the pancreas is completed. Empirically we have found that if this time interval exceeds 1.5 h, the survival of a human fetal pancreatic explant in tissue culture is much reduced.

3.1.2. Human Fetal Pancreas Dissection and Procurement

1. Commence the dissection of the fetal pancreas as early as possible. Prior to dissection, maintain the aborted fetus at +4°C. For dissection, use sterile surgical gloves and sterilized instruments. Preferably dissection is performed in a sterile hood with laminar air flow.
2. Place the fetus on its right side on a sheet of filter paper, and wash the trunk thoroughly with 70% ethanol.
3. Incise the skin and the abdominal wall gently, and open the peritoneal cavity, using separate instruments for each step of dissection. Locate the pancreas adjacent to the spleen, and either dissect it free or remove the gland together with the spleen for further separation in a sterile Petri dish.
4. Place the pancreatic gland in a sterile solution containing 2 mL of Hank's balanced salt solution (HBSS) *(8)* supplemented with antibiotics (100 U/mL benzylpenicillin + 0.1 mg/mL streptomycin). Remove nonpancreatic tissue, such as fat, lymph nodes, and connective tissue membranes.
5. At this stage, the gland is either further processed or put in a sterile test tube containing medium RPMI 1640 *(9)* and antibiotics supplemented with 10% (v/v) of fetal bovine serum (FBS) and stored at 4°C for up to 6–8 h without a significant loss of viability. The latter aspect is important if the dissected gland needs to be transported.

3.1.3. Tissue Culture of Human Fetal Pancreas

3.1.3.1. CULTURE OF FREE-FLOATING PANCREATIC FRAGMENTS

1. Chop the tissue in cold HBSS into fragments, about 2–3 mm^3 in size, using a pair of scissors.
2. Transfer the fragments to sterile plastic nontissue-culture-treated dishes (Sterilin®, 50-mm diameter, Bibby Sterilin, Stone, Staffs, UK) containing 4.0–4.5 mL of culture medium RPMI 1640 supplemented with antibiotics (*see* Section 3.1.2., step 5) 0.5–1.0 mL of FBS. The culture media used, giving comparable results in

terms of insulin content of the explanted tissue (10), are RPMI 1640 (9) or Medium 199 (11). The amount of tissue obtained from a single human fetal pancreas varies markedly. On average, from a fetus with a crown–heel length of <10 cm, the pancreas is explanted into one culture dish; at a crown–heel length of 10–15 cm and at 15–25 cm, an equal distribution of the tissue into two and three dishes, respectively, is suitable.
3. Maintain the culture dishes in an incubator with automatically controlled temperature, humidity, and CO_2 concentration. The temperature is 37°C, and the atmosphere is humidified air 5% CO_2. During culture, the tissue is submerged and floating freely in the culture dishes.
4. Every second day, remove the medium by suction through a sterile needle, and add fresh medium. By this technique the explants can be kept viable in culture for about 2 wk. However, there is an increasing central cell death in the fragments probably owing to diffusion problems. Note that the pancreatic fragments contain a majority of exocrine cells and only a minor part is islet endocrine tissue ($\approx$5%).

3.1.3.2. CULTURE OF HUMAN FETAL PANCREAS AT THE AIR–LIQUID INTERPHASE

1. This culture technique was originally developed for culture of mouse thymus explants (12,13). Mandel and Georgiou subsequently adopted the technique for human fetal pancreatic preparations (14). By this method, the cultured fragments are maintained at the gas-medium interphase of the culture system, which secures an optimal exchange of gas by the tissue and at the same time a good supply of nutrients from the culture medium.
2. Cut blocks of surgical gel foam (about 2 × 2 cm; Upjohn, Kalamazoo, MI), and soak the dry gel pieces for 10 min in culture medium, for thorough absorption with culture medium. Put a block in a culture dish (Sterilin; see Section 3.1.3.2.) containing culture medium.
3. The culture conditions and the media are otherwise identical to those adopted for the free-floating cultures. We found, however, a more sustained insulin production of the explants with Medium 199 compared to RPMI 1640 with the air–liquid interphase culture technique (10).
4. Place two Millipore® filters (pore size 8.0 µm), which have been washed and sterilized for 20 min in boiling redistilled water, on top of a gel foam block.
5. Pipet 5–10 pancreatic fragments, about 5–10 mm³ in size, onto the filters.
6. Perform medium exchange cautiously, because the fragments have a tendency to slide off the filters down into the gel foam.
7. One advantage with this culture technique is the ease of recovering the tissue from the filters. Moreover, the filters with the explants can be used directly for examination of hormone secretion employing perifusion techniques.

3.1.3.3. CULTURE OF HUMAN FETAL ISLET-LIKE CELL CLUSTERS

1. The culture methods described in Sections 3.1.3.1. and 3.1.3.2. are not specifically focused on the endocrine islets cells or on isolation of free islets. In the human fetal pancreas at gestational weeks 10–22, the number of clearly delineated islets is very small. The endocrine cells are instead scattered either as single

cells or gathered in minor irregular clumps of endocrine cells, an arrangement that precludes conventional islet isolation methods. We therefore aimed to develop a culture method where formation of islets occurred in vitro from explants of human fetal pancreas *(15).*

2. Use pancreatic preparations dissected and procured as in Section 3.1.2..
3. Mince the gland in HBSS into fragments about 4 mm^3 in size, and then transfer the tissue to sterile glass vials containing 5 mg collagenase from *Clostridium histolyticum* (Boehringer Mannheim, Mannheim, Germany), dissolved in 2 mL of Ca^{2+}- and Mg^{2+}-free HBSS supplemented with antibiotics (*see* Section 3.1.2., step 5). The Ca^{2+}- and Mg^{2+}-free solution promotes a more rapid disintegration of the tissue and prevents reaggregation during washing.
4. Shake the tissue vigorously in a water bath for 5–10 min at 37°C, and interrupt the collagenase digestion when tissue fragments are nearly completely disintegrated and can be barely seen by visual inspection.
5. Transfer the digested tissue into plastic tubes, and wash twice by centrifugation at 500*g* for 6 min.
6. Resuspend the pelleted material in a small volume of culture medium containing 10% FBS and antibiotics.
7. Explant the tissue into culture dishes allowing tissue attachment (Nunclon®, 50 mm; Nunc, Roskilde, Denmark), containing 5 mL of medium RPMI 1640 with 10% FBS.
8. Keep the cultures at 37°C in air +5% CO_2 for 7 d, with medium exchange every second day. Note that this procedure does not result in isolation of well-defined islets, as is the case for adult islets (*see* Section 3.2.1.).
9. The viable culture preparations show characteristic behavior. During the initial 2–3 d, there is a massive outgrowth of a confluent monolayer of fibroblasts until the bottom of the culture dishes becomes covered. On top of this layer, there is a gradual appearance of spherical cell aggregates, with a diameter of 200–400 μm, that are loosely attached to the fibroblast cells. The cell aggregates are easily detached by flushing with culture medium via a syringe and harvested by a pipet. Under the stereomicroscope, the cell aggregates resemble isolated pancreatic islets in shape and size. The number of cell aggregates per cultured gland ranges from about 100–2000, depending on the size of the gland. When examined under the light microscope, sections of fixed cell aggregates show a predominance of well-preserved, round or polygonal cells, and sometimes duct-like structures can be seen. Frequently, replicating cells are observed. Specific stainings for the islet hormones, on day 7 of culture, have revealed that hormone-positive cells constitute a minority (10–20%) of the cells. Since this is a staining pattern different from that observed in adult pancreatic islets, we have chosen to designate the fetal cell aggregates as islet-like cell clusters (ICC). Cells of the ICC seems to have a high capacity to replicate and mature/differentiate into hormone-positive cells on further culture or after transplantation into nude mice *(15–18).*
10. The effect of different putative factors that can affect growth and differentiation are discussed in Notes 1–4. Alternative culture methods for both fetal and adult β-cells are referred to in Note 5. Aspects of insulin secretion of fetal β-cells are discussed in Note 6.

3.2. Collection, Transport, and Culture of Adult Human Pancreatic Islets

3.2.1. Collection of Human Pancreatic Islets

As mentioned in Section 1., methods for adult human islet isolation are not discussed herein. It must be stressed, however, that isolation of human pancreatic islets is a major undertaking, requiring a close cooperation between transplantation surgeons (who will retrieve the gland from heart-beating organ donors) and cell biologists, who will isolate the islets for the actual cell culture. The previous experience from this laboratory with small-scale human islet isolation *(19)* suggested that the best strategy to obtain reliable and reproducible results in the field is to engage in a cooperative effort with other groups. As an example, we have been participating since 1989 in an European Concerted Action for the Treatment of Diabetes, under the coordination of D. G. Pipeleers, Brussels Free University, Brussels, Belgium. The project includes participating centers from 10 European countries, and it has established a network of clinical centers, assuring a constant supply of glands removed from heart-beating organ donors, for islet isolation at the Central Unit in Brussels. Against this background, the Central Unit has already processed more than 800 pancreata, and established a reproducible and reliable methodology for human islet isolation and for evaluation of quality and purity of the material. A similar arrangement, although in a minor scale, is also present in the United States, having the Human Islet Transplantation Center, Washington University Medical Center, St. Louis, MO, as the isolation center.

It is important that information on gender and date of birth of the donor, culture medium used in the isolating unit, and time of culture before shipment is recorded in conjunction with each islet isolation. Moreover, information on the islet cell preparation must be provided. Ideally, a morphological characterization at the light microscopical (for further details, *see* Section 3.3.1.) and electron microscopical levels should be performed on each preparation *(20,21)*; *see also* Note 7. If this is not possible, at least information on the insulin content per DNA must be provided (*see* Section 3.3.3.).

3.2.2. Transport of Adult Human Islets

1. The transport of human pancreatic islets is performed at room temperature in 50-mL sealed conical sterile Falcon® tubes. The transport medium is HAM F10 *(22)* containing 6.1 mM glucose, and supplemented with 0.5% BSA, 0.08 mg/mL benzylpenicillin, and 0.1 mg/mL streptomycin (these conditions have been developed at the Central Unit of the β-Cell Transplant, Brussels).
2. In our experience, islets endure well up to 10–12 h under such conditions, but they must be maintained in culture (*see* Section 3.2.3.) for at least 24–48 h after arrival to recover their function completely.

Table 1
Function of Adult Pancreatic Human Islets Cultured for 1–3 Wk[a]

DNA content, ng/10 islets	Insulin content ng/10 islets	Insulin release		IR[b]
		\multicolumn{2}{c}{Glucose concentration, ng insulin/10 islets x 1 h}	
		1.7 mM	16.7 mM	
409 ± 27	708 ± 101	5.0 ± 0.9	31.1 ± 6.6	7.4 ± 0.9

[a]Islets were isolated at the Central Unit of the β-Cell Transplant, Brussels, sent to Uppsala, and cultured as described in Section 3.2.3. The preparations contained <6% dead cells or exocrine cells, as judged by electron microscopy. The prevalence of insulin-positive cells, using light microscopy of immunocytochemically stained islets (performed in Brussels), was 51 ± 3%. Batch-type insulin release experiments were performed as described in Section 3.3.4. The results are means ± SEM of 31 observations, each performed in triplicate.

[b]The increase ratio (IR) was determined in each experiment by dividing the insulin release at 16.7 mM glucose by that obtained at 1.7 mM glucose.

3.2.3. Culture of Adult Human Islets

Perform long-term culture of free floating isolated adult human islets in nonattaching Sterilin Petri dishes. We use 50-mm dishes containing 5 mL of medium for up to 200 islets, and 90-mm dishes containing 10 mL medium for 200–500 islets. Culture the islets in medium RPMI 1640 containing 5.6 mM glucose supplemented with 10% FBS, benzylpenicillin (100 U/mL), and streptomycin (0.1 mg/mL) (as described for fetal human islets; see Section 3.1.3.1.). Change the medium every 2 d, and at each medium change, use a needle (0.4 × 20 mm) to separate mechanically islets that become attached. Indeed, human islets tend to aggregate more in culture than rodent islets, and if large aggregates are allowed to form, there is central necrosis owing to lack of diffusion of nutrients and oxygen. The glucose concentration in the medium is an important issue: β-cell function of human islets decreases when these cells are exposed for prolonged periods of time to high glucose concentrations (11 or 28 mM glucose) in medium RPMI 1640, and the islets should be kept at 5.6 mM glucose *(20)*. Using the culture conditions mentioned (and at a glucose concentration of 5.6 mM glucose), we have maintained human islets in culture for more than 4 wk without loss of function; see also Note 6. Information on insulin content and release of 31 shipments of human islets received in our laboratory from the Central Unit of the B-Cell Transplant, Brussels, in 1993, and cultured for 1–3 wk in Uppsala, is provided in Table 1.

Additional information concerning supplementation needs of the cultured human islets is provided in Note 8. Moreover considerations on islet insulin secretion and statistical computations on data obtained with human islets are given in Notes 9–11.

3.3. Evaluation of Fetal and Adult Human Islet Structure and Function

3.3.1. Fixation, Embedding, and Staining of Islets for Light Microscopy

1. Fix the islets in 10% formalin in conical glass centrifuge tubes.
2. After 4–6 h, spin down the islets, gently remove the formalin with a Pasteur pipet, and add 70% ethanol.
3. Add a drop of eosin solution (0.75% dissolved in 70% [w/v] ethanol; Eosin gelblich, Merck, Germany). The eosin makes the islets visible for further handling.
4. Dehydrate the islets with increasing concentrations of ethanol (70, 80, 90, 95, and 99.9%) by incubation for a few minutes. The islets are spun down between these steps, and can be seen owing to a slight red staining. Finally, clearing solution is added.
5. Add a small volume of melted paraffin (<1 mL) to the glass tubes. Allow the paraffin to solidify, and then cover it with another 3 mL of melted paraffin. The islets are thus infiltrated with paraffin overnight at 56°C.
6. Remove the glass tube without shaking the islets, insert a corkscrew about 0.5 cm into the paraffin, and cool the tube in running water for 2 min.
7. Finally, heat the tube in hot water ($\approx 50°C$), and exactly when the paraffin loosens from the glass (within a couple of seconds), pull the paraffin cone out from the tube. The islets are now collected close to the tip of the cone and can be mounted in a block for sectioning.
8. Check the sections (5–7 µm thick) under a microscope until the red stained islets appear. From here on, different specific staining protocols may be used. To obtain a general impression of the islet structure, hematoxylin and eosin staining can be employed. If the culture condition has been suboptimal, centrally located necrotic areas are frequently seen.
9. For estimation of the β-cell content, immunocytochemical staining for insulin must be performed, using the PAP technique *(23)*.

3.3.2. Islet Insulin and DNA Content

3.3.2.1. INSULIN CONTENT

1. Ultrasonically disrupt groups of 30–50 islets in 200 µL of redistilled water.
2. Mix a 50-µL fraction of the water homogenate with 125 µL of acid ethanol (0.18M HCl in 96% [v/v] ethanol), and extract insulin overnight at 4°C.
3. The insulin concentration is measured by radioimmunoassay (RIA). A large number of commercially available kits for assay of insulin can be used according to the manufacturers' instructions. Preferably use human insulin for the standard curve.

3.3.2.2. DNA CONTENT

1. DNA is measured based on previously described methods *(24,25)*.
2. Put two 50-µL aliquots/sample of the islet water homogenate prepared in Section 3.3.2.1. in the wells of white plastic micotiter plates (96 positions; Perkin Elmer, Beaconsfield, UK), and freeze at –20°C before assay.

3. Place the microtiter plates at 56°C overnight prior to assay in order to dry the sample.
4. Add 50 µL of 17.8 mM 3,5-diaminobenzoic acid dihydrochloride solution, which has been cleaned at least three times with 10–20 mg activated charcoal to each sample.
5. Incubate the samples in a water bath at 60°C for 45 min, and then add 200 µL of 1M HCl.
6. Prepare a standard curve (0, 0.3125, 0.0625, 0.125, 0.25, 0.50, and 1.0 µg/50 µL) of DNA, by dissolving salmon testes DNA in 1M NH$_3$, and run it in parallel.
7. Read the fluorescense of the samples in a fluorophotometer at 405 nm (excitation wavelength) and 520 nm (emission wavelength).

3.3.3. Islet Glucose Oxidation Rate

1. By this method, the metabolism of glucose, which is a key event for normal insulin secretion, is assessed.
2. Use triplicate samples if possible. Place groups of 10 islets in glass vials, which should ideally be J-shaped.
3. Each vial contains 100 µL of Krebs-Ringer buffer, D-[U-^{14}C]glucose and 1.7 or 16.7 mM nonradioactive glucose with a specific radioactivity of ≈0.5 mCi/mM. Three small aliquots are taken from the radioactive solution for direct counting, in order to convert the radioactivity of the sample to the rate of metabolism of glucose.
4. The shaft of the vial is squeezed about 1 cm through a small hole in a round rubber membrane.
5. Install the vials into 20-mL glass scintillation vials, gas them with CO_2:O_2 (5:95), tightly seal the vials with the rubber membrane, and lock with a plastic ring (Fig. 1).
6. Run in parallel vials without islets as blank samples.
7. Subsequently, incubate the vials at 37°C in a slow-shaking water bath (30 strokes/min) for 90 min. Then inject 100 µL of 0.05 mM antimycin A (dissolved in 99.9% ethanol) into the inner vial by a needle perforating the membrane, which immediately stops islet metabolism.
8. Inject into the outer scintillation vial 0.25 mL of Hyamine 10-X (Packard Instruments, Meriden, CT).
9. Promote the release of CO_2 by injection of 100 µL of 0.4M Na_2HPO_4 (pH 6.0) into the inner vial. The CO_2 is trapped in Hyamine 10-X during a further 2-h incubation at 37°C.
10. Finally, remove the inner vials and the membranes, add scintillation fluid, and count the samples in a liquid scintillator.

3.3.4. Islet Glucose-Stimulated Insulin Release in Batch-Type Incubations

1. Incubate groups of 10 islets (in triplicate per sample) in 250 µL of Krebs-Ringer buffer containing BSA, using the same vials and experimental conditions

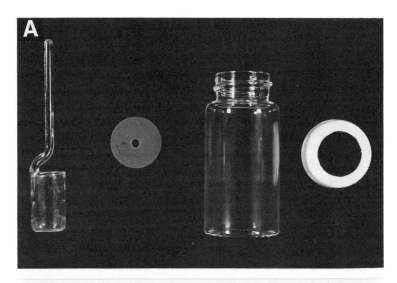
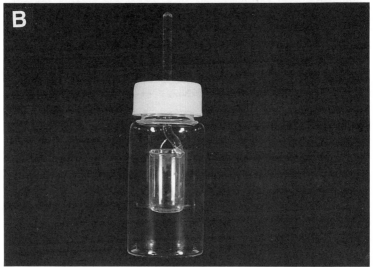

Fig. 1. Photograph of (from left to right) the inner vial, the rubber membrane, the outer vial, and the plastic ring **(A)**. The four items mounted together **(B)**.

described in Section 3.3.3. During the first hour, the medium is supplemented with 1.7 mM glucose.
2. Remove the inner vials and retrieve the medium gently with a micropipet under a stereomicroscope, avoiding touching the islets, and store the medium.
3. Then add medium containing 16.7 mM glucose, and incubate the islets for a second hour.

4. Collect the incubation media, and determine the insulin concentration by RIA. By this procedure, the stimulatory effect of glucose on islet insulin secretion can be calculated (see Table 1).

3.3.5. Islet (Pro)Insulin and Total Protein Biosynthesis Rate

3.3.5.1. LABELING AND HOMOGENIZATION

1. Incubate groups of 10 islets in duplicate for 2 h in 100 µL of Krebs-Ringer buffer containing BSA and 50 µCi/mL of L-[4.5-^{3}H]leucine for 2 h at 37°C in air + 5% CO_2.
2. After the incubation, wash the islets in HBSS containing 10 mM nonradioactive leucine, and sonicate them in 200 µL redistilled water. The samples can now be frozen at –20°C before assay.

3.3.5.2. ANALYSIS OF ISLET (PRO)INSULIN BIOSYNTHESIS RATE

1. The procedure is adapted from Halban et al. (26).
2. From each sample of the water homogenate, transfer 4 × 10-µL aliquots to Eppendorf tubes, and add 100 µL of glycine buffer with Triton X-100.
3. Add to two of the tubes 10 µL anti-insulin serum, and to the other two tubes, 10 µL guinea pig serum (nonspecific antibody binding).
4. Mix the tubes vigorously, and incubate for 1 h at room temperature.
5. Mix the samples again, and add 100 µL of protein A-Sepharose solution. Incubate the tubes for 15 min at room temperature during slow shaking.
6. Centrifuge the samples, and wash twice with 0.5 mL of glycine buffer containing Triton X-100.
7. Resuspend the pellet in 2 × 250 µL of acetic acid solution, and transfer the contents to scintillation vials.
8. Add scintillation liquid, and count the samples in a liquid scintillation counter.
9. Calculate a mean for the two anti-insulin serum tubes, and subtract the mean of the guinea pig serum tubes, which gives a measure on the (pro)insulin biosynthesis rate.

3.3.5.3. ANALYSIS OF ISLET TOTAL PROTEIN BIOSYNTHESIS RATE

1. From the water homogenate of the labeled islets, add 10 µL to two Eppendorf tubes that contain 250 µL of glycine solution and 250 µL of trichloroacetic acid solution, and mix the samples thoroughly.
2. Centrifuge the samples, remove the supernatant, and dissolve the pellet with 2 × 250 µL of 0.15M NaOH.
3. Then treat the samples for scintillation counting as in Section 3.3.5.2. This gives a measure of the total biosynthesis rate. A ratio can thus be calculated between the values obtained in Section 3.3.5.2. and in this section, which gives the relative fraction of newly synthesized (pro)insulin as compared to the total pool labeled proteins. In cultured fetal human ICC, this fraction was about 2% at 1.7 mM glucose and 5% at 16.7 mM (16), and the corresponding values for cultured adult human islets were 6 and 9% (20).

4. Notes

1. Not until a very late gestational, or even at postnatal state, the fetal human β-cell attains the hallmark of the β-cell, i.e., the capacity to respond properly to an elevated glucose concentration with an increased insulin secretion *(27)*. Much attention has been focused on factors that might regulate the differentiation and replication of fetal β-cells in vitro. Indeed, the fetal β-cell with its high capacity to replicate may be a source to generate new β-cells in vitro for clinical islet transplantation, since the adult β-cell replicates at an extremely low rate *(28,29)*.
2. We have found that human adult serum supplementation (10%) is preferable to FBS in that the number of ICC obtained is higher and the amount of fibroblasts developing in the culture dishes is limited *(16)*. However, the human serum does not accelerate β-cell differentiation. A similar effect on the number of ICC formed to that observed with human serum can also be obtained with human amniotic fluid *(30)*.
3. Addition of human growth hormone (GH) (1 μg/mL), in the presence of 10% serum, for 6–8 d to explant cultures of human fetal pancreas also promotes formation of ICC, and stimulates insulin secretion and insulin gene expression *(17,31,32)*. This effect appears not to be mediated by IGF-1, but rather appears to be a direct effect by GH *(31)*. Recently, it was also demonstrated that hepatocyte growth factor/scatter factor (HGF/SF; 25 ng/mL) might support the growth of fetal human islet cells in culture *(33)*.
4. Up to now the most potent stimulator of human fetal β-cell differentiation in vitro appears to be the vitamin B_3 derivative nicotinamide. Thus, addition of nicotinamide (5–10 mM) to cultures of human fetal pancreas enhances formation of ICC, as well as their insulin and insulin mRNA content and glucose-stimulated insulin secretion *(18,34)*. The mechanism for this action is not fully elucidated, but it might be related to inhibition of the enzyme poly(ADP-ribose) polymerase *(35)*.
5. Human fetal and adult islet cells may also be cultured in monolayers *(36–38)* or after fluorescence-activated cell sorting (FACS) of purified β-cells *(39)*. Although there are several detailed reports on the use of these techniques in rodent islets, their use for human islet preparations is still limited *(40)*.
6. The fetal islets do not normally respond to an elevation of the glucose concentration alone. In this case, 5–10 mM theophylline might be added together with the high glucose concentration. To prevent the fetal islets from "leaking" insulin owing to damage, it is advisable to incubate the islets at a low glucose concentration after the stimulatory phase. During this latter phase, the insulin secretory rate should be downregulated.
7. The common procedure of determining human adult islet "purity" after isolation by staining the islets with dithizone is unacceptable as a parameter of endocrine cell composition of the preparation. Dithizone is a small molecule with chelating properties, useful for staining of zinc, a component of the insulin molecule. Although dithizone staining can be helpful during the procedure of islet isolation, it does not provide reliable information on the actual number of β-cells per islet.

According to our experience, the percentage of β-cells per islet can vary between 15 and 60% in different human adult islet preparations, as judged by morphometrical determinations, but they will all stain positively with dithizone. Indeed, even fetal preparations containing as little as 3% β-cells stain positively with dithizone, and would thus be labeled as a "highly pure preparation," obviously an erroneous description of the sample. Finally, recent observations suggest that dithizone is toxic to islets obtained from different species *(41)*.

8. During the development of the presently described conditions for adult human islet culture, the following variables have been examined: type of culture medium (RPMI 1640 vs HAM F10), type of serum (FBS vs pooled human serum), and serum concentration (1% vs 10%). The data obtained in different functional tests (including islet retrieval, islet insulin and DNA content, insulin release in response to glucose, and rates of glucose oxidation) suggested that the best outcome was obtained with RPMI 1640 containing 5.6 mM glucose (*see* Section 3.2.3.). Islets cultured for 7 d in HAM F10 medium plus 10% FBS presented similar insulin and DNA content as islets cultured in RPMI 1640, but the absolute rates of insulin release at 1.7 and 16.7 mM glucose were 50% lower than that observed in islets cultured in RPMI 1640. Use of 10% pooled human serum did not directly affect human islet function, but we noted an increased tendency for the islets to attach to each other.

It should be remembered that use of FBS or pooled human serum introduces a noncontrolled variable in the culture system. Thus, each new batch of serum must be systematically tested in comparison with the previous one to avoid unexpected modifications in islet function (*see* description of functional islet tests in Section 3.3.). Hopefully, in the future it will be possible to perform human islet culture using serum-free conditions, with inclusion of defined medium supplements (BSA, growth factors, and so forth; *42*).

9. When assessing insulin release as a parameter of preserved adult human islet function, supplementation of the glucose load with arginine or generators of cAMP should not be used. Indeed, rodent islets damaged by β-cell toxins lose their insulin response to glucose, but still respond to arginine or cAMP generators *(43,44)*, suggesting that response to these secretagogs is not enough to define normal islet function.

10. If a more detailed monitoring of the dynamics of islet insulin secretion is needed, perifusion experiments might be performed *(45)*. However, this procedure is more cumbersome, and at least 75 islets are normally required.

11. Owing to difficulties in isolating adult human islets, it is a common practice that preparations obtained from a single individual pancreas be divided in several Petri dishes and considered as different observations (i.e., an experimental n of 5 may be obtained from a single individual). This is not acceptable, since all these "independent" observations originate from the same individual. It must be kept in mind that there is large variability in the organ donors for human islet isolation, including differences in age, sex, race, period and type of disease and/or trauma preceding brain death and subsequent organ removal, period of cold and

warm ischemia, and so forth. This further emphasizes the need for large experimental groups (i.e., with islets obtained from several individuals) in order to obtain meaningful biological information. If experiments are performed in triplicate, a mean of these three observations should be obtained and counted as one observation.

Acknowledgments

We thank Professor D. G. Pipeleers, Coordinator of β-Cell Transplant, Brussels, for providing human adult islet preparations and detailed information regarding the cell composition of the human islet preparations. Our own work was supported by financial grants from the Swedish Medical Research Council, Swedish Diabetes Association, the Juvenile Diabetes Foundation International, the Novo-Nordisk Fund, and the European Concerted Action for Treatment of Diabetes.

References

1. Orci, L. (1982) Macro- and micro-domains in the endocrine pancreas. *Diabetes* **31**, 538–565.
2. Hellerström, C. (1964) A method for microdissection of intact pancreatic islets of mammals. *Acta Endocrinol. (Copenh.)* **45**, 122–132.
3. Moskalewski, S. (1965) Isolation and culture of the islets of Langerhans of the guinea pig. *Gen. Comp. Endocrinol.* **5**, 342–353.
4. Lacy, P. E. and Kostianovsky, M. (1967) Method for isolation of intact islets of Langerhans from the rat pancreas. *Diabetes* **16**, 35–39.
5. Ricordi, C., Lacy P. E., Finke E. H., Olack, B. J., and Scharp, D. W. (1988) Automated method for isolation of human pancreatic islets. *Diabetes* **37**, 413–420.
6. Warnock, G. L., Ellis, D. K., Cattral, M., Untch D., Kneteman, N. M., and Rajotte, R. V. (1989) Viable purified islets of Langerhans from collagenase-perfused human pancreas. *Diabetes* **38(Suppl. 1)**, 136–139.
7. Krebs, H. A. and Henseleit, K. (1932) Untesuchungen über die Harnstoffbildung im Tierkörper. *Hoppe-Seylers Z. Physiol. Chem.* **210**, 33–66.
8. Hanks, J. H. and Wallace, R. E. (1949) Relation of oxygen and temperature in the preservation of tissue by refrigeration. *Proc. Soc. Exp. Biol. Med.* **71**, 196–200.
9. Moore, G. E., Gerner, R. E., and Franklin, H. A. (1967) Culture of normal human leukocytes. *JAMA* **199**, 87–92.
10. Andersson, A., Christensen, N., Groth, C.-G., Hellerström, C., and Sandler, S. (1984) Survival of human fetal pancreatic explants in organ culture as reflected in insulin secretion and oxygen consumption. *Transplantation* **37**, 499–503.
11. Parker, R. C., Healy, G., and Fisher, D. (1954) Nutrition of animal cells in tissue culture. VII. Use of replicate cell culture in the evaluation of synthetic media. *Can. J. Biochem.* **32**, 306–311.
12. Auerbach, R. (1960) Morphometric interactions in the development of the mouse thymus gland. *Dev. Biol.* **2**, 271–284.

13. Juhlin, R. and Alm, G. V. (1976) Morphologic and antigenic maturation of lymphocytes in the mouse thymus in vitro. *Scand. J. Immunol.* **5,** 497–503.
14. Mandel, T. E. and Georgiou, H. M. (1983) Insulin secretion by fetal human pancreatic islets of Langerhans in prolonged organ culture. *Diabetes* **32,** 915–920.
15. Sandler, S., Andersson, A., Schnell A., Mellgren A., Tollemar, J., Borg, H., Petersson, B., Groth, C.-G., and Hellerström, C. (1985) Tissue culture of human fetal pancreas: development and function of B-cells in vitro and transplantation of explants to nude mice. *Diabetes* **34,** 1113–1119.
16. Sandler, S., Andersson, A., Schnell Landström, A., Tollemar, J., Borg, H., Petersson, B., Groth, C.-G., and Hellerström, C. (1987) Tissue culture of human fetal pancreas: effects of human serum on development of islet and endocrine function of isletlike cell clusters. *Diabetes* **36,** 1401–1407.
17. Sandler, S., Andersson, A., Korsgren, O., Tollemar, J., Petersson, B., Groth, C.-G., and Hellerström, C. (1987) Tissue culture of human fetal pancreas: growth hormone stimulates the formation of islet-like cell clusters. *J. Clin. Endocrinol. Metab.* **65,** 1154–1158.
18. Sandler, S., Andersson, A., Korsgren, O., Tollemar, J., Petersson, B., Groth, C.-G., and Hellerström, C. (1989) Tissue culture of human fetal pancreas: effects of nicotinamide on insulin production and formation of islet-like cell clusters. *Diabetes* **38(Suppl. 1),** 168–171.
19. Andersson, A., Borg, H., Groth, C.-G., Gunnarsson, R., Hellerström, C., Lundgren, G., Westman, J., and Östman, J. (1976) Survival of isolated human islets of Langerhans maintained in tissue culture. *J. Clin. Invest.* **57,** 1295–1301.
20. Eizirik, D. L., Korbutt, G. S., and Hellerström, C. (1992) Prolonged exposure of human pancreatic islets to high glucose concentrations in vitro impairs the β-cell function. *J. Clin. Invest.* **90,** 1263–1268.
21. Eizirik, D. L., Sandler, S., Welsh, N., Cetkovic-Cvrlje, M., Nieman, A., Geller, D. A., Pipeleers, D. G., Bendtzen, K., and Hellerström, C. (1994) Cytokines suppress human islet function irrespective of their effects on nitric oxide generation. *J. Clin. Invest.* **93,** 1968–1974.
22. Ham, R. G. (1963) An improved nutrient solution for diploid Chinese hamster and human cell lines. *Exp. Cell Res.* **29,** 515–226.
23. Sternberger, L. A., Hardy, P. H., Cuculis, J. J., and Meyer, H. G. (1970) The unlabeled antibody enzyme method of immunocytochemistry. Preparation and properties of soluble antibody complex (horseradish peroxidase anti-horseradish peroxidase) and its use in identification of spirochetes. *J. Histochem. Cytochem.* **18,** 315–333.
24. Kissane, J. M. and Robins, E. (1958) The fluorometric measurement of deoxyribonucleic acid in animal tissues with special reference to the central nervous system. *J. Biol. Chem.* **233,** 184–188.
25. Hinegardner, R. T. (1971) An improved fluorometric assay for DNA. *Anal. Biochem.* **39,** 197–201.
26. Halban, P. A., Wollheim, C. B., Blondel, B., and Renold, A. E. (1980) Long-term exposure of isolated pancreatic islets to mannoheptulose: evidence for insulin degradation in the β-cell. *Biochem. Pharmacol.* **29,** 2625–2633.

27. Tuch, B. E. and Turtle, J. R. (1985) Maturation of the response of human fetal pancreatic explants to glucose. *Diabetes* **28**, 28–31
28. Hellerström, C. and Swenne, I. (1985) Growth pattern of pancreatic islets in animals, in *The Diabetic Pancreas* (Volk, B. W. and Arquilla, E. R., eds.), Plenum, New York, pp. 53–79.
29. Bonner-Weir, S. and Smith, F. E. (1994) Islet cell growth and the growth factors involved. *Trends Endocrinol. Metab.* **5**, 60–64.
30. Andersson, A., Sandler, S., Hellerström, C., Petersson, B., Schnell, A., Tollemar, J., Mellgren, A., and Groth, C.-G. (1986) Effects of amniotic fluid on the development of human fetal pancreatic B-cells in tissue culture. *Transplant. Proc.* **18**, 57–59.
31. Otonkoski, T., Knip, M., Wong, I., and Simell, O. (1988) Effects of growth hormone and insulin-like growth factor I on endocrine function of human fetal islet-like cell clusters during long-term tissue culture. *Diabetes* **37**, 1678–1683.
32. Formby, B., Ullrich, A., Coussens, L., Walker, L., and Peterson, C. M. (1988) Growth hormone stimulates insulin gene expression in cultured human fetal pancreatic islets. *J. Clin. Endocr. Metab.* **66**, 1075–1079.
33. Otonkoski, T., Beattie, G. M., Rubin, J. S., Lopez, A. D., Baird, A., and Hayek, A. (1994) Hepatocyte growth factor/scatter factor has insulinotropic activity in human fetal pancreatic cells. *Diabetes* **43**, 947–953.
34. Otonkoski, T., Beattie, G. M., Mally, M. I., Ricordi, C., and Hayek, A. (1993) Nicotinamide is a potent inducer of endocrine differentiation in cultured human fetal pancreatic cells. *J. Clin. Invest.* **92**, 1459–1466.
35. Ueda, K. and Hayashi, O. (1985) ADP-ribosylation. *Annu. Rev. Biochem.* **54**, 73–100.
36. Rabinovitch, A., Sumoski, W., Rajotte, R. V., and Warnock, G. L. (1990) Cytotoxic effects of cytokines of human pancreatic islets in monolayer culture. *J. Clin. Endocrinol. Metab.* **71**, 152–156.
37. Beattie, G. M., Lappi, D. A., Baird, A., and Hayek, A. (1991) Functional impact of attachment and purification in the short term culture of human pancreatic islets. *J. Clin. Endocrinol. Metab.* **73**, 93–98.
38. Simpson, A. M., Tuch, B. E., and Vincent, P. C. (1991) Characterization of endocrine-rich monolayers of human fetal pancreas that display reduced immunogenicity. *Diabetes* **40**, 800–808.
39. Pipeleers, D. G., In't Veld, P. A., Van de Winkel, M., Maes, E., Schuit, F. C., and Gepts, W. (1985) A new in vitro model for the study of pancreatic A and B cells. *Endocrinology* **117**, 806–816.
40. Eizirik, D. L., Pipeleers, D. G., Ling, Z., Welsh, N., Hellerström, C., and Andersson, A. (1994) Major species differences between humans and rodents in the susceptibility to pancreatic β-cell injury. *Proc. Natl. Acad. Sci. USA* **91**, 9523–9526.
41. Clark, S. A., Borland, K. M., Sherman, S. D., Rusack, T. C., and Chick, W. L. (1994) Staining and in vitro toxicity of dithizone with canine, porcine, and bovine islets. *Cell Transplant.* **3**, 299–306.
42. Ling, Z. and Pipeleers, D. G. (1994) Preservation of glucose-responsive islet β-cells during serum-free culture. *Endocrinology* **134**, 2614–2621.

43. Eizirik, D. L., Sandler, S., Welsh, N., and Hellerström, C. (1988) Preferential reduction of insulin production in mouse pancreatic islets maintained in culture after streptozotocin exposure. *Endocrinology* **122,** 1242–1249.
44. Eizirik, D. L., Sandler, S., Hallberg, A., Bendtzen, K., Sener, A., and Malaisse, W. J. (1989) Differential sensitivity to β-cell secretagogues in rat pancreatic islets exposed to human interleukin 1β. *Endocrinology* **125,** 752–759.
45. Lacy, P. E., Walker, M. M., and Fink, C. J. (1972) Perifusion of isolated rat islets in vitro. Participation of the microtubular system in the biphasic release of insulin. *Diabetes* **21,** 987–998.

30

Isolation and Culture of Human Renal Cortical Cells with Characteristics of Proximal Tubules

Vicente Rodilla and Gabrielle M. Hawksworth

1. Introduction

The kidney is an extremely heterogeneous organ from a morphological and a functional point of view. In spite of this heterogeneity, several methods have been described in an attempt to isolate and culture homogeneous populations of renal epithelial cells, seeking to maintain normal differentiated characteristics of the nephron segment of origin.

Several techniques have been used to isolate and culture human proximal tubular cells (PTC). Detrisac and coworkers first devised a simple method of culturing human renal epithelial cells from cortical tissue explants that retain the characteristics of PTC after several passages *(1)*. Other methods used involved the enzymatic digestion of the cortical tissue with collagenase *(2–5)*, removal of contaminating glomeruli and larger fragments by filtration, further purification of the cells by isopycnic centrifugation using Nycodenz *(6)* or Percoll *(7)*, or the more sophisticated technique of microdissection without collagenase *(8)* or after the proteolytic digestion of the tissue *(9–11)*.

The method presented here is a modification of that of McLaren and coworkers *(7)*. It is based on enzymatic digestion using a two-step collagenase digestion followed by mechanical disruption of the tissue, further purification by filtration through a 75-μm sieve, and sedimentation of the cells and fragments obtained in a continuous density gradient formed with Percoll. This technique allows for the separation of a cellular fraction that is highly enriched in proximal tubular cells and fragments. This method produces a far better yield than microdissection, and because the cells are purified, primary cultures can be grown in the presence of serum without fibroblast overgrowth being a significant problem. Isolating cells with this method typically produces a yield of

6–12 × 10⁶ PTC/g of cortical tissue with relatively high viability (between 70 and 90%).

Cultures initiated from this fraction can be maintained for several passages, expressing functional characteristics of proximal tubules, such as the formation of polarized monolayers and domes, parathyroid hormone responsiveness, and expression of brush border enzymes (γ-glutamyl transpeptidase, alkaline phosphatase), and do not show characteristics of other renal cells, such as responsiveness toward calcitonin and vasopressin.

Cells cultured in this manner can be used for a wide range of purposes (biochemical, physiological, or toxicological studies) where human PTC are required, thus eliminating the need for extrapolation from animal cell cultures or cell lines of uncertain origin.

2. Materials

1. Balanced salt solution (BSS): Prepare 1 L containing 5.37 mM KCl, 0.44 mM KH$_2$PO$_4$, 137 mM NaCl, 0.34 mM Na$_2$HPO$_4$, 1.35 mM NaHCO$_3$, 5.56 mM D-glucose, 25 mM (N-[2-Hydroxyethyl]piperazine-N'-[2-ethanesulfonicacid]) (HEPES), 0.5 mM ethylene glycol-bis-(β-aminoethyl ether) N,N,N',N'-tetraacetic acid (EGTA), and 0.5% bovine serum albumin (BSA). Adjust the pH to 7.0–7.2, sterilize by filtration through a 22-μm filter, and store at 4°C.
2. Phosphate-buffered saline (PBS): Prepare 1 L containing 2.68 mM KCl, 1.47 mM KH$_2$PO$_4$, 137 mM NaCl and 8.12 mM Na$_2$HPO$_4$. Adjust the pH to 7.0–7.2, sterilize by filtration through a 22-μm filter, and store at 4°C.
3. Cell-culture medium: Dulbecco's modified Eagle's medium/nutrient mixture Ham's F-12 (DMEM/Ham's F-12) mixture (1:1) with 15 mM HEPES and 14.28 mM sodium bicarbonate obtained from Sigma (Poole, UK). Store in the dark at 4°C.
4. Fetal calf serum (FCS) obtained from Gibco-BRL (Paisley, UK). On receipt, aliquot the serum and store at –20°C.
5. Penicillin/streptomycin, 50 U/50 μg/mL. The stock solution is obtained from Gibco-BRL (5000 U/mL/5000 μg/mL). Store aliquots at –20°C to avoid repeat freeze–thawing (see Note 2).
6. Supplements for defined medium: 5 μg/mL insulin from bovine pancreas, 5 μg/mL human transferrin (iron-free), 5 ng/mL sodium selenite (Na$_2$SeO$_3$), and 18 ng/mL hydrocortisone are all obtained from Sigma. Insulin, transferrin, and sodium selenite are reconstituted in medium, whereas hydrocortisone is dissolved in a small volume of absolute ethanol and brought to the desired volume with medium. Store aliquots of the stock solution at –20°C for no longer than 3 mo.
7. Versene: 0.2% (w/v) EDTA in PBS. Sterilize by filtration and store at 4°C.
8. Trypsin/EDTA solution: Obtained from Gibco-BRL as a 10X stock solution containing 0.5% porcine trypsin and 0.2% EDTA. This solution is aliquoted and stored at –20°C. Prior to use, combine an aliquot of the trypsin/EDTA solution with an equal volume of Earle's BSS without calcium and without magnesium obtained from Gibco-BRL.

Fig. 1. Photograph showing the equipment used for the isolation of PTC in the cell-culture cabinet. (A) Water-jacketed incubation vessel with magnetic bar. The tubing is connected to a water bath located outside the cabinet via a Watson-Marlow pump; the arrows on the tubing indicate the direction of the water flow. (B) Magnetic stirrer. (C) Disposable Universal containers. (D) Ice bucket. (E) Nalgene 50-mL centrifuge tubes. (F) Petri dish. (G) Stainless-steel sieves (*see* text). (H) Plunger of a 20-mL syringe.

9. Percoll solution: Percoll is obtained from Sigma, autoclaved in 18-mL aliquots, and stored at 4°C. Prior to use, combine 18 mL of sterile Percoll with 42 mL of culture medium.
10. Collagenase A solution: Just before use, prepare a solution of collagenase (from *Clostridium histolyticum*, Boehringer Mannheim, Lewes, UK) in medium (0.2% [w/v]) and filter through a 22-µm pore filter. Typically, we use 100 mg of collagenase (activity 0.5 U/mg) in a total volume of 50 mL of medium/each 10 g of cortex to be digested.
11. Incubation vessel: This consists of a glass-jacketed flask (200 mL) with a magnetic bar in it (Fig. 1A). The temperature of the collagenase solution in the inner reservoir is maintained at 37°C by circulating warm water through the jacket of the flask. The flask and the magnetic bar are autoclaved prior to use.
12. Stainless-steel test sieves obtained from Endecotts (London, UK) with mesh sizes of 300, 150, and 75 µm (Fig. 1G). The sieves are autoclaved to ensure sterility.
13. 50-mL Autoclavable centrifuge tubes (Fig. 1E) obtained from Nalgene (Rochester, NY).
14. Trypan Blue solution: Stock solution (0.4% sterile Trypan Blue) is obtained from Sigma. Just before use, this is diluted to 0.2% in PBS.

3. Methods

3.1. Isolation of Human Proximal Tubular Cells

Pieces of human renal tissue are obtained from nephrectomy specimens removed owing to renal cell carcinoma. The tissue is always obtained from a fragment as distant from the neoplastic lesion as possible and is confirmed as being normal by the Department of Pathology, University of Aberdeen. The method described here is ideally suited for the isolation of PTC from approx 10 g of cortex.

1. Transport the tissue under sterile conditions to the tissue-culture cabinet, and place it in a sterile Petri dish containing BSS. Peel off the renal capsule using sterile forceps, and dissect the cortical tissue from the adjacent medulla with a sterile scalpel.
2. Make incisions on the surface of the cortical tissue to facilitate penetration of the buffer into the tissue. Place the cortex in a preweighed disposable and sterile Universal container with BSS.
3. Weigh the Universal container.
4. Bring the cortex back to the cabinet, and place it in a plastic Petri dish containing fresh BSS. Chop the tissue coarsely with a scalpel.
5. Dissolve the required amount of collagenase in approx 50% of the final volume of medium, sterilize it by filtration, and place it in the incubation vessel.
6. Transfer the tissue fragments to a sterile Universal, and wash them thoroughly with several changes of BSS. Then cut the tissue to obtain pieces of approx 3 mm^3. Wash the cortex with several changes of BSS until the solution remains clear.
7. Wash the cortical fragments three times in tissue-culture medium. During each wash shake the Universal vigorously (*see* Note 5).
8. Resuspend the cortical fragments in prewarmed (37°C) medium, and combine it with the collagenase solution in the incubation vessel. Adjust the volume of the collagenase mixture by the addition of further prewarmed medium.
9. Incubate the cortical fragments in the collagenase solution for a variable period of time (between 20 and 30 min) depending on the consistency of the tissue.
10. Pour off the digested mixture onto the first sieve (300 µm), and force it through with the plunger of a 20-mL syringe (Fig. 1H). Then wash the sieve through with fresh medium. The same procedure is then applied to the second sieve (150 µm). Wash the material collected onto the third sieve with medium, and if necessary, stir the contents gently with a glass rod (*see* Note 6).
11. Pipet out the suspension thus collected into the Petri dish (Fig. 1F) into a series of Universals, and rinse the Petri dish with medium.
12. Centrifuge the Universals at 400*g* in a bench centrifuge to pellet the digested material.
13. Resuspend the pellets obtained in cold medium, and combine them into one or two Universals, which are centrifuged again as before.

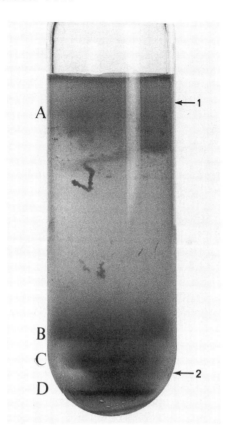

Fig. 2. Nalgene tube showing the distribution of bands after the isolation in the gradient formed with Percoll. The first band (A) is very diffuse and contains mostly cell debris, nonviable cells, and cells of uncertain origin. Bands B and C contain tubular cells and tubular fragments. Band C is enriched in proximal tubular cells and fragments. Band D contains blood cells. The density of the gradient formed was measured using density marker beads (Pharmacia); the arrows show Percoll gradients densities of (1) 1.019 g/mL and (2) 1.062 g/mL.

14. Pour off the medium, and resuspend the pellet in 10 mL of medium (5 mL/tube to be used during the isopycnic separation). Load each of the two 50-mL centrifuge tubes with 5 mL of this cell suspension and 30 mL of the Percoll mixture.
15. Centrifuge the tubes at 21,500g_{max} for 30 min in a 34° angle rotor at 4°C. This separates the material into four distinct bands (Fig. 2), three bands (A, B, and C) composed of renal fragments and cells, and a fourth band (D) that contains the blood cells. Band C contains a mixture of cells and tubules, composed mainly of proximal tubules and proximal tubular cells (*see* Notes 7 and 8).

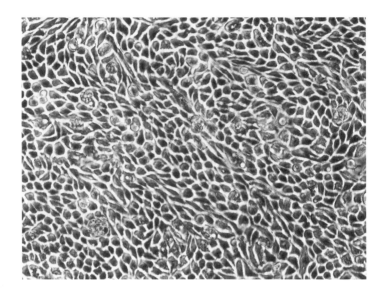

Fig. 3. Photograph of a 7-d-old confluent culture of PTC during the first passage and growing in defined medium. Note the typical cobblestone aspect of epithelial cells (original magnification 10×).

16. Pipet out the contents of band C, and wash it in medium. Centrifuge this suspension as in step 12 to remove traces of Percoll and then resuspend the pellet in a given volume of medium (*see* Note 9). Add an aliquot to the Trypan Blue solution, and estimate cell viability and number with an improved Neubauer haemocytometer (*see* Note 10).

3.2. Culture of Proximal Tubular Cells

1. Initiate the primary cultures by seeding the cells into plastic culture dishes or flasks at an approximate density of 1×10^5 cells/cm^2 in medium supplemented with 10% FCS to promote attachment.
2. Change the medium after approx 24 h and at 48-h intervals thereafter, changing, if desired, to defined medium once attachment of the cells has taken place.
3. When seeded and cultured under these conditions, the cultures reach confluence in approx 5–6 d, displaying by then the typical cobblestone appearance characteristic of epithelial cells (Fig. 3).

3.3. Subculture of Proximal Tubular Cells

1. PTC cultures can be subcultured at approx 7-d intervals.
2. Wash the cell monolayer three times with PBS. This step is of particular importance if the cells have been cultured in medium containing FCS in order to remove serum.

3. Incubate the cultures for 3–5 min with Versene solution (0.2% EDTA in PBS) to weaken cell adhesion through the action of the chelating agent.
4. After removal of the Versene, incubate the cultures at 37°C in a trypsin solution for approx 10 min or until the cells have detached from the plate.
5. Inactivate the trypsin by the addition of an equal volume of medium supplemented with 10% FCS.
6. Centrifuge the cell suspension at 400g for 5 min, and resuspend the pellet in a given amount of medium.
7. Determine viability and cell number as described using the Trypan Blue solution.
8. Seed the cells onto tissue-culture plates or dishes at a density of 1×10^4 cells/cm^2.

4. Notes

1. Although some of the solutions are stable for a long time, we use them within 3 mo of preparation.
2. The medium supplements and the antibiotics should be added to the medium prior to use for tissue culture and used within 2 wk of preparation, otherwise poor cell growth may result.
3. It is important to maintain the tissue and all solutions (other than those used for collagenase digestion) in ice at all times. This increases cell viability and helps to prevent clumping of the cells, particularly at the later stages of the isolation procedure.
4. The enzymatic digestion of the tissue using collagenase is a two-step procedure. First, the tissue is washed thoroughly with BSS, a solution containing EGTA as a chelating agent. Then the tissue is digested in a solution of collagenase containing calcium, which is essential for optimal enzymatic activity of the collagenase. The advantages of this two-step collagenase digestion are thought to lie in the loosening of tight junctions between the cells via the removal of calcium by the chelating agent and therefore producing a better digestion by collagenase *(12)*.
5. It is essential to rinse the tissue thoroughly with medium before the collagenase digestion. The EGTA present in the BSS is an inhibitor of collagenase activity, whereas the calcium present in the medium activates the enzyme.
6. Sieving the digested material through the 300- and 150-µm sieves further disrupts the tissue, whereas the function of the 75–µm sieve is primarily to remove contaminating glomeruli. This is the reason for not forcing the digested material through this latter sieve.
7. Band A contains mainly damaged cells and debris together with a number of viable cells of uncertain origin. Band B is composed of cells and tubular fragments, and although some of them are proximal tubules, the vast majority are not. Band C is mainly composed of proximal tubules and cells.
8. During the isopycnic separation of the tubular cells, the centrifuge tubes should not be overloaded with the digested material. Otherwise distinguishing between bands B and C can become extremely difficult.
9. The cells and fragments in band C are sometimes clumped together. Normally, these clumps can be dislodged by gently pipetting the pellet after resuspension in

medium. However, on some occasions, this is not sufficient to produce a homogeneous cell suspension. In these cases, we found that the suspension can be further homogenized by forcing it through a 19-gage needle, without causing a significant reduction in the viability of the suspension. These clumps may also be prevented by treating the suspension with DNase (0.05 mg/mL).

10. The assessment of viability and cell number in the isolated suspensions can only be approximate, since the preparation contains a substantial amount of tubular fragments and the number of cells in each fragment can only be estimated under the microscope.

11. In some instances, the primary cultures may be slightly contaminated with fibroblast-like cells. It is therefore convenient to culture the cells in defined medium as soon as attachment to the substratum has taken place, thus greatly reducing the problem of fibroblast overgrowth *(13,14)*. Fibroblast growth can be completely eliminated by substituting L-valine and arginine by D-valine and ornithine, respectively, in the culture medium. Epithelial cells, and therefore PTC, can convert D-valine into L-valine and ornithine into arginine, whereas fibroblasts cannot *(15,16)*.

12. With this technique, there is a risk that cell types other than proximal tubules may grow in culture. A better preparation may be obtained if the cultures are grown in glucose-deficient medium. Within the renal cortex, gluconeogenesis primarily occurs in the convoluted proximal tubules *(17,18)*. This method has been successfully used to culture rabbit PTC *(19)*.

13. When cultured on a plastic substratum, the cultures can be passaged three times before they dedifferentiate and senesce. However, by culturing these cells on a collagen matrix, the number of passages may be increased *(1)*.

14. Both the isolated cells and/or fragments and the cell suspension obtained after trypsinization of monolayers can be cryopreserved in cell-culture medium containing 10% FCS and 10% dimethyl sulfoxide (DMSO) using standard methods.

References

1. Detrisac, C. J., Sens, M. A., Garvin, A. J., Spicer, S. S., and Sens, D. A. (1984) Tissue culture of human kidney epithelial cells of proximal tubule origin. *Kidney Int.* **25**, 383–390.
2. States, B., Foreman, J., Lee, J., and Segal, S. (1986) Characteristics of cultured human renal cortical epithelia. *Biochem. Med. Metab. Biol.* **36**, 151–161.
3. Kempson, S. A., McAteer, J. A., Al-Mahrouq, H. A., Dousa, T. P., Dougherty, G. S., and Evan, A. P. (1989) Proximal tubule characteristics of cultured human renal cortex epithelium. *J. Lab. Clin. Med.* **113**, 285–296.
4. Trifillis, A. L., Regec, A. L., and Trump, B. F. (1985) Isolation, culture and characterisation of human renal tubular cells. *J. Urol.* **133**, 324–329.
5. Yang, H. A., Gould-Kostka, J., and Oberley, T. D. (1987) In vitro growth and differentiation of human kidney tubular cells on a basement membrane substrate. *In Vitro Cell. Dev. Biol.* **23**, 34–46.

6. Boogaard, P. J., Nagelkerke, J. F., and Mulder, G. J. (1990) Renal proximal tubular cells in suspension or in primary culture as in vitro models to study nephrotoxicity. *Chem. Biol. Interact.* **76,** 251–291.
7. McLaren, J., Whiting, P. H., and Hawksworth, G. M. (1990) Maintenance of glucose uptake in suspensions and cultures of human renal proximal tubular cells. *Toxicol. Lett.* **53,** 237–239.
8. Wilson, P. D., Dillingham, M. A., Breckon, R., and Anderson, R. J. (1985) Defined human renal tubular epithelia in culture: growth, characterization, and hormonal response. *Am. J. Physiol.* **248,** F436–F443.
9. Blaehr, H., Andersen, C. B., and Ladefoged, J. (1993) Acute effects of FK506 and Cyclosporine-A on cultured human proximal tubular cells. *Eur. J. Pharmacol. Environ. Toxic.* **228,** 283–288.
10. Blaehr, H. (1991) Human renal biopsies as source of cells for glomerular and tubular cell cultures. *Scand. J. Urol. Nephrol.* **25,** 287–295.
11. McAteer, J. A., Kempson, S. A., and Evan, A. P. (1991) Culture of human renal cortex epithelial cells. *J. Tissue Cult. Methods* **13,** 143–148.
12. Seglen, P. O. (1976) Preparation of isolated rat liver cells. *Methods Cell. Biol.* **13,** 29–83.
13. Taub, M. and Sato, G. H. (1979) Growth of kidney epithelial cells in hormone-supplemented, serum free medium. *J. Supramol. Struct.* **11,** 207–216.
14. Taub, M. and Livingston, D. (1981) The development of serum-free hormone-supplemented media for primary kidney cultures and their use in examining renal functions. *Ann. NY Acad. Sci.* **372,** 406–421.
15. Leffert, H. and Paul, D. (1973) Serum dependent growth of primary cultured differentiated fetal rat hepatocytes in arginine-deficient medium. *J. Cell. Physiol.* **81,** 113–124.
16. Gilbert, S. F. and Migeon, B. R. (1975) D-valine as a selective agent for normal human and rodent epithelial cells in culture. *Cell* **5,** 11–17.
17. Guder, W. G. and Ross, B. D. (1984) Enzyme distribution along the nephron. *Kidney Int.* **26,** 101–111.
18. Castellino, P. and De Fronzo, R. A. (1990) Glucose metabolism and the kidney. *Semin. Nephrol.* **103,** 458–463.
19. Jung, J. C., Lee, S. M., Kadaia, M., and Taub, M. (1992) Growth and function of primary rabbit kidney proximal tubule cells in glucose-free serum-free medium. *J. Cell. Physiol.* **150,** 243–250.

31

Glomerular Epithelial and Mesangial Cell Culture and Characterization

Heather M. Wilson, Keith N. Stewart, and Alison M. MacLeod

1. Introduction

The advent of in vitro culture techniques has allowed us to culture homogeneous populations of glomerular mesangial and epithelial cells to aid our understanding of the development of glomerular disease at the cellular level. Advances in our knowledge of the pathogenic mechanisms have made it clear that the response of intrinsic glomerular cells to external stimuli plays an important role in glomerular injury *(1,2)*. Glomerular cells from several mammalian species have been isolated, propagated, and in some instances, cloned *(3–5)*.

The four major glomerular cell types, epithelial, contractile mesangial, bone marrow-derived mesangial (akin to macrophages), and endothelial cells, have long been recognized as distinct entities because they occupy defined anatomical locations in vivo and have distinguishable morphological and cytochemical features. This compartmentalization, however, is lost in culture, as are several of the anatomical characteristics, such as endothelial fenestrations and epithelial pedicels *(6,7)*; in addition cultured cells may undergo dedifferentiation. Despite these limitations, the study of glomerular cells in culture has proven useful, and valuable information has been obtained about their physiology and pathophysiology. Human glomeruli are usually obtained from the normal pole of kidneys surgically removed from patients with renal carcinoma or from donor kidneys that cannot be used for transplantation for technical reasons. Glomeruli are isolated using sieves such that they can be rendered virtually free of tubule contamination *(7,8)*. Thereafter, glomeruli can either be seeded into culture flasks, and after a week, cells can be seen growing out of the glomerular core *(5)*. Alternatively, glomeruli can be dissociated by incubation with an enzyme, such as collagenase *(9)*.

From: *Methods in Molecular Medicine: Human Cell Culture Protocols*
Edited by: G. E. Jones Humana Press Inc., Totowa, NJ

Table 1
Main Cell Markers Used to Differentiate Between Glomerular Cell Types

MAbs to	Epithelial cells	Mesangial cells	Endothelial cells	Monocytic cells
Cytokeratin	+	−	−	−
cALLA[a]	+	−	−	−
α-Smooth muscle actin	−	+	−	−
Myosin	−	+	−	−
Thy-1	−	+	−	−
von Willbrand (factor VIII antigen)	−	−	+	−
HLA-DR	−	−	+	+
Monocyte (CD14)	−	−	−	+
LCA[b]−	−	−	+	

[a]cALLA, common acute lymphoblastic leukemia antigen.
[b]LCA, leukocyte common antigen.

The first cells to emerge from explanted glomeruli are epithelial cells, which have a distinctive "cobblestone" appearance (9) and, for the first 7–10 d of outgrowth, are the most common cell type in the mixed population of glomerular cells. If Bowman's capsule is not stripped from the glomeruli, parietal epithelial cells are the dominant cell type (10); endothelial cells, mesangial cells, and monocytes may also be present at this stage. Mesangial cells become more evident later in culture and have a stellate appearance. They grow vigorously, in multilayers, whereas epithelial cells grow in a monolayer and are subject to contact inhibition. Mesangial cells, therefore, outgrow the epithelial cells, and after 30 d of growth, the cultures are nearly pure glomerular mesangial cells (11). The difference in growth potential of primary cells in culture can be explained by the "Mosaic Theory" (12), which states that separate populations of cells can be obtained from a mixed population of cells based on their different growth rates or culture requirements.

For most experimental purposes, a homogeneous population of cells is required. It is therefore important to assess the purity of the isolated cells, since even a small population of contaminating cells can affect the experimental results. It is now clear that morphology alone is not sufficiently discriminating (13), and to ensure that a homogenous population of cells has been isolated, antibodies to specific cell-surface and cytoskeletal markers should be used to confirm their identity. Table 1 shows the main markers used to differentiate between the glomerular cell types present (5,11,14–16).

Epithelial and Mesangial Cell Culture

The outlined methods are routinely used in our laboratory to obtain and characterize pure populations of glomerular epithelial and mesangial cells (see Note 1). It should be noted that glomerular endothelial cells have proven difficult to isolate and maintain in homogeneous culture (17–20).

2. Materials

1. Wash medium: RPMI 1640 Dutch modification supplemented with 2 mM L-glutamine, 100 U/mL penicillin, 100 µg/mL streptomycin, and 2.5 µg/mL amphotericin B (all reagents from Gibco-BRL, Paisley, Scotland). Store at 4°C, preferably for not more than 1 mo.
2. Culture medium: RPMI 1640 Dutch modification or MEM-d-valine (Gibco-BRL) supplemented as in item 1 with the addition of 10% fetal calf serum (Gibco-BRL) and insulin–transferrin–sodium selenite (ITS) media supplement (Sigma, Poole, Dorset, UK, I-1884). Reconstitute 1 vial of ITS in 50 mL sterile distilled water, and add 1 mL of this stock to 100 mL of medium to give a final concentration of 5 µg/mL insulin, 5 µg/mL transferrin, and 5 ng/mL sodium selenite. Store the stock solution of ITS at 4°C and protect from light.
3. Fibronectin (Sigma F-2006): reconstitute to 1 mg/mL in sterile distilled water, and dilute to 10 µg/mL in RPMI 1640 wash medium. Store at –20°C in aliquots. Coat flasks at a concentration of 5 µg/cm^2, and leave solution to bind for 30 min at 37°C or until flasks are required.
4. Collagen Type I (tissue-culture-grade, Sigma; C-7661): reconstitute to 1 mg/mL in 0.1M acetic acid, leave for 1–3 h until it is dissolved. Transfer this to a glass bottle with chloroform at the bottom. Do not shake or stir the collagen after this point. Store in the dark at 4°C. Coat the flasks at a concentration of 10 µg/cm^2, and leave to bind for 4 h at 37°C or overnight at 4°C. Expose the coated flask to UV radiation if you suspect the solution is no longer sterile—do not filter sterilize.
5. Phosphate-buffered saline (PBS): 1.5 mM (0.2 g/L) KH$_2$PO$_4$, 8.1 mM (1.15 g/L) Na$_2$HPO$_4$, 2.7 mM (0.2 g/L) KCl, 140 mM (8.0 g/L) NaCl. Filter sterilize through a 0.22-µm filter before use. Store at room temperature.
6. Trypsin/EDTA ×10 (Sigma T-4174): Dilute trypsin/EDTA 1:10 with sterile PBS, and store in aliquots at –20°C.
7. Tris-buffered saline (TBS): 0.05M Tris-HCl, pH 7.3, containing 0.15M sodium chloride. Store at room temperature.
8. Veronal acetate buffer: Dissolve 1.47 g sodium barbitone and 0.97 g sodium acetate (trihydrate) in 200 mL water, pH to 9.2, using 0.1M HCl, and make up to 250 mL with water. Store at 4°C for no longer than 1 mo.
9. Scot's tap water substitute: Dissolve 2 g of sodium bicarbonate and 20 g of magnesium sulfate in 1 L of distilled water, and store at room temperature.
10. Stainless-steel sieves of mesh size 250, 150, 106, and 63 µm (Endecotts Ltd., London, UK). A diameter size of 7.5 cm is best for processing large amounts of tissue.
11. Lux chamber slides (8-well Permanox) from Gibco-BRL.
12. Rabbit antimouse Ig binding antibody (Dako [High Wycombe, Bucks, UK] Z0259).
13. APAAP complexes (Dako D0651).

14. Naplhol-AS-MX-phosphate, disodium salt (Sigma N-5000).
15. Levamisole (Sigma L-9756).
16. Fast red TR salt (Sigma F-2768).

3. Methods
3.1. Isolation of Whole Glomeruli

Human glomeruli are most often isolated from the nonaffected pole of nephrectomy specimens from patients with renal cell carcinoma (*see* Note 2). Aseptic conditions in a laminar flow hood should be adopted throughout.

1. Place the tissue in a sterile Petri dish, and cover it in RPMI 1640 wash medium; no fetal calf serum should be used during the isolation procedure, since this may initiate clotting owing to blood products present in the tissue. Remove surrounding capsule and any fat using a scalpel.
2. Cut the cortex away from the medulla, and chop the cortex into 1–2 mm^2 pieces. Press this through a sieve of mesh size 250 µm, into a Petri dish, using the barrel from a 5-mL syringe. This results in the separation of glomeruli from renal tubules, interstitium, and vasculature. Wash the retained tissue with a generous amount (50–100 mL) of RPMI 1640 wash medium. Collect the glomerular filtrate from the Petri dish into sterile containers on ice.
3. Separate the glomeruli from the tubular fragments by passing it through a 150-µm sieve. This also strips the Bowman's capsule from much of the glomeruli. A further 50 mL of RPMI wash medium are used to rinse the tissue retained on the sieve. Collect the filtrate into sterile containers.
4. Pass the filtrate through a 106-µm sieve that retains the glomeruli at the surface of the sieve. Since there is a substantial volume, it is advantageous to hold the sieve above a beaker to allow the filtrate to pass through quickly. Pour the filtrate through the sieve into a funnel inserted in the beaker to prevent any "splashback" that may occur. Rinse the retained glomeruli with approx 50 mL RPMI 1640 wash medium to eliminate any tubular fragments that may still be present.
5. Collect the glomeruli retained on the 106-µm sieve by inverting the sieve over a Petri dish and washing with RPMI culture medium. Transfer the glomeruli at a concentration of approx 15–20 glomeruli/mL to a fibronectin-coated culture flask (*see* Note 3).
6. Culture the glomeruli at 37°C in a 5% CO_2 incubator (*see* Note 4).

3.2. Isolation of Glomerular Epithelial Cells

1. Isolate and culture the glomeruli as described in Section 3.1. Once the glomeruli have adhered to the flask, change the medium (RPMI 1640 culture medium) every 4–5 d.
2. Epithelial cells can be seen growing out of the glomeruli around 7–10 d. This time may vary and generally takes longer to see cellular outgrowth in glomeruli prepared from kidneys from older patients (*see* Note 4).

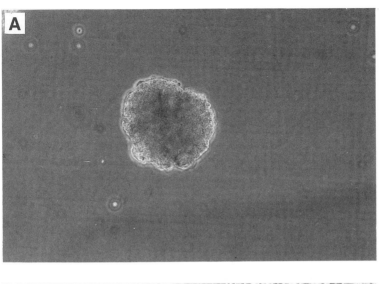

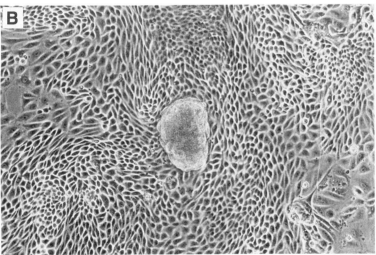

Fig. 1. Glomerular cell culture. (**A**) Decapsulated glomerulus at d 0. (**B**) Cellular growth at d 7; polygonal cells characteristic of epithelial cells. (*See* **1C,D** on p. 424)

3. Once sufficient epithelial cells are identified (these have a polygonal, cobblestone appearance) (*see* Fig. 1B), pour off the unbound glomeruli and rinse the bound glomeruli/cells in PBS. Trypsinize the cells/glomeruli off the flask (*see* Section 3.4.) and pass them through a 63-μm sieve. Collect the epithelial cells in the filtrate; glomeruli are retained on the sieve. Rinse the sieve several times with a volume of around 10 mL RPMI culture medium to stop the action of trypsin.

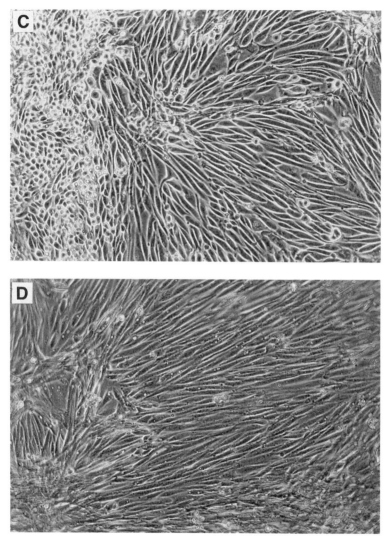

Fig. 1. *(continued)* **(C)** Stellate-shaped cells growing in multilayers characteristic of contractile mesangial cells at d 14; a few epithelial cells are still present. **(D)** Mesangial cell outgrowth at d 28; no epithelial cells are evident.

4. Pellet the cells that have passed through the sieve by centrifugation at 200g for 5 min, and then resuspend in RPMI culture medium. Plate the cells onto type I collagen at a concentration of 10 µg/cm^2 (*see* Section 2.1. and Note 5).
5. After approx 1 wk when the epithelial cells have reached confluence (*see* Note 6), passage using trypsin/EDTA (*see* Section 3.4.). Plate the trypsinized cells

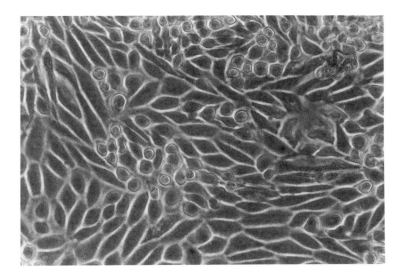

Fig. 2. Epithelial cell preparation grown on type I collagen. At confluence, epithelial cells display a cobblestone-like appearance.

onto plastic tissue-culture-grade dishes or flasks without collagen at this point. At confluence, the cells will be homogeneous and display a cobblestone appearance (Fig. 2).
6. Characterize the cells before use (*see* Section 3.5. and Note 7).

3.3. Glomerular Mesangial Cell Culture

1. Glomeruli are isolated and cultured as described in Section 3.2. Once the glomeruli have adhered to the flask, the medium should be changed every 4–5 d.
2. The first cells to grow out of the glomerular core are epithelial cells (Fig. 1B). After a further 2 wk in culture, a mixed population of glomerular cell types is observed (Fig. 1C). Continue cultures for 2–3 wk to allow the mesangial cells to overgrow the glomerular epithelial cells (Fig. 1D). From wk 3 onward, grow the cells in MEM-d-valine culture medium to inhibit any fibroblast growth (*see* Note 8) *(21)*.
3. Once the cells have grown to confluence, split them 1:1 using trypsin/EDTA (*see* Section 3.4.), and continue to culture in uncoated flasks.
4. Cells should be used between passages 3 and 7 (*see* Notes 4 and 9) after being fully characterized.

3.4. Trypsinization of Adherent Cells

To remove the epithelial or mesangial cells from tissue-culture flasks:

1. Pour the medium off the cells, and wash out the flask with sterile PBS.
2. Pour off the PBS, add 5 mL of 1% trypsin/EDTA solution, and then incubate the flask at 37°C for 4–7 min.

3. Tap the flask on the bench to loosen the cells before adding fresh medium. Split the cell suspension into two flasks to continue the culture, or centrifuge and count the cells before using for experiments (see Note 10).

3.5. Validation of Glomerular Cell Cultures

3.5.1. Cell Morphology

Cell morphology should be assessed throughout the culture period using an inverted microscope with phase-contrast illumination. Cell morphology at different stages of glomerular culture is shown in Fig 1.

1. Epithelial cells: Epithelial cells are homogenous in appearance (9), polygonal, and form cobblestone-like monolayers (Fig. 2). They are closely packed, and adhere tightly to each other at the edge of the growing monolayer. Epithelial cells are subject to contact inhibition (see Note 6) and are sensitive to puromycin amino nucleotide.
2. Mesangial cells: Mesangial cells are elongated and stellate-shaped, and are not subject to contact inhibition (Fig. 1D). They grow in multilayers to form characteristic hills and valleys (5,22).

3.5.2. Alkaline Phosphatase Antialkaline Phosphatase (APAAP) Technique

The morphological assessment of cell cultures should be confirmed by immunohistochemistry using currently available monoclonal antibodies (MAbs) (Table 1). The APAAP immunohistochemical technique is a very sensitive method (23) and is particularly useful when using MAbs. In this technique, soluble complexes of alkaline phosphatase and mouse monoclonal antialkaline phosphatase are used to amplify the primary antibody–antigen interaction via a bridging antibody (rabbit antimouse) that links the primary antibody (mouse) to the APAAP complex. An intense immunohistochemical staining with low nonspecific backgrounds can be obtained. The enzyme conjugate is detected using a napthol–phosphatase derivative as substrate, and the naphthol compound produced is visualized by forming a diazonium salt with Fast red to give a colored product. Levamisole® is included in the substrate to inactivate the endogenous phosphatase activity of cells.

1. The cells to be tested should be passaged, and 0.4 mL of cells at 1×10^6/mL seeded onto 8-well permanox Lux chamber slides (Gibco-BRL). Place in the CO_2 incubator, and incubate cells for 1–3 d.
2. Remove the slides from the incubator, and without removing the upper chamber structure aspirate tissue-culture supernatant. Wash the cells twice with 0.8 mL of TBS/well. Discard the final rinse of TBS, and allow the slides to air-dry for 20 min or overnight at room temperature.
3. Fix the slides by adding 0.3 mL of acetone at 4°C and incubating slides on ice for 10 min. (Note: Do not use more than 0.3 mL of acetone, since this dissolves the upper structure of the 8-well chamber.)

Table 2
Antibodies Used to Characterize Glomerular Cell Cultures

MAbs to	Clone/source[a]	Dilution
Thy-1	F15-42-1/Serotec	1/40
Fibronectin	NH3/Serotec	1/100
Common ALL antigen (CD10)	SS2/36/Dako	1/50
von Willebrand's factor	F8/86/Dako	1/50
HLA-DR	B-F1/Serotec	1/100
Leukocyte common (CD45)	GB3/Serotec	1/100
Monocyte (CD14)	UCHM1/Serotec	1/50
Cytokeratin	LP34/Dako	1/50
Myosin	MY-21/Sigma	1/150
α-Smooth muscle actin	1A4/Sigma	1/200

[a]Locations of companies: Serotec (Oxford, UK), Dako (High Wycombe, Bucks, UK), and Sigma (Poole, Dorset, UK).

4. Remove the slides and allow them to air-dry at room temperature for 10–15 min.
5. Prepare the antibodies at this stage, diluting them to the optimal titer in 1% BSA/TBS (w/v) (see Table 2).
6. Rehydrate the slides by adding 0.8 mL of TBS to each well (two washes of 5 min). Ensure that the slide does not dry out at any stage after this point.
7. Remove the TBS from the wells, and add 200 µL of the primary MAb diluted to its optimal titer in 1% BSA/TBS. Incubate the slides for one hour at room temperature.
8. Discard the antibody, and wash the slides three times for 2 min in TBS.
9. Add 200 µL rabbit antimouse immunoglobulin bridging antibody (Dako Z0259) diluted 1/20 in 1% BSA/TBS to each well. Incubate for 30 min at room temperature.
10. Wash as in step 8.
11. Add 200 µL of mouse APAAP complexes (Dako D0651) at a 1/40 dilution, and incubate for a further 30 min at room temperature.
12. Wash the cells in TBS, three times for 2 min each, followed by a rinse in distilled water.
13. Prepare the substrate solution: Dissolve 25 mg napthol-AS-MX-phosphate disodium salt (Sigma N-5000), 12 mg of levamisole (Sigma L-9756), and 25 mg of Fast red TR salt (Sigma F-2768) in 50 mL of veronal acetate buffer, and filter. To each well of the Permanox Lux slide, add 0.5 mL of the substrate solution.
14. Incubate the slides at room temperature for 10–15 min. During this time, a red color develops if the cells are positive for that marker.
15. Rinse in distilled water.
16. Rinse in tap water, and remove the upper chamber structure from the slides according to the manufacturer's instructions.

17. Stain the cells lightly in hematoxylin for approx 10 s, and blue the nuclei in Scot's tap water substitute.
18. Wash the slides well in running tap water.
19. Mount the slides in an aqueous mounting medium (Apathy's BDH, Merck Ltd., Poole, Dorset, UK).
20. Assess slides for positive reaction (red) by microscopy, and score on a scale of negative to ++++. Also assess the percentage of cells positive for each marker to determine the purity of the epithelial or mesangial cell cultures.

4. Notes

1. It should be noted that the methods described are for isolation of human cells, and seive sizes and culture conditions should be modified should rat cells (or other mammalian cells) be required *(4,20,24,25)*.
2. Appropriate precautions should be used when handling human tissue, e.g., sterile disposable gloves should be worn at all times.
3. Two techniques can be used to initiate glomerular cell culture. In addition to the methods described herein, glomeruli can be dissociated by incubation with collagenase type I (Sigma C-0130) in a concentration of 1 mg/mL for 20 min at 37°C. After agitation with a Pasteur pipet, glomeruli remnants can be separated from single cells by passing them through a 63-μm sieve. Glomerular fragments and single cells are plated separately for the culture of mesangial and epithelial cells, respectively. Although this improves the plating efficiency of glomeruli, great care must be taken not to "overdigest" the glomeruli, since this may damage epithelial cells.
4. The age of the patient and functional capacity of the tissue will determine how quickly the cells establish themselves in culture (cells from a young, healthy kidney will grow more rapidly) and the number of passages the cells can undergo before reaching senescence.
5. The composition of the extracellular matrix may exert major effects on the phenotypic properties of cells. Attention must be given to the modulatory influences of the matrix on which cells are cultured. Although fibronectin and collagen greatly improve the initial adherence of mesangial and epithelial cells, they can, if necessary, be cultured in the absence of such matrices.
6. Epithelial cells are subject to contact inhibition and should be passaged as soon as they reach confluence to reduce cell death. Ideally, cells should be used at, or before, passage three, since their proliferative activity decreases suddenly around this time. Epithelial cells can sometimes adopt a spindle-like structure; this is usually owing to dedifferentiation, and their use for experimentation should be considered carefully.
7. It is debatable whether the epithelial cells isolated are parietal or visceral in origin *(5,10)*. Visceral epithelial cells have a very low proliferative capacity and are unlikely to lead to confluent cultures that can be passaged *(10)*.
8. One easy and reliable way to check the cultures for fibroblast contamination is by growing cells in medium containing D-valine substituted for L-valine, a

condition in which fibroblasts cannot grow. Fibroblasts do not contain the enzyme (D-amino acid oxidase) necessary to convert the D-amino acid to its essential L-form *(21)*.

9. Phenotypic changes may occur in cultured mesangial cells after about 10 passages with the loss of angiotensin II receptors. The morphology of the cells may change from stellate-shaped cells to large, flat cells with the development of stress fibers. In addition, cells at high passage number will no longer contract isotonically to vasoactive hormones *(26)*.
10. If mesangial or epithelial cell matrix proteins are required, the cells should firstly be dislodged in 1% EDTA/PBS (w/v). The matrix should then be removed in a volume of detergent (e.g., 0.5% SDS) with vigorous scraping using the barrel of a 1-mL syringe.

References

1. Sterzel, R. B. and Lovett, D. H. (1988) Interactions of inflammatory and glomerular cells in the response to glomerular injury, in *Contemporary Issues in Nephrology*, vol. 18 (Wilson, C. B., Brenner, B. M., and Stein, J. H., eds.), Churchill Livingstone, New York, p. 137.
2. Striker, L. J., Doi, T., Elliot, S., and Striker G. E. (1989) The contribution of glomerular mesangial cells to progressive glomerulosclerosis. *Semin. Nephrol.* **9**, 318–328.
3. Scheinman, J. I., Fish, A. J., Brown, D. M., and Michael A. J. (1976) Human glomerular smooth muscle (mesangial) cells in culture. *Lab. Invest.* **34**, 150–158.
4. Holdsworth, S. R., Glasgow, E. F., Atkins, R. C., and Thomson, N. M. (1978) Cell characteristics of cultured glomeruli from different animal species. *Nephron* **22**, 454–459.
5. Kreisberg, J. I. and Karnovsky, M. J. (1983) Glomerular cells in culture. *Kidney Int.* **23**, 439–447.
6. Andrews, P. M. and Coffey, A. K. (1980) In vitro studies of kidney glomerular epithelial cells. *Scanning Electron Microsc.* **2**, 179–184.
7. Burlington, H. and Cronkite, E. P. (1973) Characteristics of cell cultures derived from renal glomeruli. *Proc. Soc. Exp. Biol. Med.* **142**, 143–149.
8. Krakower, C. A. and Greenspon, S. A. (1954) Factors leading to variation in concentration of "nephrotoxic" antigens of glomerular basement membranes. *Arch. Pathol.* **58**, 401–432.
9. Striker, G. E. and Striker L. J. (1985) Biology of disease. Glomerular cells in culture. *Lab. Invest.* **53**, 122–131.
10. Norgarrd, J. O. (1987) Rat glomerular epithelial cells in culture: parietal or visceral epithelial origin? *Lab. Invest.* **57**, 277–282.
11. Stewart, K. N., Roy-Chaudhury, P., Lumsden, L., Jones, M. C., Brown, P. A. J., MacLeod, A. M., Haites, N. E., Simpson J. G., and Power D. A. (1991) Monoclonal antibodies to cultured human glomerular mesangial cells. Reactivity with normal kidney. *J. Pathol.* **163**, 265–272.
12. Soukupova, M. and Holeckova, E. (1964) The latent period of explanted organs of newborn, adult, and senile rats. *Exp. Cell. Res.* **33**, 361–367.

13. Fish, A. J., Michael, A. F., Vernier, R. L., and Brown, D. M. (1975) Human glomerular cells in tissue culture. *Lab. Invest.* **33**, 330–341.
14. Abbott, F., Jones, S., Lockwood, C. M., and Rees, A. J. (1989) Autoantibodies to glomerular antigens in patients with Wegner's granulomatosis. *Nephrol. Dial. Transplant.* **4**, 1–8.
15. Rees, A. J. (1989) Proliferation of glomerular cells, in *New Clinical Applications in Nephrology: Glomerulonephritis* (Catto, G. R. D., ed.), Kluwer, London, pp. 163–193.
16. Müller, G. A. and Müller, C. (1983) Characterisation of renal antigens on distinct parts of the human nephron by monoclonal antibodies. *Klin. Wochenschr.* **61**, 893–902.
17. Striker, G. E., Soderland, C., Bowen-Pope, D. F., Gown, A. M., Schmer, G., Johnston, A., Luchtel, D., Ross, R., and Striker, L. J. (1984) Isolation and characterization, and propagation in vitro of human glomerular endothelial cells. *J. Exp. Med.* **160**, 323–328.
18. Savage, C. O. (1994) The biology of the glomerulus: endothelial cells. *Kidney Int.* **45**, 314–319.
19. Scheinman, J. I. and Fish, A. J. (1978) Human glomerular cells in culture: three subcultured cell types bearing glomerular antigens. *Am. J. Pathol.* **92**, 125–146.
20. Ringstead, S. and Robinson, G. B. (1994) Cell culture from rat renal glomeruli. *Vichows Arch.* **425**, 391–398.
21. Gilbert, S. F. and Migeon, B. R. (1987) D-Valine as a selective agent for normal human and rodent epithelial cells in culture. *Cell* **5**, 11–17.
22. Sterzel, R. B., Lovett, D. H., Foellmer, H. G., Perfetto, M., Biemesderfer, D., and Kashgarian, M. (1986) Mesangial cell hillocks: nodular foci of exaggerated growth of cells and matrix. *Am. J. Pathol.* **125**, 130–140.
23. Cordell, J. L., Falini, B., Erber, W. N., Ghosh, A. K., Abdulaziz, Z., MacDonald, S., Pulford, K. A. F., Stein, H., and Mason, D. Y. (1984) Immunoenzymatic labelling of monoclonal antibodies using immune complexes of alkaline phosphatase and monoclonal anti-alkaline phosphatase (APAAP complexes). *J. Histochem. Cytochem.* **32**, 219–229.
24. Harper, P. A., Robinson J. M., Hoover, R. L., Wright, T. C., and Karnovsky, M. J. (1984) Improved methods for culturing rat glomerular cells. *Kidney Int.* **26**, 875–880.
25. Mendrick, D. L. and Kelly, D. M. (1993) Temporal expression of VLA-2 and modulation of its ligand specificity by rat glomerular epithelial cells *in vitro*. *Lab Invest.* **69**, 690–702.
26. Kreisberg, J. L., Venkatachalam, M. A., and Troyer, D. A. (1985) Contractile properties of cultured glomerular mesangial cells. *Am. J. Physiol.* **249**, F229–F234.

32

Enzymatic Isolation and Serum-Free Culture of Human Renal Cells

Retaining Properties of Proximal Tubule Cells

John H. Todd, Kenneth E. McMartin, and Donald A. Sens

1. Introduction

1.1. Isolation Procedure

Procedures for the culture of the human renal proximal tubule (HPT) cell-utilizing explanted tissue have been previously reported by this laboratory *(1)*. Several other investigators have also reported the isolation and culture of human renal tubule cells *(2*, and references therein). Although explantation of tissue fragments remains an effective way to initiate cell cultures, cell outgrowth and the attainment of confluent cultures may take several weeks. The cell-culture methodology described in this chapter results in a high yield of confluent primary cultures in 7–10 d. The technique involves the digestion of minced cortical tissue with collagenase, followed by a filtering step to remove tissue fragments. The filtrate is centrifuged, and the cell pellet is resuspended in serum-free growth medium and dispensed onto prepared growth surfaces.

1.2. Overview

The growth medium is a serum-free formulation of a 1:1 mixture of Dulbecco's modified Eagle's medium (DMEM) and Ham's F-12 medium supplemented with insulin (5 µg/mL), transferrin (5 µg/mL), selenium (5 ng/mL), hydrocortisone (36 ng/mL), triiodothyronine (4 pg/mL), epidermal growth factor (EGF) (10 ng/mL), and L-glutamine (2 mM). Penicillin–streptomycin and Fungizone® may also be added to the medium. Growth surfaces are covered with a matrix consisting of bovine type 1 collagen with adsorbed fetal bovine serum (FBS) components.

Cell dissociation is accomplished by stirring minced cortical tissue at 37°C in Hank's balanced salt solution (HBSS) containing collagenase and DNase. The resulting single cells and tubule segments are placed into 75-cm² T-flasks and maintained in an incubator at 37°C with a 5% CO_2:95% air atmosphere. The following day, the growth medium is replaced with fresh growth medium to remove blood cells, debris, and floating cells. Subsequently, cultures are fed every third day. Primary cultures generally reach a confluent state in 7–10 d, and may then be treated with a cryopreservative, slowly frozen, and stored under liquid nitrogen. Those cells not frozen, as well as cells thawed at a later date, are typically passaged on a weekly basis at a 1:2 ratio.

2. Materials

2.1. Growth Surface

1. Vitrogen 100 collagen (Celtrix Laboratory, Santa Clara, CA).
2. FBS (Gibco-BRL, Grand Island, NY).
3. Phosphate-buffered saline (PBS): 8.5 g NaCl, 0.12 g Na_2HPO_4, 0.05 g NaH_2PO_4. Made up to 1 L.

2.2. Serum-Free Growth Medium

1. DMEM 1 × 5 L powder (Gibco-BRL). Low glucose, with L-glutamine, with 100 mg/mL sodium pyruvate, without sodium bicarbonate.
2. F-12 nutrient mixture (Ham) 1 × 5 L powder (Gibco-BRL), with L-glutamine, without sodium bicarbonate.
3. 7.5% (w/v) Sodium bicarbonate (Gibco-BRL).
4. Fungizone (Gibco-BRL).
5. 10,000 U/mL Penicillin–streptomycin (Gibco-BRL).
6. 0.2-µm Acrocap filter (Gelman Sciences, Ann Arbor, MI).

2.3. Growth Factor Supplements

1. 3,3',5-Triiodo-L-thyronine (T_3) (Sigma, St. Louis, MO). Final concentration is 4 pg/mL.
2. ITS Premix—insulin, transferrin, selenious acid (Collaborative Biomedical Products, Bedford, MA). Final concentrations are 5 µg/mL insulin, 5 µg/mL transferrin, and 5 ng/mL selenium.
3. Hydrocortisone (HC) (Sigma). Final concentration is 36 ng/mL.
4. Human recombinant epidermal growth factor (EGF) (Gibco-BRL). Final concentration is 10 ng/mL.
5. 200 mM L-Glutamine (Gibco-BRL). Final concentration is 2 mM.
6. Medium supplements HC, ITS, and T_3 are prepared as 1000X solutions in sterile water and added to the 1:1 mixture of DMEM and F-12. Glutamine is commercially supplied as a 100X solution. EGF is prepared as a 100X solution in PBS. Glutamine and EGF are added to the otherwise complete media just prior to use. After preparation, each supplement is stored frozen. Medium supplements may be aliquoted as single-use portions to avoid repeated freezing and thawing.

2.4. Cell Isolation

1. HBSS (Gibco-BRL).
2. Collagenase (EC 3.4.24.3) (Sigma).
3. Deoxyribonuclease 1 (DNase) (EC 3.1.21.1) from bovine pancreas (Sigma).
4. Add 700 mg collagenase and 500 mg DNase to 500 mL HBSS. Stir thoroughly in a 37°C water bath, and filter sequentially through 0.8-, 0.45-, and 0.2-µm filters.
5. 50-mL Trypsinizing flask (Bellco, Vineland, NJ).
6. White Swiss Nitex™ (T3-163T, 100% polyamide nylon, Martin, Baltimore, MD).
7. Gentamicin (US Biochemicals, Cleveland, OH). Add gentamicin (50 µg/mL) to an additional 500 mL HBSS.

2.5. Cell Passage

1. Trypsin–EDTA (0.05%, 0.53 mM) (Gibco-BRL).
2. FBS (Gibco-BRL).
3. PBS.

2.6. Cryopreservation

Dimethyl sulfoxide (DMSO) (Fisher Scientific, Pittsburgh, PA).

3. Methods
3.1. Growth Surface Preparation

1. Spread collagen evenly over the growth surface, and allow to dry under a hood (*see* Note 1). Drying may take 1–2 h.
2. Add a sufficient amount of FBS so that the surface of the collagen is covered (*see* Note 1).
3. Store overnight at 4°C to allow adsorption of serum components to the collagen.
4. The following morning, aspirate serum and rinse the surface with PBS to remove residual serum. Aspirate PBS. The growth surface is then ready for the addition of cells or may be stored at 4°C for future use. T-flasks, dishes, multiwell plates, or filter inserts may all be treated in the same manner.

3.2. Preparation of Serum-Free Growth Medium

1. Prepare DMEM and Ham's F-12 medium according to the manufacturer's instructions using tissue-culture-grade water. Make volume allowances for the addition of sodium bicarbonate and optional antimycotic/antibiotic agents (*see* Note 2).
2. Add 246.5 mL sodium bicarbonate (7.5% [w/v]) to DMEM and 78.5 mL sodium bicarbonate to F-12, and bring volume of each mixture to 5 L.
3. Sterilize by pressure filtration using peristaltic pump and Acrocap filter (Gelman Sciences).

3.3. Enzymatic Cell Dissociation

1. Using aseptic techniques, remove fibrous renal capsule, and dissect cortical tissue from the medulla (*see* Note 3).

2. Mince cortical tissue into 2–3 mm² pieces and place into a 50-mL trypsinizing flask with 30 mL HBSS containing collagenase and DNase.
3. Stir in 37°C water bath for 15 min.
4. Decant, pouring liquid through nitex cloth (*see* Note 4).
5. Centrifuge decanted liquid (150 g, 7 min). Aspirate supernatant and resuspend pellet in HBSS containing gentamicin.
6. Add 30 mL HBSS containing collagenase and DNase to remaining tissue fragments.
7. Repeat steps 3–6 as necessary. Complete digestion of tissue usually requires 4–5 cycles.
8. Combine resuspensions from each digestion and centrifuge (150g, 7 min).
9. Resuspend final pellet in growth medium, and dispense into culture flasks (*see* Note 5).
10. The following day, after gently bumping the side of the flask, aspirate the medium and replace with fresh medium. The medium change serves to remove red blood cells and dead or floating cells.

3.4. Routine Cell Maintenance

1. Cultured HPT cells are fed fresh growth medium every 3 d, and normally reach confluence 7–10 d following isolation and 7 d following subculture.
2. Stock cultures are maintained in 75-cm² T-flasks in an incubator at 37°C with a 5% CO_2:95% air atmosphere.
3. Cultures are typically passaged on a weekly basis at a 1:2 ratio.

3.5. Subculture Procedure

1. Aspirate medium from the confluent monolayer culture, and rinse the cells with PBS.
2. Aspirate PBS, and pipet trypsin–EDTA (3 mL for a 75 cm² flask) onto the monolayer. Allow trypsin solution to remain on the cells at 37°C until the cells begin to float off the culture surface (5–10 min). Bumping the side of the flask against the palm of the hand helps to detach the cells (*see* Note 6).
3. Transfer trypsin and detached cells to a centrifuge tube. Rinse the culture vessel with PBS to collect remaining cells, and add to centrifuge tube. Trypsin action is neutralized by adding FBS, at about one-half the volume of trypsin used.
4. Centrifuge the mixture of cells, trypsin, PBS, and FBS (150g, 7 min).
5. Discard supernatant and resuspend the cell pellet in an appropriate volume of fresh growth medium for delivery onto pre-prepared growth surfaces.

3.6. Cryopreservation

1. Confluent cultures are detached from the growth surface as described for subculture (*see* Section 3.5.).
2. Following centrifugation (150g, 7 min), resuspend the cell pellet in complete growth medium containing 10% DMSO (1 mL of DMSO-containing media/culture vessel), and pipet 1 mL of the cell suspension into each cryovial.
3. Place cryovials into a Styrofoam container (to slow freezing), and place container in –70°C for 1 h.
4. Transfer vials to liquid nitrogen (*see* Note 7).

4. Notes

1. Although the commercially prepared collagen is stored at 4°C, it is best to coat the growth surface with collagen warmed in a water bath. This results in a thinner, more consistent layer. Pipet in an excess to cover the bottom of the flask, aspirate the excess (to be used for the next flask), and then tilt the flask for several minutes to remove residual collagen. Use the same technique to cover the surface of the dried collagen with FBS, except in this case, leave some of the excess (for a 75-cm^2 flask, 0.5 mL FBS is sufficient).
2. Penicillin–streptomycin (50 mL/5 L) and fungizone (5 mL/5 L) may be added at this point.
3. If tissue is received late in the afternoon, it may be kept refrigerated overnight in DMEM (with penicillin–streptomycin and fungizone). The capsule is removed, and the cortex trimmed into manageable strips and placed into the holding medium.
4. Cut circles of Nitex cloth with 11.0-cm filter paper as a template. Autoclave plastic funnel and Nitex filters (3–4 needed/isolation) in a small amount of water. Use as a paper filter to remove tissue fragments from decanted mixture of tissue and HBSS.
5. The initial seeding density for the primary cultures may, of course, vary between low cell densities, which would require a longer time to become confluent and "use up" more of the generational life-span of these mortal cells, and high cell densities, which become confluent more rapidly, yet result in fewer flasks. In this laboratory, the volume of the final pellet is used to determine how many culture flasks are to be plated with cells. Each milliliter generates three 75-cm^2 flasks. For instance, a 10-mL pellet would yield 30 flasks of primary cell cultures.
6. Cell isolates differ in the relative ease of removing the cell monolayer from the growth surface using trypsin. Following trypsin treatment, remaining adherent cells may sometimes be dislodged by forceful pipeting. Although trypsin treatment is successful in most cases, it may be necessary in some instances to scrape cells from the surface. Cell viability remains high if cells are gently scraped from the surface. Scraping may be preferable in some cases, if exposure to trypsin and serum is undesirable.
7. When additional cultures are needed, a cryovial is removed from the liquid nitrogen dewar and placed into room temperature water. On thawing, the DMSO-cell solution is transferred to a centrifuge tube containing about 10 mL growth medium, which serves to dilute the DMSO. The cells are centrifuged at 150g for 7 min, and the supernatant is aspirated and discarded. The cell pellet is resuspended in fresh growth medium and pipeted into a culture vessel. Thawed cells are typically replated at the prefrozen density.

References

1. Detrisac, C. J., Sens, M. A., Garvin, A. J., Spicer, S. S., and Sens, D. A. (1984) Tissue culture of human kidney epithelial cells of proximal tubule origin. *Kidney Int.* **25**, 383–390.
2. Kempson, S. A., McAteer, J. A., Al-Mahrouq, H. A., Dousa, T. P., Dougherty, G. S., and Evan, A. P. (1989) Proximal tubule characteristics of cultured human renal cortex epithelium. *J. Lab. Clin. Med.* **113(3)**, 285–296.

33

Culture of Human Renal Medullary Interstitial Cells

Klara Tisocki and Gabrielle M. Hawksworth

1. Introduction

The renal medullary interstitial cell (RMIC) is a unique lipid containing cell found mainly in the renal medulla between the tubular and vascular structures *(1)*. These highly specialized fibroblast-like cells are abundant in the inner medulla and papilla *(2)*, and are thought to be responsible for the production of both collagenous and noncollagenous extracellular matrix, secretion of vasoactive prostaglandins, and also for degradation of certain components of extracellular material, such as hyaluronic acid *(3)*. The lipid droplets of these cells contain polyunsaturated fatty acids, which can serve as precursors for prostaglandins and other lipid-derived hormones *(4)*. Ultrastructural studies show that, compared with cortical fibroblasts, the renal medullary interstitial cells are highly differentiated, with abundant rough endoplasmatic reticulum, free ribosomes, and a characteristic mitochondrial profile, suggesting an active secretory function.

Cultured renal medullary interstitial cells derived from various animal species have been found to synthesize prostaglandins in vitro *(5)*. Under basal (unstimulated) conditions, or after addition of angiotensin II, bradykinin *(6)*, or arginine vasopressin *(7–9)*, the major prostaglandin formation was attributed to PGE_2 secretion.

Receptors for bradykinin *(10)*, angiotensin II *(11)*, atrial natriuretic factor *(12)*, endothelin 1 *(13)*, gluco- and mineralocorticoids *(14)*, and nitric oxide *(15)* have been demonstrated by specific binding assays and immunocytochemical studies. In contrast to renal epithelial cells, these receptors are functional over several passages. Based on these findings, it has been suggested that RMICs participate in the maintenance of the medullary microcirculation through modulation of the renin-angiotensin system and that prostaglandin syn-

From: *Methods in Molecular Medicine: Human Cell Culture Protocols*
Edited by: G. E. Jones Humana Press Inc., Totowa, NJ

Table 1
Techniques Applied to Culture RMIC

Species	Cell-culture method	Refs.
Rabbit	Cells derived from sc autotransplants of the renal medulla	5
Rat	Cells derived from sc isogeneic (injected into litter mate) transplants of the renal papilla	7
Dog	Selective growth in hypertonic medium from collagenase-dispersed papillae, inner medulla	18
Rat	Selective growth in hypertonic medium from collagenase-dispersed papillae, inner medulla	19
Rabbit	Selective overgrowth from collagenase-dispersed papillae	20
Rat	Selective overgrowth from collagenase-treated medulla in medium conditioned with 3T3 Swiss albino mouse fibroblasts	12

thesis in these cells is also regulated by vasoactive peptides, such as endothelin *(13)*, atrial natriuretic peptide *(12)*, or angiotensin II *(16)*.

Several techniques have been applied to culture RMIC (Table 1) in the last two decades. None of these was found to be suitable for the culture of human RMIC, and the method described here is a modification of the method used by Rodemann and Müller *(17)* to culture human renal cortical fibroblasts.

2. Materials

1. Human renal medullary interstitial cells are cultured in a 1:1 mixture of Dulbecco's modified Eagle's medium (DMEM)/Ham's Nutrient Mixture F-12 (F12) containing 15 mM HEPES and 14.3 mM NaHCO$_3$, supplemented with fetal bovine serum (FBS) at various concentrations, and a mixture of antibiotics and antimycotic agents (penicillin 100 U/mL, streptomycin 100 µg/mL, and amphotericin B 0.25 µg/mL). Glass-distilled, purified (ElgaStat water purifying system) water is used to prepare the medium, and after adjustment of the pH to 7.4 with few drops of 1M NaOH, the medium is filter sterilized through a 0.22-µm Millipore filter.
2. Sterile Hank's balanced salt solution (HBSS), containing penicillin (100 U/mL), streptomycin (100 µg/mL), and amphotericin B (0.25 µg/mL), is used for washing the renal medullary tissue and during trypsinization of renal medullary cell cultures.
3. Sterile HBSS without Ca^{2+} and Mg^{2+}, but containing 0.5 mM EDTA, is used during trypsinization of renal medullary cell cultures.
4. Sterile 0.5% (v/v) trypsin solution is prepared by dilution of 2.5% stock solution of trypsin with Ca^{2+}- and Mg^{2+}-free HBSS containing 0.5 mM EDTA.
5. Sterile 0.4% (w/v) trypan blue solution prepared with phosphate-buffered saline is used for determination of cell number and viability.

6. 0.2% (w/v) solution of collagenase A (*Clostridiopeptidase A*, EC 3.4.24.3, Boehringer Mannheim, Mannheim, Germany) is prepared freshly for enzymatic digestion of minced tissue in serum-free DMEM/F12 medium.
7. Normal human renal tissue is collected from nephrectomy samples, no more than 1 h postoperatively. The tissue is assessed macroscopically by a pathologist.
8. Forceps, scissors, scalpels, and all other equipment (i.e., beakers, bottles, pipet tips, etc.) are sterilized by autoclaving. A sterile 100-mm tissue-culture Petri dish is used for mincing tissue, and sterile disposable plastic Universal containers used for washing and dispersing tissue explants or cultured cells.
9. For cryopreservation of cells, DMEM/F12 medium is supplemented with 10% FBS, 10% dimethyl sulfoxide, penicillin (100 U/mL), streptomycin (100 µg/mL), and amphotericin B (0.25 µg/mL).
10. Storage of solutions: Medium and balanced salt solutions should be stored in a dark place at 4°C for no longer than 3 mo. FBS, antibiotic/antimycotic solutions, and trypsin solution can be stored at –20°C in aliquots suitable for making up small volumes of medium or trypsin solution, in order to avoid subsequent freezing and thawing of the stock solutions. Once supplements are added to the medium and/or solutions, store at 4°C and use within 2 wk.

3. Method

The principles of the method are summarized in Fig. 1.

1. Dissect normal renal tissue from the nephrectomy specimen, and transfer it in a sterile Universal container on ice, from the pathology laboratory to the tissue-culture facility.
2. Prepare renal medullary tissue explants in a vertical laminar air-flow hood under aseptic conditions.
3. Remove the renal capsule. Then separate the cortex and outer medulla from the inner medulla and papilla using sterile forceps and a scalpel. Dissect all fatty tissue away with sharp scissors or a scalpel. Place the inner medulla and papilla (at least 2.0 g tissue) in ice-cold sterile HBSS containing penicillin (100 U/mL), streptomycin (100 µg/mL), and amphotericin B (0.25 µg/mL), and keep it on ice.
4. Transfer the medullary and papillary tissue fragments to a sterile Petri dish containing a small amount of HBSS, and cut into 6–8 mm^3 cubes (*see* Note 1), ensuring that the tissue is covered with fluid at all times. Transfer these small fragments to a Universal, and wash at least three times with ice-cold HBSS or until most of the blood has been removed.
5. Transfer the tissue fragments into 15 mL 0.2% Collagenase A solution in a sterile plastic Universal, and incubate at 37°C for 20 min. During this time, gently agitate the contents of the Universal in every 5 min. There is no need for continuous stirring. Short enzymatic digestion (20 min) of tissue fragments results in better outgrowth of medullary cells from explants.
6. At the end of the incubation period, inactivate Collagenase A by the addition of 10 mL ice-cold Ca^{2+}- and Mg^{2+}-free HBSS containing 0.5 mM EDTA. Wash the

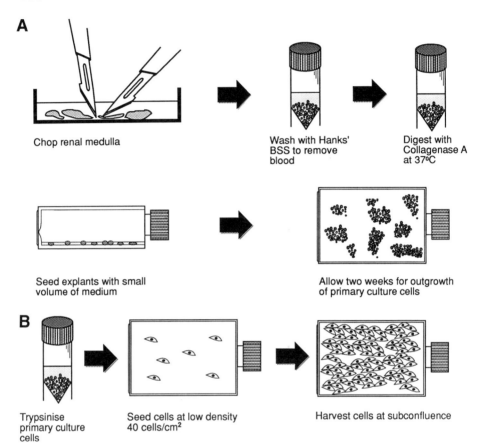

Fig. 1. Culture of medullary interstitial cells. **(A)** Primary explant technique. **(B)** Selective overgrowth of interstitial cells.

tissue fragments twice by resuspending the fragments in the same solution allowing them to settle under gravity each time before removal of HBSS.
7. After the final removal of HBSS, add a small volume of DMEM/F12 medium supplemented with 20% FBS, penicillin (100 U/mL), streptomycin (100 µg/mL), and amphotericin B (0.25 µg/mL).
8. Transfer the tissue suspension to a tissue-culture flask (10 fragments/20 cm^2), with 1–1.5 mL medium/20 cm^2 growth area (*see* Note 2). Tilt the flask gently to spread the pieces evenly over the growth surface. The medium should just barely cover the tissue fragments (*see* Note 3).
9. Place the tissue-culture flasks in a humidified incubator at 37°C in a 5% CO_2/95% air environment.
10. Do not move tissue-culture flasks for 72 h to allow attachment of explants (*see* Note 4).

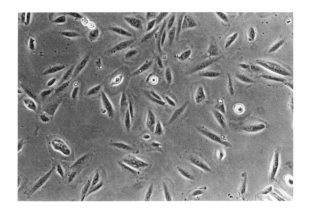

Fig. 2. Photomicrograph of a culture of human renal medullary interstitial cells. Human renal medullary interstitial cells displaying spindle-shaped morphology, passage 2. Magnification 27×.

11. Change the medium after 72 h by gently pouring off the medium and any unattached tissue pieces. Gently pipet a small volume of fresh medium down the side of the tissue-culture flasks, so as not to disturb the adhered explants. Replace the medium with fresh medium every 48 h. Gradually increase the volume of medium after the first week, up to 3–4 mL/20 cm^2 growth area.
12. After approx 2 wk, or when a mixed primary culture of renal medullary cells has spread over 60–75% of the growth area, the cells may be subcultured.
13. Subculture: Remove the medium from the cell monolayer, and wash it with Ca^{2+}- and Mg^{2+}-free HBSS containing 0.5 mM EDTA. Gentle tapping of the flask during this washing step will remove any remaining tissue explants.
14. Add 0.5% trypsin solution in a volume of 5 mL/20-cm^2 growth area. Place flasks in the incubator for 6–8 min to allow detachment of cells from the surface (*see* Note 5).
15. Centrifuge suspensions of the dispersed cells at 400g to remove the trypsin, and then resuspend them in ice-cold HBSS. Keep the cell suspension on ice.
16. Mix a small aliquot of the cell suspension (50–200 µL) with 0.4% trypan blue solution, and determine the number of viable cells by counting cells in a hemocytometer.
17. Seed the dispersed primary culture cells at 40 cells/cm^2 density (*see* Note 6), and incubate with DMEM/F12 supplemented with 20% FBS and antibiotic/antimycotic agents. Passage 1 cells may reach near confluence 14–18 d after subculture.
18. Individually attached cells form loosely connected colonies within a few days of seeding, and selective overgrowth by these cells results in a highly homogenous cell population (Fig. 2) displaying characteristics of renal medullary interstitial cells (*see* Note 8 and Figs. 3 and 4).
19. Cells at this stage can be subcultured further with a seeding density of 1×10^4 cells/cm^2 and grown in DMEM/F12 medium containing 10% FBS (*see* Note 7).

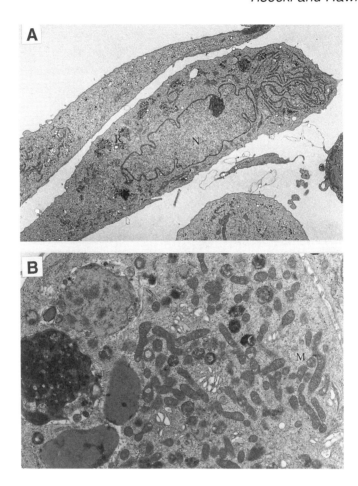

Fig. 3. Electron micrographs of a cultured human renal medullary interstitial cell demonstrating **(A)** (5000×) irregularly shaped nucleus (N) with nucleolus, and free ribosomes (R) in the cytoplasm; **(B)** (10,000×) elongated or round mitochondrial profile (M).

20. Alternatively, the cell suspension may be cryopreserved in DMEM/F12 medium (supplemented with 10% FBS) containing 10% DMSO as a cryopreservative.
21. Cryopreservation: Sediment cell suspension by centrifugation after determination of cell count. Remove supernatant, and add medium with cryopreservative to the cell pellet to give a final cell concentration of $2–5 \times 10^6$ cells/mL. Resuspend the cells, and keep them on ice for 10 min. Transfer 1-mL aliquots of the cell suspension into cryotubes, and place in a –20°C freezer for 4 h. After 4 h, move cryotubes into the storage freezer. Cells should be stored at –80°C or below (*see* Note 10).

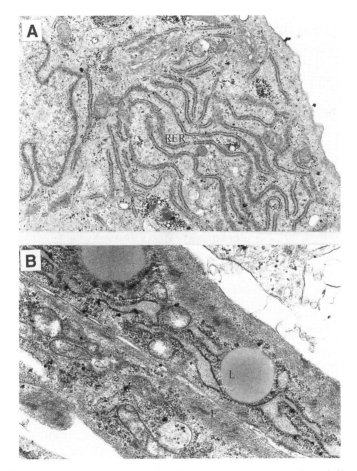

Fig. 4. Electron micrographic details of a cultured human renal medullary interstitial cell demonstrating **(A)** (12,500×) abundance of rough endoplasmatic reticulum (RER); **(B)** (18,750×) lipid droplets (L) and bundles of filament adjacent to cell membrane (F).

4. Notes

1. When cutting up the renal tissue, try to avoid fine mincing of tissue, since very small tissue pieces will float in medium and not attach to the growth surface.
2. In our laboratory, we found that rectangular tissue-culture flasks were superior for growing explants when compared with Petri dishes.
3. During the seeding of explants, it is essential to add the correct volume of medium to the tissue fragments, i.e., only a thin film of medium should cover the pieces in order to enhance attachment of renal medullary fragments.

4. In the first 72-h period, place tissue-culture flasks on a little-used shelf in the incubator to facilitate attachment of tissue fragments.
5. Since epithelial cells are relatively resistant to trypsin, during trypsinization of mixed primary cultures of renal medullary cells, cells with nonepithelial characteristics will detach from the growth surface in the first 10 min. Therefore, it is not necessary to wait for the dispersion of all the cells.
6. When preparing very high dilutions (40 cells/cm^2) of cells for seeding in passage 1, use ice-cold solutions to prevent clumping of cells, and disperse them well by drawing up the diluted cell suspension several times into a 10-mL sterile pipet before seeding.
7. Human renal medullary interstitial cells can be subcultured up to passage 4. After this passage number, cell morphology and growth characteristics are significantly changed owing to senescence of the cells.
8. We have characterized these cells using light and electron microscopy and immunocytochemistry. Lipid droplets (stained with Sudan Black or Oil Red O) are constant characteristics of RMIC. Immunocytochemical analysis (technique described in *21*) revealed that RMIC stained positively with antihuman antifibroblast antibody, and negatively for epithelial, endothelial, and mesengial cell markers. Ultrastructural details of cultures hRMIC are shown in Figs. 3 and 4.
9. If the method described here is followed, it should be adequate to confirm the identity and homogeneity of the cultures using light microscopy.
10. Human RMIC can be stored under these conditions for up to 1 yr, and then thawed and cultured successfully.

References

1. Bulger, R. E., Tisher, C. C., Myers, C. H., and Trump, B. F. (1967) Human renal ultrastructure. *Lab. Invest.* **16,** 124–141.
2. Bohman, S. O. (1974) The ultrastructure of the rat renal medulla as observed after improved fixation methods. *J. Ultrastruct. Res.* **47,** 329–360.
3. Lemley, K. V. and Kriz, W. (1991) Anatomy of the renal interstitium. *Kidney Int.* **39,** 370–381.
4. Comai, K., Prose, P., and Farber, S. J. (1974) Correlation of renal medullary prostaglandin content and renal interstitial cell lipid droplets. *Prostaglandins* **6,** 375–379.
5. Muirhead, E. E., Germain, G., Leach, B. E., Pitcock, J. A., Stephenson, P., Brooks, B., Brosius, W. L., Daniels, E. G., and Hinman, W. J. (1972) Production of renomedullary prostaglandins by renomedullary interstitial cells. *Circ. Res.* **Suppl. II.,** II-161–II-172.
6. Zusmann, R. M. and Keiser, H. R. (1977) Prostaglandin E$_2$ biosynthesis by rabbit renomedullary interstitial cells in tissue culture. *J. Biol. Chem.* **252,** 2069–2071.
7. Beck, T. R., Hassid, A., and Dunn, M. J. (1980) The effect of arginine vasopressin and its analogs on the synthesis of prostaglandin E$_2$ by rat renal medullary interstitial cells in culture. *J. Pharm. Exp. Ther.* **215,** 15–19.
8. Ausiello, D. A. and Zusman, R. M. (1984) The role of calcium in the stimulation of prostaglandin synthesis by vasopressin in rabbit renal-medullary interstitial cells in tissue culture. *Biochem. J.* **220,** 139–145.

9. Craven, P. A. and DeRubertis, F. R. (1991) Calcium and prostaglandin E_2 in renomedullary interstitial cells. *Hypertension* **17**, 303–307.
10. Fredrick, M. J., Abel, F. C., Rightsel, W. A., Muirhead, E. E., and Odya, C. E. (1985) B2 bradykinin receptor-like binding in rat renomedullary interstitial cells. *Life Sci.* **37**, 331–338.
11. Zhuo, J., Alcorn, D., Allen, A. M., and Mendelsohn, F. A. (1992) High resolution localization of angiotensin II receptors in rat renal medulla. *Kidney Int.* **42**, 1372–1380.
12. Fontoura, B. M., Nussenzweig, D. R., Pelton, K. M., and Maack, T. (1990) Atrial natriuretic factor receptors in cultured renomedullary interstitial cells. *Am. J. Physiol.* **258**, C692–C699.
13. Wilkes, B. M., Ruston, A. S., Mento, P., Girardi, E., Hart, D., Molen, M. V., Barnett, R., and Nord, E. P. (1991) Characterization of endothelin 1 receptor and signal transduction mechanisms in rat medullary interstitial cells. *Am. J. Physiol.* **260**, F579–F589.
14. Farman, N., Oblin, M. E., Loabes, M., Delahaye, F., Westphal, H. M., Bonvalet, J. P., and Gasc, J. M. (1991) Immunolocalization of gluco- and mineralo-corticoid receptors in rabbit kidney. *Am. J. Physiol.* **260**, C226–C233.
15. Ujiie, K., Hogarth, L., Danziger, R., Drewett, J. G., Yuen, P. S. T., Pang, I.-H., and Star, A. (1994) Homologous and heterologous desensitization of a guanylyl cyclase-linked nitric oxide receptor in cultured rat reno-medullary interstitial cells. *J. Pharmacol. Exp. Ther.* **270**, 761–767.
16. Kugler, P. (1983) Angiotensinase A in the renomedullary interstitial cells. *Histochemistry* **77**, 105–115.
17. Rodemann, P. E. and Müller G. A. (1990) Abnormal growth and clonal proliferation of fibroblasts derived from kidneys with interstitial fibrosis. *Proc. Soc. Exp. Biol. Med.* **195**, 57–63.
18. Kuroda, M., Ueno, H., Sakato, S., Funaki, N., and Takeda, R. (1979) A unique affinity adaptation of renomedullary interstitial cells for hypertonic medium. *Prostaglandins* **18**, 209–220.
19. Benns, S. E., Dixit, M., Ahmed, I., Ketley, C., and Bach, P. H. (1985) The use of renal cells as alternatives to live animals for studying renal medullary toxicity, in *Alternative Methods in Toxicology, Vol. 3: In Vitro Toxicology* (Goldburg, A. M., ed.), Mary Ann Liebert Inc., New York, pp. 211–234.
20. Rodemann, H. P., Müller G. A., Knecht, A., Norman J. T., and Fine L. G. (1991) Fibroblasts of rabbit kidney in culture I. Characterization and identification of cell specific markers. *Am. J. Physiol.* **261**, F238–F291.
21. Hawksworth, G. M., Cockburn, E. M., Simpson, J. G., and Whiting, P. H. (1993) Primary culture of rat renal medullary interstitial cells, in *Methods in Toxicology, In Vitro Biological Systems*, vol. 1, part A (Tyson, C. A. and Frasier, J. M., eds.), Academic, New York, pp. 385–396.

34

Culture Technique of Human Pituitary Adenoma Cells

Masanori Kabuto

1. Introduction

Cell cultures are indispensable to the study of human pituitary adenomas. Nevertheless, in vitro studies of human pituitary adenomas have been hindered by the small amount of tissue available and by difficulties in establishing monolayer cultures using standard culture techniques owing to poor cell attachment to ordinary plastic flasks. It is well known that pituitary adenoma cells attach poorly to plastic culture flasks, form floating aggregations in the medium, and are easily washed out by medium change (1). Pituitary adenoma cells do not grow rapidly enough for subcultures, so the primary culture is maintained for a relatively long period. However, such cells are occasionally overgrown by fibroblasts, which eventually become free floating in the medium (1,2). Plastic dishes coated with extracellular matrix have been used recently for culturing of pituitary adenoma cells (3,4), but good results are not necessarily obtained using this method in our experience. Therefore, in this chapter, I describe simple techniques for obtaining an in vitro model system of human hormone-producing pituitary adenoma, which has relatively firm cell attachment and well-preserved hormonal function (5). In brief, human pituitary adenoma cells are cultured on a microporous membrane (6) coated with a basement membrane extract under the control of seeding medium volume (5).

2. Materials

1. Falcon 3091 or 3092 cell-culture inserts (Becton Dickinson Labware, Lincoln Park, NJ) with 3.0-µm pore-sized microporous membrane at the bottom, which are available for use in a six-well tissue-culture plate (Falcon 3502).
2. Matrigel (Collaborative Biomedical Products, Becton Dickinson Labware, Bedford, MA), which is a basement membrane extract of EHS mouse sarcoma

(3). This is stable for at least 9 mo when stored at −20°C. Matrigel gels rapidly at 22–35°C, so it must be thawed at 4°C overnight and kept on ice before use, and prepared by using precooled pipets, tubes, and other ware for use.
3. Coon's Modified Ham's F-12 medium (Ham's Nutrient Mixture F-12) (JRH Bioscience, Lenexa, KS). Coon's Modified Ham's F-12 medium powder is dissolved with distilled water, supplemented with 10% (or 15%) fetal bovine serum, 0.2% glucose, 100 U/mL penicillin, 100 µg/mL streptomycin, 10 mM HEPES, and 14 mM NaHCO$_3$, and buffered to pH 7.1–7.3 (complete medium). This complete medium must be stored at 4°C and incubated at 37°C when used for culture. It must be renewed after 1 mo (see Note 1).
4. Tris buffer for lysis of contaminated blood cells.
   a. Solution A: 0.83% NH$_4$Cl solution.
   b. Solution B: 20.6 g of Tris (Hydroxymethyl) amino methane is dissolved with about 700 mL of distilled water, and this is then buffered to pH 7.65 with 1N HCl solution. Distilled water is further added until the total volume is 1 L.
   Solutions A and B are mixed at a ratio of 9:1. This buffer is passed through a 0.45-µm filter unit (Millipore Products Division, Bedford, MA) for sterilization and can be stored at 4°C for up to 3 mo.
5. 0.25% Trypsin solution (0.25% trypsin in Hank's balanced salt solution) (ICN Biomedical Inc., Costa Mesa, CA).

3. Methods

3.1. Coating of Membrane Insert with Matrigel (see Note 2)

This preparation is carried out 1–2 d before use.

1. Dilute Matrigel 1:8 in cold distilled water, and vortex until Matrigel is completely dissolved. Do this at 4°C or on ice.
2. Add 400 µL (–700 µL) of the diluted Matrigel to the inside of each membrane insert in a six-well cell-culture plate.
3. Gently shake the cell-culture plate until the diluted Matrigel evenly coats the surface of the membrane insert.
4. Air-dry overnight in a laminar flow hood with the cell-culture plate cover slightly ajar.
5. Keep the coated membrane inserts sterile in the six-well cell-culture plate at room temperature until use.

3.2. Sampling of Pituitary Adenoma Tissues

1. Provide a labeled container of the medium to your clinical collaborator. A 50-mL sterile centrifuge tube with cap containing about 15 mL of complete medium is ideal.
2. If it is not possible to culture the obtained tissues soon after sampling, provide an ice pail filled with ice to keep the container at 0–4°C. A successful culture will be obtained unless the tissues are kept for more than 6 h.
3. Have an operator put surgically removed adenoma tissues into sterile physiological saline in a sterile plastic dish and wash them once to remove contaminated blood cells.

4. Transfer several pieces of the adenoma tissues to the medium in a container, and deliver them to the culture laboratory as soon as possible.

3.3. Primary Culture on Microporous Membrane

1. Transfer the all of the adenoma tissue to a 60-mm Petri dish using a pair of forceps or pipet, and mince and crush them using scalpels.
2. Add adequate 0.25% trypsin solution to the Petri dish to suspend the crushed tissues, and then pipet them several times (*see* Note 3).
3. Incubate for 20 min at 37°C, and pipet several times during the incubation.
4. After incubation, transfer to a new 50-mL centrifuge tube, and pipet rapidly about 10 times to disperse adenoma tissues into the cell suspension containing small pieces of the cell aggregate.
5. Filter the cell suspension through a cell strainer (Falcon 2350) into a new 50-mL centrifuge tube to obtain a single-cell suspension. Press all filtered off pieces with a cotton swab or a pipet tip against a mesh of the cell strainer, and rinse the pressed pieces from the cell strainer and the cotton swab with complete medium (*see* Note 4).
6. Centrifuge the filtered cell suspension into a cell pellet at 1500 rpm for 10 min. Resuspend the pellet in 5 mL of serum-free medium, and transfer to a 15-mL centrifuge tube. Then centrifuge at 1000 rpm for 5 min and repeat.
7. Add 2–3 mL of Tris buffer containing NH_4Cl to the final pellet for lysis of contaminating red cells, and resuspend this in the solution. Leave these at room temperature for 2–5 min. The solution will change to a light red wine color owing to the hemolysis. Then centrifuge at 1000 rpm for 5 min (*see* Note 5).
8. Resuspend the cell pellet in 5 mL of serum-free medium, and centrifuge at 1000 rpm for 5 min. Repeat this process to rinse the Tris buffer containing NH_4Cl thoroughly.
9. Resuspend the cell pellet in adequate complete medium to obtain the required cell density (or the required number of membrane inserts). About 5 mL of complete medium are required to obtain a density of $1-5 \times 10^5$ cells/mL (*see* Notes 6 and 7).
10. Seed about 1 mL of complete medium containing dispersed cells at a density of $1-5 \times 10^5$ cells/mL in each coated membrane insert that has been set up in advance as follows. The coated membrane insert is placed into the six-well cell-culture plate into which about 0.5 mL of complete medium has already been poured to moisten the outside of the microporous membrane so that the medium can pass through the membrane easily (Fig. 1A) (*see* Note 8).
11. Incubate the seeded cells at 37°C in an atmosphere of 5% CO_2 and 95% air. The medium added to the membrane insert will gradually pass through the microporous membrane from the inside to the outside during incubation, and consequently the condition depicted in Fig. 1B will be observed within several hours of seeding (*see* Note 9).
12. Change the medium in the well of the six-well cell-culture plate every 3–4 d. Pour about 1.5 mL of fresh complete medium into a new well of a six-well cell-culture plate, and quickly transfer the incubating membrane insert to it. If you need to

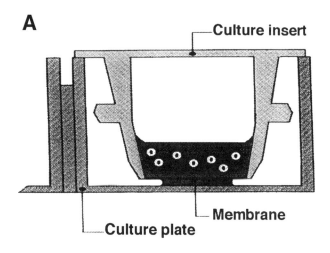

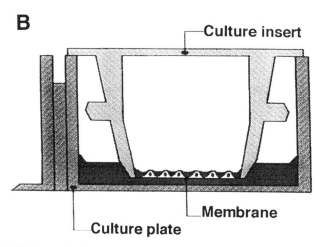

Fig. 1. (A) Schema depicting cell culture on a coated microporous membrane immediately after seeding. A membrane insert is placed into a six-well culture plate into which about 0.5 mL of complete medium has already been poured to moisten the outside of the microporous membrane so that the medium can pass through the membrane easily. About 1 mL of complete medium containing dispersed cells is seeded into the inside of the membrane insert. (B) Schema depicting cell culture on a coated microporous membrane several hours after the start of incubation. Culture medium in the membrane insert is diminished because of its passing through membrane to the outside, forcing the seeded cells to remain attached to the coated membrane both by surface tension and lack of medium buoyancy.

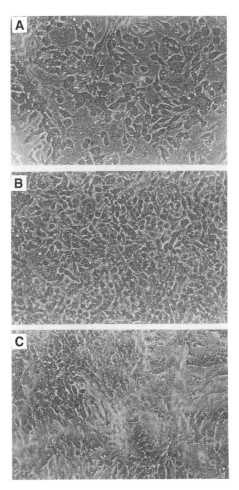

Fig. 2. Phase-contrast photomicrographs of growth hormone-secreting pituitary adenoma cells cultured on a microporous membrane coated with basement membrane extract for 2 wk **(A)**, 2 mo **(B)**, and 3 mo **(C)** (×50).

save cell culture plates, remove the used medium in the well of the six-well cell-culture plate using a Pasteur pipet, and replace it with about 1.5 mL of fresh complete medium. Do not add the medium to the inside of the membrane insert. Using this procedure, the culture condition shown in Fig. 1B can be maintained (see Note 10) (Fig. 2A).

3.4. Subculture on Microporous Membrane (see Note 11)

1. When the culture reaches confluence (Fig. 2B), replace the medium in the well of the six-well cell-culture plate with about 1.5 mL of 0.25% trypsin

solution, and add about 1.0 mL of 0.25% trypsin solution to the inside of the membrane insert.
2. Incubate at 37°C for 5–10 min or until the cells loosen from the membrane.
3. Add about 1.0 mL of complete medium to the inside of the membrane insert, and pipet up and down until all cells are freed from the membrane and a single-cell suspension is obtained.
4. Transfer to a 15-mL centrifuge tube, and centrifuge at 1500 rpm for 10 min. Resuspend the cell pellet in the serum-free medium, and centrifuge again at 1000 rpm for 5 min.
5. Resuspend the cell pellet in adequate complete medium for reseeding at a subcultivation ratio of 1:4 to 1:2. Seed 1.0 mL of complete medium containing the dispersed cells as described in step 10 of Section 3.3.
6. Renew the medium every 3–4 d as described in step 12 of Section 3.3. until confluence is reached again.

4. Notes

1. If you cannot obtain Coon's Modified Ham's F-12 medium, Ham's F-10 or 12, Dulbecco's Modified Eagle's Medium and Dulbecco's Modified Eagle's Medium/Ham's Nutrient Mixture F-12 (1:1 DME/F12) are available for this culture. The supplementation of 0.2% glucose is not necessarily needed for the medium, which already contains a sufficient quantity. The pH of the medium is very important. The optimal pH range is 7.1–7.3. If you incubate a stock bottle of the medium in a water bath repeatedly, the pH of the medium will gradually rise (especially when the air space in the bottle increases as the medium diminishes). Therefore, you must transfer an appropriate quantity of the medium from the stock bottle to 50-mL centrifuge tubes and incubate them at 37°C with a water bath whenever you use fresh medium. The osmotic pressure of the medium is also important, so you should measure the osmotic pressure of your medium once after preparation if possible.
2. Precoated cell-culture inserts (Falcon 40548) have recently become available from Becton Dickinson Labware (Lincoln Park, NJ). However, they are relatively expensive.
3. The trypsin may damage pituitary adenoma cells to a certain degree. It may be desirable to avoid the addition of the trypsin solution. Pituitary adenoma tissues generally contain only a small number of connective tissues and few junctional complexes among the adenoma cells, and are soft, so you can easily disperse minced adenoma tissues to a cell suspension composed of single cells, cell aggregates, and small tissue fragments by rapid and repeated pipeting.
4. If you do not need a single-cell suspension and an exact seeding cell density, it may be desirable to omit this filtration with a cell strainer. Better functional preservation can be obtained when many aggregated cells rather than single cells are seeded *(5)*. Cell aggregations may resemble the in vivo physiological system more closely than do cultures of single cells *(7)*. However, larger cell aggregates will be accompanied by central necrosis over the time of cultivation.

5. If the number of contaminating red cells is small, it may be desirable to omit the addition of the Tris buffer for lysis of the contaminating red cells. This solution may damage pituitary adenoma cells to a certain degree. However, many contaminating red cells also disturb the cell attachment, functions, and growth.
6. In general, compared with other kinds of cells, human pituitary adenoma cells do not grow rapidly enough for subcultures, so a primary culture must be maintained for a relatively long period. Therefore, you must prepare the number of membrane inserts required for your research before the primary culture. From the viewpoint of the hormonal function of pituitary adenoma cells, higher seeding cell density is desired because the opportunity of cell aggregation may be high (*see* Note 4). If you need many membrane inserts for your research, you should use membrane inserts available for use in 12- or 24-well cell-culture plates.
7. If you want to remove as many fibroblasts from pituitary adenoma tissue as possible, seed about 5 mL of complete medium containing the dispersed cells (the single-cell suspension) at a density of $1–5 \times 10^5$ cells/mL in a 25-cm^2 ordinary tissue-culture flask, and incubate for several days before seeding them in the membrane insert. After this, harvest the floating cells and the cells detached from the plastic flask by shaking and vigorously pipeting without trypsin, and then centrifuge at 1500 rpm for 10 min. Resuspend the cell pellet in adequate complete medium for your research as described in Note 6, and then seed the cells in membrane inserts as described in Section 3.3., step 10. By using this procedure, you can remove the fibroblasts firmly attached to plastic flask. This cell suspension will contain many aggregated cells.
8. Section 3.3., step 10 is very important and a unique point for this culture. It is important to control the quantity of the medium poured into the membrane insert. About 1 mL is appropriate for obtaining the culture condition depicted in Fig. 1B when a membrane insert for a six-well culture plate is used. With the culture conditions shown in Fig. 1B, seeded cells are forced to remain attached to the coated membrane both by surface tension and the lack of medium buoyancy, resulting in a higher rate of cell attachment. The pituitary adenoma cells will not float free in the medium even if the fibroblasts overgrow the pituitary adenoma cells in this condition.
9. If the coating of the membrane insert with Matrigel is thick, it may take the medium overnight to pass through the coated membrane as shown in Fig. 1B.
10. Changing of the medium in a laminar flow hood must be performed quickly to avoid drying of the membrane surface. If it is necessary to increase the quantity of the medium in the insert, add the required quantity of the medium to the well (the outside of the insert), causing it to pass through the membrane from the outside to the inside of the insert. This may prevent the loosely attached cells from becoming detached.
11. In general, human pituitary adenoma cells do not grow rapidly enough for subcultures, so the primary culture must be maintained for a relatively long period. It may be desirable to maintain the primary culture (Figs. 2C and 3) even if the growth of cultured adenoma cells is observed, because there is a report *(8)* stating

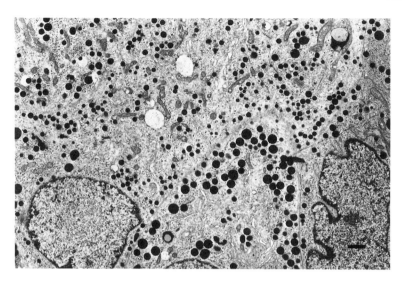

Fig. 3. Electron micrograph of growth hormone-secreting pituitary adenoma cells cultured on a coated microporous membrane for 3 mo showing numerous secretory granules which indicate well-preserved endocrinological activity (see Note 12). Bar = 1 µm.

that hormonal production continued to decline rapidly with increasing subculture number. It is generally difficult to establish cell lines with well-preserved hormonal functions in human pituitary adenoma cells (8).
12. Electron microscopic preparation of cultured cells (5,9,10) has been published, so I have omitted a description of this technique in this chapter.
13. A pituitary cell perifusion system (11) may be useful for analysis of the hormonal functions in human pituitary adenoma cells in the short term.
14. As the number of centrifugations increases, the final collection rate of the pituitary adenoma cells from the cell suspension generally decreases. Many cells may be lost when the cell collection is performed from the cell suspension, especially if it contains serum, by low-speed centrifugation. However, centrifugation at high speed may damage the pituitary adenoma cells. Therefore, when the available amount of tissue is small, it is desirable to decrease the total number of centrifugations.

References

1. Kletzky, O. A., Marrs, R. P., Rundall, T. T., Weiss, M. H., and Beierle, J. W. (1980) Monolayer and suspension culture of human prolactin-secreting pituitary adenoma. *Am. J. Obstet. Gynecol.* **138**, 660–664.
2. Kohler, P. O., Bridson, W. E., Rayford, P. L., and Kohler, S. E. (1969) Hormone production by human pituitary adenomas in culture. *Metabolism* **18**, 782–788.

3. Kleinman, H. K., McGarvey, M. L., Hassell, J. R., Star, V. L., Cannon, F. B., and Laurie, G. W. (1986) Basement membrane complexes with biological activity. *Biochemistry* **25**, 312–318.
4. Westphol, M., Jaquet, P., and Wilson, C. B. (1986) Long-term culture of human corticotropin-secreting adenomas on extracellular matrix and evaluation of serum-free condition. *Acta Neuropathol. (Berl.)* **71**, 142–149.
5. Kabuto, M., Kubota, T., Kobayashi, H., and Nakagawa, T. (1993) Modified procedure of monolayer culture of human pituitary adenoma cells on microporous membrane coated with extracellular matrix. *Neurol. Res.* **15**, 304–309.
6. Pitt, A. M., Gahriels, J. E., Badmington, F., McDowell, J., Gonzales, L., and Waugh, M. E. (1987) Cell culture on a microscopically transparent microporous membrane. *Bio-Techniques* **5**, 162–171.
7. Denef, C. and Baes, M. (1982) β-Adrenergic stimulation of prolactin release from superfused pituitary cell aggregates. *Endocrinology* **111**, 356–358.
8. Kikuchi, Y., Seki, K., Momose, E., Kizawa, I., Oomori K., Shima, K., Mukai, K., and Kato, K. (1985) Establishment and characterization of a new human cultured cell line from a prolactin-secreting pituitary adenoma. *Cancer Res.* **45**, 5722–5727.
9. Gray, M. E. and Morris, F. (1994) *Preparation of Falcon Cell Culture Inserts for Transmission Electron Microscopy.* Technical Bulletin 406, 1-3. (Available from Becton Dickinson Labware.)
10. Neumuller, J. (1990) Transmission and scanning electron microscope preparation of whole cultured cells, in *Methods in Molecular Biology, Vol. 5: Animal Cell Culture* (Pollard, J. W. and Walker, J. M., eds.), Humana Press, Clifton, NJ, Ch. 34, pp. 447–471.
11. Kotsuji, F., Winters, S. J., Keeping, H. S., Attardi, B., Oshima, H., and Troen, P. (1988) Effects of inhibin from primate sertoli cells on follicle-stimulating hormone and luteinizing hormone release by perifused rat pituitary cells. *Endocrinology* **122**, 2796–2802.

35

Preparation of Human Trophoblast Cells for Culture

Suren R. Sooranna

1. Introduction

The difficulty in obtaining a preparation of pure trophoblast cells for culture can be appreciated by understanding the structure of the placenta. The outer surface of the chorionic villi is covered by the syncytiotrophoblast, underlying which is a single cell layer of cytotrophoblast cells that sits on the basement membrane. Microvascular vessels connect this cell layer to the umbilical arteries and vein. The apical membrane of the syncytiotrophoblast is folded into numerous microvilli, and this layer forms a syncytium. Disaggregation of this villous tissue results in a broken syncytial membrane, releasing not only the cytotrophoblastic cells, but also the rest of the villous cell population (i.e., Hofbauer cells or macrophages, fibroblasts, giant cells, some adhering decidual and endothelial cells) as well as DNA from the nuclei of the syncytium. Separation of trophoblast from this heterogeneous cell population has proven to be a challenge.

Disaggregation of tissue by trypsinization is the favored method for obtaining isolated cells from the placenta for subsequent culture and was first attempted by Thiede *(1)*. Variations in the period of trypsinization and concentrations of trypsin used by several workers *(2–5)* resulted in the same three main cell types as described by Thiede (morphologically assessed as spindle-shaped, epitheloid, and multinucleated cells). The use of specific immunocytochemical markers *(6)* was a major step toward evaluating potential methods for purifying trophoblast cells for culture.

Kliman and coworkers *(7)* used a Percoll gradient for separating trophoblast cells from the mixture of placental cells obtained after trypsinization. The method described is a modification of Kliman's technique and yields more than 95% pure trophoblast cells. Subsequently, a further purification step based

From: *Methods in Molecular Medicine: Human Cell Culture Protocols*
Edited by: G. E. Jones Humana Press Inc., Totowa, NJ

on the fact that trophoblast is one of the few tissues to lack surface major histocompatibility antigens *(8)* was described by Douglas and King *(9)*. The HLA-positive contaminating cells can be removed by incubating the cell mixture with monoclonal HLA antibodies and then with immunomagnetic microspheres coated with goat antimouse IgG. Rosetted cells can be immobilized using a magnet, freeing nonrosetted trophoblast cells for subsequent culture.

2. Materials

1. PBS without calcium and magnesium salts: Prepare from tablets (Oxoid, Basingstoke, UK). Add D-glucose to a final concentration of 1 mM, and sterilize the solution by autoclaving.
2. Ham's F12 culture medium (Sigma, Poole, UK): Add 100 U/mL penicillin, 100 μg/mL streptomycin, 100 U/mL nystatin, and 15% fetal calf serum (FCS) to single-strength liquid medium.
3. 10X concentrated Hank's balanced salt solution (HBSS; Sigma) is diluted with sterile water.
4. Percoll (Pharmacia Biotech AB, Uppsala, Sweden) gradient: The gradient is made from 70 to 5% Percoll (v:v) in fourteen 5% steps of 3 mL each by dilutions of 90% Percoll (9 parts Percoll:1 part 10X HBSS) with calcium- and magnesium free HBSS and is layered using a pump at a flow rate of approx 1 mL/min. The gradient is formed in a 50-mL conical polystyrene centrifuge tube (Sterilin, Feltham, UK; *see* Note 1).
5. 1M HEPES is adjusted to pH 7.4 and filter-sterilized.
6. Bovine pancreatic deoxyribonuclease I (Sigma) is added at a concentration of approx 2000 Kunitz U/mL when needed.
7. 0.45 μm Millicell-HA (12-mm diameter) nitrocellulose filters (Millipore, Watford, UK): Rinse and preincubate with culture medium for at least 30 min prior to use.
8. 24- and 96-Well plates (Sterilin): Rinse with culture medium prior to use.
9. HLA-ABC (W6/32) and HLA-DR monoclonal antibodies (MAbs) (Dako, High Wycombe, UK) and Dynabeads M-450 coated with goat antimouse IgG (Dynal, Wirral, UK): Dilute with culture as needed.
10. Antibodies to keratin and vimentin for determination of purity of cultures (Labsystems, Basingstoke, UK).

3. Method

1. Term placenta from spontaneous vaginal delivery or Caesarean section is obtained within a few minutes of delivery (*see* Note 2).
2. The basal plate is heat sterilized with a hot scalpel blade.
3. 8–10 × 1 cm^3 pieces of tissue are removed and placed in cold PBS containing 100 U/mL penicillin, 100 μg/mL streptomycin, and 100 U/mL nystatin.
4. This is washed several times to remove as much blood as possible (*see* Note 3).

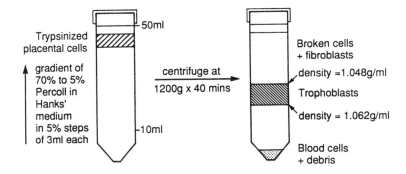

Fig. 1. Percoll gradient centrifugation of the placental cell mixture.

5. The basal plate is clearly distinguishable from the villous tissues after the washing procedure and it is removed with scissors. The placental villi are teased into pieces that are as small as possible, carefully removing any obvious blood vessels and clots (see Note 4).
6. This tissue is rewashed several times with PBS to remove blood (see Note 3).
7. Approximately 30 mL of settled tissue are made up to 75 mL with PBS, and 75 mL of HBSS are then added. Further additions are 3 mL of $1M$ HEPES, 0.4 mL of 5% EDTA (if calcium- and magnesium-free HBSS is not available), and 3 mL of 10% trypsin (see Note 5).
8. The tissue is magnetically stirred (speed 3 on a Voss HV65 stirrer) at 37°C in a conical flask for 90 min.
9. After incubation, tissue fragments are allowed to settle for 2 min, and the suspension is poured through a double layer of gauze (1.5 × 3 mm).
10. The filtrate (approx 150 mL) is then layered over 1.5 mL of FCS in each of 3 × 50 mL conical polystyrene centrifuge tubes and spun at 2500 rpm (1000g) for 10 min in a bench centrifuge (see Note 6).
11. The resultant pellets are dispersed in 2–3 mL of HBSS by repeated gentle aspiration using a 19-gage needle and a 5-mL syringe, and filtered through a metal gauze (mesh 0.5 × 0.5 mm). The pooled suspension of cells is kept on ice.
12. The remaining placental tissue from step 9 is further trypsinized by the addition of 100 mL PBS containing 1 mL of 10% trypsin for 45 min at 37°C.
13. Steps 9–11 are repeated with this second digest.
14. The resuspended pellets obtained are pooled with step 11, made up to 50 mL with HBSS, and spun at 2500 rpm (1000g) for 10 min in a benchtop centrifuge.
15. The resultant pellet is resuspended in 5 mL of HBSS and layered over the preformed 70 to 5% Percoll gradient.
16. After centrifugation (3000 rpm for 40 min on a benchtop centrifuge; 1500g), three regions are visible (Fig. 1):
    a. Top—containing connective tissue elements, small vessels, and villous fragments. If this layer is washed and cultured, it contains a high percentage of fibroblasts.

Fig. 2. Phase-contrast photograph of trophoblast cells after 30 min in 96-well plate (×150).

b. Middle (corresponding to densities of between 1.048 and 1.062 g/mL)—containing trophoblasts.
c. Bottom—containing red blood cells and some polymorphonuclear leukocytes.
17. The middle layer is passed through a disposable cell strainer with a 70-μm nylon mesh (Falcon 2350) and spun at 2500 rpm for 10 min on a benchtop centrifuge (1000g; *see* Note 7).
18. The pellet is dispersed into Ham's F12 culture medium at a concentration of 2×10^6 live cells/mL.
19. Cell viability is determined by mixing two drops of the cell suspension with two drops of 0.5% Trypan Blue and after 2–5 min, counted in a Neubauer chamber. The average number of cells that exclude the dye in each large square is multiplied by 2×10^4 to give the number of cells/mL. Those that take up the dye are dead cells (*see* Note 8).
20. 0.2 mL of cell suspension (4×10^5 cells) is dispensed into 96-well Sterilin plates and incubated at 37°C in an atmosphere of 5:95% (CO_2:air). Figures 2 and 3 show the cells after 30 min and 16 h at 37°C.
21. 0.5 mL of cell suspension (1×10^6 cells) is placed in Millicell-HA inserts for ligand transport experiments. 0.75 mL of culture medium is added to the outer chamber in 24-well plates (Figs. 4 and 5).
22. Medium should be changed on d 1, 3, 5, and 7, but cells are best used on the first or second day of culture.
23. During culture, cells are crossreacted with keratin and vimentin MAbs to determined the proportion of trophoblast cells, fibroblasts, and macrophages present. Cells in 96-well plates are fixed with methanol:acetone (1:1) for 5-min at room

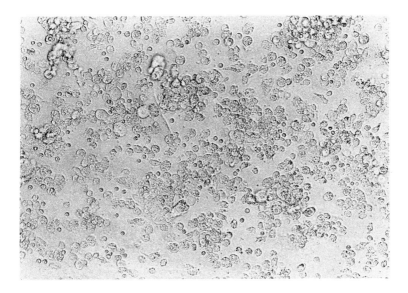

Fig. 3. Phase-contrast photograph of trophoblast cells after 16 h in 96-well plate (×150).

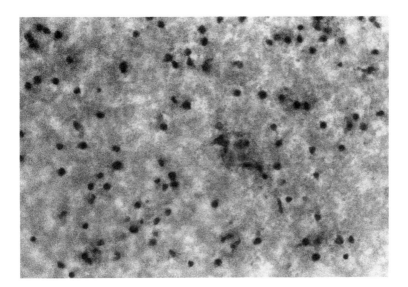

Fig. 4. Trophoblast cells on a filter stained with Hematoxylin (×260).

temperature, washed with PBS, and incubated with the different MAbs for 90 min. A further wash with PBS is followed by incubation with goat-antimouse-FITC for 90 min. Cells are then viewed and counted under fluorescence (Fig. 6; see Note 9).

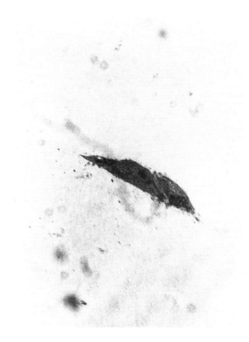

Fig. 5. Cross-section of trophoblast cells grown on a filter for 1 d and stained with toluidine blue (×660).

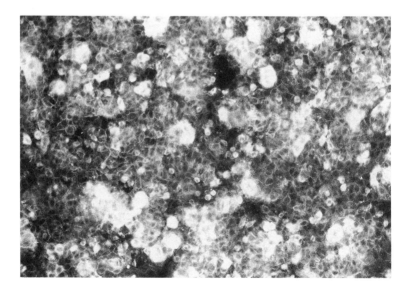

Fig. 6. Day one cultured trophoblast cells stained with keratin MAb (×150).

24. This procedure yields >95% trophoblast cells, which can be increased to more than 99% by the following procedure.
25. At step 18, cells are made up to 4×10^6 live cells/mL in ice-cold culture medium. 0.02 mL of monoclonal anti-HLA-ABC and 0.04 mL of monoclonal anti-DR antibodies are added and the mixture incubated at 4°C for 45 min with occasional gentle shaking. After spinning at 2500 rpm ($1000g$) for 10 min in a benchtop centrifuge, the pellet is resuspended in HBSS containing 1% FCS in a sterilized borosilicate tube (12×75 mm). The washing procedure is repeated twice more with HBSS containing 1% FCS. 0.25 mL of M-450 goat antimouse IgG-coated Dynabeads (a ratio of 5 beads/cell) is then added and incubated at 4°C for 30 min with occasional gentle shaking. The tube is placed on a magnet for 5 min, and the supernatant collected for steps 18–23 (*see* Note 10).

4. Notes

1. Special care must be taken in the preparation of the Percoll gradient, which when formed is stable at room temperature, but must be used within 5 d.
2. It is important to use the placenta as soon as possible because of the protease-rich nature of the tissue.
3. Washing two or three times is usually sufficient to remove most of the blood, but the tissue may require several more washes because contaminating red blood cells are a problem particularly during plating of cells.
4. After washing in PBS, the basal plate is clearly distinguished from the rest of the tissue. It has a dark brown to dark blue appearance and is much tougher than the villi; the latter can be scraped away easily.
5. The proportion of tissue to medium is crucial. Approximately 30 mL of tissue to 150 mL of medium has been found to be optimal. More tissue results in large amounts of extracellular DNA and the need to add DNase, whereas too little tissue can result in a low cell yield. If a large number of cells are needed, it is better to use two separate digests rather than alter these proportions.
6. The addition of FCS stops the action of trypsin.
7. Prior to removal of the middle layer, mark the boundaries on the outside of the tube and then carefully removed the top layer with a Pasteur pipet. A generous amount of the middle layer can be removed, but care must be taken since the lower layers contain an increasing number of red blood cells.
8. A large portion of the cell suspension (93–97%) consists of live cells, and approx $1–3 \times 10^7$ cells are obtained from 30 mL of settled placental villous tissue.
9. A typical d 1 cell population is composed of approx 95% keratin-positive cells. For most purposes, this preparation is adequate.
10. Although these steps result in a population of >99% cytokeratin-positive cells, there is a loss of recovery and viability of cells. The extra time, effort, cost, and the resultant loss of cells makes this additional procedure not worthwhile for most studies.

    Percoll gradients need to be made at least the day before cells are to be prepared. After sterilization of plastic tubing, it is probably best to form several gradients *en block* (approx 45 min each). When the placenta is obtained, the whole procedure to obtain cells will take 4–5 h without step 25. The last step will add an additional 2 h.

This procedure was also used for separation of trophoblasts from trypsinized first-trimester placental villi (obtained from 8–12 wk terminations of pregnancy). However, on d 1 of culture, only 49% of the cell population was positive to keratin antibody. This result is in agreement with Kliman and coworkers (personal communication). The lower yield of trophoblasts obtained from Percoll gradient centrifugation of early placental cells indicates their densities to be similar to the contaminating cells.

References

1. Thiede, H. A. (1960) Studies of the human trophoblast in tissue culture. I. Cultural methods and histochemical staining. *Am. J. Obstet. Gynecol.* **79,** 636–647.
2. Thiede, H. A. and Rudolph, J. H. (1961) A method for obtaining monolayer cultures of human fetal cells from term placentas. *Proc. Soc. Exp. Biol. Med.* **107,** 565–569.
3. Soma, H., Ehrmann, R. L., and Hertig, A. T. (1961) Human trophoblast in tissue culture. *Obstet. Gynecol.* **18,** 704–718.
4. Taylor, P. V. and Hancock, K. W. (1973) Viability of human trophoblast *in vitro*. *J. Obstet. Gynecol. Br. Commonw.* **80,** 834–838.
5. Roy, P. W., Ryan, G. E., and Bransome, E. D. (1976) Primary culture and maintenance of steroid-secreting human placental monolayers. *In Vitro* **12,** 115–119.
6. Contractor, S. F., Routledge, A., and Sooranna, S. R. (1984) Identification and estimation of cell types in mixed primary cell cultures of early and term human placenta. *Placenta* **3,** 41–54.
7. Kliman, H. J., Nestler, J. E., Sermasi, E., Sanger, J. M., and Strauss J. F., III (1986) Purification, characterization and *in vitro* differentiation of cytotrophoblasts from human term placentae. *Endocrinology* **118,** 1567–1582.
8. Sunderland, C. A., Naiem, M., Mason, D. Y., Redman, C. W. G., and Stirrat, G. M. (1981) The expression of major histocompatibility antigens by human chorionic villi. *J. Reprod. Immunol.* **3,** 323–329.
9. Douglas, G. C. and King, B. F. (1989) Isolation of pure villous cytotrophoblast from term human placenta using immunomagnetic microsperes. *J. Immunol. Methods* **119,** 259–268.

36

Superfused Microcarrier Cultures of BeWo Choriocarcinoma Cells

A Dynamic Model for Studies of Human Trophoblast Function

Suren R. Sooranna and Bryan M. Eaton

1. Introduction

The human placenta is a unique and still very poorly understood organ. During gestation, it is responsible for performing most synthetic and transport functions that are necessary to ensure the normal growth and development of the fetus *in utero*. Subsequently, at birth it is discarded. Current understanding of the role of the placenta in maintaining fetal homeostasis is based primarily on studies using either artificially perfused term placental lobules, placental slices or explants, or isolated plasma membrane fractions *(1–3)*. Studies with placental tissue are complicated by the presence of serial tissue layers in the maternofetal barrier, i.e., syncytiotrophoblast, cytotrophoblast, and fetal endothelium, as well as by the presence of other cell types within the placental tissue, such as the Hofbauer cells or placental macrophages.

It is now well established that most synthetic and transport activity is restricted to the trophoblastic tissue *(4,5)*. Although syncytiotrophoblast plasma membrane fractions are of obvious use in characterizing receptors and transport systems (*see* Eaton and Contractor *[6]* for review), they are unable to provide information on trophoblast metabolism. Since the mature trophoblast exists as a true syncytium, it cannot be isolated intact from placental tissue. The syncytiotrophoblast is believed to be formed in vivo by fusion of cytotrophoblast cells. These cytotrophoblast cells are still present in substantial quantities even at term and can now be isolated with a high degree of purity using gradient techniques (*see* Chapter 35). Unfortunately, cytotrophoblasts are extremely difficult to grow in culture and have a limited life-span. Although they will fuse to form multinucleated cells, they will not form large sheets of syncytial trophoblast. In order to overcome the limitations imposed by the use of normal cytotrophoblast, several investigators have turned to the use of human choriocarcinoma cell lines, of which the BeWo line has been the most extensively characterized *(7)*. In culture, these cells grow, divide, and fuse to form large syncytial masses with many characteristics of normal syncytiotrophoblast *(8,9)*. The rapid growth of these cells in culture and their ability to fuse and form syncytia make them ideal cells to culture on microcarrier beads. Microcarrier culture of anchorage-dependent cells has been routinely employed for many years, because this culture procedure results in a very large surface area to volume ratio and hence large numbers of cells can be cultured economically in small volumes. The methodology has been also been employed when harvesting of cell products is required. We here report a technique for growing BeWo cells on microcarrier beads that provides a new, dynamic in vitro system for studying trophoblast function without the complications associated with the presence of other cell types. An advantage to growing cells on microcarrier beads is that the beads can be packed into columns and superfused, thus allowing sequential studies to be carried out on cell-surface receptors, transport systems, hormone release, and so forth. The use of rapid, single-passage tracer techniques *(10)* permits measurements to be made under control conditions and also allows the effects of varying substrate concentrations, inhibitors, hormones, or other agents to be determined **in the same preparation**.

2. Materials

1. PBS without calcium and magnesium salts: Prepare from tablets (Oxoid, Basingstoke, UK). Add D-glucose to a final concentration of 1 mM and sterilize the solution by autoclaving.
2. PBS containing a final concentration of 0.25% trypsin and 0.02% EDTA: Prepare by diluting stocks of 10% trypsin in PBS and 5% EDTA in PBS, using PBS also prepared from tablets.
3. Ham's F12 culture medium (Sigma, Poole, UK): Add 100 U/mL penicillin, 100 μg/mL streptomycin, and 10% fetal calf serum (FCS) to single-strength liquid medium.
4. BeWo cells are purchased from the European Collection of Animal Cell Cultures (Porton Down, Salisbury, UK, ECACC No. 86082803). After culture, BeWo cells are frozen in Ham's F12 medium containing 20% FCS and 20% dimethyl sulfoxide (DMSO) Grade I (Sigma).
5. Ninety-six-well plates (Sterilin, Feltham, UK) are rinsed with culture medium prior to use. Seventy-five-milliliter culture flasks (Falcon, Becton Dickinson Ltd., Cowley, Oxford, UK) are also rinsed before use.
6. Cytodex-3 microcarrier beads (crosslinked dextran covered with covalently linked denatured collagen, diameter range: 130–215 μm) and Biosilon microcarrier beads (solid plastic coated with gelatin, diameter range: 150–210 μm) are

purchased from Pharmacia (Uppsala, Sweden) and Sigma, respectively. One gram of Cytodex is rehydrated according to the manufacturer's instructions and autoclaved. One gram of beads swells to 14 mL on rehydration. One milliliter of solid Biosilon beads (which do not swell) is mixed with an equivalent volume of PBS and autoclaved. Prior to use, beads are rinsed twice with approx 10 mL of culture medium.
7. Techne stirrer flasks (125 mL; Techne, Duxford, Cambridge, UK) are siliconized, washed, and autoclaved. Prior to use, flasks are rinsed twice with about 20 mL of PBS, followed by two additional washes with culture medium.
8. Fluorochrome-labeled antibodies (Miles Laboratories, Stoke Poges, Bucks, UK).
9. Monoclonal antibody (MAb) to keratin (Labsystems, Basingstoke, UK). MAb to trophoblasts is from our laboratory *(11)*.

3. Method

1. Thawed aliquots of BeWo cells are initially grown to near confluence in 75-mL flasks in Ham's F12 medium containing FCS and antibiotics (*see* Note 1).
2. When the cells in the 75-mL flasks have reached about 80% confluence (about 7 d), the culture medium is removed and the cells are washed twice with 20 mL PBS (at 37°C). Ten milliliters of PBS/trypsin/EDTA are added and incubated with the cells for 1 min at room temperature. The trypsin/EDTA solution is removed by aspiration and replaced with 10 mL PBS. Cells are continually monitored by phase-contrast microscopy, and after approx 5 min, substantial numbers of cells will be seen to have detached from the surface of the flask into the medium. The flask can be tapped vigorously 2–3 times to accelerate the release of cells (*see* Note 2).
3. The cell suspension is transferred into a 10-mL polystyrene centrifuge tube and spun at 2500 rpm (1000g) for 10 min in a benchtop centrifuge. A further 10 mL of PBS are added to the 75-mL flask and left for 10 min, with occasional vigorous tapping to release the remaining cells. This cell suspension is similarly centrifuged, and the two cell pellets are combined and resuspended in 4 mL of Ham's F12 medium.
4. Cell viability can be determined by mixing two drops of the cell suspension with two drops of 0.5% Trypan Blue in PBS, and after 2–5 min, cells are counted in a Neubauer chamber. The average number of cells that exclude the dye in each large square is multiplied by 2×10^4 to give the number of cells/mL. Cells that take up the dye are dead and should be excluded from the count. A 75-mL flask of confluent cells will provide approx 1×10^7 cells. The cell concentration should the be adjusted to 3×10^6 cells/mL with Ham's F12 medium.
5. One-milliliter aliquots of this cell suspension are mixed with 1 mL of Ham's F12/FCS/DMSO medium in a 2-mL cryovial (Camlab, Cambridge, UK). Vials are cooled over liquid nitrogen vapor at a rate of 1°C/min over a period of approx 5 h and are stored in liquid nitrogen (*see* Note 3).
6. For microcarrier culture, one vial (containing approx 3×10^6 cells) is rapidly thawed, over a period of 5 min in a water bath (37°C), keeping the vial in con-

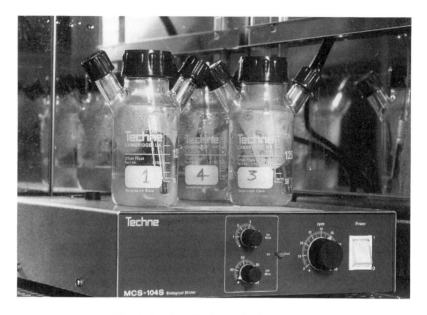

Fig. 1. Techne flasks and stirrer base.

stant motion. The cell suspension is transferred into a 10-mL polystyrene centrifuge tube, 8 mL of PBS are added, and the tube spun at 2500 rpm (1000g) for 10 min in a benchtop centrifuge.

7. The resulting cell pellet is resuspended in 2 mL of Ham's F12 medium, and the cells are gently dispersed by aspiration. The cell suspension is then transferred into a 75-mL flask, and 20 mL of culture medium added. The flask is incubated (with the cap loosened) at 37°C in an atmosphere of 5% CO_2:95% air (see Note 4).

8. The culture medium is changed after 24, 48, and 72 h, at which stage the cells are almost confluent and can be trypsinized as described in steps 2 and 3. The resulting cell pellet (about 8×10^6 cells) is dispersed in 4 mL of culture medium.

9. Forty milliliters of Ham's F12 culture medium are added to prewashed siliconized Techne flasks (Fig. 1) together with 1 mL of microcarrier beads (cell-to-bead ratio of about 8:1) and the medium stirred at 20 rpm for approx 30 min in the incubator to allow equilibration. After this period, 2 mL of the cell suspension are added to the flask and the stirring rate adjusted to 40 rpm for 5 min every 10 min (see Notes 5 and 6).

10. After 24 h, half the medium is removed by decanting and replaced with fresh medium. The stirring conditions are then changed to 60 rpm for 5 min every 5 min.

11. A further 48 h later, half of the medium is again replaced, and the cells are then stirred continuously at 50 rpm. Half the medium is subsequently replaced at 24-h intervals for a further 3–4 d by which time the cells are ready for use.

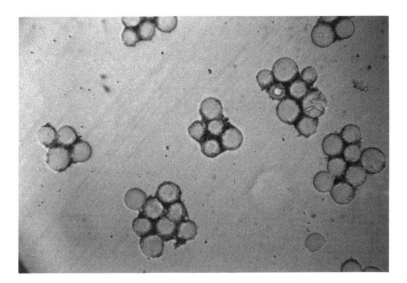

Fig. 2. Phase-contrast photograph of cells growing on microcarrier bead (magnification ×60).

12. At this stage, a large number of aggregates of between two and six beads, with a covering of multinucleated syncytium (Fig. 2) are present. A sample is fixed and examined by light microscopy, after staining with Giemsa stain (Fig. 3) as follows: 0.5 mL of the cell suspension from the stirrer flask is placed in a 1.5-mL Eppendorf tube and spun for 30 s at 4000g in a bench-top microfuge. The supernatant is removed by aspiration, and the cells are fixed for 10 min with 70% methanol. The pellet is washed twice with PBS and incubated with Giemsa stain (Raymond Lamb Lab., North Acton, UK) for 2 min and then washed with PBS at least four times. Some of the stained beads can then be placed on a glass microscope slide, air-dried, dipped in xylol, and mounted with styrolite mounting medium (Raymond Lamb Lab.). Cell samples can also be processed and examined by transmission electron microscopy (Fig. 4, p. 471). For the latter procedure, 0.5 mL of bead suspension from the stirrer flask is placed into a 1.5-mL Eppendorf tube and spun at 4000g for 30 s in a benchtop microfuge. The medium is removed by aspiration, and the cells fixed with 2.5% glutaraldehyde in 0.1M sodium cacodylate buffer, pH 7.4, for 2 h. After two washes with cacodylate buffer, the sample is postfixed in 2% osmium tetroxide for 30 min. Following a further two washes with distilled water, cells are incubated with 2% uranyl acetate for 45 min and then dehydrated in graded alcohols and embedded overnight in Spurr's resin (*see* Note 6).
13. The two- to six-bead aggregates can be readily packed into small columns for superfusion (Fig. 5, p. 472). Superfusion is carried out in a thermostatically controlled hot-box at 37°C. Three-millimeter diameter circles of nylon gauze

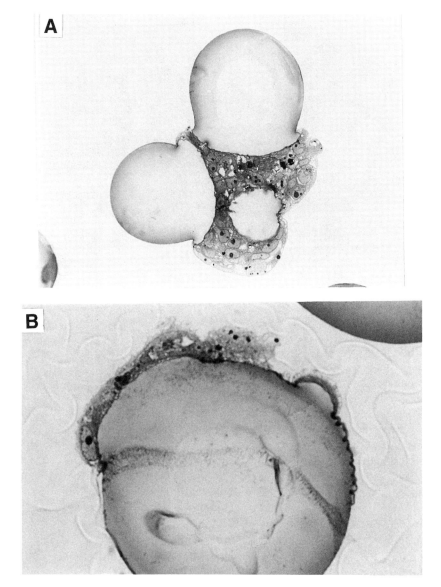

Fig. 3. Giemsa-stained cells on microcarrier beads. Magnification **(A)** ×150 and **(B)** ×660.

(0.1-mm mesh size) are prepared and one is placed to seal off the Luer tip of a 1-mL syringe barrel (Monoject). The mesh is wetted with medium, and the microcarrier bead-cell aggregates are then packed into the syringe barrel, using a Pasteur pipet, to a volume of 0.3–0.5 mL. Another circle of gauze is fixed to the tip of the

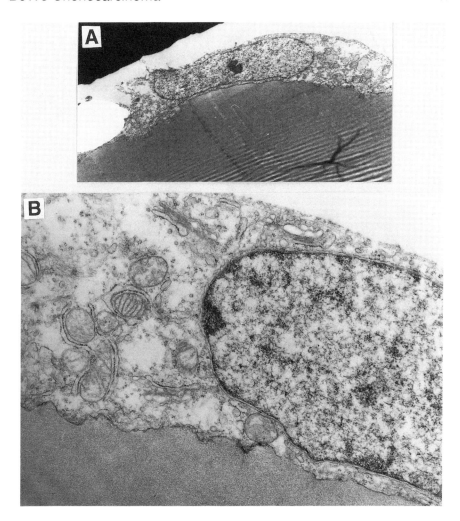

Fig. 4. Electron micrograph of BeWo cells on microcarrier beads. Magnification (**A**) ×2700 and (**B**) ×16,000.

    syringe plunger, which is then gently lowered to the top of the packed beads. Plastic tubing (diameter 0.5 mm) inserted through the rubber tip of the plunger is connected to a pump. The column can be superfused from below at flow rates of 0.3–1.2 mL/min with the perfusate being drawn through the column by a peristaltic pump (Multipuls 3, Gilson, Anachem; *see* Note 7). The column effluent can be collected for assay either by hand or by using a fraction collector.
14. BeWo cells can also be grown in 96-well plates if required, since this is more convenient for studies involving, for example, immunofluorescence (Fig 6). Following trypsinization (steps 2 and 3), 0.2 mL of a cell suspension (adjusted to

Fig. 5. Superfusion of microcarrier cultures in a column.

contain 5×10^5 cells/mL) is dispensed into each well and incubated at 37°C in an atmosphere of 5% CO_2:95% air. Medium should be changed every 24 h. When the cells reach confluence (about 4 d), they are fixed with methanol:acetone (1:1) for 5 min at room temperature, washed with PBS, and incubated with a mouse MAb for 90 min. A further wash with PBS is followed by incubation with goat antimouse FITC-labeled secondary antibody for 90 min. Cells are then viewed and counted using a fluorescence microscope.

Preliminary data obtained using the single-passage paired-tracer technique in superfused microcarrier cultures have shown the binding of iron-saturated, but not iron-free transferrin (measured relative to a column void volume marker, ^{131}I-IgG) *(12)*, suggesting that these cells behave like normal syncytiotrophoblast *(13)*.

4. Notes

1. Fungizone® or Nystatin is not needed in the culture medium. If present, the cultures tend to show decreased cell viability and growth.
2. A large number of the "multinucleate masses" separate into single cells following trypsinization, and it is these cells that will attach and grow.

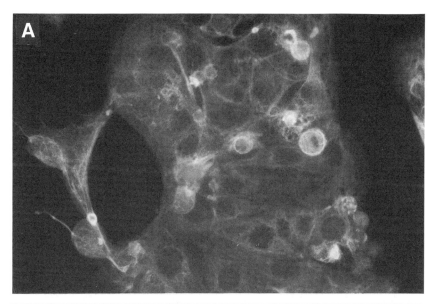
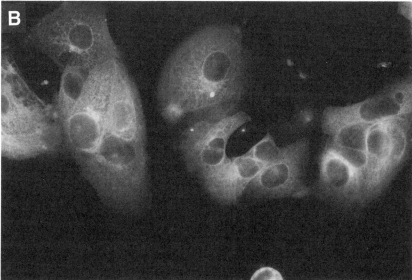

Fig. 6. Immunofluorescence staining of BeWo cultures grown in 96-well plates using (A) keratin MAb and (B) trophoblast-specific MAb. Magnification ×150.

3. BeWo cells cannot be stored at −70°C for longer than 24 h. They have to be kept in liquid nitrogen; otherwise they will be permanently damaged.
4. The culture medium rapidly becomes acidic (as indicated by the yellow color imparted by the pH indicator dye). This, however, reflects healthy growth of the

BeWo cells. The medium must therefore be partially replaced every 24 h to compensate for this effect.
5. The conditions for stirring the microcarrier cultures (Section 3., steps 9–11) are important to ensure that the cells have an opportunity to attach to the beads at an early stage in the microcarrier culture, but are prevented from forming overly large bead-cell aggregates.
6. The same culture conditions apply to both types of microcarrier beads. The techniques for light and electron microscopy also apply to both types of beads. However, cutting thin sections of the plastic Biosilon beads for the electron microscope resulted in many of the beads shattering, and hence, the Cytodex beads were preferred for EM studies.
7. When superfusing columns of bead-cell aggregates, the perfusate should be drawn through the column by the pump rather than pumped through to prevent compacting the aggregates.

Acknowledgment

This study was funded by the Wellcome Trust.

References

1. Contractor, S. F., Eaton, B. M., Firth, J. A., and Bauman, K. F. (1984) A comparison of the effects of different perfusion regimes on the structure of the isolated human placental lobule. *Cell Tiss. Res.* **237**, 609–617.
2. Contractor, S. F. and Sooranna, S. R. (1993) *In vitro* assessment of trophoblast receptors and placental transport mechanisms, in *The Human Placenta. A Guide for Clinicians and Scientists* (Redman, C. W. G., Sargent, I. L., and Starkey, P. M., eds.), Blackwell Scientific Publication, Oxford, UK, pp. 504–526.
3. Eaton, B. M. and Oakey, M. P. (1994) Sequential preparation of highly purified microvillous and basal syncytiotrophoblast membranes in substantial from a single term human placenta: inhibition of microvillous alkaline phosphatase activity by EDTA. *Biochim. Biophys. Acta* **1193**, 85–92.
4. Sibley, C. P. and Boyd, R. D. H. (1988) Control of transfer across the mature placenta, in *Oxford Reviews of Reproductive Biology*, vol. 10 (Clarke, J. R., ed.), Oxford University Press, Oxford, UK, pp. 382–435.
5. Dearden, L. and Ockleford, C. D. (1983) Structure of the human trophoblast: correlation with function, in *Biology of the Trophoblast* (Loke, Y. W. and Whyte, A., eds.), Elsevier, Amsterdam, pp. 69–110.
6. Eaton, B. M. and Contractor, S. F. (1993) *In vitro* assessment of trophoblast receptors and placental transport mechanisms, in *The Human Placenta. A Guide for Clinicians and Scientists* (Redman, C. W. G., Sargent, I. L., and Starkey, P. M., eds.), Blackwell Scientific Publication, Oxford, pp. 471–503.
7. Pattillo, R. O. and Gey, G. O. (1968) The establishment of a cell line of human synthesising cells *in vitro*. *Cancer Res.* **28**, 1231–1236.
8. van der Ende, A., du Maine, A., Simmons, C. F., Schwartz, A. L., and Strous, G. J. (1987) Iron metabolism in BeWo chorion carcinoma cells. *J. Biol. Chem.* **262**, 8910–8916.

9. Wice, B., Menton, D., Geuze, H., and Schwartz, A. L. (1990) Modulators of cyclic AMP metabolism induces syncytiotrophoblast formation *in vitro*. *Exp. Cell Res.* **186,** 306–316.
10. Eaton, B. M., Toothill, V. J., Davies, H. A., Pearson, J. D., and Mann, G. E. (1991) Permeability of human venous endothelial cell monolayers perfused in microcarrier cultures: effects of flow rate, thrombin, and cytochalasin D. *J. Cell. Physiol.* **149,** 88–99.
11. Contractor, S. F. and Sooranna, S. R. (1986) Monoclonal antibodies to the cytotrophoblast and syncytiotrophoblast of human placenta. *J. Dev. Physiol.* **8,** 277–282.
12. Sooranna, S. R. and Eaton, B. M. (1994) Perfused microcarrier cultures of choriocarcinoma cells: a dynamic model for studying endocytosis of macromolecules. *Placenta* **15,** A66.
13. Contractor, S. F. and Eaton, B. M. (1986) Role of transferrin in iron transport between maternal and fetal circulations of a perfused lobule of human placenta. *Cell Biochem. Funct.* **4,** 6974.

37

Corneal Organ Culture

Robert W. Lambert, Normand R. Richard, and Janet A. Anderson

1. Introduction

Long-term organ culture of human corneas offers a tool for conducting a variety of biochemical, histological, and wound-healing studies on human tissue in vitro. A requirement of any organ culture system is that cell differentiation and structural integrity of the tissue be maintained or even improved during culture. The latter is particularly relevant in corneal organ culture, since available donor corneas are sometimes of poor quality. These corneas are usually unsuitable for transplantation and are from older donors (>65 yr), from donors with known bacterial or viral infections at the time of death (e.g., septicemia, hepatitis), or are rejected because of a low endothelial cell count, such as might occur following ocular surgery.

Previous corneal organ culture techniques have described a procedure whereby corneal tissue is fully immersed in culture medium, with the epithelial side down *(1)*. Using this technique, tissue can be maintained in culture for up to 21 d *(2)*. However, structural integrity appears to be compromised, as evidenced by a reduction in the thickness of the epithelial cell layer and loss of cellular differentiation, the presence of epithelial and stromal edema, keratocyte deterioration, and thinning of the endothelium *(3,4)*.

The air/liquid culture technique described in this chapter differs from the immersion procedure described by keeping the corneal epithelium up and alternately exposed to air and liquid, while the endothelium has continuous access to the culture medium. The technique is a modification of a procedure described by Trowell *(5)*. The cornea is placed on a circular retainer, the Richard ring, which is immobilized to the bottom of a culture dish. Culture medium is added, and the dish is then placed on a rocker platform in the incubating chamber. The culture dish is gently rocked at 10 cycles/min, so that at the

From: *Methods in Molecular Medicine: Human Cell Culture Protocols*
Edited by: G. E. Jones Humana Press Inc., Totowa, NJ

highest point of the upswing, the cornea is exposed to air, and at the lowest point, it is fully immersed in culture medium. This episodic exposure of the corneal epithelium to an air interface mimics the in vivo situation in which the oxygen supply to the cornea is continuously replenished. The medium is changed every 24–48 h, depending on the experimental protocol.

Using this technique, Richard et al. demonstrated superior ultrastructural morphology after 3 wk in culture *(6)*. When compared to previous culture techniques, reduced epithelial and stromal edema and improvements in the endothelial morphology were observed. The model has been used to examine corneal wound healing after epithelial debridement *(7)* and stromal wounds, such as keratectomy, excimer laser photoablation *(8,9)*, and thermal burns *(10)*.

Using the air/liquid culture technique, we found epithelial wound-healing rates were comparable to those reported in animal models *(7)*. The model has also been used to examine the effects of pharmaceutical compounds on corneal wound-healing responses in tissue cultured in serum-free medium *(7)*, and to examine the nutritional requirements of the human cornea for vitamin A *(11)* (*see* Note 1). The air/liquid organ culture model provides a valuable tool for experimental manipulation of donor tissue. The culture model and techniques are continuously undergoing modification and improvement.

2. Materials
2.1. Tissue Procurement

1. Eye banks: One of the functions of an eye bank is to provide corneas for transplant surgery. Corneas are procured from donor eyes at the time of death. Since not all donor corneas are suitable for transplant, tissues are usually available for research. Inquire from eye banks in your region about their procedures for providing research tissues. You may be asked to submit a description of your research needs, so that eye bank personnel will be alerted to call you when suitable tissue is available. There are usually nominal charges associated with tissue handling and administrative costs.

    Patient information provided by the eye bank should include the location of the eye (right or left), the donor age and sex, the cause of death, the date and time of death, and the time at which the enucleation was performed by the eye bank technician. A list of drugs the donor was taking just prior to death may also be available. The eye bank technician can determine whether the donor has had previous ocular surgery by examining the eye and from medical histories if they are available.

    The integrity of the cornea, including the degree of epithelial attachment, the condition of the endothelium, and the presence or absence of corneal scarring, will usually be determined by eye bank personnel prior to releasing the tissue for either transplant or research purposes. There are no current routinely used tests

for determining corneal viability. Serology is usually performed on the blood of donors whose corneal tissue is earmarked for transplantation to ascertain HIV and HBV. These tests are usually not performed for research tissues because of the costs involved; however, they can be requested by the researcher. Corneal tissue may be provided in a whole globe or the cornea may have already been dissected from the globe.
2. Other sources of research tissues: Another source of research tissues is from ophthalmic surgeons, who may have access to pathologic corneal tissues during transplant procedures. Tissue obtained from this source generally requires the donor to sign a consent form and the permission of the internal review board of the institution where the surgery is performed. Obtaining permission of the internal review board may require a written submission to the board and can take several months for approval, depending on how often the board meets.

2.2. Tissue Handling

1. Tissue transport: Whole globes are usually transported from eye banks in sealed containers in a moist environment. Excised corneas are immersed in sterile corneal storage solutions, such as Optisol® or Dexsol® (Chiron IntraOptics, Irvine, CA). Tissues should be transported at 4°C in a container cooled with ice; however, tissues should not be in direct contact with the ice. Express mail services may be used to transport tissues. In the United States, several airline carriers will deliver human tissue at no charge if it is delivered to the point of embarkation and collected when it arrives. This is a quick, minimal cost alternative and is useful during weekends and vacation time when personnel are not usually in the laboratory to receive and process packages.
2. Evaluation: Independent evaluation of the corneal tissue should be performed by the researcher after the tissue has been received from the eye bank. The researcher should be familiar with the use of universal precautions for handling potentially biohazardous tissues. These include the wearing of a disposable face mask, gloves, and either a washable or disposable gown. Immunization for HBV antigen is strongly recommended for all researchers in contact with human tissues.

  Characteristics of the corneal surface should be examined under the dissecting microscope. Depending on the requirements of the study, damaged or irregularly attached epithelium may not be acceptable. Scarring and edema in the stromal tissue may be evident to the naked eye as irregular opacities, but a more thorough examination under the dissecting microscope is recommended. Corneas with these characteristics are suggestive of ocular disease (e.g., ocular herpes, pseudophakic bullous keratopathy) or improper handling, and should be eliminated from most studies. Death-to-utilization time is also an important factor to achieve optimal corneal culture. Human corneal tissue that is in a whole globe and not excised or immersed in a preservation solution is best used within 72 h of death of the donor. After 72 h, corneal morphology and function may be compromised. Excised corneas stored in a corneal preservation solution such as Optisol or Dexsol, may be used up to 7 d after donor death.

Donor sepsis at the time of death and some causes of death may preclude using tissues for organ culture. The experience of our laboratory has shown that corneas that have been immersed in fluids other than preservation medium, for example, tissue from drowning victims, show poor response in tissue culture. Also we have observed corneal edema and poor epithelial attachment in corneas from donors that have been comatose for a period of time before death. In addition, neurological disorders, such as Parkinson's disease or Alzheimer's disease, which usually disqualify a donor's corneas for transplantation, may also make them unsuitable for your requirements. Drugs given to a donor just prior to death may pose some problems in corneal organ culture. Careful consideration should be give to these aspects, since corneal function might be compromised.

3. Recording information: Maintaining accurate and up-to-date donor records is an essential component of long-term organ culture studies. The date of receipt, source of the cornea, age and sex of the donor, and cause of death should all be permanently recorded in a bound laboratory notebook. After the cornea has been examined and accepted into a study, it should be given a unique laboratory number. The original paperwork from the eye bank or surgeon who has donated the tissue is marked with this number and retained in a separate loose leaf notebook that can be readily accessed.

4. Storage prior to organ culture: Whole globes and corneas are stored in the containers in which they are received at 4°C. After removal of the tissues, the containers and their remaining contents should be autoclaved to eliminate any potential biohazard, and returned to the eye bank. An area of the refrigerator should be dedicated for storing biohazardous materials. Tissues should be stored away from toxic chemicals and fixatives. Since most experimental protocols require use of the corneas within 72 h following death or surgical removal, storage times will be limited.

All biological tissues removed from either the donor whole globe or the dissected cornea that are not required for the study should be placed in a "morgue" for subsequent disposal. A simple morgue may consist of an appropriately labeled plastic pail with an air-tight lid containing between 1 and 1.5 L of a 10% formalin solution. The morgue should be kept at 4°C and should only be opened in a fume hood because of the toxic nature of the formalin. The contents should be disposed of as biohazardous waste when the container is full.

2.3. Equipment and Supplies

1. Storage and organ culture: The equipment required for tissue storage and organ culture are as follows: steam autoclave, 4°C refrigerator, laminar flow hood containing access to vacuum and gas, binocular dissecting microscope, 37°C incubator, and a rocker platform. To minimize risk of contamination and to maintain sterility, disposable, plastic culture supplies are recommended. These include culture dishes (60 × 15 mm), 10- and 25-mL pipets, sterile gauze squares, 5-cc syringes, and 0.20-µm filters. Polysulfone 15 × 3-mm Richard rings (discussed in step 2) are available from J. M. Specialities (San Diego, CA).

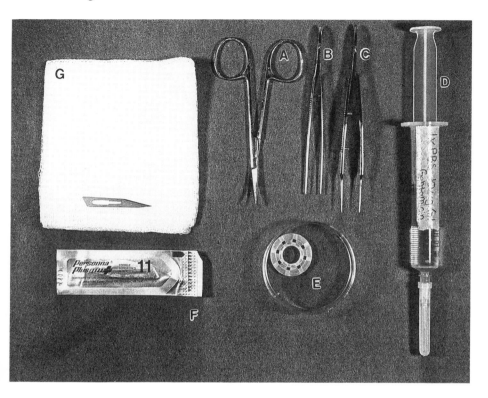

Fig. 1. Surgical instruments required for corneal excision and air/liquid organ culture include **(A)** 4¾-in. curved corneal scissors with sharp 30-mm blades; **(B)** 4-in. straight, microdissecting forceps, serrated 0.8-mm wide tips; **(C)** 4-in. straight, microdissecting forceps, 1 × 2 teeth, 0.8-mm wide tips; **(D)** 1X PBS with 10 µg/mL gentamicin; **(E)** Richard ring and 60 × 15 mm disposable culture dish; **(F)** #11 scalpel blade (Personna, American Safety Razor Co., Staunton, VA); and **(G)** sterile 4 × 4 in. cotton gauze.

Instruments required for corneal dissection and handling should be sterile prior to use (*see* Note 2). These are shown in Fig. 1 and include:
a. 4¾-in. curved corneal scissors with sharp 30-mm blades.
b. 4-in. straight, microdissecting forceps, serrated 0.8-mm wide tips.
c. 4-in. straight, microdissecting forceps, 1 × 2 teeth, 0.8-mm wide tips.
d. Plastic syringe containing 1X PBS with 10 µg/mL gentamicin.
e. Richard ring and 60 × 15 mm disposable culture dish.
f. #11 Scalpel blade (Personna, American Safety Razor Co., Staunton, VA).
g. Sterile 4 × 4 in. cotton gauze.

Forceps and scissors can be obtained from Biomedical Research Instruments Inc. (Rockville, MD). Comparable instruments can usually be obtained from most

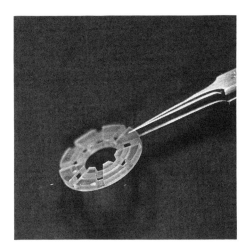

Fig. 2. Corneas are retained on the bottom of the culture dish using Richard rings. Richard rings are constructed from two concentric polysulfone rings joined by radial struts cut into the body of both rings. The struts form channels that allow the culture medium access to the endothelial side of the cornea. The corneal-scleral rim sits in a moat that holds the cornea onto the retainer.

surgical instrument companies. While handling corneal tissue, all instruments can be repeatedly flame sterilized using a solution of 95% ethanol. However, care should be taken using an open flame in the laminar flow hood.
2. Richard rings: Richard rings were designed at the National Vision Research Institute to support and hold donor corneas during organ culture. The rings are produced from polysulfone and can be autoclaved for reuse. They consist of two concentric rings joined by radial struts cut into the body of the rings to form channels. These channels are designed to allow culture medium to flow under the cornea during culture (Fig. 2). The outer ring is approx 21 mm in diameter and is separated from the inner ring by a 2-mm moat. The inner ring is open in the center to allow culture medium unrestricted access to the corneal endothelium. The rim of the cornea is fitted into the moat between the two rings to retain the tissue and avoid damage from tissue movement (Fig. 3). The Richard ring is fixed to the bottom of the culture dish adjacent to one wall using sterile molten paraffin.

2.4. Culture Medium and Buffers

All solutions are prepared under sterile conditions and should be kept at 4°C unless otherwise stated. Culture medium can be prepared and stored for 2–3 mo. However, retinol acetate and a protein carrier, such as bovine serum albumin (BSA), should be added to the medium just prior to using. Because retinoids are light-sensitive, culture dishes and medium containers should be kept

Fig. 3. Top view of a human donor cornea that has been positioned on a Richard ring.

covered in aluminum foil, and culture procedures should be conducted at low levels of light.

1. Serum-free culture medium: The use of serum or serum-free medium is dependent on the requirements of a particular study (see Note 3). The serum-free culture medium consists of a mixture of Dulbecco's modified Eagle's medium (DMEM) and Ham's F12 culture medium (1:1, Sigma, St. Louis, MO) with additional supplements. The medium is prepared by dissolving 15.6 g of the DMEM/F12 culture medium (Sigma) in 900 mL of distilled deionized water. While stirring, add 2.2 g sodium bicarbonate and 10 g dextran 40 (Sigma). Bring to 1000-mL vol, and filter sterilize, on ice, through a 0.20-μm pore size membrane filter (Fisher Scientific, Pittsburgh, PA). After filtration, add 13.5 g of chondroitin sulfate C (mixed isomer, Sigma) to 900 mL of the solution. Vigorously mix until 80% of the chondroitin sulfate is in solution. Using a magnetic stir bar and plate, dissolve the remaining chondroitin sulfate. Adjust pH to 7.1 with $1N$ NaOH or $1N$ HCl, and bring up the volume to 1000 mL with remaining sterilized medium. Resterilize the medium at 4°C using a 0.20-μm filter.

After sterilization, supplement the medium with 20 nM progesterone, 20 nM hydrocortisone, 100 μM putrescine, 30 nM selenium, 5 μg/mL transferrin, 5 μg/mL insulin, and 10 ng/mL epidermal growth factor (EGF). These chemicals may be obtained from Sigma. In addition, the medium is further supplemented with 100 U/mL/100 μg/mL penicillin–streptomycin (Flow Laboratories, McLean, VA), 2.5 μg/mL fungizone, and 10 μg/mL gentamicin (Irvine Scientific, Irvine, CA). Aliquot the supplemented medium into sterile screw-capped 100-mL plastic containers, flame sterilize the neck of the containers and the interior of the caps before resealing, and store at 4°C.

Prior to use, warm the culture medium to 37°C, and supplement with retinol acetate (vitamin A) and BSA. To supplement the medium with vitamin A, place 100 mg (2.9×10^6 USP/g, Sigma, stored at $-20°C$) into 10 mL absolute ethanol in a foil-wrapped centrifuge tube. Vortex until dissolved, and use this stock solution the same day as prepared to make a final dilution of $5 \times 10^{-9} M$ depending on the requirements of the culture *(11)*. To supplement the medium with 0.05% BSA, dissolve 50 mg in a 10-mL aliquot of the culture medium. Filter sterilize the dilution through a 0.20-µm filter into 100 mL of the culture medium.

2. Serum-plus culture medium (10% FCS): Serum-plus culture medium consists of DMEM supplemented with either 10% fetal calf serum (Irvine Scientific Co., Irvine, CA), 10% human serum (Irvine Scientific Co.) or human plasma in 4% sodium citrate solution (1:9 dilution, Interstate Blood Bank Inc. Memphis, TN).

   The DMEM medium is prepared by dissolving 15.6 g of the powdered premixed culture medium in 900 mL of distilled deionized water. While stirring, add 2.2 g sodium bicarbonate and 10 g dextran 40. Bring to 1000-mL vol, and filter sterilize, on ice, through a 0.20-µm pore size membrane filter. The addition of chondroitin sulfate, BSA, vitamin A, or dextran is not required in this medium. However, 100 U/mL penicillin, 100 µg/mL streptomycin, 2.5 µg/mL fungizone, and 50 µg/mL gentamycin should be added.

   Medium can be stored at 4°C for up to 6 mo after preparation. Prior to use in culture, the medium should be brought to 37°C.

3. Phosphate-buffered saline (10X PBS, stock solution): 2.0 g KCl, 2.0 g KH_2PO_4, 80 g NaCl, 11.5 g Na_2HPO_4. Dissolve ingredients in 900 mL distilled water, adjust volume to 1000 mL, and filter sterilize using a 0.20-µm pore membrane filter. Store the stock solution at room temperature in sterile 500-mL resealable containers. To prepare a working solution (1X PBS), dilute the stock solution 1:9 in sterile distilled, deionized water. Gentamicin (10 µg/mL) can be added to the working solution of 1X PBS for rinsing corneal tissue.

3. Methods

1. All culture work should be performed in the laminar flow hood. Prior to working in the hood, the area should be wiped with a 3.0% solution of Amphyl® (National Laboratories, Montvale, NJ) followed by a further wipe with 70% ethanol. The researcher should wear gloves, mask, and a long-sleeve washable or disposable gown to handle the biohazardous material and to avoid contaminating the cornea. The laminar flow cabinet should contain a clear protective screen that provides a working barrier between the researcher and the corneal tissue.

2. If the donor cornea is supplied in a preservative solution, such as Optisol, it can be rinsed off with 1X PBS containing 10 µg/mL gentamicin just prior to being placed in culture. If the donor tissue is supplied as a whole globe, the cornea must be excised before it is placed in culture.

3. Before excising the cornea from a whole globe, the exterior of the globe should be rinsed in 1X PBS containing 10 µg/mL gentamicin. A small puncture is made in the sclera of globe, using a sterile scalpel blade. The tips of the corneal scissors are then introduced into this puncture, and the cornea is cut away from the globe with a 2–3 mm scleral collar intact. In most instances, the initial cut will be made behind the iris, which can be subsequently pulled off using forceps. After removing the iris, the cornea can be rewashed in 1X PBS containing 10 µg/mL gentamicin by placing the tissue into a 60 × 15-mm culture dish containing 15 mL of the buffer for 5 min, using sterile forceps.
4. In a separate 60 × 15-mm culture dish, a Richard ring should be affixed to the bottom of the dish with sterile paraffin, which has been placed at 1–2 mm from the edge of the dish. Culture medium (12.5 mL) is added to this dish, and the cornea lowered onto the Richard ring using forceps. The cornea should be handled with sterile forceps at the scleral rim. The cornea is placed in culture medium at a slight angle to avoid trapping air bubbles under the endothelial surface. If air bubbles persist while the cornea is in culture, the endothelium will be damaged.
5. After the cornea is seated onto the Richard ring (Fig. 3), the lid of the culture dish is replaced, and the dish is placed onto a rocker platform that has been previously wiped with a 70% ethanol solution and placed in the incubator. The cornea is placed at the edge of the rocker platform, and the culture is rocked at 10 cycles/min. At the lowest arc of the down cycle, the cornea should be fully immersed in culture medium. In contrast, the corneal epithelium should be exposed to a 95% air:5% CO_2 environment in the incubator at the highest arc of the up cycle. The rocking motion also moves the culture medium around the dish to provide an exchange of nutrients to the cornea. Care should be taken not to overfill the culture dish to avoid spilling the medium during the rocking motion, risking contamination (see Note 4). If insufficient culture medium is used, the corneal epithelial surface will not be adequately bathed, and epithelial drying with subsequent cell damage may occur.
6. Tissue should be incubated at 37°C under humidified conditions. If the incubation chamber does not regulate humidity, a dish of sterile distilled deionized water containing 10 µg/mL gentamicin should be placed at the bottom of the chamber. Culture medium should be replaced every 24–48 h depending on the requirements of the experimental protocol. The cornea should be kept moist at all times during handling and photography to avoid tissue damage.

4. Notes

1. The serum-free air/liquid organ culture model provides an ideal vehicle to examine the effects of pharmaceutical compounds on corneal wound-healing responses. Anderson et al. (7) used the model to examine the effect(s) of an RGD-containing compound on corneal wound-healing responses after an epithelial scrape wound. They determined the rates of re-epithelialization in the presence and the absence of the compound, and concluded that the addition of the compound had significant effects on these rates.

The air/liquid organ culture model has also been used to compare corneal wound-healing responses to identical treatments in the presence and the absence of serum in the culture medium. Anderson et al. *(8,9)* examined endothelial cell loss in excimer laser-treated corneas cultured for 7 d after wounding. They observed that endothelial cell loss occurred in corneas cultured in serum-free medium, but did not occur in corneas cultured in the presence of serum, suggesting that excimer laser-related effects on endothelial cells are repaired by a serum factor(s).

The model can also be used to examine other aspects of endothelial structure and function, including wound-healing responses in human corneas. Studies have shown that following a touch wound to the surface of the corneal endothelium, cells spread and migrate to reestablish a confluent monolayer within 24–48 h *(12,13)*. Further analysis of such responses can be used to determine the nature of the interactions between the corneal endothelium and the extracellular matrix.

2. All instruments should be scrubbed in a mild detergent and soaked in a disinfectant, such as 70% ethanol, 10%, H_2O_2, or a 3% solution of Amphyl (National Laboratories). Instruments should not be left longer than 10 min in these solutions because of the corrosive properties of the chemicals. After soaking in disinfectant, the instruments should be rinsed in three changes of distilled deionized water, dried, and autoclaved at between 120 and 130°C at 15 psi for 30 min in a steam autoclave. All nondisposable items, such as glass containers, should be similarly washed and then autoclaved.

    Solutions, such as buffers and culture medium, should be sterilized by filtration through a 0.20-μm filter into sterile containers. Disposable items, such as tissue-culture dishes, pipets, gauze swabs, and centrifuge tubes, should be purchased sterilized. All containers should be opened and resealed under the laminar flow hood. Any instrument or item brought from under the laminar flow hood into the environment of the tissue-culture room is assumed to be nonsterile and should not be subsequently reintroduced to the hood or used for tissue culture, before cleaning and resterilization.

    Care should be taken to avoid contact of the polysulfone Richard rings with solutions containing methanol, toluene or xylene, aromatic or halogenated hydrocarbons, or ketones, e.g., acetone, esters, and/ or aldehydes. The Richard rings should be cleaned in mild detergent and water before autoclaving. Use of Amphyl should be avoided because of the phenol derivative contained in this preparation.

3. Eliminating serum from the culture medium allows the researcher to control factors that might influence the performance of the cornea in culture. Because of the absence of humoral and neuronal responses, the serum-free organ culture model can be used to monitor the effect(s) of biological mediators and pharmaceutical agents.

    Recent work by Anderson et al. *(11)* has shown that adult human corneas cultured in the air/liquid culture model for 21 d with serum-free DME:F12 medium exhibit dose-related responses to vitamin A. In the absence of vitamin A, corneas presented the morphologic characteristics of vitamin A deficiency, including multilayering of squamous cells, bundling of tonofilaments to form keratofibrils, increased desmosomes, and loss of microplicae on the epithelial cell surface. Epithelial morphology was responsive to the addition of concentrations of vitamin A ranging from 5×10^{-9} to $5 \times 10^{-6} M$. At the lowest concentration of vitamin A added, cell differentiation and morphology appeared normal. Corneal epithelium developed secretory morphology at the higher concentrations of vitamin A, with intracellular vesicles, basement membrane overproduction, and reduced cell–cell and cell–substrate attachments. Keratocytes and endothelial cells showed increased intracellular vacuoles in the presence of higher concentrations of vitamin A.

4. Contaminated air/liquid corneal cultures can be identified by examination: the culture medium should be clear, and the cornea should remain clear and translucent during the time it is in organ culture. If either the culture medium or the cornea becomes cloudy, then the culture is probably contaminated and should be discarded. A gram-negative stain for identification of bacteria or, alternatively, plating of the culture medium on agar gels would assist in identifying the type of contamination.

If a culture is discarded owing to contamination, the following procedures should be performed before further organ culture is attempted:

a. A solution of 70% ethanol should be used to wipe down all accessible parts of the incubator, laminar flow hood, and rocker platform as well as the walls, floor, and working area around the incubator. This should be done at least twice before attempting another culture.

b. All solutions, including culture medium and buffers, should be discarded and freshly made for subsequent culture. Glassware and other nondisposable equipment, including surgical instruments, should be rewashed and autoclaved. Opened and resealed packages of sterile disposable culture equipment should be put aside and relegated to nonsterile purposes.

c. Re-examine the patient data sheet to determine if the corneal tissue might have been compromised before the culture. If possible, identify the compromising factor and avoid using similar tissue in future experiments.

d. Tissue-culture techniques should be examined by the researcher performing the culture and by a second individual to determine possible sources of contamination in the procedure.

Acknowledgments

The encouragement and advice of Perry S. Binder were most invaluable during the development of these procedures. The help of Carol J. Sutherland and Max M. Moore in the preparation of this manuscript was greatly appreciated. Funding for the work was provided by the National Vision Research Institute, the San Diego Eye Bank, and the Sharp Hospital Foundation.

References

1. Doughman, D. J. (1980) Prolonged donor cornea preservation in organ culture long term clinical evaluation. *Trans. Am. Ophthalmol. Soc.* **LXXCVIII**, 567–628.
2. Cintron, C., Covington, H. I., and Gregory, J. D. (1988) Organ culture of rabbit cornea: morphologic analysis. *Curr. Eye. Res.* **7**, 303–312.
3. Doughman, D. J., Van Horn, D., Harris, J. E., Miller, G. E., Lindstrom, R., and Good, R. A. (1974) The ultrastructure of the human organ-cultured cornea. I. Endothelium. *Arch Ophthalmol.* **92**, 516–523.
4. Van Horn, D., Doughman, D. J., Harris, J. E., Miller, G. E., Lindstrom, R., and Good, R. A. (1975) The ultrastructure of the human organ-cultured cornea. II. Stroma and Epithelium. *Arch Ophthalmol.* **93**, 275–277.
5. Trowell, O. A. (1953) A modified technique for organ culture in vitro. *Exper. Cell. Res.* **6**, 246–248.
6. Richard, N. R., Anderson, J. A., Weiss, J. L., and Binder, P. S. (1991) Air/liquid organ culture: a light microscopic study. *Curr. Eye. Res.* **10**, 739–749.
7. Anderson J. A., Sipes, N. J., and Binder, P. S. (1993) Healing rates of corneal wounds in the presence of an RGD-containing peptide conjugate. *J. Cell. Biochem.* **17(Suppl.)**, 131.
8. Binder, P. S., Anderson, J. A., and Lambert, R. W. (1993) Endothelial cell loss associated with excimer laser. *Ophthalmology* **100**, 107.
9. Anderson, J. A., Lambert R. W., Sutherland C. J., and Binder, P. S. (1994) Endothelial cell loss in excimer laser-treated cultured human corneas. *Invest. Ophthalmol. Vis. Sci.* **35(4)**, 2016.
10. Collin, H. B., Anderson, J. A., Richard, N. R., and Binder, P. S. (1995) *In vitro* model for corneal wound healing: organ cultured human corneas. *Curr. Eye Res.* **14**, 331–339.
11. Anderson, J. A., Richard, N. R., Rock, M. E., and Binder, P. S. (1993) Requirement for vitamin A in long-term culture of human cornea. *Invest. Ophthalmol. Vis. Sci.* **34**, 3442–3449.
12. Weiss, J. L., Richard, N. R., Anderson, J. A., and Binder, P. S. (1990) Endothelial wound healing in bi-phasic organ culture and the effects of epidermal growth factor. *Invest. Ophthalmol. Vis. Sci.* **31**, 53.
13. Lambert, R. W., Weiss, J. L., Binder, P. S., and Anderson, J. A. (1993) Co-localization of F-actin and fibronectin in wounded human corneal endothelium. *J. Cell. Biol.* **17E**, 116.

38

Keratocyte Cell Culture

Normand R. Richard, Robert W. Lambert, and Janet A. Anderson

1. Introduction
1.1. Keratocyte Definition and Function

Keratocytes, or corneal fibroblasts, are the primary cell type of the corneal stroma. They lie between and are oriented parallel to the orthogonally arranged collagen lamellae, forming a continuous interconnecting cellular network that has been hypothesized to transmit information throughout the cornea concerning the status of the tissue *(1)*. Under normal conditions, the keratocyte in the adult cornea is a relatively quiescent cell. However, in the event of a corneal injury or trauma, the keratocytes near the injured area differentiate into active, synthesizing cells and rapidly replace damaged stromal matrix *(2,3)*.

The adult corneal stromal matrix consists of collagens and proteoglycans so structured as to create a transparent stroma *(4)*. Type I is the major structural collagen of the cornea *(5)*, and colocalizes with types III and V in the striated collagen fibrils *(6)*. Type VI constitutes a large proportion of the stromal collagen *(5)* and is found in the fine filaments of the interfibrillar matrix *(6)*. Dermatan sulfate- and keratan sulfate-containing proteoglycans interact with the collagens to form the structural basis of the stroma *(7)*.

1.2. In Vitro Studies

In vitro studies of keratocytes have examined the signals that stimulate these cells to synthesize corneal matrix *(8)*, as well as those disorders of the stroma, such as keratoconus *(9)* and herpes simplex *(10)*, that lead to breakdown and/or scarring of the corneal tissue. Since the nature of the matrix materials is important to ensure corneal transparency *(4)*, cultured keratocytes are also studied to examine the effects of environmental factors on the matrix molecules synthesized *(11)*.

From: *Methods in Molecular Medicine: Human Cell Culture Protocols*
Edited by: G. E. Jones Humana Press Inc., Totowa, NJ

Cell-culture studies indicate that the nature of the cell–matrix relationship influences the orientation of the newly synthesized extracellular matrix. When keratocytes are cultured on a planar substrate, they flatten and randomly spread out in a disorganized manner and, after several days, become confluent *(12,13)*. Type I collagen synthesized by plated keratocytes is laid down in a network over the surface of the plate *(12)*. Keratocytes plated on culture plates that have linearly grooved surfaces orient themselves parallel to the direction of the grooves *(12)*. The type I collagen synthesized by these cells is also distributed in a linear fashion, parallel to the linear grooves *(12)*.

Human keratocytes cultured in a three-dimensional, hydrated collagen gel more closely resemble keratocytes in the cornea with regard to cell shape, nuclear configuration, contact with adjacent cells through pseudopodia, and spatial relationships to other cells than do the same cells grown on planar substrates *(13–16)*. Collagen gel-grown keratocytes also display the characteristic locomotion seen in vivo, but not observed when these same cells are grown on planar substrates *(17,18)*. Gel-grown keratocytes synthesize and secrete type I collagen, which is dispersed throughout the gel *(19,20)*.

Since varying culture conditions of keratocytes can alter the distribution of newly synthesized matrix materials, it would be of interest to determine the effect of culture conditions on the nature of the materials synthesized by these cells. The ability of newly synthesized extracellular materials to form a transparent matrix apparently depends both on the nature of the material synthesized and on its orientation *(4)*.

2. Materials
2.1. Corneal Tissue

1. Tissue procurement: Whole globes and excised corneas can be procured directly from local eye banks. Contact the eye banks to determine the procedures for obtaining tissue. Usually there are some nominal tissue-handling fees. Indicate your tissue requirements, e.g., no tissue older than 48 h postmortem, no evidence of corneal disease or surgery, no eyes from donors with a known history of diabetes, and so forth (*see* Note 1).

    Another source of corneal tissue is corneal surgeons. At the time of corneal transplant, tissues that are usually scarred and/or diseased are removed from the patient. If you are studying corneal disorders, such as bullous keratopathy, herpes simplex, keratoconus, Fuchs' dystrophy, or lattice dystrophy, you may find it useful to work with a surgeon. A portion of each corneal button may need to be taken for examination by the pathologist, but you can usually obtain sufficient tissue for cell culture. Tissue from this source requires application to the hospital's Internal Review Board.

2. Tissue handling: Because of the possibility of exposure to pathogens when handling human tissue, biohazardous material precautions should always be

employed. A washable or disposable lab coat should be worn along with a disposable face mask and surgical latex gloves, which should be changed frequently. All bench surfaces, pipets, containers, and surgical instruments should be decontaminated with a 10% solution of chlorine bleach. Reusable items should be cleaned and autoclaved at 120°C for 30 min at 15 psi immediately after use. Tissue processing and cell-culture work should be conducted under a laminar flow biological cabinet (*see* Note 2).

2.2. Instruments for Tissue Dissection

Ophthalmic instruments can be obtained from Storz Ophthalmics (St. Louis, MO), Biomedical Research Instruments (Rockville, MD), or Katena Eye Instruments, Katena Products, Inc. (Denville, NJ). Instruments should be autoclaved or flame sterilized prior to use. To ethanol flame-sterilize instruments and equipment, dip the item in 95% ethyl alcohol, ignite using a Bunsen burner flame, and let the flame self-extinguish away from the Bunsen burner. Repeat twice, taking care not to ignite the ethanol container.

1. Corneal trephine holder and blades, 6–10 mm in diameter (Fig. 1A): A trephine holder with a disposable circular blade is used to excise corneal buttons. Trephine blades with varying diameters can be purchased to accommodate different size donor corneas.
2. Curved corneal scissors (Fig. 1B): Curved corneal scissors, 4¾ in. in length with 30-mm blades, are required to excise the cornea, including a 3-mm scleral rim, from whole globes.
3. Straight, microdissecting 4-in. forceps with 1 × 2 teeth, standard pattern 0.8-mm wide tips (Fig 1C).
4. Straight, microdissecting 4-in. forceps with serrated 0.8-mm tips (Fig. 1D).
5. Scapel holders and sterile #10 or #11 blades (American Safety Razor Co., Staunton, VA) (Fig 1E).
6. Sterile 4 × 4 in. cotton gauze (Fig. 1F).
7. Sterile cotton-tipped swabs (Fig. 1G).
8. Sterile plastic syringe containing 1X PBS with 10 µg/mL gentamicin (Fig. 1H).

2.3. Tissue-Culture Solutions and Media

All solutions and media used in culturing procedures must be sterile. Sterile, ready-made solutions can be purchased, or solutions and media can be freshly prepared and sterilized by filtering through a 0.22-µm pore membrane filter.

1. Hank's balanced salt solution (HBSS) without calcium and magnesium, pH 7.4: Weigh the following ingredients and place in a 1-L flask or beaker: 0.4 g KCl, 0.06 g KH_2PO_4, 8.0 g NaCl, 0.35 g $NaHCO_3$, 0.0475 g Na_2HPO_4, 1.0 g glucose, 0.017 g phenol red. Dissolve ingredients in 900 mL of distilled water, adjust pH to 7.4, bring the final volume to 1000 mL, and filter sterilize using a 0.22-µm pore membrane filter. Divide into usable quantities and store at 4°C in sterile, screwtop bottles. Discard after 1 yr.

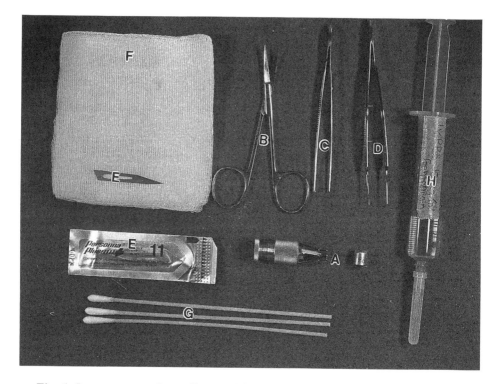

Fig. 1. Instruments and supplies used in corneal dissection: **(A)** corneal trephine holder and disposable blade, **(B)** curved corneal scissors, **(C)** toothed forceps, **(D)** fine-tip forceps, **(E)** scalpel blade #11, **(F)** sterile cotton gauze, **(G)** sterile cotton swabs, and **(H)** syringe filled with gentamicin in PBS (10 µg/mL).

2. Dulbecco's phosphate-buffered saline (PBS) without calcium and magnesium, stock solution (10X): 2.0 g KCl, 2.0 g KH_2PO_4, 80.0 g NaCl, 11.5 g Na_2HPO_4. Dissolve ingredients in 900 mL of distilled water, adjust final volume to 1000 mL, and filter sterilize using a 0.22-µm pore membrane filter. The sterilized salt solution can be kept at room temperature for 6 mo to 1 yr. To make a working solution, dilute 10 times in sterile distilled water. The pH of the final solution will be 7.2.
3. Medium 199 with Earle's salts, 25 mM HEPES and L-glutamine (Gibco-BRL, Gaithersburg, MD): Weigh out enough powdered medium to make 1 L. Add 900 mL of distilled water. While stirring, add to this mixture 2.2 g NaH_2CO_3 and adjust pH to 7.3 with 1M NaOH. Bring the volume up to 1000 mL and filter sterilize with 0.22-µm pore membrane filters. Store unsupplemented media in 500-mL bottles at 4°C for up to 6 mo.

   Before use, supplement 1 L of medium as follows:
   a. Add 10 mL of 10,000 U/mL–10,000 µg/mL penicillin–streptomycin stock solution giving final concentrations of 100 U/mL–100 µg/mL. The stock

solution can be divided into usable amounts and stored below −20°C for up to 1 yr.

    b. Add 10 mL of 250 µg/mL fungizone stock solution (final concentration 2.5 µg/mL) if mold contamination is a problem in growing cultures. Store stock fungizone in usable aliquots below −20°C for up to 1 yr. Store fungizone in nonbreakable containers and fill containers half full; fungizone has a tendency to expand and crack storage containers when freezing.

    c. Add 5 mL of 10 mg/mL gentamicin stock solution (final concentration 50 µg/mL) if bacterial contamination in the cell cultures is not kept in check by penicillin–streptomycin. Gentamicin can be stored at room temperature for up to 1 yr.

    d. Supplemented media should be discarded after 1 mo at 4°C. Warm medium to 37°C before adding to the cell culture.

4. Fetal calf serum (heat-inactivated): Aliquot in usable amounts, and store below −20°C for up to 1 yr. Supplement Medium 199 with 10% fetal calf serum (*see* Note 3).

5. Alseivers Trypsin Versene (ATV): This trypsin solution is used for dissociating cells from growth surfaces prior to passage. The solution is prepared by mixing 0.5 g trypsin (1:250, Irvine Scientific, Santa Ana, CA), 8.0 g NaCl, 0.4 g KCl, 0.2 g EDTA, and 1.0 g dextrose in 1000 mL of doubly distilled water. Store filter-sterilized material in usable aliquots below −20°C for up to 1 yr.

6. Collagenase (0.1%), pH 7.2: Prepare collagenase solution by mixing 0.1% (w/v) of dry powder in HBSS, filter sterilize, warm to 37°C, and use immediately (*see* Note 4).

7. Freezing medium: Freezing medium is used to preserve cells for long-term storage either in liquid nitrogen or at −70°C. The medium consists of Medium 199 supplemented with 20% fetal calf serum, 10% sterile dimethyl sulfoxide, and 100 U/mL–100 µg/mL penicillin–streptomycin.

8. Rat tail collagen or Vitrogen 100 (bovine collagen): Either rat tail collagen (Sigma, St. Louis, MO) or Vitrogen 100 (bovine collagen, Collagen, Palo Alto, CA) can be used to culture keratocytes effectively within a three-dimensional collagen matrix (*see* Section 3.5., step 2).

9. Trypan Blue (0.4%): This vital stain solution can be obtained from Sigma.

2.4. Tissue-Culture Supplies

The following items are needed for processing corneal tissue and culturing human keratocytes. They can be purchased sterile from any general laboratory supplier, such as Fisher Scientific (Pittsburgh, PA) or Baxter Scientific Products (McGaw Park, IL) (*see* Note 5).

1. 75 and 25 cm^2 Tissue-culture flasks, 60 × 15 and 35 × 15 mm culture plates, and 6- and 24-well plates (*see* Note 5).
2. Freezing vials (1–1.5 mL) intended for complete immersion in liquid nitrogen.
3. 15-mL Sterile, plastic centrifuge tubes.
4. Sterile 0.22-µm pore filters and filter-sterilization apparatus for large volumes.

3. Methods
3.1. Preparation of Primary Cultures

1. Corneal tissue can be obtained from whole eyes or excised corneas maintained in preservation media. If whole eyes are used, aseptically remove the whole globe from its packaging, rinse with sterile PBS containing 10 µg/mL gentamicin, and excise the cornea with curved corneal scissors, leaving a 2–3 mm rim of sclera surrounding the corneal edge.
2. Pick up the excised corneas by the scleral rim with a sterile, toothed forceps, and place in a dry, sterile culture dish with the anterior, epithelial side down.
3. Set up a dissecting microscope under the laminar flow hood, first wiping the microscope with 70% ethanol. Place the cornea under the microscope and gently peel away any iris that may be attached to the posterior cornea using sterile toothed forceps.
4. Remove the endothelial cell monolayer, and the attached Descemet's membrane from the posterior side of the cornea with a sterile cotton-tipped swab. To do this, position the sterile cotton swab at the inside edge of the cornea, and twist while pushing in a downward motion with moderate pressure. This will result in tearing of Descemet's membrane, which can then be easily wiped away using another sterile cotton swab.
5. Using a trephine and attached blade, excise the largest possible corneal button from the center of the corneal scleral rim, and place it into 0.1% collagenase-A solution, prewarmed to 37°C, using 3 mL of solution for each corneal button; incubate at 37°C in a culture incubator for 45 min.
6. After the initial enzyme digestion, aseptically remove the button(s), and place into a dry, sterile culture plate. Gently wipe off the epithelial layer from the anterior cornea(s) using light to moderate pressure applied by a sterile cotton-tipped swab.
7. Transfer the remaining stromal button(s) to fresh prewarmed 0.1% collagenase-A solution, 3 mL for each button, and mince tissue into 1–2 mm pieces with stainless-steel surgical scissors.
8. Incubate the minced stroma for 1.5 h at 37°C in the 0.1% collagenase solution; to facilitate digestion, place the culture plates on the rocker platform positioned inside the incubator, and rock at a moderate speed. At the end of the incubation period, the stromal fragments should appear "slimy" in consistency.
9. With a sterile, 10-mL pipet, transfer the slurry into a sterile 15-mL centrifuge tube and triturate 15 times.
10. Centrifuge at 100g for 10 min at room temperature, remove the collagenase-containing supernatant, and resuspend the cell pellet in HBSS.
11. Triturate again 15 times with a 10-mL pipet, centrifuge at same speed and time as indicated in step 10, and remove the supernatant.
12. Resuspend the cell pellet into prewarmed, supplemented Medium 199, and plate into appropriate culture vessels. Incubate the culture plates for 2 d undisturbed at 37°C under humidified conditions and 5% CO_2; 95% air.
13. Evaluate the cultures daily by inverted phase-contrast microscopy. This type of microscopy provides the best contrast and resolution for daily evaluations, when

Keratocyte Cell Culture

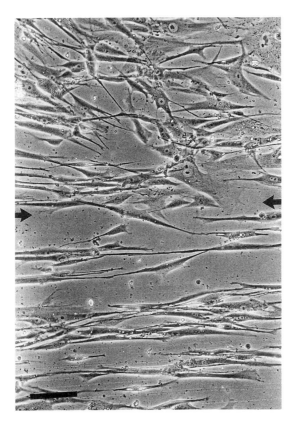

Fig. 2. Inverted phase-contrast microscopy of plated human keratocytes. Keratocytes taken from a 58-yr-old donor, passage 5, were plated at a concentration of 10^4 cells/mL on a culture plate, a portion of which had been etched with a cotton swab. Three days after plating, cells in the unetched portion of the plate, above the arrows, show the dorsal-ventrally flattened, multilayered, and random distribution typical of plated fibroblasts. In the etched portion of the plate, below the arrows, the cells are bipolar, spindle-shaped, and have aligned along the etched lines. Bar = 240 μm.

compared to the conventional inverted light microscope. Figure 2 shows a phase-contrast micrograph of a growing keratocyte culture, note the typical flattened shaped, fibroblastic-like morphology of these cells.

3.2. Maintenance and Subculture of Established Cell Lines

1. After an initial 2-d incubation period subsequent to plating, the culture medium should be changed every other day. To do so, aspirate the expired medium from the culture vessel, and replace with the same volume of prewarmed, supplemented Medium 199.

2. When the monolayer of cells is 90% confluent, cells will need to be subcultured. Remove the expired medium and rinse the monolayer with enough sterile PBS to cover the cells fully, rocking the culture vessel several times to wash the cells (*see* Note 6).
3. Remove the PBS by aspiration, and add ATV to cover the monolayer. Rock the culture three times, and immediately remove most of the ATV. Leave a small amount of ATV in the culture vessel, e.g., for a 75-cm^2 culture flask, leave behind 0.5–1.0 mL of ATV.
4. Incubate at 37°C for 1–3 min. During this time, periodically examine the cultures under an inverted microscope for detaching cells, which will appear rounded up.
5. When the majority of the cells appear rounded up, gently tap the bottom of the culture vessel to an edge of a hard surface, e.g., lab bench, to dislodge the cells from the growth surface.
6. For a 1:1 split, resuspend the cells in a volume equal to the original volume of culture medium, and transfer one-half of the cell suspension into a similar-size culture vessel. Add enough medium to bring both cultures up to the original volume, and reincubate at 37°C.

3.3. Cryopreservation

1. Trypsinize monolayers of keratocytes at 90% confluence using the methods described in Section 3.2.; freeze cultures that have been passaged only a few times.
2. Resuspend the detached cells in a 15-mL, sterile centrifuge tube in complete Medium 199, and centrifuge at 100g at room temperature for 10 min.
3. Remove the supernatant by aspiration, and gently resuspend the cell pellet in freezing medium; use 1 mL of freezing medium/75-cm^2 culture flask.
4. Pipet aliquots of the cell suspension into freezing vials. To minimize cell damage from freezing too rapidly, place the vials overnight in a freezer at –70°C.
5. On the following day, transfer the frozen vials to a liquid nitrogen-filled canister.
6. To replate cryopreserved cells, first quick-thaw the vials containing the frozen cell suspension by gentle agitation in a 37°C water bath.
7. Wipe off the capped ends of the vials with 70% ethanol, transfer the thawed cell suspension to a 15-mL conical centrifuge tube, and centrifuge for 10 min at 100g at room temperature.
8. Remove the supernatant, and gently resuspend the cell pellet into supplemented Medium 199.
9. Replate the cell suspension at a density twice that of the original prefrozen cell density, e.g., use two vials of cells for each 75-cm^2 culture flask.

3.4. Plating Cells for Microscopic Procedures

Cells grown on coverslips can be easily processed for a variety of microscopic methods of evaluation. The following is a suggested method for plating cells onto 22 × 22 mm coverslips placed into 35 × 15 mm culture dishes at a density of 5×10^5 cells/mL, which is suitable for most experimental purposes.

1. Dissociate cells grown in one or more 75-cm² flasks using procedures described in Section 3.2. Resuspend cells in one-half volume supplemented Medium 199, and transfer to a 15-mL centrifuge tube.
2. Remove 0.2 mL from the cell suspension, and mix it with an equal volume of 0.4% Trypan Blue.
3. Using a small-bore Pasteur pipet, completely fill both chambers of a coverslipped hemocytometer with the cell–Trypan Blue mixture. Trypan Blue is a vital stain that will distinguish dead cells from live ones; dead cells will be stained dark blue, whereas live cells will remain unstained.
4. Using a hemocytometer, determine cell density, and dilute cells to a concentration of 10^5 cells/mL with supplemented Medium 199.
5. Plate diluted cell suspension in 2-mL amounts in 35-mm plates containing ethanol flame-sterilized coverslips.
6. Incubate dishes, and maintain as indicated in the above steps until cells are ready for evaluation and analysis.

3.5. Alternative Culture Conditions

In addition to culturing the keratocytes on two-dimensional substrata, such as tissue-culture plastic, the cells can be grown on etched surfaces or within a collagen matrix. These methods provide ways to study the growth characteristics of keratocytes in environments that mimic some of the characteristics of the in vivo environment.

3.5.1. Etched Cultures

1. Prior to cell dissociation, etch the bottom growth surface of 35- or 60-mm tissue-culture plates with parallel lines using a sterile cotton-tipped swab.
2. While working under a dissecting microscope set up under a laminar flow biological cabinet and place the bottom portion of the culture plate over a sheet of graph paper (3 × 3-mm grids).
3. Using light to moderate pressure applied to the cotton swab, trace each parallel line of the graph paper until the entire bottom surface of the plate is etched. Parallel lines of approx 3–21 µm apart and 0.9–21 µm wide are produced with this method.
4. Seed plates at density of 10^5 cells/mL.
5. After 3 d of culture, cells will be observed growing parallel to the etched lines (Fig. 2).

3.5.2. Collagen-Gel Cultures (2.2 mg/mL Bovine Collagen Type I)

1. Trypsinize cells growing in a 75-cm² flask, and resuspend in 5 mL supplemented Medium 199.
2. Count the number of viable cells in the cell suspension.
3. Withdraw into a 15-mL centrifuge tube enough cells for five 35-mm plates at 1 mL/plate and a cell density of 5×10^4 cells/mL.

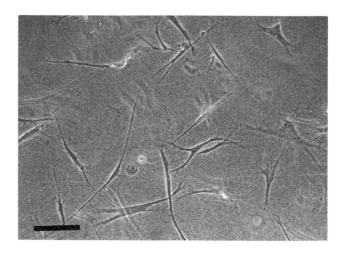

Fig. 3. Inverted phase-contrast microscopy of human keratocytes cultured in collagen gel. Keratocytes from a 69-yr-old donor, passage 11, were plated in Vitrogen 100 at a concentration of 10^5 cells/mL. After 3 d in culture, bipolar, spindle-shaped cells can be seen at various depths in the three-dimensional gel. Bar = 400 μm.

4. Centrifuge cells for 10 min at 100g.
5. Using a 1-mL pipet, gently resuspend cells in 0.26 mL sterile PBS.
6. To the resuspended cells, add 3.74 mL cold Vitrogen 100 (3.0 mg/mL collagen), and gently resuspend using a 5-mL pipet.
7. To the above mixture add 0.5 mL 10X cold Medium 199 supplemented with 20% fetal calf serum, along with 1000 U/mL–1000 μg/mL penicillin–streptomycin. Quickly but gently mix using a 5-mL pipet.
8. Add 0.5 mL cold 1N NaOH, and immediately gently resuspend the mixture. Avoid introducing air bubbles while resuspending the collagen mixture.
9. Immediately aliquot 1-mL amounts into the 35-mm plates, and gently swirl the plates so as to coat the bottom of each culture dish evenly with the collagen cell suspension.
10. Cover the plates and incubate in a humidified incubator set at 37°C and 5% CO_2 for 1 h; during this time period, the collagen mixture will congeal.
11. After the 1-h incubation, overlay the congealed collagen culture with 1 mL of Medium 199 supplemented with 2% fetal calf serum and 100 U/mL–100 μg/mL penicillin–streptomycin. Feed cultures every other day with same medium. The lower percentage of fetal calf serum is used in the culture medium and the collagen gel to prevent rapid cell proliferation, leading to contraction of the collagen gel.
12. After 48–72 h under standard incubation conditions, the keratocytes will be observed to be bipolar, with elongated cell processes interconnecting with other cells growing within the collagen matrix (**Fig. 3**).

4. Notes

1. There are many reasons why primary cultures lose viability during the first days or weeks following initial plating. The following suggestions will help you avoid some of the pitfalls encountered while learning to culture keratocytes.

    Try to avoid using corneas from donors over 80 yr old, unless they are to be used for age-related studies, since they are slow growing and do not provide many passages before reaching senescence. Generally, we have found that tissues derived from donors ranging in age from newborn to 20 yr old provide rapidly growing cells that may undergo as many as 20 passages. Cells from donors over 20 and under 80 yr of age are adequate for most studies, providing approx 12–15 passages. The number of times each keratocyte culture can be passed is also dependent on the health of the original donor tissue.

    Procure corneas within 48 h or less after death, if possible, and process the tissue for cell culture as soon as possible. Tissue that is used within 24 h following procurement will provide more viable cells than tissue sitting at 4°C for several days. Carefully examine the state of the cornea on receipt of the tissue. Healthy donor corneas should be clear and smooth, with an intact epithelium and white scleral tissue.

2. Although it is possible to request of the eye banks that blood pathogen testing be performed on donor blood each time eyes become available, not all pathogens, e.g., HIV, show up on testing. Furthermore, negative results may give you a false sense of security. It is safer to assume that all tissues contain pathogens and consistently use biohazardous materials precautions at all times.

3. Keratocyte cultures grow more rapidly when grown in Medium 199 supplemented with fetal calf serum rather than cadet or newborn serum. Although fetal calf serum is more expensive, it is worth the extra few cents. If expense is an issue, then supplementing Medium 199 with a higher percentage of a less expensive serum, e.g., 15 or 20%, may help to improve the growth rate of the cultured keratocytes.

4. Selection of the right collagenase can make a difference between success and failure when growing primary keratocyte cultures. Not all collagenases of the same type and specificity will have the same activity. Not only do collagenase activity and purity vary from supplier to supplier, but the activity can also vary between lots. Not all collagenases work well on corneal tissue. Choose one that will preferentially degrade collagen type I, the predominant collagen type in corneas. It is our recommendation to try different suppliers and lots, and then choose the one that works best. In our laboratory, we have had success with collagenase-A purchased from Boehringer Mannheim (Indianapolis, IN).

5. For collagenase treatment, use 3 mL of 0.1% collagenase-A in a 35 × 15-mm culture plate for digestion of one-half or one whole corneal button. For two corneal buttons, use 6 mL of collagenase in a 60 × 15-mm plate.

    The number of corneas used in preparing a primary keratocyte culture will determine the size of the culture vessel chosen for the initial plating. For example, the dissociated cells derived from one-half of a corneal button should be plated in a vessel with a small growth surface area, such as single well in a 24-well culture plate, with 1 mL of culture medium. After cell confluence is achieved, the cells may be passaged into a larger culture vessel, e.g., 35 × 15-mm plate, and so on. The dissociated cells from one corneal button should be suspended in 3 mL of medium and placed in a 35 × 15-mm culture plate. Cells from two corneal buttons from the same donor should be resuspended in 6 mL of medium and placed in a 60 × 15-mm plate.

6. The rate of cell growth will dictate how often the cells will need to be subcultured or passaged. For very rapid growers, when passaged 1:1, cultures may need to be subcultured every other day. To avoid frequent subculturing of rapid growers, passage cells at a lower cell density, e.g., split at 1:2 or 1:3.

Acknowledgments

The encouragement and suggestions of Perry S. Binder were most helpful during the development of these procedures. Funding for the work was provided by the Sharp Hospitals Foundation and the National Vision Research Institute.

References

1. Nishida, T., Yasumoto, K., Otori, T., and Desaki, J. (1988) The network structure of corneal fibroblasts in the rat as revealed by scanning electron microscopy. *Invest. Ophthalmol. Vis. Sci.* **29,** 1887–1890.
2. Cionni, R. J., Katakami, C., Lavrich, J. B., and Kao, W. W. (1986) Collagen metabolism following corneal laceration in rabbits. *Curr. Eye Res.* **5,** 549–558.
3. Funderburgh, J. L., Cintron, C., Covinton, H. I., and Conrad, G. W. (1988) Immunoanalysis of keratan sulfate proteoglycan from corneal scars. *Invest. Ophthalmol. Vis. Sci.* **29,** 1116–1124.
4. Cintron, C. (1989) The function of proteoglycans in normal and healing corneas, in *Healing Processes in the Cornea* (Beuerman, R. W., Crosson, C. E., and Kaufman H. E., eds.), Gulf Publishing Co., Houston, TX, pp. 99–109.
5. Zimmerman, D. R., Fischer, R. W., Winterhalter, K. H., Witmer, R., and Vaughan, L. (1988) Comparative studies of collagens in normal and keratoconus corneas. *Exp. Eye Res.* **46,** 431–442.
6. Marshall, G. E., Konstas, A. G., and Lee, W. R. (1991) Immunogold fine structural localization of extracellular matrix components in aged human cornea. II. Collagen types V and VI. *Graefes Arch. Clin. Exp. Ophthalmol.* **229,** 164–171.
7. Wollensak, J. and Buddecke, E. (1990) Biochemical studies on human corneal proteoglycans—a comparison of normal and keratoconic eyes. *Graefes Arch. Clin. Exp. Ophthalmol.* **228,** 517–523.
8. Birk, D. E., Lande, M. A., and Fernandez-Madrid, F. R. (1981) Collagen and glycosaminoglycan synthesis in aging human keratocyte cultures. *Exp. Eye Res.* **32,** 331–339.
9. Kenney, M. C., Chwa, M., Opbroek, A. J., and Brown, D. J. (1994) Increased gelatinolytic activity in keratoconus keratocyte cultures. A. Correlation to an

altered matrix metalloproteinase-2/tissue inhibitor of metalloproteinase ratio. *Cornea* **13**, 114–124.
10. Oakes, J. E., Monteiro, C. A., Cubitt, C. L., and Lausch, R. N. (1993) Induction of interleukin-8 gene expression is associated with herpes simplex virus infection of human corneal keratocytes but not human corneal epithelial cells. *J. Virol.* **67**, 4777–4784.
11. Hassell, J. R., Schrecengost, P. K., Rada, J. A., SundarRaj, N., Sossi, G., and Thoft, R. A. (1992) Biosynthesis of stromal matrix proteoglycans and basement membrane components by human corneal fibroblasts. *Invest. Ophthalmol. Vis. Sci.* **33**, 547–557.
12. Richard, N. R., Anderson, J. A., and Binder, P. S. (1991) Contact guidance modulates cultured human keratocyte morphology and distribution of expressed cellular proteins. *Invest. Ophthalmol. Vis. Sci.* **32(Suppl.)**, 876.
13. Nishida, T., Ueda, A., Fukuda, M., Mishima, H., Yasumoto, K., and Otori, T. (1988) Interactions of extracellular collagen and corneal fibroblasts: morphologic and biochemical changes of rabbit corneal cells cultured in a collagen matrix. *In Vitro Cell Dev. Biol.* **24**, 1009–1014.
14. Richard, N. R., Anderson, J. A., Duzman, E., Kolb, B., and Binder, P. S. (1989) Keratocytes in collagen gel: morphology and collagen synthetic capability. *Invest. Ophthalmol. Vis. Sci.* **30(Suppl.)**, 203.
15. Ueda, A., Nishida, T., Otori, T., and Fujita, H. (1987) Electron-microscopic studies on the presence of gap junctions between corneal fibroblasts in rabbits. *Cell Tissue Res.* **249**, 473–475.
16. Anderson, J. A., Richard, N. R., Rock, M. E., and Binder, P. S. (1990) *Keratocyte Extracellular Matrix Interactions in Hydrated Collagen Gel Culture and in Human Cornea.* Third International Conference on the Molecular Biology and Pathology of Matrix, II-9, Philadelphia, PA.
17. Bard, J. B. L. and Hay, E. D. (1975) The behavior of fibroblasts from the developing avian cornea. Morphology and movement *in situ* and *in vitro*. *J. Cell Biol.* **67**, 400–413.
18. Heath, J. P. and Hedlund, K.-O. (1984) Locomotion and cell surface movements of fibroblasts in fibrillar collagen gels. *Scanning Electron Microscopy* **IV**, 2031–2043.
19. Richard, N. R., Anderson, J. A., and Binder, P. S. (1990) Immunohistochemical demonstration of synthetic capabilities of human keratocytes grown in collagen gel. *Invest. Ophthalmol. Vis. Sci.* **31(Suppl.)**, 539.
20. Doane, K. J., Babiarz, J. P., Fitch, J. M., Linsenmayer, T. F., and Birk, D. E. (1992) Collagen fibril assembly by corneal fibroblasts in three-dimensional collagen gel cultures: small-diameter heterotypic fibrils are deposited in the absence of keratan sulfate proteoglycan. *Exp. Cell Res.* **202**, 113–124.

39

Conjunctiva

Organ and Cell Culture

Monica Berry, Roger B. Ellingham, and Anthony P. Corfield

1. Introduction
1.1. Anatomy

The conjunctiva is the mucous membrane of the eye. It is a delicate transparent epithelium with its own stroma overlying the tough white sclera, which forms the ball of the eye. Only at the circular limbus, where the sclera meets the transparent cornea, and at the eyelids, is the conjunctiva firmly attached, thus permitting free movement of the eye. The loose conjunctiva between these points of attachment folds into a blind sac deepest under the upper and lower lids. Removed in one piece from the eye, the conjunctiva is a flimsy sheet (usually with adherent fascial tissue called Tenon's capsule) with an 11-mm circular defect in the center.

Three main cell types make up the conjunctiva:

1. Epithelial cells whose morphology ranges from cylindrical in the stratified columnar forniceal epithelium to flattened in the stratified squamous limbal and lid margin areas *(1)*.
2. The mucus-secreting epithelial goblet cells.
3. The fibroblasts of the stroma.

Also within the stroma are melanocytes, fine blood vessels, and lymphatics.

There are several layers of epithelial cells. The deepest layer is a germinal layer with actively dividing cells. At the conjunctival-corneal limbus, there is a concentration of stem cells. Stem cells are assumed to exist in all self-renewing tissues. They are characterized by asymmetric division. One daughter cell maintains stem-cell characteristics, and the other differentiates *(2)*. It is debatable whether limbal stem cells are capable of differentiating into corneal epithelial or conjunctival epithelial cells according to their orientation on the globe. There are sites in the conjunctiva other than the limbus where stem cells reside. The rabbit has stem cells in the conjunctival fornix *(3)*. It is believed that those which become conjunctival may differentiate into either goblet cells or cells of the germinal layer. Conjunctival epithelium may transdifferentiate and grow to re-epithelialize a corneal wound: phenotypic transformation depends on limbal integrity *(4,5)*.

For successful growth of epithelium in tissue culture, conjunctival explants should be chosen that are likely to contain germinal, or better still, stem cells *(6)*. For successful growth of fibroblasts, such care is not required. All fibroblasts appear capable of mitosis and are less fastidious than epithelial cells in culture. The cells of the conjunctival vasculature need special conditions to survive in tissue culture.

1.2. Applications of Conjunctival Tissue Culture

Culture of conjunctival tissue and cells has a wide application in the investigation of basic cell biological properties, changes resulting from disease, and in the clinical field.

1.2.1. Epithelial Cells

Cultured conjunctival cells can be used for microscopic and histochemical analysis of cell phenotype *(7)*, pathogenic inclusions *(8,9)*, and cell secretions in health and disease (e.g., mucins *[10,11]*, cytokines *[12]*, and growth factors). Primary cell culture may provide a source of material for biochemical analysis of conjunctival cell products *(13,14)*.

Tissue-culture assays of toxicity and wound healing are being used increasingly to evaluate eye irritation and toxicity *(15,16)*. Autologous conjunctival cell grafts have been used to help repopulate severely damaged corneal surface *(17–20)*.

1.2.2. Fibroblasts

Fibroblasts have been shown to modulate the growth and differentiation of human conjunctival epithelial cells grown on collagen gels. Stratification and differentiation of goblet cells occurred when the carrier was populated with fibroblasts, and depended on the source of stromal cells *(21,22)*.

2. Materials
2.1. Source of Tissue

It is a prerequisite to have satisfied all legal requirements for the use of human tissue in experimentation. With the appropriate approval and consent, both cadaver material and surgical specimens may be available. Readers should inquire about the legal requirements that apply to them.

2.2. Tissue-Culture Media

A variety of defined media have been used to grow epithelial cells *(6)*. We have had most success with a 1:1 (v/v) mixture of: RPMI 1640 (21875-034 Gibco [Paisley, UK]) and Ham F10 (N 6013 Sigma [St. Louis, MO]).

To 1 L of mixture add:

1. 50 mL Fetal bovine serum (FBS) (EC approved 10106-078 Gibco).
2. 5 mL Antibiotic/antimycotic mixture (Sigma). Final dilutions are: 50 U/mL penicillin, 50 µg/mL streptomycin, and 125 ng/mL amphotericin.
3. 10 mL Nonessential amino acids (RPMI-1640 [50X] R-7131 Sigma).
4. 5 mL L-glutamine (200 mM, G7513, Sigma).
5. Epidermal growth factor (EGF) (Human recombinant; E1264 Sigma), final concentration 5 µg/L.

Prepare all solutions in a laminar flow hood with an aseptic technique and filter-sterilize (0.2-µm filter) all additives except EGF.

2.3. Dissociation Media

1. CaMg-free phosphate buffered saline (Dulbecco's PBS D5527, Sigma).
2. Trypsin–EDTA (0.25 and 0.02%, respectively, T4538 Sigma)—aliquot aseptically and filter sterilize (0.2-µm filter) before use.

2.4. Cryopreservation Medium

This consists of RPMI 1640 (21875-034 Gibco) with 20% FBS (EC-approved 10106-078 Gibco), 10% dimethyl sulfoxide (DMSO) (Sigma D8779), and antibiotic/antimycotic mixture (Sigma). Final dilutions of antibiotics are: 50 U/mL penicillin, 50 µg/mL streptomycin, and 125 ng/mL amphotericin.

2.5. Collagen Gels

1. 8 mL Collagen type I acid-soluble rat tail collagen (Sigma C8897), dissolved in 0.01% acetic acid (at 4°C) to final concentration between 2 and 4 mg/mL. Filter-sterilize dissolved collagen through a 0.45-µm filter.
2. 1.8 mL of tissue culture medium (*see* Section 2.2).
3. 200 µL of 1M sodium hydroxide.

2.6. Histochemistry

2.6.1. Cell Fixation

1. Ice-cold methanol (BDH [Poole, UK] 10158) Analar.
2. 10% Formalin + 0.5% cetylpyridinium chloride

2.6.2. Tissue and Collagen Gel Fixation

1. Periodate lysine paraformaldehyde (PLP). Prepare by mixing the following:
   a. 100 mL Lysine phosphate buffer, pH 7.4.
   b. 0.428 g Sodium periodate.
   c. 3.15 mL Paraformaldehyde (8%) in dextrose.
2. 70% Ethanol.
3. 90% Ethanol.
4. Ethanol absolute.
5. Histo-Clear (National Diagnostics).
6. Paraffin wax, congealing point 51–53°C (BDH).

2.6.3. Anticytokeratin Staining

1. Blocking buffer: bovine serum albumin (Fraction V, A4325 Sigma), 3% (w/v) made in PBS, pH 7.4, and containing 0.05–0.1% (v/v) polyoxyethylene-sorbitan monolaurate (Tween-20, P1379 Sigma).
2. Rabbit antimouse IgG (H + L) chain FITC conjugated (ICN [High Wycombe, UK] 65-171) made up 1:20 in PBS.
3. Monoclonal mouse antiepithelial keratin AE5 (ICN 69-143) made up 1:50 in PBS.
4. Monoclonal antiepithelial keratin AE1 and AE3 (ICN 69-145) made up 1:50 in PBS.

2.6.4. Periodic Acid-Schiff (PAS) Stain

1. Periodic acid (Sigma).
2. Schiff reagent (Sigma).
3. 0.1% Sodium metabisulfite (Sigma S9000) in 1 mM HCl.

2.6.5. Alcian Blue Stain

This consists of Alcian Blue 8GX (0.5%) (Sigma A3157) dissolved in 3% acetic acid adjusted to pH 1.0 with HCl.

2.6.6. Mountant for Histological Sections

This consists of Aquamount (Emscope [Watford, UK] C334).

2.7. Plastics

1. Sterile pipets:
   a. 1 mL Plugged pipets (Sterilin [Barking, UK] 40101).
   b. 10 mL Plugged pipets (Sterilin 03123).
   c. 25 mL Plugged serological pipets (Falcon [Cowley, UK] 7525).
   d. 230 mm Disposable glass Pasteur pipets (John Poulten Ltd. D812).
   e. 230 mm Preplugged disposable glass Pasteur pipets (John Poulten Ltd. D812/PP).
2. Sterile culture dishes:
   a. Sterile tissue-culture dishes 35 × 10 mm (Falcon 3001).
   b. Sterile tissue-culture dishes 60 × 15 mm (Falcon 3004).

c. Sterile organ culture dishes 60 × 15 mm (Falcon 3037).
d. Sterile 25-cm^2 vented tissue-culture flasks (Falcon 3108).
e. Sterile 75-cm^2 vented tissue culture flasks (Falcon 3110).
f. Sterile 12-well tissue-culture plates (Falcon 3503) for use with 12-mm inserts.
g. Sterile 24-well tissue-culture plates (Falcon 3504) for use with 9-mm inserts.
h. Sterile cell-culture inserts, 0.45-µm pore size, 12-mm diameter (Falcon 3180).
i. Sterile cell-culture inserts, 0.45-µm pore size, 9-mm diameter (Falcon 3095).
j. Lab-tek Chamber Slide Culture Chambers (Nunc [Naperville, IL] 177437).
3. Syringes and filters:
a. Disposable luer lock syringes.
b. Disposable luer lock syringe filter holders, 0.2-µm pore size (Sartorius [Göttingen, Germany] 165 34 K).
c. Disposable luer lock syringe filter holders, 0.45-µm pore size (Sartorius 165 55 K).
d. Sterile 0.2-µm pore size 100-mL bottle filter (Costar [High Wycombe, UK] 8310).
4. 50-mL Polypropylene screw-top conical tubes (Falcon 2070).

2.8. Laboratory Infrastructure

1. Laminar flow hood.
2. Humidified CO_2 incubator.
3. Inverted microscope with phase optics.
4. Heated water bath.
5. Suction pump.
6. Storage facilities at 4, –20, –70°C, and liquid nitrogen.

3. Methods
3.1. Collection of Tissue (see Note 1)

1. Ensure all appropriate donation and consent forms have been obtained.
2. To collect conjunctiva from a cadaver donor, first insert an eyelid speculum to part the eyelids. Use iris scissors to cut the conjunctiva all around the anatomical limbus and detach it from adherent Tenon's fascia. Make a further incision into the conjunctiva away from the limbus, and cut around to obtain an annulus of conjunctiva (see Note 1).
3. Place the conjunctiva in sterile saline or tissue-culture medium for transport.

3.2. Organ Culture (see Notes 2–6)

Organ culture is used to follow the metabolic fate of radioactive precursors in tissue fragments (see Note 2 and ref. 23).

1. Place 2 mL tissue-culture medium in the central well of an organ culture dish.
2. Cut the conjunctiva in pieces 2–4 mm^2.
3. Place lens tissue squares on triangular stainless-steel grids, and saturate the lens squares with medium. Place the grid over the well, and check that it is in contact with the medium (see Note 3).

4. Place up to six pieces of tissue on the grid.
5. Add radioactive precursor(s) to medium in central well. Add 1 mL distilled sterile water to the outside (circular) well (*see* note 4).
6. Incubate for at least 24 and up to 96 h in 5% CO_2 incubator at 35°C (*see* Notes 5 and 6).

3.3. Primary Cell Culture

Various methods have been described for the primary culture of epithelial cells. Some authors prefer to dissociate the epithelium enzymatically *in situ* and plate single cell suspensions *(9)*; others remove the epithelial sheet by intrastromal injection of protease followed by blunt dissection *(24)*. Our best results were obtained by plating explants of full thickness conjunctiva, and this method is described next.

3.3.1. Epithelial Cell Culture (see Notes 7–12)

1. Rinse the conjunctiva in tissue-culture medium for 10 min. Change medium twice (*see* Note 7).
2. Cut into explants 1–2 mm^2 on a Petri dish (*see* Note 8).
3. Plate on tissue-culture dishes (35 × 10 mm Easy Grip, Falcon) (*see* Note 9).
4. Allow to adhere for at least 15 min in humidified 5% CO_2 incubator at 37°C before adding 1 mL tissue-culture medium.
5. Leave explants in the dish until epithelial growth has been established (3–5 d) (*see* Note 10 and Fig. 1A).
6. Remove explants to minimize fibroblast contamination.
7. Explants can be replated by transfer with a needle (*see* Note 11).
8. Feed twice a week until confluent (*see* Note 12 and Fig. 1B).

3.3.2. Fibroblast Culture

This method is identical to the aforementioned, but do not remove explants. For initial outgrowth, allow at least 7 d (*see* Fig. 2). Stromal cells do not require EGF and grow more quickly in a medium containing 10% FBS.

3.4. Epithelial Cultures on Stromal Carriers

3.4.1. Collagen Gels (see Notes 13 and 14)

1. Dissolve collagen in 0.01% acetic acid, overnight at 4°C, to a final concentration of 2–4 mg/mL. Sterilize by filtering through a 0.45-µm filter *(25)*.
2. Prepare cold pipets, universals, plates, inserts, and ice pack.
3. Mix, keeping cold: 8 mL ice-cold collagen solution, 1.8 mL tissue-culture medium, and 0.2 mL of 1*M* NaOH.
4. Dispense 0.5 mL/well in a cold 24-well plate or 0.25 mL/well in 9-mm inserts.
5. Incubate at 37°C for at least 1 h.
6. Detach gels carefully with a Pasteur pipet, and equilibrate with tissue-culture medium.

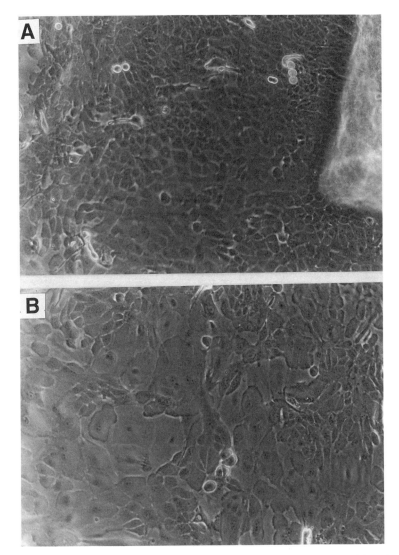

Fig. 1. (A) Primary epithelial growth from conjunctival explant after 5 d in culture. **(B)** Epithelial cells after 11 d in culture. The enlarged cells may be or may become senescent.

7. Plate fibroblasts in 100 µL medium.
8. Leave for approx 14 d feeding when necessary (*see* Notes 13 and 14).

3.4.2. Epithelialization (see Notes 15 and 16)

1. Plate epithelium on gels in a minimal volume of medium. Plate a small aliquot in a well as a control for growth (*see* Note 15).

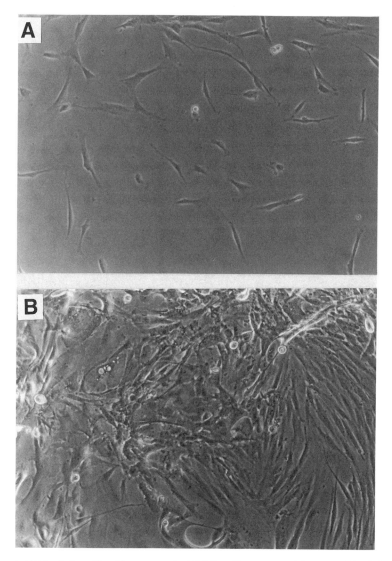

Fig. 2. (**A**) Incipient fibroblast growth. (**B**) Fibroblasts overtaking an epithelial growth.

2. When the epithelium in the control well has reached confluence, the gels can be lifted to air to improve epithelial differentiation. This is best done by placing the gel or the insert on a thick layer of sterile filters soaked in tissue-culture medium (*see* Note 16).

3.5. Passage (see Notes 17 and 18)

Both epithelial and stromal cells preserve contact inhibition of mitosis. Cultures can be amplified by dissociation and replating (*see* Note 17).

1. Rinse with cold (4°C) CaMg-free PBS (Dulbecco's PBS D5527, Sigma).
2. Add 0.5 mL filter-sterilized trypsin (0.25% trypsin–EDTA, T4538 Sigma).
3. Place at 37°C for 5 min.
4. Under the microscope, check that cells are rounding up and detaching. Agitation of the medium may help displace the cells.
5. When the cells are rounded and detached, stop trypsin activity by adding 3 mL of medium containing 10% FBS.
6. Collect in a sterile centrifuge tube.
7. Spin for 10 min at 1000 rpm.
8. Discard the supernatant.
9. Add 1 mL tissue-culture medium, and pipet up and down repeatedly to obtain a single-cell suspension.
10. To estimate cell density and viability, take a 50-µL aliquot, dilute as required with PBS and Trypan Blue, and apply the suspension to a hemocytometer. Dead cells will stain blue (*see* Note 18).
11. Plate as required. Cells from one confluent dish should provide enough cells for three new dishes.

3.6. Cryopreservation (see Note 19)

Dissociated cells can be preserved in liquid nitrogen until use.

1. Pellet dissociated cells.
2. Resuspend in 1–1.5 mL cold freezing medium in a cryotube at a density of 10^6 cells/mL.
3. Place cryotubes in a polystyrene box at –70°C overnight, and then transfer quickly to a freezing cane in liquid nitrogen (*see* Note 19).
4. To defrost for culture, heat the cryovial quickly in a 37°C bath. When thawed, quickly add 5 mL tissue-culture medium containing 10% FBS and mix.
5. Centrifuge for 10 min at 1000 rpm.
6. Resuspend the pellet in tissue-culture medium as appropriate.

4. Histology and Cytology

For this purpose, cells are best grown on well inserts or chamber slides.

4.1. Fixing and Embedding (see Note 20)

4.1.1. Gels

1. To fix gels, use PLP (*see* Note 21). Fix at 4°C for 4 h or overnight.
2. To dehydrate gels, carry out the following procedures. All solutions and incubations are at 4°C.
   a. PBS: three times for 5 min.
   b. 70% Ethanol: 45 min.
   c. 90% Ethanol: 45 min.
   d. 100% Ethanol: twice for 30 min.
   e. Histo-Clear: twice for 30 min.

3. Following dehydration, vacuum embed at 56°C in wax for 20–30 min.
4. Orientate and block in wax.
5. Leave at least 20 min on cold plate. Keep at –20°C until sectioning. Cut 5–10-μm sections, and place on polylysine-coated slides.
6. Deparaffinize before staining.

4.1.2. Cells

Fixing for mucin immunocytochemistry (*see* Note 22).

1. Discard tissue-culture medium.
2. Fix with cold methanol for 30 min at least, at –20°C (*see* Note 22).

Fixing for cytokeratin determination (*see* Notes 23 and 24).

Epithelial cells can be identified by their polygonal morphology and, more specifically, by the intermediate filaments, the cytokeratins, in their cytoskeleton. Cytokeratins have been described which are ontgenesis specific, tissue specific, and area specific. Antibody AE1 recognizes most acidic-type keratins, whereas antibody AE3 recognizes all known basic cytokeratins. AE5 has been described as cornea specific, and binds to corneal epithelium and a transitional zone on the conjunctiva *(3)*. Antibodies are not available against cytokeratins that are specific to conjunctival epithelium alone.

1. Discard tissue-culture medium.
2. Briefly rinse in PBS.
3. Fix in methanol at –20°C for 5 min.
4. Drain excess methanol, and quickly dip in acetone at –20°C (*see* Note 24).
5. Dip for 5 min each into three changes of PBS.

Cells are now ready for immunohistochemical detection of intermediary filaments.

4.2. Histological Stains

4.2.1. PAS (see Note 25)

1. Incubate sections in 1% periodic acid in 3% acetic acid for 30 min.
2. Rinse twice in 0.1% sodium metabisulfite in 1 mM hydrochloric acid for 2 min.
3. Dip in Schiff's reagent for 15 min.
4. Rinse in 0.1 % sodium metabisulfite for at least 5 min.

Counterstaining with hematoxylin–eosin can be used if desired.

4.2.2. Alcian Blue (see Note 26)

1. Best fixed in 10% formalin with 0.5% cetylpyridinium chloride.
2. Rinse with 3% acetic acid or 3% acetic acid adjusted to pH 1 with hydrochloric acid.
3. Stain with 0.5% Alcian Blue in 3% acetic acid (pH 1 or 2.5) for 30–45 min (*see* Note 26).
4. Rinse briefly in 3% acetic acid and then wash in running tap water for 3 min.

4.2.3. Detection of Cytokeratins (see Notes 27–29)

When cells have been fixed and delipidized:

1. Incubate with blocking buffer for 1 h at room temperature.
2. Incubate with 1:50 anticytokeratin antibody (in blocking buffer) for 60 min at room temperature.
3. Wash three times in PBS.
4. Incubate with 1:20 FITC-linked antibody for 30 min at room temperature in the dark.
5. Wash three times in PBS.
6. Remove upper structure of a slide chamber or well insert (*see* Note 28).
7. Mount with water-soluble mountant, e.g., Aquamount (*see* Note 29).

5. Notes

1. To remove the whole globe with its conjunctiva attached, first evert the upper lid and cut the conjunctiva with a scalpel blade from its insertion into the tarsal plate. With scissors, free the conjunctiva from underlying tissue at the deep part of the upper and lower eyelids, and leave it as a collar attached to the globe. Remove the globe according to your usual method.
2. This method is a modification of the organ-culture method described by Corfield and Paraskeva *(23)*. We adapted it for the study of ocular mucins.
3. Nonessential amino acids may be omitted from the tissue-culture medium for organ culture.
4. ^{14}C-glucosamine (185 kBq/dish) and ^{3}H-ethanolamine (1850 kBq/dish) were incorporated in both secreted and membrane-bound human conjunctival glycoproteins, but only a small part of ^{35}S-sodium sulfate (925 kBq/dish) radioactivity was recovered in mucin-like material *(10)*.
5. We have observed that goblet cells migrate off human conjunctival explants for at least 4 d. In dogs, maximal migration was observed after 48 h (S. D. Carrington, personal communication). Maintenance of organ culture is limited by media depletion and tissue availability of nutrients.
6. The isolation of mucins from organ culture is extensively described by Corfield and Paraskeva *(23)*. Several antibodies to nonconjunctival mucin crossreact with partially purified human conjunctival mucins *(26)* and goblet cells *(27)*. Unfortunately, these antibodies are not commercially available.
7. The methods described were developed for human cadaver conjunctiva collected within 24 h of death and stored at 4°C until processed.
8. It is not important to place the explants epithelium up or down. We have observed good growth irrespective of orientation.
9. Establishing primary cultures in Petri dishes has the advantage of easy access to explants and the disadvantage of increased risk of contamination. Cells grow equally well on tissue-culture flasks or cell inserts.
10. Epithelial cells will usually migrate off the explant and onto the dish before stromal cells (Fig. 1). The speed of migration and difference in onset vary. The cultures, therefore, should be inspected daily after d 3, and explants removed if any

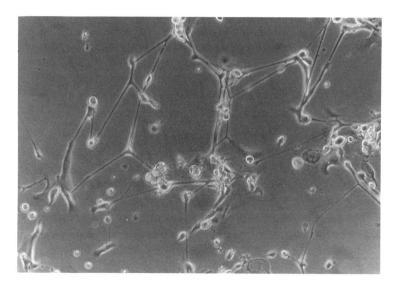

Fig. 3. "Unhealthy" epithelial growth, which may benefit from an increased concentration of fetal calf serum in the medium.

fibroblasts appear. A small number of fibroblasts can be eliminated by rubbing the area of growth with a small swab moistened with ethanol (Fig. 2).
11. Replated explants may continue to grow epithelium. Increased care is needed to remove them before the onset of vigorous stromal growth.
12. If the cells are lifting or looking excessively rounded, it may help to increase temporarily the amount of FBS in the medium (Fig. 3).
13. If fibroblasts are growing, the gels contract a lot.
14. If the gels are not on inserts, it may be easier to place them now on sterile filters in preparation for lifting to the air/liquid interface.
15. Gels become opaque with contraction, and epithelium cannot be directly visualized.
16. Care should be taken not to allow filters to dry.
17. Dissociation of preconfluent dishes is sometimes successful in restoring senescent cultures.
18. The details of Trypan Blue staining and hemocytometer use can be found in the current Sigma catalog.
19. If a controlled freezing instrument is available, a 10-min incubation with 10% DMSO at 0°C followed by cooling at 1°/min should be used.
20. This fixing and embedding method is designed to preserve cell-surface epitopes.
21. The fixative is labile, and it should be mixed just before use. Lysine-phosphate buffer and paraformaldehyde in dextrose solutions can be kept refrigerated for up to a month.
22. Fixing with methanol also inhibits endogenous peroxidases, an advantage if peroxidase-linked antibodies are used for staining.

23. Cells can be grown on tissue-culture inserts or in slide chambers. Preconfluent cultures give best results.
24. A longer (30-min) incubation with methanol at −20°C can be used to delipidize membranes instead of dipping in acetone. Some culture dishes become opaque on contact with acetone.
25. Several general histological stains are available to detect carbohydrate-rich structures, e.g., PAS, Alcian Blue, or high-iron diamine. The methods described here are useful as quick and easy tools for goblet cells.
26. All acid glycoconjugates are stained at pH 2.5; only sulfated glycoconjugates are stained at pH 1 *(28)*.
27. Incubation times and antibody concentrations have to be optimized for the tissue and antibody source used, but the titers and schedule given here are a good starting point.
28. An easy way of removing the upper structure of an insert is to trephine the filter directly onto a polylysine-coated slide.
29. If a permanent record is needed, the slides should be photographed within 24 h of staining because of the decay in fluorescence.

References

1. Setzer, P. Y., Nichols, B. A., and Dawson, C. R. (1987) Unusual structure of rat conjunctival epithelium: light and electron microscopy. *Invest. Ophthalmol. Vis. Sci.* **27,** 531–537.
2. Zieske, J. D. (1994) Perpetuation of stem cells in the eye. *Eye* **8,** 163–169.
3. Wei, Z. G., Wu, R. L., Lavker, R. M., and Sun, T. T. (1993) In vitro growth and differentiation of rabbit bulbar fornix and palpebral conjunctival epithelia: implications on conjunctival epithelium transdifferentiation and stem cells. *Invest. Ophthalmol. Vis. Sci.* **34,** 1814–1828.
4. Huang, A. J. W., Tseng, S. C. G., and Kenyon, K. R. (1990) Alteration of epithelial paracellular permeability during corneal epithelial wound healing. *Invest. Ophthalmol. Vis. Sci.* **31,** 429–431.
5. Chen, J. J. Y. and Tseng, S. G. (1991) Abnormal corneal wound healing in partial thickness removal of limbal epithelium. *Invest. Ophthalmol. Vis. Sci.* **32,** 2219–2233.
6. Freshnay, R. I. (1992) Introduction, in *Culture of Epithelial Cells* (Freshnay, R. I., ed.), Wiley-Liss, New York, pp. 1–23.
7. Garcher, C., Bara, J., Bron, A., and Oriol, R. (1994) Expression of mucin peptide and blood group ABH and Lewis related carbohydrate antigens in normal human conjunctiva. *Invest. Ophthalmol. Vis. Sci.* **35,** 1184–1191.
8. Patton, D. L., Chan, K. Y., Kuo, C. C., Casgrove, Y. T., and Langley, L. (1990) In vitro growth of Chlamidia trachomatis in conjunctival and corneal epithelium. *Invest. Ophthalmol. Vis. Sci.* **29,** 1087–1095.
9. Rapoza, P., Tahija, S. G., Carlin, J. P., Miller, S. L., Padilla, M. L., and Byrne, G. I. (1991) Effect of interferon on a primary conjunctival cell model of trachoma. *Invest. Ophthalmol. Vis. Sci.* **30,** 2919–2923.
10. Corfield, A. P., Myerscough, N., Berry, M., Clamp, J. R., and Easty, D. L. (1991) Mucins synthesised in organ culture of human conjunctival tissue. *Biochem. Soc. Trans.* **19,** 352S.

11. Frescura, M., Corfield, A. P., and Berry, M. (1993) Cultured human conjunctival cells secrete mucins. *Biochem. Soc. Trans.* **2**, 460(S).
12. Bernauer, W., Wright, P., Dart, J. K., Leonard, J. N., and Lightman, S. (1993) Cytokines in the conjunctiva of acute and chronic mucous membrane pemphigoid. *Invest. Ophthalmol. Vis. Sci.* **34**, 857.
13. Chen, M. and Wolosin, J. M. (1992) An organ culture model to study glycocalyx synthesis in corneal epithelium; evidence of inhibition of terminal glycosylation by retinoic acid. *Invest. Ophthtal. Vis. Sci.* **33**, 1239.
14. Sullivan, D. A. (ed.) (1994) *Lacrimal Gland, Tear Film and Dry Eye Symptoms.* Plenum, New York.
15. Simmons, S. J., Jumblatt, M. M., and Neufeld, A. H. (1987) Corneal epithelial wound closure in tissue culture: an in vitro model of ocular irritancy. *Toxicol. Appl. Pharmacol.* **88**, 13–23.
16. Hobson, D. W. and Blank, J. A. (1991) In vitro alternative methods for the assessment of dermal irritation and inflammation, in *Dermal and Ocular Toxicology— Fundamentals and Methods* (Hobson, D. W., ed.), CRC, London, pp. 323–368.
17. Lindberg, K., Brown, M. E., Chaves, H. V., Kenyon, K. R., and Rheinwald, Y. G. (1993) In vitro propagation of human ocular surface epithelial cells for transplantation. *Invest. Ophthalmol. Vis. Sci.* **34**, 2672–2679.
18. Clinch, T. E., Goins, K. M., and Cobo, L. M. (1992) Treatment of contact lens-related ocular surface disorders with autologous conjunctival transplantation. *Ophthalmology* **99**, 634–638.
19. He, Y.-G. and McCulley, J. P. (1991) Growing human corneal epithelium on collagen shield and subsequent transfer to denuded cornea in vitro. *Curr. Eye. Res.* **10**, 851–863.
20. Thoft, R. A. (1982) Indications for conjunctival transplantation. *Ophthalmology* **89**, 335–339.
21. Chen, W. Y. W. and Tseng, S. C. G. (1992) Differential intrastromal invasion by cultured corneal and conjunctival epithelium: an event also modulated by different stromal fibroblasts. *Invest. Ophthalmol. Vis. Sci.* **33**, 1175.
22. Tsai, R. J.-F., Ho, Y.-S., and Chen, J.-K. (1994) The effects of fibroblasts on the growth and differentiation of human bulbar conjunctival epithelium in an in vitro conjunctival equivalent. *Invest. Ophthalmol. Vis. Sci.* **35**, 2865–2875.
23. Corfield, A. P. and Paraskeva, C. (1993) Secreted mucus glycoproteins in cell and organ culture, in *Methods in Molecular Biology, Vol. 14: Glycoprotein Analysis in Biomedicine* (Hounsell, E. F., ed.), Humana Press, Totowa, NJ, pp. 119–210.
24. Geggel, H. S. and Gipson, I. K. (1985) Removal of viable sheets of conjunctival epithelium with Dispase II. *Invest. Ophthalmol. Vis. Sci.* **26**, 15–22.
25. Fusening, N. E. (1992) Cell interactions and epithelial differentiation, in *Culture of Epithelial Cells* (Freshnay, R. I., ed.), Wiley-Liss, New York, pp. 25–37.
26. Berry, M., Corfield, A. P., and Easty, D. L. (1994) Partial purification of human ocular mucins. *Invest. Ophthalmol. Vis. Sci.* **35**, 1795.
27. Ellingham, R. B., Berry, M., Corfield, A. P., and Easty, D. L. (1994) Mucins in impression cytology of human conjunctiva. *JERMOV Abstr.* **205**.
28. Kiernan, J. A. (1992) *Histological and Histochemical Methods*, 2nd ed., Pergamon, Oxford.

40

Establishment and Maintenance of In Vitro Cultures of Human Retinal Pigment Epithelium

Eva L. Feldman, Monte A. Del Monte, Martin J. Stevens, and Douglas A. Greene

1. Introduction

The retinal pigment epithelium (RPE) is a layer of multipotential cells of neural ectoderm origin lying between Bruch's membrane and the neural retina. The RPE subserves several essential ocular functions, including phagocytosis of shed photoreceptor outer segments, maintenance of the blood–retinal barrier, absorption of stray light, regulation of the biochemical, metabolic, and ionic composition of the subretinal space, and induction of embryonic differentiation of adjacent neural retina and choroid *(1)*. Experimental evidence indicates that early in embryonic life, the neural retina can regenerate from the pigment epithelium *(2)*. In vitro cultures of pure RPE provide a vehicle for studying RPE function in both normal and diseased states, and may also serve as a model for other neural cells *(3,4)*. Multiple techniques have been described for culturing human RPE *(5–12)*. The authors describe here a modification of the technique of Del Monte and Maumenee *(10)*, which is simple and effective in establishing primary cultures and extended cell lines of human RPE for research.

2. Materials

1. Hank's balanced salt solution (HBSS) without Ca^{2+} and Mg^{2+}, and Earle's balanced salt solution are supplied by Gibco (Grand Island, NY), and are stored at 4°C or room temperature.
2. Ham's F12 Nutrient Mixture (HAM's F12) and Dulbecco's minimum essential medium (DMEM) are supplied by Gibco and stored at 4°C.
3. Bovine calf serum (defined and iron-supplemented) and fetal calf serum (defined) are supplied by Hyclone Laboratories (Logan, UT), sterilely divided into 50-mL aliquots in 50-mL tubes, and stored at –20°C.

4. L-Glutamine, penicillin/streptomycin, and sterile sodium bicarbonate solutions are supplied by Gibco and stored at –20°C.
5. Establishment of primary cultures of human RPE is accomplished with the following recipe: Ham's F12 Nutrient Mixture, 16% fetal bovine serum, 0.02 mM L-glutamine, 100 U/mL penicillin/100 µg/mL streptomycin, 0.075% (w/v) sodium bicarbonate.
6. Trypsin-EDTA is supplied by Gibco, sterilely divided into 10-mL aliquots in 15-mL tubes, and stored at –20°C.
7. Papain (cat. no. P3125) and L-cysteine-HCl (cat. no. C7880) are supplied by Sigma (St. Louis, MO). Papain stock solution is prepared in advance in Ca^{2+}- and Mg^{2+}-free HBSS by adding 3 mM L-cysteine HCl, 1 mM EDTA, and 10 µL/mL papain. The stock solution is stable for 2 wk when stored at 4°C.
8. Acid-soluble collagen (Sigma type III) is supplied by Sigma and stored at –20°C.
9. Silane (cat. no. M6514) for treating pipets is supplied by Sigma. Pipets are coated by rinsing in 0.2% silane, then chloroform, ethanol, and several water rinses, or by coating with silane vapor in a desiccator and then autoclaved to sterilize.

3. Methods

3.1. Preparation of Collagen-Coated Culture Vessels (see Note 1)

1. Coating of dishes with acid-soluble type I collagen is easily accomplished as follows: 0.5% solution of acid-soluble collagen in 0.1M acetic acid is painted on the surface of the culture vessels with a fine brush.
2. The thin collagen coating is gelled by exposure to NH_3 fumes from ammonium hydroxide for 12 h.
3. The excess ammonia is neutralized with buffered saline.
4. Dishes are sterilized by overnight exposure to UV light.

3.2. Initial Dissection, Cell Isolation, and Primary Culture (see Note 2)

1. Postmortem eyes are obtained from an eye bank within 96 h of death, and the anterior segment of the eye is removed by circumferential incision 5 mm posterior to the limbus.
2. The vitreous and neurosensory retina are separated from the pigment epithelium, cut free at the optic nerve, and removed.
3. The eyecup is rinsed three times with HBSS without Ca^{2+} or Mg^{2+}. It is important to keep the cyccup moist throughout the isolation procedure.
4. The eyecup is filled with stock HBSS/papain solution and incubated for 40 min at 37°C or until the RPE cell layer appears marbled.
5. The papain reaction is stopped by the addition of 50 µL of fetal bovine serum to the eyecup.
6. The HBSS/papain solution is removed from the eyecup and replaced with Ham's F12-medium without serum.
7. The RPE cells are dislodged and suspended with a stream of medium from a fire-polished Pasteur pipet treated with 0.2% silane to prevent sticking. Remaining adher-

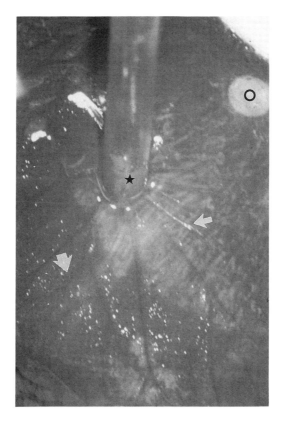

Fig. 1. Dissecting microscope view of harvesting human RPE from the posterior pole of the eye using a fire-polished Pasteur pipet. Velvety-appearing RPE (large arrow) is being gently vacuumed into the pipet (star) leaving behind the shiny Bruch's membrane (small arrow). Optic disk is marked with ○ for orientation.

ent RPE is gently vacuumed free from Bruch's membrane using the fire-polished Pasteur pipet under direct observation with a dissecting microscope as shown in Fig. 1.
8. The cell/papain solution is removed from the eyecup, placed in a sterile tube, and centrifuged at 50g for 3 min to pellet the cells.
9. The cells are resuspended and washed in HBSS.
10. The washed pellet is resuspended in complete media consisting of Ham's F12-M media supplemented with 16% fetal calf serum, 0.075% sodium bicarbonate, 0.02 mM glutamine, 100 U/mL penicillin, and 100 µg/mL streptomycin.
11. Cells are distributed evenly into sterile uncoated or extracellular matrix-coated 35-mm plastic culture dishes or the wells of 6- or 24-well culture plates. Twenty-four-well PRIMARIA culture plates (Falcon Plastics, Franklin Lakes, NY) provide excellent cell adhesion and colony formation characteristics, and are recommended for most primary culture applications. Other dishes or extracellu-

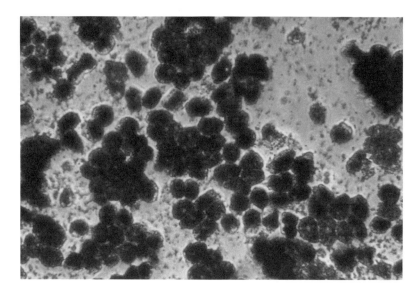

Fig. 2. Primary explants of human RPE as examined by phase-contrast microscopy immediately after harvesting (140×).

lar matrix coatings are fine for established cultures. Fresh RPE explants consist of single or small clusters of polygonal, deeply pigmented epithelial cells, which appear much as they do in vivo as seen in Fig. 2.

12. Cultures are incubated at 37°C in a humidified 95% air, 5% CO_2 atmosphere. After 2 d, the media are aspirated and replaced. Cell attachment and initiation of colony formation are monitored by phase-contrast microscopy.
13. Alternatively, at step 6, medium with serum is added to the eyecup, and small patches of cells are dislodged and plated directly into culture dishes or plates without the centrifugation step. This technique is especially useful for obtaining cultures from specific locations of the globe, such as the macular area, midperiphery, or periphery, to study regional differences.

3.3. Establishment and Maintenance of RPE Cell Lines (see Notes 3–6)

1. The medium is aspirated, and the cells are rinsed with 5 mL HBSS without Ca^{2+} or Mg^{2+} for 5 min. Cells can then be removed from the dishes by addition of either 2 mL 0.05% trypsin in 0.53 mM EDTA or 2 mL of stock HBSS/ papain solution. The cells are then incubated at 37°C for 2–15 min until they are rounded and begin to detach from the plastic.
2. Trypsinization is terminated by addition of 1 vol of medium supplemented with 10–20% calf serum.
3. The cell suspension is centrifuged at 50g for 5 min and resuspended in medium. Established cell lines can be maintained in simpler media, such as DMEM with

**Table 1
Donor Factors Influencing Human RPE Culture Success**

Donor characteristics	Success/attempts	%
Age		
<40	3/3	100
40–60	6/7	85
60–80	6/8	75
>80	1/3	33
Sex		
Male	6/8	75
Female	10/13	77
Time from death, h		
<24	5/5	100
24–48	9/10	90
48–72	2/4	50
>72	0/2	0

Adapted from ref. *13* with permission.

16% fetal bovine serum replaced by as low as 2–5% calf serum or even, for short defined periods of time, with defined serum-free medium.
4. Cells are subcultured at a density of 10,000–20,000 cells/cm^2 in T-25 or T-75 tissue-culture flasks.
5. Cells are maintained at 37°C and humidified 95% air, 5% CO_2 atmosphere with fresh medium changes three times weekly.

4. Notes

1. Cell attachment and colony formation are enhanced, especially in eyes from older donors and those obtained later after death, by coating the culture vessels with collagen or other artificial extracellular matrix *(10,14)*. However, with the development of specially treated tissue-culture plastics, such as PRIMARIA (Falcon Plastics), cell adhesion and colony formation are excellent. Therefore, the use of artificial extracellular matrix is usually unnecessary today unless required for specific experimental conditions.
2. Donor characteristics influence the establishment of successful cultures: The highest rates of viable cultures are found if patients are <60 yr of age and deceased <48 h. Table 1 summarizes donor characteristics that correlated with viable cultures in 21 consecutive pairs of eyes.
3. After initiation of primary culture, cell adhesion occurs during the first 24–48 h. Viable cells begin to divide and establish heavily pigmented primary colonies within 5–14 d as seen in Fig. 3. Dividing cells become large, flattened, and polygonal with pigment granules clumped around a prominent nucleus with one or two nucleoli (Fig. 4). The peripheral cytoplasm is thin and relatively transparent with prominent and often refractile cell/cell boundaries.

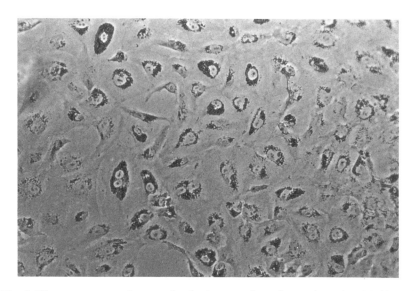

Fig. 3. Phase-contrast micrograph of primary colony formation after 14 d in culture showing pigmented, epithelioid cells, some with dividing nuclei (100×).

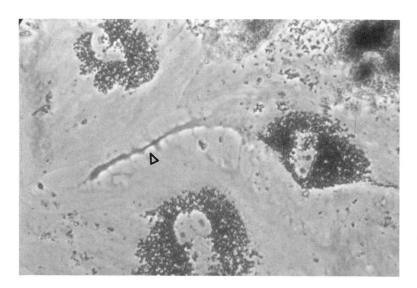

Fig. 4. Higher power view of primary culture of human RPE after 21 d in culture showing dividing epithelioid cells with dense pigment surrounding the nucleus, attenuated transparent peripheral cytoplasm with refractile cell/cell boundaries (phase contrast 220×).

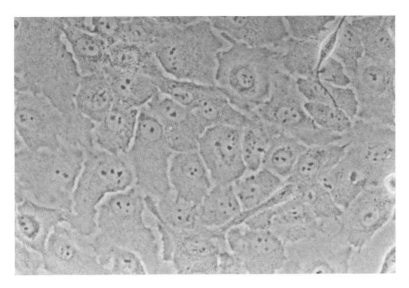

Fig. 5. Established cell line from human RPE after 13th passage. Epithelioid features are maintained in spite of total loss of pigment by dilution during cell division (phase contrast 160×).

4. In successful cultures, dense colonies expand to fill 25–50% of the culture surface within 14–28 d. Cell lines can be established and maintained for over 20 passages.
5. Improved cell adhesion and colony formation in established cultures and cell lines are encouraged by using dishes coated with artificial extracellular matrix, such as collagen Type I (Sigma), as described above (10), or collagen Type IV, laminin, and Matrigel (all available from Collaborative Research, Bedford, MA) (14).
6. Established human RPE cell lines from adult donors produced in this way maintain polygonal epithelioid morphological characteristics and many in vivo functional characteristics, making them excellent models for study of the cell biology of these important neural-derived cells under normal and pathological conditions. In addition, future refinements and improvements in culture conditions should augment the utility of these cells in vision and neurobiology research. However, melanin synthesis generally does not occur after the fetal period and cells in established cultures become slowly depigmented by pigment granule dilution during successive cell divisions, such that third or fourth passage cells are almost completely devoid of pigment as shown in Fig. 5.

Acknowledgments

These studies were supported by USPHS Research Grants R29 NS32843 (E. L. Feldman), RO1-DK38304 (D. A. Greene), and the Michigan Diabetes Research and Training Center P60-DK20572 (E. L. Feldman, M. J. Stevens, and D. A. Greene), and a grant from the Skillman Foundation (M. A. Del Monte).

References

1. Zinn, K. M. and Benjamin-Henkind, J. (1991) Retinal pigment epithelium, in *Biomedical Foundations of Ophthalmology*, vol. 1 (Duane, T. D., ed.), Harper & Row, Philadelphia, PA, ch. 21, pp. 1–20.
2. Coulombre, J. L. and Coulombre, H. A. (1965) Regeneration of neural retina from the pigmented epithelium in the chick embryo. *Dev. Biol.* **17,** 79–83.
3. Newsome, D. A. (1983) Retinal pigmented epithelium culture: current applications. *Trans. Ophthalmol. Soc. UK* **103,** 458–466.
4. Boulton, M. E., Marshall, J., and Mellerio, J. (1982) Human retinal pigment epithelial cells in tissue culture: a means of studying inherited retinal diseases. *Birth Defects* **19,** 101–118.
5. Albert, D. M. and Buyukmihci, N. (1979) Tissue culture of the retinal pigment epithelium, in *The Retinal Pigment Epithelium* (Zinn, K. M. and Marmor, M. F., eds.), Harvard University Press, Cambridge, MA, ch. 16, pp. 277–292.
6. Albert, D. M., Tso, M. O. M., and Rabson, A. S. (1972) *In vitro* growth of pure cultures of retinal pigment epithelium. *Arch. Ophthalmol.* **88,** 63–69.
7. Mannaugh, J., Dhaarmendra, V. A., and Irvine, A. R. (1973) Tissue culture of human retinal pigment epithelium. *Invest. Ophthalmol. Vis. Sci.* **12,** 52–64.
8. Edwards, R. B. (1982) Culture of mammalian retinal pigment epithelium and neural retina. *Methods Enzymol.* **81,** 39–43.
9. Flood, M. T., Gouras, P., and Kjeldbye, H. (1980) Growth characteristics and ultrastructure of human retinal pigment epithelium *in vitro*. *Invest. Ophthalmol. Vis. Sci.* **19,** 1309–1315.
10. Del Monte, M. A. and Maumenee, I. H. (1981) *In vitro* culture of human retinal pigment epithelium for biochemical and metabolic study. *Vis. Res.* **21,** 137–142.
11. Pfeffer, B. A., Clark, V. M., Flanery, J. G., and Bok, D. (1986) Membrane receptors for retinol-binding protein in cultured human retinal pigment epithelium. *Invest. Ophthalmol. Vis. Sci.* **27,** 1031–1040.
12. Mircheff, A. K., Miller, S. S., Farber, D. B., Bradley, M. E., O'Day, W. T., and Bok, D. (1990) Isolation and provisional identification of plasma membrane populations from cultured human retinal pigment epithelium. *Invest. Ophthalmol. Vis. Sci.* **31,** 863–878.
13. Del Monte, M. A. and Maumenee, I. H. (1980) New techniques for *in vitro* culture of human retinal pigment epithelium, in *Birth Defects: Original Article Series*, vol. XVI, number 2, pp. 327–338.
14. Dutt, K., Scott, M. M., Del Monte, M., Brennan, M., Harris-Hooker, S., Kaplan H. J., and Verly, G. (1991) Extracellular matrix mediated growth and differentiation in human pigment epithelial cell line 0041. *Curr. Eye Res.* **10,** 1089–1100.

41

Demonstration of Mycoplasma Contamination in Cell Cultures by a Mycoplasma Group-Specific Polymerase Chain Reaction

Frank van Kuppeveld, Jos van der Logt, Karl-Erik Johansson, and Willem Melchers

1. Introduction

Cell cultures are widely used in both medical and biotechnical research centers, industries, and also as diagnostic tools in a clinical setting. It has been reported that up to 50% of cell cultures are contaminated with mycoplasmas *(1)*. Mycoplasma contamination may alter cellular growth characteristics, enzyme patterns, and cell membrane composition, and can induce chromosomal abnormalities and cytopathogenic changes *(2–5)*. In experimental results being published, it is now becoming standard practice to show that mycoplasma-free cell cultures have been used. For industrial production of biological materials derived from mammalian cells and intended for diagnostic or therapeutic use in humans, regulatory guidelines are emerging that are intended to guarantee the quality and safety of these products. Obligatory testing for mycoplasmas is therefore increasing *(6)*.

Several methods have been described for the detection of mycoplasma contamination in cell cultures, but all methods have their drawbacks. Mycoplasma culture requires specific growth conditions and is often time consuming *(7,8)*. Results obtained with DNA fluorescent staining and biochemical assays may be difficult to interpret *(9–11)*. Molecular detection assays with probes directed to the mycoplasmal rRNA gene are commercially available, but lack sufficient sensitivity for the detection of low-level infection. This assay is also hampered by aspecific hybridization to gram-positive bacteria *(6,12,13)*. A combination of these different methods is often necessary to obtain reliable results *(1)*. Recently, attention has focused on the polymerase chain reaction (PCR) for the

From: *Methods in Molecular Medicine: Human Cell Culture Protocols*
Edited by: G. E. Jones Humana Press Inc., Totowa, NJ

detection of mycoplasma contamination in cell cultures. The high sensitivity of this method is provided by the in vitro amplification of the target sequence to be analyzed *(14)*. The principle of the polymerase-mediated amplification reaction is shown in Fig. 1. A pair of synthetic oligonucleotide primers of about 20 nucleotides in length, complementary to sequences flanking the particular region of interest, are used to direct DNA synthesis in opposite and overlapping directions. Repeated cycles of DNA denaturation, primer annealing, and elongation with a thermostable DNA polymerase are used. The cycle is repeated about 40 times, and the final result is a more than 10^{13}-fold amplification of the target sequence. Theoretically, a single target sequence (e.g., one single mycoplasma) can be detected.

Several research groups have described PCR assays with the mycoplasma 16S rRNA as a common target *(15–20)*. The use of ribosomal RNA as a target in PCR has several advantages. In addition to the unique sequence features *(21)*, which facilitate the selection of highly specific primers, rRNA molecules form part of all ribosomes, and the RNA can therefore be used as a PCR target independent of gene expression. Since rRNA is naturally present in high copy numbers (up to 10,000 molecules/cell *[22]*), it provides a target for a highly sensitive PCR assay *(23)*. Five mycoplasmas, *Acholeplasma laidlawii, Mycoplasma arginini, Mycoplasma fermentans, Mycoplasma hyorhinis,* and *Mycoplasma orale,* are considered to be responsible for 98% of cell-culture contaminations *(6,9)*. The remaining 2% include other mollicute species. It should, therefore, be possible to detect this broad range of organisms specifically by PCR.

Three research groups have evaluated their PCR assay with conventional mycoplasma detection methods. All concluded that PCR is superior to any other method or combination of methods *(17–19)*. However, in these assays, either a combination of primers or a nested-PCR reaction had to be used to obtain either sufficient specificity or sufficient sensitivity, thereby making the assays less applicable for routine screening *(17–19)*.

Recently, in our laboratory, a 16S rRNA-based mycoplasma group-specific PCR assay has been developed, which requires the use of a single primer set to detect all mycoplasma species *(20)*. It appeared to be possible to select a mycoplasma-specific primer set after systematic computer alignments of the known mycoplasmal 16S rRNA sequences with sequences of other prokaryotes *(23,24)*. In vitro amplification by the PCR with this primer set resulted in amplification of the sequences of the members of *Mycoplasma, Ureaplasma, Spiroplasma,* and *Acholeplasma* genera, but not in amplification of sequences of any other pro- or eukaryote *(20)*. The suitability of this PCR assay for the detection of mycoplasma contamination has been investigated in a large-scale study by a comparison of the results with those obtained by microbiological

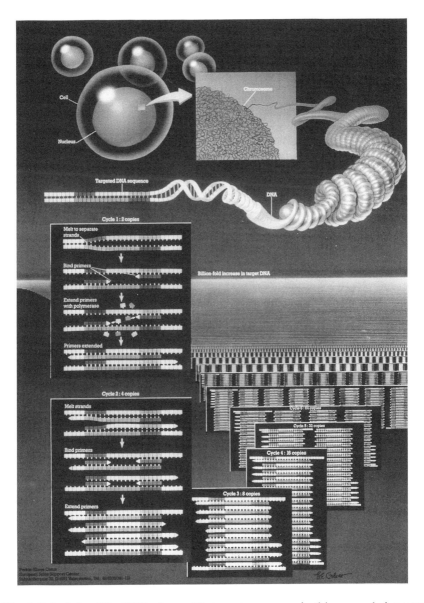

Fig. 1. Principle of the PCR. A specific target sequence, in this example integrated proviral HIV-1 DNA, is amplified by repeated cycles of denaturation, primer binding, and primer extension as indicated in the figure (reproduced with permission of Perkin Elmer Cetus, Germany).

culture, DNA fluochrome staining, and DNA–rRNA hybridization *(20)*. It appeared that the PCR was the method of choice for the routine screening of cell cultures for the presence of contaminating mycoplasmas. The method is easy to perform, requires no specific experience, and can therefore be incorporated in any laboratory. It is relatively fast, and results can be obtained within 2 d. The method is specific, sensitive, and superior to any other mycoplasma detection method. It does not have to be used in combination with other methods and produces highly reliable results. The amplification can be completely automated with specially designed equipment. Since the cost of such a PCR thermocycler is nowadays relatively low, the investment can be recommended. Because no radioactive step is involved, no specific laboratory conditions are required, except to avoid contamination (*see* Section 3.2.). The PCR assay as described in this chapter is ready to use, but it is expected that modifications, such as sample preparation, the use of carryover prevention, and the direct colorimetric detection systems, will further simplify the method in the near future.

2. Materials
2.1. DNA Extraction

1. DNA to be purified: Cell-culture sample containing 10^5 cells in 0.5 mL culture medium (*see* Note 1).
2. 10% Sodium dodecyl sulfate (SDS).
3. 5 mg Proteinase K/mL distilled water. Proteinase K is unstable and must be prepared fresh with each use.
4. Phenol: Ultrapure, ready to use (Boehringer Mannheim Biochemica, Mannheim, Germany).
5. 25:24:1 (v/v/v) Phenol/chloroform/isoamyl alcohol.
6. 24:1 (v/v) Chloroform/isoamyl alcohol.
7. 3*M* Sodium acetate buffer, pH 5.2.
8. 96% Ethanol, ice-cold.
9. 70% Ethanol, room temperature.
10. Distilled water.

2.2. PCR: Amplification Procedure

1. Purified DNA sample (*see* Section 2.1.).
2. 10X PCR buffer: 500 m*M* potassium chloride, 100 m*M* Tris-HCl, pH 9.0, 15 m*M* magnesium chloride, 1% Triton X-100, 0.1% gelatin.
3. 5X Deoxynucleotide triphosphate mix (dNTP): 1 m*M* dATP, 1 m*M* dTTP, 1 m*M* dCTP, 1 m*M* dGTP, in 50 m*M* Tris-HCl, pH 8.3. Aliquot in small volumes, store at –20°C, and use only once.
4. Oligonucleotide primers (Note 2):
    GPO-3: 5'-GGGAGCAAACAGGATTAGATACCCT-3'
    MGSO: 5'-TGCACCATCTGTCACTCTGTTAACCTC-3'

Detection of Mycoplasmas by PCR

Work solution is 50 pmol primer/μL distilled water. Stock solution must be kept apart and stored at −20°C.

5. SuperTaq DNA polymerase (5 U/μL stock stored at −20°C in work solutions, HT Biotechnology, Cambridge, UK): The specific content of the PCR buffer may alter significantly when using other *Taq* DNA polymerases. Normally, the manufacturers supply their specific reaction buffer and/or conditions.
6. Mineral oil.

2.3. Identification of Amplified DNA

2.3.1. Agarose Gel Electrophoresis

1. Electrophoresis-grade agarose.
2. 10 mg Ethidium bromide/mL distilled water.
3. 10X TBE buffer: 890 mM Tris-HCl, pH 8.2, 890 mM boric acid, 20 mM EDTA.
4. 6X Loading buffer: 0.25% bromophenol blue, 0.25% xylene cyanol, 15% Ficoll 400 in water.
5. 100-bp Ladder size marker (Pharmacia Biotech, Uppsala, Sweden).

2.3.2. Southern Blot Analysis

1. Agarose gel containing samples to be analyzed.
2. 0.2M HCl.
3. Alkaline buffer: 0.5M sodium hydroxide, 1.5M sodium chloride.
4. Nylon membrane (Hybond, Amersham Int., Amersham, UK).

2.4. Oligonucleotide Probe Labeling

1. 5X TdT buffer: 1M potassium chloride, 125 mM Tris-HCl, pH 6.6, 1.25 mg/mL bovine serum albumin (BSA).
2. 25 mM Cobalt chloride.
3. Terminal dideoxynucleotidyl transferase (TdT, 50 U/μL Boehringer Mannheim Biochemica).
4. Oligonucleotide probe: 5'-CTTAAAGGAATTGACGGGAACCCG-3' 30 pmol/μL distilled water.
5. 1 mM DIG-11-ddUTP (Boehringer Mannheim Biochemica).
6. 4M Lithium chloride.
7. 100% Ethanol (ice-cold).
8. 70% Ethanol (ice-cold).
9. Distilled water.

2.5. Hybridization

1. Nylon membranes containing transferred DNA.
2. 20X SSC: 3M sodium chloride, 300 mM sodium citrate, pH 7.0.
3. 100X Denhardt's solution: 2% BSA, 2% Ficoll, 2% polyvinylpyrrolidone.
4. 10% SDS.
5. 10 mg Sonicated, denatured salmon sperm DNA/mL distilled water.
6. Labeled oligonucleotide probe (30 pmol/20 μL, see Section 2.4.).

2.6. Detection of the Hybridized Probe

Use DIG Luminescent Detection Kit for nucleic acids on nylon membranes (Boehringer Mannheim Biochemica).

3. Methods
3.1. DNA Extraction

In rats infected with *M. hominis*, a higher sensitivity was obtained by rRNA amplification than by rDNA amplification *(25)*. However, RNA amplification requires an additional reverse transcriptase step, which makes the procedure more complex and expensive. In a large comparative study between rRNA and rDNA amplification of cell-culture samples, we have found no differences in the number of positive samples, indicating that rDNA amplification in most cases yields sufficient sensitivity for mycoplasma detection in tissue-culture samples. We therefore use rDNA as target for the amplification.

1. Add 25 µL SDS and 50 µL proteinase K to 0.5 mL cell-culture sample in a 1.5-mL microcentrifuge tube.
2. Gently mix the solution, and incubate for 2 h at 37°C.
3. Add 0.5 mL phenol.
4. Shake thoroughly on a shaker for 10 min and microcentrifuge for 10 min.
5. Carefully transfer the upper aqueous phase to a new tube.
6. Add 0.5 mL phenol/chloroform/isoamyl alcohol.
7. Shake thoroughly on a shaker for 10 min, and microcentrifuge for 10 min.
8. Transfer the upper aqueous phase to a new tube.
9. Add 0.5 mL of chloroform/isoamyl alcohol.
10. Shake thoroughly on a shaker for 5 min, and microcentrifuge for 10 min.
11. Transfer the upper aqueous phase containing the extracted DNA to a new tube.
12. Add 50 µL (0.1 vol) $3M$ sodium acetate buffer (pH 5.2) to the solution of DNA and mix briefly.
13. Add 1.25 mL (2.5 vol) 96% ice-cold ethanol, and gently mix the solution.
14. Precipitate the DNA either for 2 h at –80°C or overnight at –20C.
15. Microcentrifuge for 15 min, and discard the supernatant.
16. Add 1 mL of 70% ethanol to the DNA pellet, and gently revolve the tube.
17. Microcentrifuge for 10 min, and carefully discard the supernatant.
18. Remove the remaining ethanol by air- or vacuum drying of the pellet. Do not completely dry the DNA pellet; otherwise the DNA will be difficult to dissolve.
19. Dissolve the purified DNA in 50 µL of distilled water. Either use directly in PCR or store at –20°C until use.

3.2. PCR

An example of mycoplasma detection in cell cultures by PCR is shown in Fig. 2. In general, clear-cut and easy to interpret results are obtained directly after agarose gel electrophoresis. However, it is also generally considered that

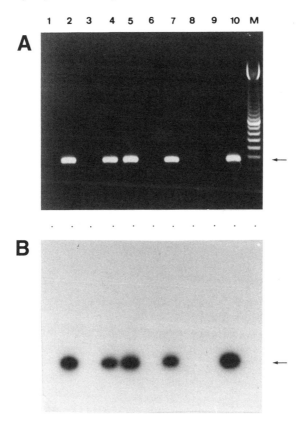

Fig. 2. Detection of mycoplasma contamination in cell cultures by PCR. In this example, eight different cell cultures (lanes 1–8) were analyzed for the presence of mycoplasma contamination. In four cell cultures (lanes 2, 4, 5, and 7), mycoplasma DNA was detected, resulting in an amplification product of 280 bp (arrow). The positive amplification was confirmed by hybridization with a <DIG>-labeled oligonucleotide probe. **(A)** Agarose gel electrophoresis; **(B)** Southern-blot analysis; lane 9: negative control (distilled water); lane 10: positive control (*M. pneumonia* DNA); lane M: Size marker (100-bp ladder).

PCR results must be confirmed by hybridization with a specific oligonucleotide probe that is directed to the internal portion of the amplified fragment. Because of the high sensitivity of the PCR, hybridization is not primarily performed to increase the sensitivity, although samples negative by gel electrophoresis can still become positive after hybridization *(26)*. The main reason for hybridization is to confirm the gel results and to increase the specificity. Although many hybridization assays are still based on the use of radiolabeled probes (as in our previous studies with mycoplasma), we are now using nonradioactive digoxigenin-labeled probes, by which the same sensitivity can be obtained as with radioactive probes. This facilitates the introduction of the PCR procedure in any laboratory where cell cultures are used.

PCR is highly sensitive to contamination, resulting in false-positive results or interpretation difficulties. The main sources of contamination are the amplified PCR products themselves *(27)*. A method that prevents the carryover of PCR products has been developed (Note 3). The transfer of infectious agents from positive to negative samples must also be avoided. Strict separation of the different technical steps involved in PCR amplification at different spatially separated locations and working with aliquoted reagents are absolute requirements for successful PCR analysis. Relevant negative controls should be included in each assay. In addition to this, the precautions as described by Kwok and Higuchi *(28)* should be taken. In case contamination occurs, one should remove all working solutions and start again.

3.2.1. Amplification Reaction

1. Location 1, mix for each single reaction:
    a. 10 μL 10X PCR buffer.
    b. 20 μL 5X deoxynucleotide triphosphate mix (dNTP mix).
    c. 1 μL primer GPO-3.
    d. 1 μL primer MGSO.
    e. 0.2 U of SuperTaq DNA polymerase.
2. Location 2:
    a. Add 25 μL of the extracted DNA and 43 μL of distilled water to the reaction mixture to a final volume of 100 μL. Half of the sample can be stored at −20°C in case the assay has to be repeated.
    b. Mix the components and spin down for 5 s.
    c. Overlay the samples with two drops of mineral oil to prevent evaporation.
3. Location 3:
    a. Start with an initial denaturation of the DNA for 5 min at 94°C.
    b. Subject the reactions to 40 cycles of amplification: 1 cycle: 1 min 94°C (DNA denaturation), 1 min 55°C (primer annealing), and 2 min 72°C (chain elongation).
    c. After the last cycle, include a final step of 10 min at 72°C. Amplification will result in the generation of a 280-bp product *(see* Note 4).

3.2.2. Identification of Amplified DNA

3.2.2.1. AGAROSE GEL ELECTROPHORESIS

1. Add 6 g agarose to 300 mL of 0.5X TBE buffer.
2. Melt the agarose by heating, and swirl to ensure even mixing.
3. Cool the melted agarose to approx 50°C.
4. Add 3 μL of ethidium bromide solution, and mix gently, avoiding air bubbles.

5. Prepare the gel frame, and pour the agarose onto the gel platform. Let the solution gel.
6. Add 4 µL of 6X loading buffer to 20 µL of amplified DNA. Include a molecular-size marker per series of samples.
7. Separate the DNA fragments by electrophoresis in 0.5X TBE buffer.
8. Turn off the power supply when the bromophenol blue dye from the loading buffer has migrated approx 8 cm into the agarose gel.
9. Analyze the reactions by visualization on a UV transilluminator, and document the results by photography.

3.2.2.2. SOUTHERN-BLOT ANALYSIS

The principle of Southern blotting is that DNA fragments are transferred from a nonsolid carrier (agarose gel) and immobilized to a solid carrier (membrane). The DNA becomes strongly bound to the membrane, which is necessary for hybridization. Primarily, it is performed by diffusion or capillary blotting. The technique was described in 1975 by Southern *(29)*, and is nowadays a standard technique in molecular biology *(30)*.

For Southern-blot analysis:

1. Depurinate DNA in the agarose gel by gently shaking the gel in 0.2M HCl. The solution must cover the gel. Depending on the thickness of the gel, this will take approx 20 min. As an indicator, we use the color change of the bromophenol blue dye present in the sampling buffer to a yellow color.
2. Pour off the HCl solution, and rinse the gel in the alkaline blot buffer to denature the DNA. Gently shake until the yellow color has changed to blue again. The solution must cover the gel.
3. Blot the gel to a nylon membrane (for example, Hybond, Amersham) of exactly the gel size in alkaline blot buffer by capillary blotting. The transfer will be completed after approx 12 h (*see* Note 5).
4. Dissassemble the transfer pyramid.
5. Mark the positions of the wells on the membrane to identify the specific lanes and the size marker.
6. Bake the membrane for 2 h at 80°C to dry the membrane completely.
7. Store the membrane between Whatmann 3-MM paper until use.

3.3. Labeling of the Oligonucleotide Probe

Two major choices for the labeling of the oligonucleotide probe and subsequent hybridization of the Southern blot are possible: radioactive labeling and nonradioactive labeling. Radioactive labeling is generally performed by the addition of a labeled phosphorous analog (γ-^{32}P-dCTP) at the 3'-end of the specific oligonucleotide probe (3'-end labeling). It is the most widely used method for the hybridization and confirmation of PCR amplification reactions. However, since the amount of amplified product is in such an order that sensitivity of the reaction is in principle of less relevance, a nonradioactively labeled oligonucleotide probe can easily be used instead of a radioactive label. In fact, it appears that nonradioactive hybridization is as sensitive as radioactive hybridization and detection. We therefore recommend the use of nonradioactive labels, since it will provide the opportunity to perform the complete assay in any laboratory working with cell cultures. The choice of nonradioactive labels is overwhelming. We limit ourselves in this chapter to the method we use in our laboratory, namely, the DIG system developed by Boehringer Mannheim *(31)*. Using this system, one digoxigenin residue is added per oligonucleotide with the enzyme terminal deoxynucleotidyl transferase. Probes labeled with this method retain their high degree of specificity. For the DIG labeling of the oligonucleotide probe, complete ready to use kits are commercially available (DIG Oligonucleotide 3'-End Labeling Kit, Boehringer Mannheim GmbH, Biochemica).

1. Mix for each labeling reaction in a 1.5-mL microcentrifuge tube: 4 µL 5X Tdt buffer, 4 µL 25mM CoCl$_2$-solution, 1 µL oligonucleotide probe, 1 µL DIG-11-ddUTP, 1 µL TdT, and 9 µL distilled water.
2. Incubate for 15 min at 37°C.
3. Add 2.5 µL of 4M LiCl and 75 µL 100% ethanol, and mix the solution.
4. Precipitate the labeled probe either for 30 min at –80°C or for 2 h at –20°C.
5. Microcentrifuge for 15 min, and discard the supernatant. Nonincorporated DIG-11-ddUTP nucleotides will be removed with the precipitation.
6. Add 50 µL of 70% ethanol to the labeled probe, and gently revolve the tube.
7. Microcentrifuge for 5 min, and carefully discard the supernatant.
8. Remove the remaining ethanol by air- or vacuum drying the pellet.
9. Dissolve the dried pellet in 20 µL of distilled water.

3.4. Hybridization
3.4.1. Prehybridization of the Membrane

1. Prepare prehybridization solution at room temperature. Per 100 mL, add: 25 mL 20X SSC, 5 mL 100X Denhardt's solution, 1 mL 10% SDS, 25 µL denatured salmon sperm DNA (heat the sonicated salmon sperm DNA for 5 min at 100°C and chill on ice until use), and 69 mL distilled water.
2. Place the nylon membrane with the DNA side up in a hybridization tube, and add 10 mL of prehybridization solution.
3. Place the membrane in a hybridization oven (Salmon and Kipp, BV, Breukelen, The Netherlands), and incubate for 2 h at 42°C.

3.4.2. Hybridization of the Membrane

1. Prepare the hybridization solution at room temperature. Per 20 mL, add: 5 mL 20X SSC, 200 µL 100X Denhardt's solution, 200 µL 10% SDS, 5 µL denatured salmon sperm DNA, 5 µL <DIG> labeled oligonucleotide probe (to prevent secondary

structures in the oligonucleotide probe, heat the probe for 5 min at 100°C, and chill on ice until use), and 14.5 mL distilled water.
2. Discard the prehybridization solution.
3. Add 10 mL of hybridization solution to the membrane.
4. Allow the probe to hybridize overnight at 42°C.

3.4.3. Washings of the Membrane

1. Prepare wash solutions:
    a. Wash solution 1. Per 200 mL, add: 50 mL 20X SSC, 2 mL 10% SDS, and 148 mL distilled water.
    b. Wash solution 2. Per 200 mL, add: 10 mL 20X SSC, 2 mL 10% SDS, and 188 mL distilled water.
    Preheat both wash solutions at 42°C.
2. Remove the membranes from the hybridization tube to a preheated wash container.
3. Add 100 mL wash solution 1, and wash the membrane under continuous shaking for 15 min at 42°C.
4. Replace wash solution 1 by 100 mL fresh wash solution 1, and repeat the wash procedure.
5. Replace wash solution 1 by 100 mL wash solution 2, and repeat the wash procedure.
6. Replace wash solution 2 by 100 mL fresh wash solution 2, and repeat the wash procedure.
7. Air-dry the membrane, and store it between Whatmann 3 MM paper until use for the detection with the hybridized probe.

3.4.4. Detection of the Hybridized Probe

The DIG system offers several different detection assays. In principle, an antidigoxigenin antibody conjugated with alkaline phosphatase is allowed to bind to the hybridized probe. The signal is subsequently detected with colorimetric or chemiluminescent alkaline phosphatase substrates. If a colorimetric substrate is used, the signals develop directly on the membrane. When a chemiluminescent substrate (Lumigen™PPD) is used, the signal is detected on an X-ray film. The advantage of the latter detection method is its higher sensitivity (0.03 pg homologous DNA) over the colorimetric detection (0.1 pg). A disadvantage is the use of X-ray films, which requires equipment to develop the films. A DIG Luminescent (or Multicolor) Detection Kit is commercially available (Boehringer Mannheim Biochemica), and the detection method and principles are extensively described in The DIG System User's Guide for Filter Hybridization. We perform the method exactly as described in this manual *(31)*.

4. Notes

1. Cell-culture samples to be analyzed must contain approx 10^5 cells and 0.5 mL culture medium for the analysis of both intracellular and extracellular mycoplasmas. Suspension cell cultures can be directly used for the test after careful mix-

ture. Monolayer cells must be scraped off and suspended in their media. There are in principle no limitations in the cell-culture origin to be analyzed. Analysis can be performed on freshly prepared samples, on stored samples (frozen or cooled), or in fact, lyophilized cell cultures *(20)*. The latter facilitates transport of the samples. Detection can be performed on crude cell-culture samples, or the nucleic acids can be purified for the analysis. In this chapter, a commonly used protocol to extract and purify DNA from liquid samples is described. In fact, any protocol can be used. As indicated, the analysis can even be performed on crude samples.
2. The primers used in this PCR assay are described in detail in refs. *20* and *23*. Nowadays, it is very cheap to let primers be synthesized commercially. Normally it will cost about US $50 for an amount of primers that is sufficient for at least 1000 reactions.
3. A method has been described to ensure that PCR products from a previous amplification reaction cannot be reamplified to produce false-positive results *(32)*. In this system, the nucleotide triphosphate dTTP is substituted by dUTP, which will be incorporated during the amplification reaction as effective as dTTP. PCR products containing dUTP are susceptible to degradation by treatment with uracil *N*-glycosylase, which cleaves the PCR product directly after each incorporated dUTP. Treating samples with uracil *N*-glycosylase prior to the amplification reaction will degrade any PCR contaminant possibly present and will therefore prevent false-positive results. A commercial system is available (GeneAmp^r PCR Carry-over Prevention Kit, Perkin Elmer Cetus, Norwalk, CT). False-positive results will be avoided, and the complete amplification and determination reaction can be performed in a single room. The system is, however, rather expensive.
4. In general, a clear-cut 280-bp amplification product will be obtained in the case of a mycoplasma-contaminated cell culture. On rare occasions, some aspecific background fragments of different sizes might be generated that might influence the interpretation of the results. We therefore recommend that Southern-blot analysis of the gel results be incorporated as a standard in the assay.
5. Transfer of DNA to the nylon membrane is performed with a pyramid setup for Southern blotting via upward capillary transfer. We recommend *Current Protocols in Molecular Biology (30)* in which several setups are described.

Generally, the gels are neutralized after denaturation in sodium hydroxide. However, because the PCR fragments are so small, direct rehybridization will occur. Therefore, the blotting is performed under denaturing conditions. It is, therefore, essential to use nylon membrane instead of nitrocellulose. Nitrocellulose will break and become porous in sodium hydroxide. The transfer of small fragments is normally a very fast process and takes at the most 2 h. In our hands, however, the best results are obtained after blotting overnight.

References

1. Uphoff, C. C., Brauer, S., Grunicke, D., Gignac, S. M., Macleod, R. A. F., Quentmeier, H., Steube, K., Tümmler, M., Voges, M., Wagner, B., and Drexler, H. G. (1992) Sensitivity and specificity of five different Mycoplasma detection assays. *Leukemia* **6**, 335–341.

2. McGarrity, G. J., Vanaman, V., and Sarama, J. (1984) Cytogenetic effects of mycoplasma infection of cell cultures: a review. *In Vitro* **20,** 1–18.
3. McGarrity, G. J. and Kotani, H. (1985) Cell culture mycoplasmas, in *The Mycoplasmas*, vol. IV (Razin, S. and Barile, M. F., eds.), Academic, New York, pp. 353–390.
4. Barile, M. F. (1979) Mycoplasma tissue cell culture interactions, in *The Mycoplasmas* (Tully, J. G. and Whitcomb, R. F., eds.), Academic, New York, pp. 425–474.
5. Stanbridge, E. J. (1971) Mycoplasmas and cell cultures. *Bacteriol. Rev.* **35,** 206–227.
6. Johansson, K.-E., Johansson, I., and Göbel, U. B. (1990) Evaluation of different hybridization procedures for the detection of mycoplasma contamination in cell cultures. *Mol. Cell. Probes* **4,** 33–42.
7. Del Giudice, R. A. and Hopps, H. E. (1978) Microbiological methods and fluorescent microscopy for the direct demonstration of mycoplasma infection of cell cultures, in *Mycoplasma Infections in Cell Cultures* (McGarrity, G. J., Murphy, D. G., and Nichols, W. W., eds.), Plenum, New York, pp. 57–69.
8. Hopps, H. E., Meyer, B. C., Barile, M. F., and Del Giudice, R. A. (1973) Problems concerning "non-cultivable" mycoplasma contaminants in tissue cultures. *Ann. NY Acad. Sci.* **225,** 265–276.
9. Hay, R. J., Macy, M. L., and Chen, T. R. (1989) Mycoplasma infection of cultured cells. *Nature* **339,** 487,488.
10. McGarrity, G. J., Sarama, J., and Vanaman, V. (1985) Cell culture techniques. *ASM News* **51,** 170–183.
11. McGarrity, G. J. (1982) Detection of mycoplasmal infection of cell cultures, in *Advances in Cell Cultures* (Maramorosch, K.), Academic, New York, pp. 99–131.
12. Göbel, U. B. and Stanbridge, E. (1984) Cloned mycoplasma ribosomal RNA genes for the detection of mycoplasma contamination of tissue cultures. *Science* **226,** 1211–1213.
13. Razin, S., Gross, M., Wormser, M., Pollack, Y., and Glaser, G. (1984) Detection of mycoplasmas infecting cell cultures by DNA hybridization. *In Vitro* **20,** 404–408.
14. Saiki, R. K., Bugawan, T. L., Horn, G. T., Mullis, K. B., and Erlich, H. A. (1986) Analysis of enzymatically amplified β-globin and HLA-DQ-alfa DNA with allel-specific oligonucleotide probes. *Nature* **324,** 163–166.
15. Blanchard, A., Gautier, M., and Mayau, V. (1991) Detection and identification of mycoplasmas by amplification of rDNA. *FEMS Microbiol. Lett.* **81,** 37–42.
16. Spaepen, M., Angulo, A. F., Marynen, P., and Cassiman, J.-J. (1992) Detection of bacterial and mycoplasma contamination in cell cultures by polymerase chain reaction. *FEMS Microbiol. Lett.* **99,** 89–94.
17. Roulland-Dussoix, D., Henry, A., and Lemercier, B. (1994) Detection of mycoplasmas in cell cultures by PCR: a one year study. *J. Microbiol. Methods* **19,** 27–134.
18. Hopert, A., Uphoff, C. C., Wirth, M., Hauser, H., and Drexler, H. G. (1993) Specificity and sensitivity of polymerase chain reaction (PCR) in comparison with other methods for the detection of mycoplasma contamination in cell lines. *J. Immunol. Methods* **164,** 91–100.

19. Teyssou, R., Poutiers, F., Saillard, C., Grau, O., Laigret, F., Bové, J.-M., and Bébéar, C. (1993) Detection of mollicute contamination in cell cultures by 16S rDNA amplification. *Mol. Cell. Probes* **7,** 209–216.
20. van Kuppeveld, F. J. M., Johansson, K.-E., Galama, J. M. D., Kissing, J., Bölske, G., van der Logt, J. T. M., and Melchers, W. J. G. (1994) Detection of Mycoplasma contamination in cell cultures by a Mycoplasma group-specific PCR. *Appl. Environ. Microbiol.* **60,** 149–152.
21. Gray, M. W., Sankoff, D., and Cedergren, R. J. (1984) On the evolutionary descent of organisms and organelles: a global phylogeny based on highly conserved structural core in small subunit ribosomal RNA. *Nucleic Acids Res.* **12,** 5837–5852.
22. Waters, A. P. and McCutchan, T. F. (1990) Ribosomal RNA: nature's own polymerase-amplified target for diagnosis. *Parasitol. Today* **6,** 56–59.
23. van Kuppeveld, F. J. M., van der Logt, J. T. M., Angulo, A. F., van Zoest, M. J., Quint, W. G. V., Niesters, H. G. M., Galama, J. M. D., and Melchers, W. J. G. (1992) Genus- and species-specific identification of mycoplasmas by 16S rRNA amplification. *Appl. Environ. Microbiol.* **58,** 2606–2615. (Author's correction [1993] **59,** 655.)
24. Neefs, J.-M., van der Peer, Y., Hendriks, L., and de Wachter, R. (1990) Compilation of small ribosomal subunit RNA sequences. *Nucleic Acids Res.* **18,** 2237–2318.
25. van Kuppeveld, F. J. M., Melchers, W. J. G., Willemse, H., Kissing, J., Galama, J. M. D., and van der Logt, J. T. M. (1993) Detection of *Mycoplasma pulmonis* in experimentally infected laboratory rats by 16S rRNA amplification. *J. Clin. Microbiol.* **31,** 524–527.
26. Melchers, W. J. G., Verweij, P. E., van den Hurk, P., van Belkum, A., De Pauw, B. E., Hoogkamp-Korstanje, J. A. A., and Meis, J. F. G. M. (1994) General primer-mediated PCR for detection of *Aspergillus* species. *J. Clin. Microbiol.* **32,** 1710–1717.
27. Kwok, S. K. (1990) Procedures to minimize PCR-product carry-over, in *PCR Protocols. A Guide to Methods and Applications* (Innis, M. A., Gelfand, D. H., Sninsky, J. J., and White, T. J., eds.), Academic, San Diego, CA, pp. 142–145.
28. Kwok, S. K. and Higuchi, R. (1989) Avoiding false positives with PCR. *Nature* **339,** 237,238.
29. Southern, E. M. (1975) Detection of specific sequences among DNA fragments separated by gel electrophoresis. *J. Mol. Biol.* **98,** 503–517.
30. Ausubel, F. M., Brent, R., Kingston, R. E., Moore, D. D., Seidman, J. G., Smith, J. A., and Struhl, K. (1994) *Current Protocols in Molecular Biology,* vol. I and II, Wiley, New York.*
31. Boehringer Mannheim Biochemica (1993) *The DIG System User's Guide for Filter Hybridization.* Boehringer Mannheim GmbH, Biochemica, Mannheim, Germany.
32. Longo, M. C., Berniger, M. S., and Hartley, J. L. (1990) Use of uracil DNA glycosylase to control carry-over contamination in polymerase chain reactions. *Gene* **93,** 125–128.

*Reference *30* is strongly recommended for background information on any molecular biological method used.

Index

A

β-Aminoproprionitrile, 219, 224
Acoustic neuromas, 96
Adipocytes,
 cell differentiation, 44–46
 culture, 44
 isolation, 43, 44
Agarose gel, 529, 532
Alkaline phosphatase,
 see glomerular cells
 see PTH
Ascorbate,
 in culture systems, 246, 247
 passaging cells cultured in, 249
Astrocytes, 56

B

BeWo cells,
 media, 466
 microcarrier culture, 467–472
Biotin, 42
Bone resorption, 265, 269
 area measurement, 269
 pH optimum, 272
 volume measurement, 272
Bone wafers,
 devitalized, bovine, 274
 fixation, 264
 preparation, 264
 staining, 265
Brain, adult,
 cell culture, 64–66

Brain, fetal,
 cell culture, 55–57
 growth substrate, 56
 receptor expression, 57, 58

C

3',5' Cyclic adenosine monophosphate, see PTH
c-Kit ligand, 142, 148
Cell growth kinetics,
 of glioma, 73, 74
 of osteoclasts, 241
 of vascular smooth muscle, 325, 327, 330
Cervical fibroblasts,
 biopsy collection, 337, 340
 cryopreservation, 341
 differentiation, 338
 media, 337
 phenotype characterization, 339, 341
 primary culture, 337, 340
 subculture, 338
Chondrocytes,
 isolation, 218, 220
 matrix protein synthesis, 219, 224
 media, 218
 mRNA analysis, 220, 225, 228
 subculture, 218, 220, 227, 228
 suspension culture, 219, 223
 transient transfection assays, 220, 226, 229
Clonogenic assays, 75, 76
Collagen,
 chain, made by osteoclasts, 243, 244
 coating, 518

gels, 2, 505, 508
lattice, 205
Conditioned media,
  MRC-5, 308, 309, 313
Conjunctiva,
  cryopreservation, 511
  epithelial cells, 504, 508
  epithelialization, 509
  fibroblasts, 504, 508
  histochemistry, see
      Histochemistry media,
  cryopreservation, 505
  dissociation, 505
  growth, 505
  primary culture, 508, 513
  subculture, 510
  tissue, 507, 513
Cornea,
  culture, 484
  endothelia, see Endothelia
  equipment for dissection,
      480, 486, 491
  eyebanks, 478
  media, 482, 486
  tissue, 479, 490
Cryopreservation,
  cervical epithelia, 341
  conjunctiva, 511
  corneal endothelia, 97
  cystic fibrosis airway epithelia,
      178, 181
  fetal brain, 56
  hepatocytes, 383
  keratinocytes, 6
  keratocytes, 496
  myometrial cells, 341
  myometrial smooth muscle,
      341
  osteoclasts, 253
  renal cells, 433, 434
  T-cell clones, 128, 129
  thymic epithelia, 117
  vascular smooth muscle, 327

Cystic fibrosis airway epithelial cells,
  cryopreservation, 178, 181
  culture, 175, 178, 180
  immortalization, 178, 181
  isolation, 177
  lipofection, 187
  media, 176, 186
  transduction, 187
  transfection, 186
Cystic fibrosis transmembrane
    conductance receptor (CFTR),
    173, 176, 179
  activity assessment of, 179–181
  detection of mRNA, 198
  gene transfer of, 189, 190
  genotypic DNA analysis, 198, 199
  RT-PCR, 195, 199
  sequencing of,
    dsDNA, 197
    mRNA, 196, 199
    ssDNA, 196

D

Dexamethazone, 43
Doxorubicin, 139, 140, 142, 146

E

ECM, see Extracellular matrix
Electroporation,
  buffer, 186
  of CF airway epithelial cells, 186, 189
Elutriation, centrifugal,
  of endocrine cells, 347, 351, 354
  of neurons, 358, 362
Endocrine cells, antral,
  characteristics, 352
  culture, 347, 352, 354, 355
  elutriation, 347, 351, 354
  media, 348
  mucosa, digestion of, 347, 350, 353
  stomach dissection, 346, 348
  tissue collection, 345, 348

Endothelia,
  culture from cornea, 85, 90, 97
  culture from umbilical vein,
    104–106
Endothelial cell growth factor,
  102, 108
Epidermal growth factor (EGF), 3, 11,
  23, 432
Epidermal Langerhans' cells
  short-term culture, 36
Epidermal melanocytes,
  see Melanocytes
Epinephrine,
  selection by, 179, 181
  sensitivity to, 176
  solution, 177
Epithelia,
  conjunctival, see Conjunctiva
  thymic, see Thymic
Erythrocyte lysis buffer, 42
Eurocollins buffer, 348, 359
Extracellular matrix preparation
  from bovine eyes, 82, 84–86
  from glioma, 86
Eye, bovine,
  see Extracellular Matrix

F

Feeder cells
  3T3 fibroblasts, 4
  dermal fibroblasts, 28
Fibroblasts,
  cervical, see Cervical
  conjunctival, see Conjunctiva
Fibroblast growth factor (FGF), 83, 87
Flow cytometry, 308, 311, 314

G

Gelatin, as culture substrate, 103
Glioma,
  cell cloning, 66–73, 88
  extracellular matrix, 86

Glomerular cells,
  alkaline phosphatase antialkaline
    phosphatase, 426
  epithelial cells
    culture, 425
    isolation of, 422
  medium,
    culture, 421
    washing, 421
  mesangial cells,
    culture, 425
    trypsinization, 425, 428
  morphology, 426, 429
Glomeruli, isolation of, 422
Glucose oxidation rate, 392, 399
Glycerol-3-phosphate dehydrogenase
  assay, 43, 46
Granulocyte monocyte colony-
  stimulating factor (GM-CSF),
  138, 142, 146, 154

H

Hematopoietic progenitor cells,
  colony-forming cell culture,
    137, 146
  immunoselection, 142
  isolation of, 143, 144
  long-term culture, 137
  media, 142
Hepatocytes,
  coculture, with rat epithelial cells,
    382
  cryopreservation, 383
  culture, 379
  dispersal buffer, 373
  fetal, 383
  isolation, 375–379
  media,
    attachment, 373
    culture, 373
Histochemistry, of conjunctiva
  alcian blue, 506, 512
  anticytokeratin, 506, 513

cell fixation, 505, 512
gel fixation, 511
PAS (periodic acid Schiff), 506, 512
tissue fixation, 506

I

Immunophenotyping, of hematopoietic progenitor cells, 143, 144
In situ hybridization, 64, 70, 147, 529, 530
  detection, 535
  membrane, 534
  prehybridization, 534
  washing, 535
Insulin, 43
  biosynthesis by pancreatic islet cells, 392, 401
  glucose stimulated release by pancreatic islet cells, 399
  labeling, 401
  release by pancreatic islet cells, 392, 398
  transferrin and selenious acid, 432
Interleukin-1β (IL-1β), 138, 139, 142
Interleukin-3 (IL-3), 138, 142, 154

K

Keratin, 117
Keratinocyte,
  culture from hair follicle, 25–27
  culture from skin, 5
  feeder layer, 4, 28
  media, 3, 23
Keratocyte,
  collagen gel culture, 497
  cryopreservation, 496
  definition and function, 489
  etched culture, 497
  media,
    culture, 491, 499
    freezing, 493

plating, 496
primary culture, 494, 499
subculture, 495, 497

L

Langerhans' cells
  see Epidermal

M

Macrophages, Monocyte derived, 153
  culture, 156
  isolation, 153–156
  media, 153
Matrigel, 447, 448
Medulloblastomas, 97
Melanocytes, epidermal,
  culture, 11–13
  media, 11
  surface antigen expression, 14
Meningiomas, 92, 93
Mezerein, 154, 156, 158
Microcarrier beads, 466, 474
Microcarrier culture, of BeWo cells, 467–472
MTT, 157, 159
Muscle, smooth vascular,
  see Vascular
Mycoplasma,
  DNA extraction, 528, 530
  hybridization, 529, 530
  oligonucleotide probe labeling, 529, 533
  PCR, 528, 530
    amplification, 528, 532
    southern blots, 529, 533
Myoblast,
  bulk culture, 310
  clonal culture, 313
  isolation, 309
  media,
    dissociation, 308
    freezing, 308, 315

Index

growth, 308
transport, 308
purified culture, 312
Myometrial cells,
  biopsy collection, 337, 340
  cryopreservation, 341
  differentiation, 338
  media, 337
  phenotype characteristics, 339, 341
  primary culture, 337, 340
  subculture, 338

N

Natural killer cells,
  contamination of by other PBL subsets, 170
  low recovery, 170
  media, 163
  positive/negative selection, 166
  proliferation, 169
  purification, 167
Neural cell adhesion molecule (NCAM) antibody, 308, 311, 314
Neurocytomas, 97
Neuroma, acoustic, 96
Neurons, 56
  characteristics, 363
  culture, 359–363
  isolation, 358, 360, 365, 366
  media, 359
  separation from plexus, 358–360, 364
Nylon mesh, 42

O

Oil Red O, 43, 47
Oligonucleotide probe labeling, 529, 533
Osteocalcin, 243, 245
Osteoblasts, response to PTH, 242

Osteoclasts,
  collagen chain production, 243, 244
  cryopreservation, 253
  culture, 235, 238, 253, 266, 267
  explants, 235, 236, 254–257
  fixation, 269
  growth characteristics, 241
  immunocytochemistry, 273
  isolation, 265
  long-term culture, 274
  media, 235, 253, 264
  mineralization, 236, 249–253
  phenotypic characterization, 235, 240, 244–246

P

p450,
  analysis, 378
  buffer, 373
Pancreas, protein biosynthesis by, 392, 401
Pancreas, adult,
  culture, 397
  DNA content, 398
  fixation and staining, 398
Pancreas, fetal,
  culture,
    at air liquid interphase, 394
    islet like cell clusters, 394
    pancreatic fragments, 393
  dissection, 393
  DNA content, 398
  fixation and staining, 398
  registration, 392
Pancreatic islet cells,
  see Insulin
Pantothenic acid, 42
PCR, 528, 530, 532
  of CF cells, 194, 195, 199
Peripheral blood lymphocytes,
  coculture, 165
  isolation, 164, 169
Phorbol retinoyl acetate, 154, 156, 158

Pituitary adenoma,
  media, 448, 452
  primary culture, 448
  sampling of, 448
  subculture, 451
Pituitary extract preparation, 10
Polyethyleneimine, 56
Protein biosynthesis, by pancreas, 392, 401
  labeling, 401
PTH,
  effect on alkaline phosphatase activity, 243
  effects on production of 3',5' cyclic adenosine monophosphate, 242
  osteoblast response to, 242

R

Renal cells,
  cryopreservation, 433, 434
  culture, 433, 434
  isolation, 431, 433
  media, 432
Renal cortical cells,
  culture of proximal tubular cells, 414
  isolation of proximal tubular cells, 412
  media, 410, 415
  subculture of proximal tubular cells, 414
Renal medullary interstitial cells,
  culture, 441–444
  isolation, 439, 440, 443
  media, 438
Retinal pigment epithelia,
  media, 517
  primary culture, 518
  subculture, 520
Retroviral vectors, 187, 189
  E1 deficient, 188
Rosette formation, 166, 167, 170

S

Southern blots, 529, 533

T

T-cells,
  cell cloning, 126
  media, 123
TGFβ, 24, 331
Thymic epithelia,
  culture, 114–117
  media, 113, 131
TPA, 154, 156, 158
Tracheal gland cells,
  applications, 208
  culture,
    collagen lattice, 205
    monolayer, 205
  cystic fibrosis, 211
  digestion of, 204
  isolation, 203, 213, 214
  media,
    detachment, 203
    digestion, 203
    growth, 203, 212
    transport, 203
  phenotypic expression, 207, 210
  physiology and pharmacology, 208
  subculture, 206
Triiodothyronine, 43
Trophoblasts,
  culture, 460
  immunocytochemistry, 461
  isolation, 458
  media, 458
Tumor necrosis factor α (TNFα), 138, 139, 142, 146

U

Umbilical vein,
  see Endothelia

V

Vascular smooth muscle,
  contractile proteins, 329
  cryopreservation, 327
  DNA synthesis, 330
  enzyme-dispersed cells,
    cell synchronization, 325
    comparison with explant-derived cells, 328
    establishment, 322, 331
    growth kinetics, 325, 330
    subculture, 324
  explant-derived cells,
    cell synchronization, 327
    comparison with enzyme-dispersed cells, 328
    establishment, 325
    growth kinetics, 327, 330
    subculture, 328
  growth kinetics, 330
  media,
    isolation, 320
    growth, 320, 331, 332
  morphology, 328
  tissue samples, 322, 332
  yield, 328